JUN 2 1982

Chance
change
&challenge

THE
EVOLVING
BIOSPHERE

Mimicry in some African swallowtail butterflies (see chapter 12).

The centre column shows females of *Papilio dardanus* Brown. The upper five specimens all came from Entebbe (Uganda), where the polymorphic female of this species mimics five different poisonous species (corresponding upper five butterflies in left-hand column). The yellow, tailed *dardanus* female (centre, bottom) is from Djemdjem Forest (Ethiopia), one of several areas where some or all the females are, in contrast, similar to the universal, non-mimetic male form (illustrated by a male from Entebbe, at the foot of the left-hand column). The lower four butterflies in the right-hand column are *Papilio phorcas* Cramer, the species presumed to be the closest living relative of *dardanus* and sympatric with it throughout much of Africa. The male of *phorcas* (foot of column) is always green and broad-banded. The female of *phorcas* is polymorphic, and (shown in ascending sequence) is either male-like, intermediate, or narrow-banded yellow. This last female form is similar to another African species, *Papilio constantinus* Ward (top right), probably the closest living relative of the *dardanus/phorcas* pair.

Chance
change
&challenge

General Editor: *P. H. Greenwood*

THE EVOLVING BIOSPHERE

Editor: *P. L. Forey*

BRITISH MUSEUM (NATURAL HISTORY)

CAMBRIDGE UNIVERSITY PRESS

Published by the British Museum (Natural
History), London and the Press Syndicate of the
University of Cambridge
The Pitt Building, Trumpington Street,
Cambridge CB2 1RP
32 East 57th Street, New York, NY 10022, USA
296 Beaconsfield Parade, Middle Park, Melbourne
3206 Australia

First published 1981

Printed in Great Britain at the University Press, Cambridge

**British Library Cataloguing
in Publication Data**

The evolving biosphere. – (Chance change and challenge)
I. Evolution
II. Forey, Peter Lawrence
III. British Museum (Natural History)
575 QH 366.2

ISBN 0 521 23811 0 hard covers
ISBN 0 521 28230 6 paperback

Cover illustration

The fossil diatom *Lepidodiscus elegans*
A single valve of the fossil diatom *Lepidodiscus elegans* Witt
(Eocene, Russia) photographed through a phase-contrast
microscope which produces its unnatural coloration. The
valve is made of silica and is one of two which compose the
frustule or box-like exoskeleton which is the definitive feature
of diatoms. The earliest known diatoms date from the
Jurassic; the group flourished in the Miocene and is widespread and abundant today. Diatoms are noteworthy for the
constancy of their appearance, there being little change in
form between species in the Miocene and those living today.

FOREWORD

Chance, change and challenge

The basis of natural history

When the Museum moved to South Kensington in 1881, controversy over evolution was at its height and although the reference collections were not at first employed directly to advance evolutionary ideas, the work of its staff has always contributed ultimately to evolutionary knowledge and to explaining the principles underlying natural history. It is therefore appropriate that the Museum's centenary year should be marked by the publication of two volumes on evolution. This subject not only links all living organisms, but also the organic to the inorganic world, and the Past to the Present, thereby providing a theme to which all five science departments in the Museum (Botany, Entomology, Mineralogy, Palaeontology and Zoology) can contribute.

The wide range of subjects currently studied in the Museum ensures that there are many areas and individual spheres of interest and expertise within the broad spectrum of theoretical and empirical evolutionary studies. This knowledge is tapped here in a series of very personal essays which reflect the widely differing approaches to the study of evolution.

The title summarises, in simple terms, the essential elements of evolution. CHANCE – the interplay of cosmic forces that brought the Earth into existence and the often fortuitous, random mutations and other genetical changes that collectively contribute to the evolution of life. CHANGE – the formation and destruction of continents and ocean floors as rocks and sediments are generated, eroded, and recycled, and the infinitely variable patterns of life as plants and animals diversify. CHALLENGE – the ever-changing physical and biotic environments to which living organisms respond, and in so responding, evolve.

Although, in its generally accepted sense, the concept of evolution dates back some 200 years, it did not attract great interest among naturalists (and the general public) until after the publication of the famous paper on 'Evolution by natural selection' by Darwin and Wallace in 1858, followed almost immediately by Darwin's monumental *Origin of Species by means of Natural Selection* (1859). The ensuing stream of publications on evolution and evolutionary theory eventually became a flood which as yet shows no sign of diminishing. The present volumes do not set out to review the research output of the last century, neither are they intended to serve as text-books: their aims are more modest, although not perhaps less important. They show how the evolution of life is linked to that of the Earth itself, thus providing a broader perspective than is possible with most text-books. Intelligible accounts of highly specialised and very different work should help to disseminate ideas between specialists in fields between which there is usually little contact. Finally, students should benefit from a set of essays which are not always uniform in philosophical outlook and which demonstrate the disagreements that are part of the continuing challenge of research and of the theory of evolution itself.

The first volume – *The Evolving Earth* – demonstrates the interrelated contributions of Mineralogy, Geology, Oceanography and Palaeobiology to an understanding of the changing physical and chemical backgrounds against which plants and animals arose and evolved. In so doing it also describes the environments which have supported living organisms, and shaped their adaptation and diversity during the last 3·5 thousand million years. The revolutionary changes in cosmological and geological thought during this century are reflected in the sections on the origin of the Earth, and on continental drift and plate tectonics. Controversial hypotheses, such as that of an expanding Earth, which still require further testing and evaluation, are included. Apart from those essays which review the background to continental drift, biogeographers should be interested especially in the later chapters on Mesozoic and Cenozoic palaeogeography which provide a framework for the study of present-day distributions.

The second volume – *The Evolving Biosphere* – is necessarily more selective and less comprehensive than its companion since its scope is potentially greater. It is concerned with the mechanisms and interactions which produce and account for the diversity, coexistence, coev-

olution and distribution of plants and animals in the world today. There is naturally much emphasis on speciation, the basic process underlying these phenomena, and the one subject not discussed in the *Origin of Species* since nothing was known in Darwin's day of the possible mechanisms involved. The arrangement of these essays is essentially similar to that followed by Darwin in '*The Origin*', thus serving to underline changes in thinking on evolution and evolutionary processes since the mid-nineteenth century.

Major problems, both philosophical and practical, which still hinder our understanding of evolution are not avoided. The contributions show the diversity of interpretation and opinion held by students of evolution, and highlight the dynamic state of modern evolutionary biology. They also show how taxonomists have contributed to the advance and interpretations of evolutionary theory, and how in turn a deeper appreciation of evolutionary processes, and hence of phylogeny, has influenced taxonomic theory and the practice of classification.

These volumes were conceived and planned by a group of colleagues which first met in 1977 to consider the feasibility of a book of this sort. I should like to express special gratitude to the General Editor, Dr P. H. Greenwood, and the volume editors, Drs L. R. M. Cocks and P. L. Forey, for their time-consuming and dedicated work in bringing these publications to fruition. They, I know, recognise the special help they have been given by Dr C. G. Adams and Mr J. F. Peake at various stages in their work. The Museum is grateful to all the contributors, especially those from outside who so willingly filled the gaps in our knowledge. Their help demonstrates the close links forged over the years between the Museum, other government establishments and the universities. Thanks are also due to Mr C. J. Owen (Co-ordinating Editor) and Mr E. Dent (Production Controller).

R. H. Hedley
Director, British Museum (Natural History)
January 1981

Contents

Introduction

This book lays no claim to being a comprehensive review of evolutionary theory or of evolutionary processes; there are several excellent and recent text-books which fulfil that requirement. Its objective is rather to take a closer look at certain aspects of evolution through the eyes of practising taxonomists whose work brings them into close contact with the products of those evolutionary processes. To study and to handle these products, unless one is peculiarly insensitive, is bound to prompt thoughts about evolution, to question current hypotheses, or to find corroboration of them.

Despite the range of topics ultimately discussed, each of the 21 chapters is in effect concerned with a single, albeit multifaceted feature of life: its diversity.

The diversity of form and function amongst living organisms presents to the human eye and mind the most intriguing problems of evolution. Indeed, such diversity is the basis for the Darwinian concept of evolution through natural selection, a subject that contributes a parallel theme to the book.

Since the eyes and minds of human observers are themselves diverse, and prone to a variety of biases, the results of their observations and deliberations inevitably will produce a variety of opinions. That side of diversity, too, is apparent in the individual approaches to a common problem: the analysis and interpretation of the processes which shaped and are even now changing the biosphere.

In order to bring some cohesion and, indeed, a kind of large-scale evolutionary sequence into the presentation of these ideas, the chapters are arranged in relation to three major topics – species and speciation, coexistence and coevolution, and biogeography. But, as in life, the impossibility of actually treating evolution in any way other than as a dynamic nexus will be obvious to the reader.

Each chapter is a self-contained essay stemming from its author's particular interests in the field of evolutionary biology, and often expresses the influences different plant and animal groups have on the authors' approaches to a common evolutionary problem.

The first major topic discussed is concerned mainly with one level of organic diversity – the species and, in particular, with the modes and mechanisms of speciation. It sets the scene for the second section where consideration is given to the ways in which different species interact, the results of these interactions, and their effects as evolutionary processes operating at a level beyond that of the species. Nevertheless, there is a constant return to the species (and its subdivisions) as operational units within the broader picture created by these essays.

The third group of essays is concerned with particular instances of large-scale animal and plant distribution. This subject involves not only species and speciation, but also problems of coexistence and coevolution and, of course, the geological and geomorphological background to organic evolution. That background is considered in greater detail in a companion volume, *The Evolving Earth*.

Perhaps more than elsewhere in this book, those chapters dealing with the biogeographical aspects of evolution highlight different philosophical and methodological approaches to a common problem.

No attempt has been made to round-off the volume with an all-embracing 'Conclusion'. Its subject matter and the nature of its constituent parts preclude such pretentiousness. However, each of the three sections is prefaced by an introduction broadly reviewing and integrating the various chapters, and providing a complementary commentary on the major topic under review in that section.

PART I

Species and speciation

Introduction

It is evident to the most casual observer that the variety of life on earth does not form a continuum, but rather shows a marked tendency to be packaged into discrete units. The various kinds of units are recognisable by an assortment of characteristics – morphological, behavioural, physiological, ecological, and so on – which are exhibited in common by many individuals. Early attempts to systematise and classify life-forms employed elements of Aristotelian methodology, associating similar kinds of animals or plants as groups, each called a genus, but differentiating the component kinds within each genus as species. These were originally defined by a series of Latin adjectives which listed the characteristics of each species. However, the increasing complexity of descriptions as studies became more detailed led Linnaeus, in his tenth edition of *Systema Naturae*, to develop the use of a single adjectival name or epithet (the 'trivial name') for each species. This now serves as a shorthand, in combination with a generic name, for each individual species, the characters which diagnose it, and ultimately for the sum total of man's knowledge of that species.

Our casual observer, a member of the human species, will be more perceptive of the similarities and differences in those forms of life which present themselves most conspicuously to human senses; he will readily spot many differences between a cat and a dog, and even more between himself and his close relative the gorilla, but will be inclined to the view that all flies, or all grasses, look alike. This is, of course, largely due to his anthropocentric viewpoint. Flies generally have the sensory equipment to distinguish their own kind from other species of fly, and the stigma of one grass species is generally unreceptive to the pollen of another grass species. Certain fly species can also recognise grasses on which their larvae will feed, and lay their eggs on these, rather than on other grasses which are not suitable food for their larvae. Species differentiation and recognition is not the peculiar prerogative of *Homo sapiens*, but is an essential feature of the functional organisation of life.

Linnaeus at the time of *Systema Naturae* saw species as the products of special creation, each morphologically distinct and invariable in time and space, each with a definite role in the natural system. Later he became sufficiently disturbed by the indistinctness of some of the boundaries between plant species to claim that it must be the genus, not the species, that was specially created. Species, he suggested, arose within genera by hybridisation and by adaptation to different environments. Linnaeus' genera were in many groups equivalent to the family level in present-day classifications, so this rather hazy later philosophy had to account for the major part of natural variation. Thus, Linnaeus never found the natural system that he sought so single-mindedly. Nevertheless, his belief in its existence enabled him to introduce order into the classification of living things. His teachings continued to promote the concept of fixed and invariable species for long after his death, during the time that most of the world's life forms were described and named. In the process, one or more specimens from among those originally described were often designated as 'types', to be regarded as typical examples of their species. Such 'types' were regarded and used in a fashion analogous to physical standards of measurement.

The nineteenth century saw the birth of our evolutionary concepts. Observations of geographically isolated and locally adapted populations such as those made by Darwin and Wallace were 'followed by the demonstration of Mendelian segregation of characters. Thus was postulated a mechanism permitting the origin of variation and its transmission to offspring, combined with differential survival of the best fitted, giving the idea of the species a new apparent objectivity. The taxonomist – the describer and namer of species – became a student of evolution, and endowed his classificatory structure with a new significance, that of representing the relationships and descent of species. This phylogenetic or phyletic approach has been refined, with principles, terminology and methodology of its own, by workers such as Hennig (1966).

With the acceptance of evolutionary theory, variation within species in both space and time had to be recognised as the rule rather than the exception. The old typological ideas were swept away, as there was no longer any conceptual requirement for individuals to conform to the 'typical' form of their species. Nevertheless, some of the techniques developed during the typological phase remain appropriate to the needs of the phyletic approach. 'Types' are still used, but they are now a rather different concept, often misunderstood by non-taxonomists. The continued use of the word 'type' is perhaps unfortunate, as it suggests that the type specimen is still regarded as a typical member of the species, or at least that it represents or exemplifies it in some way. Today, the type of the taxonomist is merely a nomenclatural tool, the name-bearer (onomatophore) for the taxon. It says nothing more than 'I am a part of the species *A-us b-us*, erected by Henry Bloggs, and must always remain so, no matter how much you decide in the future to change your concept of the species *A-us b-us* Bloggs'. Thus, as happens frequently, more thorough taxonomic studies may reveal that the name *A-us b-us* is being applied to more than one species; there may even have been more than one species in the original material described by Bloggs. The type-specimen removes any doubt as to which of these species should continue to have the name *A-us b-us* Bloggs.

The development of the science of population genetics in the 1920s and 1930s resulted in a new approach to species, in which the events at the local population level came to be regarded as of great evolutionary significance. Genetic change, morphological differentiation and divergence happen within and between populations, and geneticists could define, sample and study selected populations more easily than they could entire species. Species therefore came to be defined in terms of populations; most succinctly by Mayr (1969) as 'groups of interbreeding natural populations, reproductively isolated from other such groups'. This biological, or more strictly, genetical concept of species is still widely accepted by both geneticists and taxonomists, although criticised at various times by palaeobiologists (e.g. Simpson, 1961), population geneticists (Ehrlich & Raven, 1969), logicians (Hull, 1970), numerical taxonomists (Sokal & Crovello, 1970) and phylogeneticists (Løvtrup, 1979; Rosen, 1979). The most radical positions adopted by critics of the biological species concept amount almost to denial of the existence of species. Ehrlich & Holm (1963), Sokal & Crovello (1970) and Løvtrup (1979) all incline to the view that the evolutionist need only concern himself with populations, and that the taxonomist would do better to abandon or ignore the biological species concept and use a strictly phenetic approach to classification, at the species level as well as above and below it. Such views have been discussed most recently by M. J. D. White (1978), and the controversy over the phylogenetic *versus* phenetic approaches to taxonomy is covered by Hull (1970) and many other writers. Volume 24 of *Systematic Zoology* includes several papers discussing the controversy. Suffice it to say here that the biological species concept seems so far to have survived these more fundamental attacks and is still regarded as a valid conceptual basis for their work by the majority of taxonomists and population geneticists. The links so carefully forged between genetics and taxonomy by Dobzhansky and by Mayr still remain largely unbroken.

It has been claimed, mainly by those not primarily engaged in taxonomy, that the biological species concept is of little practical use to the museum taxonomist, who has to work almost entirely with dead specimens. But the usefulness of a concept lies more in its theoretical significance that in its applicability in a practical sense.

The search for biological species in a particular group of organisms defines the taxonomist's task. He can use his knowledge and experience of the group in question to identify among the available data those which are likely to be relevant to this task, and those which can be discarded as misleading or irrelevant. In erecting new species he will have to make assumptions about reproductive relationships within the group, and sometimes these assumptions will be falsified by subsequent evidence. The experienced taxonomist recognises only too well the inadequacy of the data provided by dead specimens. He is ready to accept that the material he is studying may include some 'good' biological species which he has no hope of separating on the grounds of their gross morphology alone (so-called sibling species). On the other hand, in interpreting morphological data he can usually make valid inferences based on his knowledge of the general biology of the organisms he is studying, so that the instances when he will be misled by morphological criteria are much rarer than non-taxonomists often suppose.

The biological species concept is no help at all, however, to a taxonomist who is faced with organisms in which interbreeding populations do not occur; that is, asexual or uniparental organisms. Dobzhansky excluded uniparental organisms from his concept of species, labelling them 'pseudospecies' (Dobzhansky, 1972). The taxonomist has traditionally described species in uniparental organisms as well as in biparental ones, and continues to do so, but without any peace of mind. He can regard them as distinct from sexual species, and term them 'agamospecies', but this does not solve the conceptual problem, which arises from the very fact that taxa apparently akin to biological species do seem to occur in uniparental organisms. If interbreeding is so crucial to the delimitation of biological species, why do not uniparental organisms exhibit 'a chaotic range of variation that defies analysis and classification' (M. J. D. White, 1978)? One possible answer is that in evolutionary terms uniparental organisms may be very short-lived, so that extant 'agamospecies' still bear the stamp of biparental species from which they were derived relatively recently.

The most significant limitation of the biological species concept must be the absence of the time dimension. Species are the essential units of evolution. Palaeobiologists and evolutionists legitimately ask how any species concept can be adequate if it fails even to refer to the evolutionary process. They seek to extend the species into the time dimension. Simpson (1961) conceived his evolutionary species as 'a lineage (an ancestor–descendant sequence of populations) evolving separately from others and with its own unitary evolutionary role and tendencies'. Simpson had the problem of delimiting species in time. He suggested that an evolutionary lineage would have to be divided arbitrarily into a succession of species in the fossil record, between which the morphological difference was at least as great as that between extant sister species in the same group. Students of the Hennigian school of phylogenetics,

however, have no conceptual problem in delimiting the evolutionary species; for them, species originate when one lineage branches into two, so that each species exists as a single lineage spanning the time interval between two speciation events. Wiley (1978) adopts a phylogenetic approach to the evolutionary species, and slightly modifies Simpson's concept to provide for the fact that species do not evolve at a constant rate, but go through periods of relative stasis when they nevertheless still 'maintain their identity' with respect to other species. This view of the evolutionary process has been discussed at length by Gould & Eldredge (1977).

The biological species concept and the evolutionary species concept are in no way contradictory to one another; the first is just a special case of the second, applying at one point in time to contemporaneous populations of bisexual organisms. A single evolutionary lineage at any one point in time can comprise a group of interbreeding populations, and separate lineages must be reproductively isolated from one another to the extent that is required to maintain their separate identities and evolutionary tendencies. The idea that a species should have a particular 'evolutionary role' is, however, something that is missing from Mayr's statement of the biological species concept, although it certainly has a place in the population geneticist's concept of species. Dobzhansky, for instance, referred to species (and other taxa) as arrays of gene combinations forming 'adaptive peaks', distinguishable because the 'non-adaptive valleys' of unfavourable gene combinations remain largely uninhabited. 'Adaptive peaks' can be thought of as the 'evolutionary roles' of individual species. The problem is, which is more important in maintaining the integrity (or identity) of a species; interbreeding – which can only apply to biparental organisms – or centripetal selection for gene combinations at or around a particular 'adaptive peak' or 'evolutionary role', which can apply not only to interbreeding populations but also to populations of uniparental organisms. The applicability of the evolutionary species concept to uniparental organisms has been argued by all its proponents, originally by Meglitsch (1954) but subsequently also by Simpson (1961) and by Wiley (1978). Meglitsch's words provide a simple expression of the argument: 'The species, in the case of uniparental and biparental organisms, may be visualised as a natural population, evolving as a unit in actuality, or retaining the capacity to evolve as a unit if...barriers are removed...The species population is the visible manifestation of a pool of genes which retains its character as a unified pool because, in theory, any allele present may eventually come to replace all the allelomorphic factors in the pool, either as a result of interbreeding or as a consequence of simple differential survival in the case of uniparental organisms...A species thus, is an independent and distinctive region of gene spread, regardless of the mechanisms involved in the distribution of these genes, and is applicable equally to organisms which reproduce sexually and asexually.'

Perhaps the most important implication of such ideas is not that they accommodate uniparental organisms as well as biparental organisms into our concept of species, but that they provide some insight into the species as a functional unit. They emphasise that distinct species exist not so much because of the reproductive barriers separating them, but because they occupy separate evolutionary roles or niches in the economy of Nature. The reproductive barriers follow from the need of biparental organisms to concentrate their gene combinations on or around the adaptive peaks. The biological species concept, by omitting any reference to the part played by unifying selective influences in shaping species and maintaining their identity, may have tended to overemphasise the role of reproductive isolation in delimiting species, and in the process of speciation.

Recognition of the unitary evolutionary role of a species is also underlined by the philosophical points made by Ghiselin (1974) and developed by Hull (1976), that species are not classes but individuals. The idea that species are individuals is of course implicit in the practice of taxonomy, where they are described (*not* defined) as particular entities, and given proper names. The concept of the type in modern taxonomy referred to earlier also recognises the individual status of the species taxon; the type-specimen is erected to form *part* of the species, not to serve as an instance of it. If the species is an individual, its constituent organisms should be regarded as parts of the species, not as members; and as a unique entity, the species taxon has a reality which is distinct from its categorisation (as the category, 'species'). The species category is a class, an artificial grouping together of unique species taxa which do not individually conform to any general definition. Thus the mode of origin of a species, the genetic mechanisms by which its unity is maintained, and the organisms which compose it are all unique properties of that species.

It is the mode of origin of species that is elaborated upon in the next eight chapters. M. J. D. White (1978) has recently emphasised how we need to recognise that there is a multiplicity of events contributing to the evolutionary phenomenon whereby one species splits into two. It is reasonable to argue that the particular circumstances and course of any one speciation event are unique, just as are the species themselves. The origins of speciation events lie in the genetic variation within and between populations of a species. This variation is one of the principal practical difficulties confronting the working taxonomist who wishes to classify organisms, or simply to identify them from some other worker's classification. Lane & Marshall (Chapter 1) discuss the ways in which discontinuities can arise in the spectrum of genetic variation as a result of the patterns of spatial distribution which inevitably occur within species, and the attempts which have been made, with very limited success, to fit these discontinuities into one or more generally definable infraspecific categories. G. B. White (Chapter 2) develops this theme and considers particularly the special case where speciation is completed but is accompanied by little or no morphological differentiation. 'Sibling' species are only recognisable by characters that cannot readily be observed in museum specimens such as differences in their behaviour, cytology, ecology, etc. Recognition of the existence of such species can be very important to man if the organism is of agricultural or medical importance, as exemplified by the *Anopheles gambiae* group of mosquitoes.

The earliest coherent theories of genetic disruption leading to the multiplication of species emphasised geographical isolation, and became condensed into the widely accepted allopatric model of speciation. Based primarily on observations of much-studied birds and mammals, the allopatric model has the advantage of being readily understandable. Snow (Chapter 3) outlines the concept with examples drawn from birds, for which group the model frequently provides a very satisfactory explanation of the speciation process. However, alternative hypotheses have frequently been proposed in which the initial isolation of component populations of a species is not a prerequisite of speciation. Hammond (Chapters 4 and 5) develops the view that disruption of the gene pool of a species may be the result of alternative, more subtle mechanisms operating within populations (sympatric speciation) or between contiguous populations (parapatric speciation). Such processes are not demonstrated so easily as the allopatric model, but may have equal validity, particularly for less well-studied groups of small animals such as insects.

Many population studies have profited from situations where a clear geographic correlation facilitates the identification of individual species distributions and the derivation of hypotheses. Island faunas have produced many classic examples. However, Greenwood (Chapter 6) takes us 'through the looking glass' to examine the cichlid fish of the African Rift Valley Lakes. These provide further examples of supposedly allopatric speciation events, but emphasise just how rapidly an ecologically and behaviourally diverse fauna can develop from common or closely inter-related ancestors. Such species flocks provide, in microcosm, the analogue of past continental adaptive radiations of major animal groups.

Thus far the book deals with concepts based on processes applicable to sexual, or biparental, organisms. However, sexual systems are by no means universal. Blackman (Chapter 7) places sex in its true context, as a powerful facilitator of evolution, before discussing the implications of an asexual life system. An opportunistic process but an evolutionary blind alley, parthenogenesis *per se* has arisen in many groups, and brings with it some philosophical difficulties for the biologist. However, when parthenogenesis is combined with the sexual process, in those organisms characterised by an alternation of uniparental and biparental generations, the evolutionary implications are somewhat different. Aphids include many species of great agricultural significance to man, and examples of cyclical parthenogenesis, and developments from it, are drawn from this group.

There is one type of speciation that has been recognised for over 50 years as quite distinct from any other. Speciation by polyploidy (Gibby – Chapter 8) is particularly interesting and informative because it involves a new species arising suddenly and unambiguously within the distribution area of an existing one. Successful formation of a new polyploid species almost always involves hybridisation between two parent species, so that the new form starts out with a unique, highly heterozygous (and by implication, highly adaptive) genome.

References

Dobzhansky, T. 1972. Species of *Drosophila*. New excitement in an old field. *Science* **117**: 664–669.

Ehrlich, P. & Holm, R. W. 1963. *The process of evolution*. 347 pp., New York: McGraw-Hill.

Ehrlich, P. & Raven, P. H. 1969. Differentiation of populations. *Science* **165**: 1228–1232.

Ghiselin, P. 1974. A radical solution to the species problem. *Systematic Zoology* **23**: 536–544.

Gould, S. J. & Eldredge, N. 1977. Punctuated equilibria: the tempo and mode of evolution reconsidered. *Paleobiology* **3**: 115–151.

Hennig, W. 1966. *Phylogenetic systematics*. 263 pp., Urbana: University of Illinois Press.

Hull, D. L. 1970. Contemporary systematic philosophies. *Annual Review of Ecology and Systematics* **1**: 19–54.

Hull, D. L. 1976. Are species really individuals? *Systematic Zoology* **25**: 174–191.

Løvtrup, S. 1979. The evolutionary species: fact or fiction. *Systematic Zoology* **28**: 386–392.

Mayr, E. 1969. *Principles of systematic zoology*. 428 pp. New York: McGraw-Hill.

Meglitsch, P. A. 1954. On the nature of the species. *Systematic Zoology* **3**: 49–65.

Rosen, D. E. 1978. Vicariant patterns and historical explanations in biogeography. *Systematic Zoology* **27**: 159–188.

Rosen, D. E. 1979. Fishes from the uplands and intermontane basins of Guatemala: revisionary studies and comparative geography. *Bulletin of the American Museum of Natural History*. **162**: 270–375.

Simpson, G. G. 1961. *Principles of animal taxonomy*. 247 pp., New York: Columbia University Press.

Sokal, R. R. & Crovello, T. J. 1970. The biological species concept: a critical evaluation. *American Naturalist* **104**: 127–153.

White, M. J. D. 1978. *Modes of speciation*. 455 pp., San Francisco: Freeman.

Wiley, E. O. 1978. The evolutionary species concept reconsidered. *Systematic Zoology* **27**: 17–26.

R. L. Blackman and M. C. Day
British Museum (Natural History)

CHAPTER 1

Geographical variation, races and subspecies

R. P. Lane and J. E. Marshall

Just as no two sexually reproducing individuals are identical, neither are any two populations of a given species. There are many reasons for this differentiation, one of the more important being the spatial separation of the populations. Such spatial variation is more commonly referred to as geographical variation.

Man possibly first became interested in the geographical diversity of organisms when he began his voyages of exploration, crossing oceans to unfamiliar lands inhabited by many weird and wonderful plants and animals. Although not the first to write on the subject, Darwin and Wallace incorporated their observations on the geographical distribution of organisms into their theory of the role of natural selection in evolution, ultimately presented in Darwin's (1859) *On the Origin of Species*. Wallace later (1876) published a work in which he divided the world into six more or less distinct biogeographical realms (following Sclater, 1858) each characterised by a particular assemblage of organisms. However, on a smaller scale, Darwin had already noted the effect of the local environment on just a single species or group of closely related organisms. Here we shall attempt to show how the variation is generated within a species as a result of the spatial separation of populations, how this variation is dealt with taxonomically and finally, how it contributes to our knowledge of speciation.

The study of geographical variation, like that of evolution itself, encompasses many disciplines – physiology, ecology, cytology, genetics, taxonomy, biometry, ethology, and evolutionary studies. Consequently, many diverse techniques have been and are being used. Even the basic morphological approach, on which the subject was founded is now being refined by numerical techniques. The application of biometric comparisons between populations and races is a rapidly growing subject (Thorpe, 1976). One of the advantages of these new techniques is that many attributes may be considered simultaneously, thereby incorporating correlations (biological and statistical) between structures that are the result of the composite evolution of an organism.

It is not essential to travel great distances, as did Darwin and Wallace, in order to study or recognise geographical variation in animals or plants. Of course, the differences in the behaviour, morphology or ecology of a species may be particularly obvious over great distances but it is also possible to detect variation over relatively small distances (microgeographical variation). Perhaps one of the best known studies of microgeographical variation is that of the snail *Cepaea* (for a good review see Ford, 1971). Many molluscs are relatively sedentary or have small home ranges and therefore variation over such small distances is preserved by a low level of interbreeding. In *Cepaea* a colony only 30 metres in diameter may function as a discrete interbreeding unit. In *Cepaea nemoralis* the proportion of differently coloured and banded snails within each population varies over only a few kilometres, the prevalent phenotype being the form most suitably camouflaged in the habitat in which any particular colony lives. The frequency of different phenotypes is determined principally by natural selection which, in this case, is predation by thrushes – the least conspicuous phenotypes are left alone.

Fundamentally there is no real difference between variation over such small distances and variation over continents – the effects of spatial separation are basically the same. The main causes of geographical variation in animals may be divided into two major components – the response of an organism to different environments and the limitations to the flow of genetic material between spatially separated populations. We will first consider how the environments differ and what effects these differences have on the organisms concerned.

Continuous and discontinuous variation

Where conditions such as climate and altitude vary gradually over a given area, as is often seen in continental regions, they will effect gradual changes in certain attributes of a species resulting in regular gradients of variability.

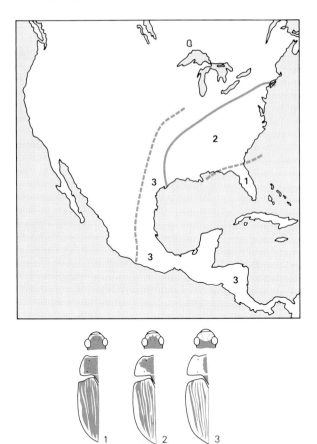

Figure 1.1 The distribution of water beetle *Tropisternus mexicanus* in North and Central America. (1: *T. m. viridis*, 2: *T. m. striolatus*, 3: *T. m. mexicanus*). After Young (1960).

Such gradients, where they are of a measurable character, such as size or colour, are termed clines. This term was proposed by Julian Huxley in 1938 to describe the expression of continuous variation and it should be noted that it refers to characters and not to populations. A population may 'belong' to as many clines as it has variable characters. The slope of the gradient is a measure of the divergence between different geographical areas, whilst the smoothness of the gradient (that is, the difference between successive variants) will be a function of several factors, including spatial arrangement of populations, gene flow between populations and selection pressure in different environments.

Clines are widespread and occur in the majority of continental species where gene flow is possible between populations in diverse environments via one or more intermediate populations. Figures 1.1 and 1.2 show the clinal variation in the pigmentation of the water beetle *Tropisternus mexicanus* in America and of the ladybird *Harmonia axyridis*, in south-east Asia (see discussion below), respectively. Clines may not be so evident in

species that have a discontinuous distribution such as island, lake or mountain populations for example. A further complication is that in the latter case the different populations may be called allopatric species, in which case a 'cline' would be revealed by a phylogenetic analysis of a super-species (a series of closely related species that are allopatric) as a transformation series.

Clines may be expressed in many characters and may be the result of any number of geographically variable conditions. The effect of temperature on size and colour for example will be discussed more fully under 'Ecogeographical Rules' below.

As implied above, not all geographical variation is demonstrated by a smooth transitional series. Often there is an irregular distribution of variation which may be attributed to a number of causes. It may simply be the result of a discontinuous distribution with the resulting cessation of gene flow between populations, these populations being subjected to differing local environments. It is a truism to say that no species exists in isolation. It has to interact with many factors in the environment, both biotic and non-biotic and either of these two factors may give rise to a discontinuous distribution. For example, competition with closely related species may lead to character displacement in areas where the two distributions overlap (this is discussed in Chapter 17). Another biotic factor that has been the subject of a considerable amount of research is variation in mimetic species (Chapter 12). Some species are polymorphic (the species is composed of several different forms), and if mimetic in addition (a common situation in butterflies), then the proportion of morphs in each population may vary discontinuously from area to area, according to the model available in each area. For example, in the Eastern Tiger Swallowtail (*Papilio glaucus*), a butterfly palatable to birds, there are populations with high frequencies of dark brown females in certain areas. These forms resemble the Aristolochia Swallowtail (*Bathus philenor*) also found in these areas. *B. philenor* is distasteful to birds. Selection apparently favours the development and maintenance of the mimetic form of *P. glaucus* in these areas.

A biotic factor of importance to many insects is the distribution of the host-plant. Where an association can be demonstrated between a particular phenotype and a host-plant preference, the effect of the host-plant distribution can be investigated. For example, in eastern North America two distinct groups of the cerambycid beetle, *Saperda inornata*, can be recognised – a light form and a dark form. The light form also tends to be a little larger with shorter elytral hairs and shallow, partly contiguous and confluent elytral punctures. *S. inornata* is found on three main host-plants, *Populus balsamifera*, *P. tremuloides* and *Salix* spp. which occur together although *P. balsamifera* has a slightly more northerly restricted range. It has been shown that the dark form of *S. inornata* is found principally on *P. balsamifera* and the light form on *P. tremuloides* and *Salix* spp. This is reflected in the distri-

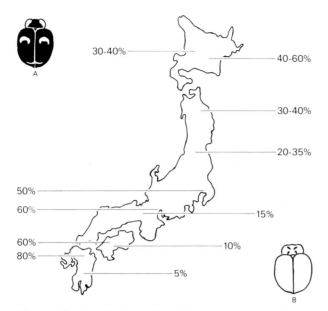

Figure 1.2 The clinal variation of the extreme dark and light forms of *Harmonia axyridis* throughout Japan (A: *H. a. conspicua*, left; B: *H. a. succinea*, right) expressed as percentage of total local population.

bution of the different forms of the beetle. The dark, a few light and a number of intermediates of various grades of *S. inornata* are only found to the north where all three hosts occur. In the south, beyond the limit of *P. balsamifera*, only light and light intermediate forms are found.

Non-biotic factors in the environment which may generate discontinuous variation may be various physical properties of the substrate such as colour or texture. Perhaps one of the most famous examples of substrate colour influencing the variation of a species is that of the peppered moth *Biston betularia*, which is the classic case of industrial melanism. An example less dependent on man's activities is that of the water beetle, *Tropisternus collaris* in America. The colour patterns of the head, pronotum and elytra differ in accordance with the colour of the mud at the bottom of the pools in which the beetles live, for example, whether the pools are sandy or muddy. The beetles are predators and camouflage themselves against the substrate before pouncing on their prey. A fuller discussion of the similarities and dissimilarities of continuous and discontinuous variation is given in Grant (1963).

The effects of other environmental factors, such as temperature and humidity, have led to the proposal of several so-called ecogeographical rules which will be discussed in the following section.

Ecogeographical rules

It has been known for some time that certain features of animals, such as size and pigmentation, vary in a regular manner, usually in association with climatic conditions. Initial observations were made on birds and mammals and it was for these that the first ecogeographical rules were proposed. One of the earliest rules was put forward by Gloger who suggested that animals from warm and humid areas were more heavily pigmented than those from cool, dry areas. The observations were originally made on vertebrates but such a phenomenon has since been recognised in insects. In butterflies, for example, it is commonly found that the cold-season phenotypes of multivoltine (many generations per year) populations resemble the single phenotype produced by univoltine (one generation per year) high altitude or latitude populations. This observation has been verified by rearing butterflies in the laboratory under temperature regimes different from those of the habitat in which they were taken. This shows that the temperature effect is principally a physiological response (through growth and melanin deposition) and not genetic.

The ladybird *Harmonia axyridis* is a species found in south-east Asia that has distinctly polymorphic elytral patterning based on black/orange pigmentation. There are four main types, or morphs, controlled by a single gene with the order of dominance, with respect to the extent of the melanic area, expressed by the named forms: *conspicua-spectabilis-axyridis-succinea*. The number of individuals of each form was measured as a percentage of the population in samples taken from Japan, North China, Korea and Taiwan and it was found that there was a distinct and regular gradient in the relative frequency of types from the north and south of Japan (Fig. 1.2, where only the percentages of the extreme forms are shown).

This clinal variation in degree of pigmentation is said to conform to Gloger's rule and is explained by the effect of humidity and, perhaps, temperature. The light form, *succinea*, was also found to exhibit hardiness in cold climates.

The best known study of variation in ladybirds was that of Dobzhansky (1933) in which he concluded (p. 108) that 'homologous varieties of different species may be more similar to each other in appearance than the different varieties of the same species'. Dobzhansky examined two groups of ladybirds (i. *Coccinella septempunctata*, *C. quinquepunctata*, *Adonia variegata* and *Synharmonia conglobata*; ii. *Anatis ocellata*, *Coccinella transversoguttata*, *C. divaricata*, *Coccinula quatordecimpustulata*, *Anisosticta novemdecimpunctata* and *Adalia bipunctata*) which exhibit similar trends in the proportional distribution of individuals with more or less pronounced pigmentation. For all members of both species groups there exist geographical centres with less pigmented populations and geographical centres with more pigmented populations. Such centres for the different species roughly coincide in geographical locations. The centre for light forms lies in central Asia in the the eastern hemisphere as in, for example, *Synharmonia conglobata* (Fig. 1.3) and in California in the western hemisphere. In the Old World a significant centre for dark

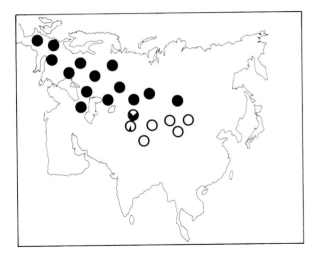

Figure 1.3 Geographical distribution of the typical form of the ladybird *Synharmonia conglobata* (black) and its variety *buphthalmus* (white) throughout Europe and Asia (Dobzhansky, 1933).

forms is situated in the Far East. However, the proportion of pigmented forms increases radially in all directions from each light centre, for example, not only from central Asia northeast to the dark centre of the Far East, but also northwest to Europe and south to Africa. Similarly in North America the dark forms increase east and north from California. According to these results, such a situation again supports Gloger's rule but only as far as humidity is concerned, the centres for light populations occurring in arid areas and pigmentation increasing with humidity. Pigmentation may not have any selective value but may merely coincide with changes in physiological processes.

Altitude has also been found to have an effect, as for example, on the thoracic and elytral coloration of the cicindelid beetle, *Cicindela depressula* which occurs from Alaska to California west of the Rockies. A study of nine populations suggested that there was a tendency for populations to be green and blue at high altitudes while at lower elevations the trend was to have brown elytra with green or coppery thoracic sides. Of course, it needs to be stated that it is difficult to suggest that altitude alone is correlated with change in colour pattern since changes in altitude are associated with changes in temperature, light regimes and probably humidity.

Perhaps the best known rules relate body size to climatic factors. Bergmann's rule states that body size in geographically variable species of warm-blooded animals is larger in the cooler parts of the range of the species. Allen extended this rule to include protruding body parts, such as tails, ears and bills, which he found to be relatively shorter in cooler regions. Classically, these rules are interpreted in terms of the relationship between body surface area and environmental temperature and are expressed in simple mathematical terms. As an object increases in size its surface becomes relatively smaller (increases as the square) than the volume (increases as the

cube). When the protruberances are relatively shorter then the surface area is reduced even further. The traditional physiological explanation is that this mechanism reduces heat radiation so that animals in colder climates have a relatively smaller surface area from which they may lose heat. However, there has been some doubt cast on whether size change with ambient temperature is really a consequence of heat conservation. The opposing view is that the relative reduction in surface area is hopelessly small for an effective reduction in heat loss, and holds that the principal mechanism for regulation of body temperature is insulation and vascular control. However, there have been no critical experiments to test which theory is correct.

These rules were founded on studies of endotherms and indeed the majority of animals studied conform to these predictions. They may be illustrated by reference to our own species in which a number of distinct 'races' are readily recognisable, each with its own distinct geographical origin. Pigmentation is an example, where the skin colour appears to be darker in the hotter regions of the earth – black in the more tropical areas and brown in deserts. This is related to the effect of humidity, since in a dry atmosphere evaporation takes place readily so there is less discomfort than in a damp atmosphere.

Recently Ray (1960) has tested whether these rules also apply to poikilotherms (excluding fish) by rearing both vertebrates and invertebrates at different temperatures. He found that body length increased between 10 and 50 per cent for a 10° F (5·5° C) decrease in temperature and that body weight was affected to an even greater extent, increasingly by well over 100 per cent. Protruding body parts showed a less dramatic difference, usually 2–9 per cent less than that shown by body length. In fishes it has been found that temperature influences the definitive number of vertebrae and fin-rays. Thus, overall, the relatively good correlation between the laboratory findings and the observations in nature show that these rules also apply to poikilotherms.

In insects, the final size of an individual may be an incidental rather than a functional response to temperature. It has been shown in mosquitoes, for example, that the size of an adult depends very much on the ambient temperature at the time of emergence from the pupa (more so than the temperatures at which the larva grew). It appears that the lower the temperature, the longer the insect takes to expand before the chitinised exoskeleton begins to harden and prevent further expansion. Thus, at higher temperatures the insect has relatively little time available for expansion before the temperature-dependent hardening process inhibits expansion, resulting in a smaller insect.

It is of interest to note that the correlation between size and temperature has been used as evidence for 'Lamarkian' evolution (inheritance of acquired characters) in different parts of the geographical range of a species. This is now viewed differently through the process known as the Baldwin effect: the adaptations of individuals of a species in particular environmental conditions may eventually,

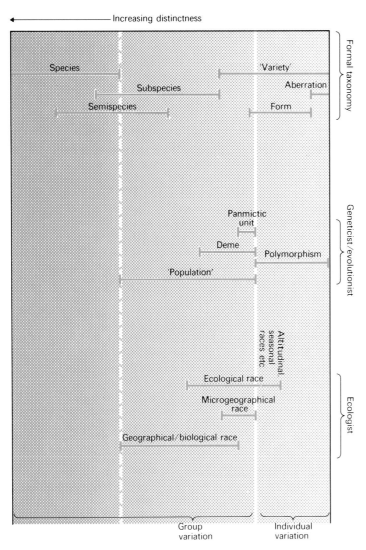

Species

Subspecies

Semispecies

'Variety'

Aberration

Form

Formal taxonomy

Panmictic unit

Deme

Polymorphism

'Population'

Geneticist/evolutionist

Altitudinal seasonal races etc.

Ecological race

Microgeographical race

Geographical/biological race

Ecologist

Increasing distinctness

Group variation

Individual variation

Figure 1.4 Diagram to show relationships of terms used in intraspecific variation studies.

ation to altitude or latitude should be made with considerable reservation. This may be the reason for the apparent exceptions to the rules that have appeared in the literature. Furthermore, these rules are based on observation and make no predictions as to the underlying mechanisms – the heat regulatory function has been associated with them to such an extent that it has become part of the conventional wisdom, despite the lack of experimental evidence. Finally, however, it should be noted that although the environment can influence size, colour etc., such phenomena do have a hereditary basis as will be discussed below ('Genetic basis of variation').

We have thus seen how organisms vary geographically and what effects the environment may have on organisms to effect such variation. We will now discuss the repercussions such variation has on taxonomy.

Geographical variation and taxonomy

The study of geographical variation was the by-product of taxonomic research when specimens of a species from geographically widely separated localities came to be examined. Were the differences merely due to geographical variation or was there more than one species involved? Could the forms be considered to be distinct subspecies or just local variants? Were the differences, in fact, due to geographical variation or simply individual variation or freak specimens? There thus evolved a nomenclatural nightmare with individuals or groups of specimens being assigned to different categories by taxonomists on the one hand and evolutionists on the other. It should be made clear that, to a certain extent, such categories (and their accompanying definitions) are essential for effective communication; that many of the discrepancies between definitions reflect differing interests, and that many of the categories are not strictly equivalent. The basic unit of the taxonomist is the species (however it is defined, if at all!) whereas for the evolutionist and geneticist it is the population or, more precisely, the deme (a local breeding population). These two attitudes are not irreconcilable, they merely reflect different objectives.

The relationships between the commonly used categories are summarised in Figure 1.4 and are discussed below. One of the main reasons for the present confusion is the acknowledgement of the polytypic species. As its name implies this type of species shows more than one form. In its first usage the polytypic species was associated with morphological differences and variation but now there are examples known where physiology, ecology, cytology and behaviour, for example, differ. There are many historical/ evolutionary causes of this polytypy but here we shall only consider differences derived from the spatial distribution of populations in a heterogeneous environment.

The term variety has been extensively used in the past to refer to any deviation from the typical or ideal of the species. In the early days of taxonomy the typical or ideal usually meant that individual(s) that was examined first. The term variety has been greatly misused and has led to

under the influence of selection, be reinforced or replaced by similar hereditary characters. Experiments on *Drosophila* fruitflies confirm this interpretation.

The widespread adherence to Bergmann's rule throughout the animal kingdom suggests a fundamental relationship between temperature and growth. Both Bergmann's and Allen's rules, however, are only widespread if expressed in their purest sense – relating size to ambient temperature and *not* directly to latitude or altitude. It must be stressed that comparisons should be made at the intraspecific and not the interspecific level since, for example, when comparing temperature and size, it is of no significance that the largest species of insects are found in the tropics. For geographic studies it is very important to realise that the ecogeographical rules express a relationship between size and temperature and therefore extrapol-

much confusion as it has included two very different sources of variation. Firstly, individual variants within a single population (due to balanced polymorphisms or freak specimens) and secondly, different populations of a polytypic species (a major reason for which is geographic variation). The term variety has now largely disappeared from use (although still retained to a certain extent in, for example, Lepidoptera and the Coccinellidae), individual variants now being termed aberrations, population variants, subspecies and forms of fixed polymorphisms.

The basic unit of the evolutionary geneticist is the deme which basically refers to the largest group of individuals in which panmictic (random) mating occurs. Numbers of individuals in a deme may vary from a few score to millions and gene flow occurs between demes which may be transitory and vary in size from season to season. The definition of a deme may vary according to the organism studied. Dobzhansky (1970), for example, considers it to be a local population composed of one or more panmictic units. There is no taxonomic equivalent of the deme but it is by no means a purely theoretical concept as demes have been demonstrated in several groups of animals.

The next category in the heirarchy is the race which has, for some taxonomists at least, a concrete taxonomic equivalent in the subspecies. The extent to which these two categories are equated varies from group to group as well as from person to person. Frequently races are placed as subordinates of a subspecies as in, for example, the European fish *Coregonus laveratus* which has 42 subspecies of which one, *C. laveratus wartmanni*, has 24 races (natios) (Berg, 1932).

The term race has been variously defined. For example Dobzhansky (1970) regards races as genetically distinct Mendelian populations which, as a rule, are allopatric and are only maintained as distinct by geographical separation. Others have defined the term as a potential subspecies or stage in a stepped cline. The term usually refers to a geographic race, a microgeographic race being a group of organisms confined to a small area (sometimes equivalent to a deme). However, many other types of race have been defined in the literature such as biological, behavioural, ecological, chromosomal, altitudinal, seasonal and host-specific races.

Once again the use of the terms race and subspecies has led to much confusion and controversy. How then does a taxonomist decide to what category a particular organism or group of organisms belongs and what criteria and methods does he use to aid his decision? The taxonomist examines available material and, on the basis of the extent of the discontinuities (usually morphological) and according to experience, he decides whether the sample is a distinct species or merely a geographical variant. In the latter case, distinct infraspecific groups may be assigned to subspecies (the lowest rank permitted in formal nomenclature). The use of a subspecific name may be quite arbitrary since the discontinuities between samples may be due to a lack of available material from intervening

regions. As collections become more comprehensive variation between two previously designated subspecies is sometimes found to be clinal and the discontinuities not so clear-cut as originally thought. However, in some cases where intermediate populations are known not to exist (for example between island, mountain and isolated forest populations) the term subspecies is often applied with more confidence.

The taxonomist's subspecies varies in its practical expression from group to group and often its use merely reflects the current status of the taxonomy of a particular group rather than a true biological entity. Most decisions are based on morphological evidence and only rarely on behavioural and genetic characters. It is important to note that the absence of the subspecies category in some groups, for example as in the Diptera, does not mean that substantial geographic variation does not exist – it merely reflects the attitude of taxonomists towards its use.

There have been some attempts to make the recognition of subspecies more 'objective' by the use of rules, a commonly used criterion being the 75 per cent rule, originally introduced by ornithologists. According to this rule zoologists (rather than botanists) recognised distinct subspecies as those populations that are sufficiently distinct to allow correct differentiation of at least 75 per cent of specimens of the one from the other. Many have pointed out that this rule is absurd, since for example the very inbred inhabitants of the more remote East Anglian villages of England would be assigned to a distinct subspecies!

There has been a recent trend towards quantifying differences between populations and towards using numerical techniques to analyse the data. However, many characters cannot be compartmentalised because they show clinal variation and these techniques have not proved very rewarding. However, despite these shortcomings, work on genetic differences between populations (allelic substitutions found by electrophoretic methods) has shown some general trends. For example, in the *Drosophila willistoni* complex in South America an average of 23 allelic substitutions per 100 loci were estimated to have occurred in the evolution of two subspecies from a common ancestor. This is, on average, 10 times greater than the differences found between local populations of each subspecies. Between 10 and 25 allelic substitutions per 100 loci are estimated to have occurred in the evolution of pairs of subspecies of such diverse organisms as *Drosophila*, sunfishes, salamanders, iguanid lizards and rodents (Ayala, 1975).

One of the commonest but erroneous assumptions in the study of geographical variation is that characters that vary geographically do so together. This is not so since different characters are affected differently by different environmental factors and a species may belong to as many clines as it has characters. For example, in the moth *Lymantria monarcha* eight characters were found to vary independently in populations from south-west England (Ford,

1971). Under these conditions, different subspecies could be proposed depending on which characters were chosen to base them on. To assign subspecies on the basis of the study of a single character would be meaningless but unfortunately this occurs all too often. The extraordinary microgeographical variation found in snails such as *Cepaea* and *Partula*, discussed above, could be used to generate a multitude of subspecific names (even using the most restrictive definition of a subspecies) if the procedures so common in studies on Lepidoptera, birds and mammals were adopted. Such a plethora of names would only serve to hinder communication of ideas, not to expedite them, as is one of the objectives of classification.

In recent years there has been a growing dissatisfaction with the subspecies concept in taxonomy, or rather the multitude of concepts resulting in no coherent concept. Wilson & Brown (1953) have expressed considerable doubt about its utility and philosophical basis and suggest that the category should be abandoned as it is 'the weakest category in systematics'. The use of subspecific names not only implies discontinuity where none may exist but also unity where there may be discontinuity. Mayr (1969) suggests that more effort should be put into describing geographical variation accurately, rather than describing new taxa.

The fact that, in reality, the entities usually referred to as subspecies are not always, or even usually, discrete and that they may be connected by transitional populations is itself of considerable interest. Opinion differs on the extent of gene flow between populations, but again there are no hard and fast rules. To think that subspecies (or races) do not exist simply because they cannot be defined, is fallacious typological thinking. No one denies that Mount Everest exists; the problem is deciding where Everest begins and the adjacent mountain ends. If taxonomy is to benefit from the advances in evolutionary biology, then a system must be developed which is capable of recognising indistinct species as well as smaller units, such as populations or geographical isolates. The traditional taxonomic system was developed prior to neoDarwinian concepts of evolution, and it is unable to incorporate potentially incipient species in a realistic manner.

In most forms of taxonomy there is one unifying concept – that species form mutually exclusive sets, which may be ordered in an hierarchical manner. Although this axiom has been the basis of most taxonomy, it does have considerable restrictions when dealing with highly variable species and with the various infraspecific categories. Usually, all taxonomic observations are fitted into this model even when the data suggest it may not be wholly appropriate. One future line of development in which this problem may be resolved is to allow subspecies, and for that matter species, to overlap in terms of morphological or any other trait. One procedure which would allow this to a limited degree, uses the statistical technique of multiple discriminant analysis and has been used experimentally for the subspecies of bumblebees with some success (DuPraw, 1964). Basically, this system uses scatter diagrams (multiple discriminant function diagrams) as the classification. Using several characters simultaneously, multiple discriminant analysis maximises the ratio of the variance between groups to the variance within groups (populations or samples). The relative contribution of each character to the analysis is therefore directed to those providing the best discrimination between taxa. Individual specimens are points plotted onto the diagrams which usually form areas of high density linked to other similar areas by intermediate zones of lower density. Traditionally, the areas of high density would be circumscribed with more or less arbitrary boundaries. This generalisation from actual specimens to the taxonomic concept of a species may introduce serious artifacts of information. In the case of distinct species with little geographic variation or intermediate specimens, there would be no problem. The method is most useful in those cases where species overlap, either morphologically, cytologically or physiologically. In such cases, the grouping of specimens is poorly defined, but the traditional taxonomic concept of the species would imply that they were sharply defined and mutually exclusive.

Identification of a population or a new sample requires a simple arithmetic calculation based on the observed features in the sample, and it is fixed exactly on the diagram. Whether this system is truly non-Linnaean, as has been claimed, is a matter of theoretical argument; it does however overcome most of the shortcomings of Linnaean taxonomy in the classification of geographical isolates.

Genetic basis of variation

We have now seen that species do vary throughout their ranges (in fact they are more likely to do so than not) and we have also seen that this can cause problems for a taxonomist; but what is the basis for this variation? On what do the effects of the differing environmental conditions act to produce the forms we see?

It has long been accepted that the environment does not directly affect the individual to produce adaptive morphological and physiological features, but rather that it acts through natural selection on the gene composition of the population as a whole. Also, as we have seen, geographical variation is principally concerned with the variation of spatially segregated populations. Most geographical variation is based on the inherent genetic variation of a population although there are instances where non-genetic factors have been found to play a significant role.

The phenotype under genetic control may respond in time and space in various ways such as in age variation, seasonal variation (summer and winter plumage and fur for example) and seasonal generations. Substrate factors such as the quality of the soil or of the water may also have an effect as seen for example in sessile animals such as freshwater bivalves. Phenotypes resulting from modification by edaphic or other ecological conditions are known as ecophenotypes and have been most intensively studied in plants.

To return to genotypic variation, every local population is adapted, through natural selection, to a specific environment. Since there must, theoretically, be an optimum genotype for a population in any particular environment so there ought, theoretically, to be a genetic uniformity within that population. This, however, is found not to be the case as variation is a fundamental and characteristic property of all biological systems. All organisms contain a great store of latent genetic variability which, even though it may not manifest itself totally in any individual phenotype, is an extremely important attribute of the population and, ultimately, the species as a whole.

The maintenance of genetic variability is of extreme importance and has numerous advantages for ensuring the survival of the species. For instance it may harbour genotypes that are able to survive catastrophes such as dramatic environmental changes which produce conditions adverse to the modal genotype that is best adapted to the original conditions. Likewise it may contain genotypes capable of colonising new and previously unfavourable environments; this allows for greater utilisation of different environments and marginal habitats and thus counteracts specialisation and gives plasticity to the species. However, too much variation can be disadvantageous as it may lead to the production of many locally inferior genotypes, so there must consequently be a balance between genetic variation on the one hand and the selective forces of the environment on the other. The action of the latter is commonly referred to as 'balancing natural selection' and it acts as a buffer in this homeostatic biological system to maintain the optimum fitness of a population while at the same time endowing it with a certain plasticity and potential to evolve.

Morphological variation as the result of genetic differences can be either continuous (for example, longer versus shorter, lighter versus darker) or discontinuous (brown eyes versus blue eyes) just as in geographical variation (for instance, clines) but for differing reasons. When variation is discontinuous it is known as polymorphism and the resultant character types are referred to as morphs. For any given polymorphic character there may be two or more morphs, the relative frequencies of which may vary from population to population. As a result of geographical variation, morph frequencies, just like gene frequencies, may vary clinally or discontinuously throughout a species' range.

According to the Hardy-Weinberg Law, in a closed population the frequency of a particular gene remains constant in the absence of selection, of non-random mating and of accidents of sampling (genetic drift). However, the occurrence of new genetic factors will result in alterations of gene frequencies and consequently give rise to genetic variability. The principal sources of this genetic variability are mutations, gene flow and recombination. Mutations caused by alterations in the molecular structure of the gene are usually deleterious and are quickly eliminated by natural selection. Some, however, may be positively advantageous but more usually they have no direct effect at all and remain as part of the latent variability of the population.

Gene flow between adjacent populations as a result of migration is a significant source of new genetic material. Populations are rarely closed systems as hypothesised in the Hardy-Weinberg Law but constantly exchange genetic material. Gene flow is also an important factor in the formation of clines and tends to smooth out differences between populations.

The most important source of genetic variability in sexually reproducing organisms is recombination. The biological significance of sexual reproduction is that it permits the exchange of genetic material between individuals. Basically, recombination of parental genetic material increases the potential variability of the offspring of any given set of parents.

These then are the sources of genetic variability; what then are the factors that reduce the genetic variation of a population? Essentially they are natural selection and genetic drift. Genetic drift has been defined by Dobzhansky (1951, p. 156) as 'random fluctuations in gene frequencies in effectively small populations'. In such small populations, especially if in total or partial isolation, particular genes may be lost completely and thus give the population certain characteristic traits.

Natural selection is, as its name implies, 'Nature's way' of ensuring that only those particular phenotypes that are best adapted for the particular environmental conditions concerned will survive and thus perpetuate that particular genotype through their descendants. Geographical variation presents a variety of environmental selective forces and thus gives rise to the predominance of certain genotypes in populations from different areas or habitat-types.

Geographical variation and the process of speciation

Geographical variation plays a central role in most discussions of speciation mechanisms, especially in the controversy concerning the allopatric–sympatric models of speciation (see Chapters 3 & 4). The main difference of opinion is as to whether populations need to be geographically separated or isolated before differentiation may take place.

One of the classic models of speciation is called allopatric or geographical speciation. The essential feature of this hypothesis is that populations become spatially separated by any one of a range of barriers such as advancing ice caps, rising water levels, interstitial areas of hostile environments (for instance, forests in deserts, mountains) and any patchy environment in general. The separated populations become genetically differentiated due to genetic drift and natural selection in the new restricted environment. If the barrier is subsequently removed after this period of independent development there are several possible out-

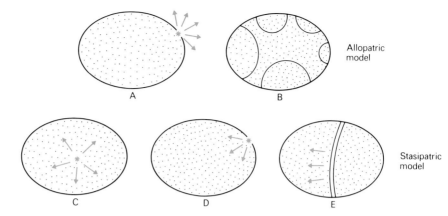

Figure 1.5 Allopatric (A, B) and stasipatric (C–E) models of geographic speciation. Allopatric model: A, an isolated peripheral deme spreads into areas previously unoccupied by the species, or B, the species range is disrupted to form several isolated groups of demes. Stasipatric model: C, the chromosomal rearrangement establishes itself in a non-peripheral local colony and spreads through the range of the species on an advancing front. D, the chromosomal rearrangement establishes itself in a peripheral colony and then spreads through the existing species population. E, the result of C or D: a narrow hybrid zone showing a slow secular movement across the territory occupied by the species until it is arrested in some way. (Based on White, 1968.)

comes: either the two populations have not diverged sufficiently and they will interbreed freely; interbreeding will occur but with low fitness of hybrids; or the divergence may be such that the populations will no longer be able to mate or produce viable offspring. The last possibility is considered to be speciation. It may be that peripheral populations become isolated (Fig. 1.5A) or that the whole range of a species is disrupted and divided into several refuges which develop independently (Fig. 1.5B). Many biologists maintain that this explanation is the only means of speciation in sexually reproducing organisms (Mayr, 1970) but others (Dobzhansky, 1970; Scudder, 1974; White, 1978) maintain that it is only one of a range of speciation processes. Although theoretically the allopatric explanation is quite plausible, this type of speciation, like all others, has never actually been observed.

In the model described above, one vital point is that populations need to be completely separated with absolutely no exchange of genetic material (gene flow). However, it is becoming clearer from recent research that speciation can take place in the face of considerable gene flow (see also Chapter 4), and under these conditions the spatial arrangement of populations, especially geographical variation in gene frequencies, is crucial.

Work on the genetics of morabine grasshoppers by M. J. D. White and collaborators (see White, 1974, 1978) has led to the proposal of an alternative method of speciation – stasipatric speciation. The ecology, distribution and evolution of several species of these wingless grasshoppers – the *viatica* species-group – have been studied in their natural habitats in southern Australia. The members of this group normally live at ground level or on low vegetation. Morphological differences between them are minimal and some characters show considerable geographical variation in individual taxa. In the *viatica*

species-group there are coastal and inland populations, but the coastal forms show (Fig. 4.2) major rearrangements of chromosomes (pericentric inversions). White has shown that different chromosomal types of grasshoppers are spatially separate with little overlap in distribution but with extensive adjoining populations (parapatry).

Stasipatric speciation involves the origin of chromosomal rearrangements (commonly found to be polymorphic in many animals) giving adaptively superior homozygotes to a particular part of the range of a species, but with inferior heterozygotes. Although most rearrangements die out by natural selection, White believes that on rare occasions one of them may spread geographically throughout the territory occupied by a species and give rise to differentiation between populations carrying the rearrangement (Fig. 1.5C–E). This is especially important in patchy environments.

The intrusion of one form into the territory of another is continuously opposed by diminished fertility of hybrids. The zones of overlap between populations of different chromosomal types is not static, they move as bands to and fro across the country. The boundary of mixed chromosomal types, as well as hybrids between them, was found in Australian grasshoppers to be as narrow as 300 metres, but introgression extends much further, up to 10 km. Although the two populations are connected via an intermediate population, there is little evidence of significant gene flow between them. The populations have minimal or non-existent behavioural reproductive isolation at the interfaces, so application of traditional biological species concepts would not regard them as different species (or even subspecies) even though there is little introgression between them. The principal difference between this hypothesis and that of allopatric speciation is that no physical barrier to the interruption of gene flow is believed to have occurred at any stage of the process (see also Chapter 4).

A particularly interesting example of the role of geographical variation in speciation concerns a number of laboratory studies of artificial clines in *Drosophila*. These experiments show that two important factors – gene flow and natural selection – have an overriding importance in speciation along clines (Endler, 1974).

The relative magnitude of selection and gene flow alter the extent to which a given deme's gene frequency is influenced by that of its neighbours. Low local selection and high gene flow will, as expected, produce poor differentiation. In laboratory experiments (see Endler, 1974, 1977 for details), 15 demes of *Drosophila* were set up, and the resulting clines for gene frequency of a marker gene over 35 generations were followed. It was found that there was no significant difference between the selective cline with 40 per cent gene flow and that without gene flow, so the effect of gene flow in the experiment was not detectable. This suggests that under strong selection pressure in each deme, differentiation is possible in spite of gene flow as high as 40 per cent.

It is an oversimplification to consider only gene flow and selection as this ignores their spatial patterns. In most circumstances pockets of individuals are often found in favourable conditions. Gene flow occurs in both directions along a cline, so if dispersal (not migration) is regular, then the net effects on differentiation will be minimal. This is because the mean gene frequency of immigrants will not differ substantially from the population itself. This neutralising effect will be the same for all values of gene flow, hence clines from smooth environmental gradients will be insensitive to the attenuating effects of gene flow as shown by experimental clines in *Drosophila*. Furthermore, it is possible for local differentiation to occur along a relatively weak environmental gradient – perhaps at a lower level than it is possible to measure in the field but which can be controlled in the laboratory. The equalising effect of gene flow along an environmental gradient can be reduced if there is some distinct spatial event or an abrupt change in the slope of the environmental gradient. Under these conditions, immigration may not be the same from both directions along the cline. This would be very important at the ends of clines or at the edges of a distribution, and may represent some form of edge-effect.

Results of experiments and theoretical models show that it is possible for local differentiation to evolve parapatrically in spite of considerable gene flow if the selection gradients are relatively uniform. Irregularities in environmental gradients increase the sensitivity of demes to differences in immigration rates (= gene flow). In some cases even a gentle environmental gradient can give rise to marked spatial differentiation along a genetically continuous series of demes. Such environmental differences may be below the practical limits of detection in the field. Any asymmetry in gene flow does not lead to dedifferentiation if the environmental gradient is smooth – it merely shifts the position of the transition zone between the differentiated areas from that which would be expected if there were no asymmetry. This has important repercussions for interpreting field studies. Abrupt changes in gene frequency, morph frequency, etc., should not therefore be interpreted necessarily to mean there is a change in the immediate local environment.

Conclusion

Geographical variation in species has many repercussions in almost all aspects of biology. An awareness of its existence and an understanding of its causes are important for the interpretation of the species and the process of speciation. Taxonomists, geneticists, evolutionists and ecologists alike are becoming increasingly aware of the need to study the geographical variation of species in greater depth. Of particular importance is the need for studies that try to correlate the level of gene flow between populations with the observed genetic and phenotypic variation between those populations. These studies will continue to be of importance as long as variation below the species level is thought to hold the key to our understanding of the origin of species.

References

Ayala, F. J. 1975. Genetic differentiation during the speciation process. *Evolutionary Biology* **8**: 1–78.

Berg, L. S. 1932. Übersicht der Verbreitung der Süsswasserfische Europas. *Zoogeografica* **1**: 107–208.

Darwin, C. 1859. *On the origin of species by means of natural selection* (6th ed. 1972). London: Murray.

Dobzhansky, Th. 1933. Geographical variation in lady-beetles. *American Naturalist* **67**: 97–126.

Dobzhansky, Th. 1951. *Genetics and the origin of species* (3rd ed.). 364 pp. New York: Columbia University Press.

Dobzhansky, Th. 1970. *Genetics of the evolutionary process*. 505 pp. New York: Columbia University Press.

DuPraw, E. J. 1964. Non-Linnean taxonomy. *Nature* **202**: 849–852.

Endler, J. A. 1974. Gene flow and population differentiation. *Science* **179**: 243–250.

Endler, J. A. 1977. *Geographical variation, speciation and clines*. 246 pp. Princeton: Princeton University Press.

Ford, E. B. 1971. *Ecological genetics* (3rd ed.). 410 pp. London: Chapman & Hall.

Grant, V. 1963. *The origin of adaptations*. 606 pp. New York: Columbia University Press.

Mayr, E. 1969. *Principles of systematic zoology*. 428 pp. New York: McGraw-Hill.

Mayr, E. 1970. *Populations, species and evolution*. 453 pp. Cambridge, Massachusetts: Belknap Press.

Ray, C. 1960. The application of Bergmann's and Allen's rules to poikilotherms. *Journal of Morphology* **106**: 85–108.

Sclater, P. L. 1858. On the general geographical distribution of the members of the class Aves. *Journal of the Proceedings of the Linnean Society*, London **2**: 130–145.

Scudder, G. G. E. 1974. Species concepts and speciation. *Canadian Journal of Zoology* **52**: 1121–1134.

Thorpe, R. S. 1976. Biometric analysis of geographic variation and racial affinity. *Biological Reviews* **51**: 407–452.

Wallace, A. R. 1876. *The geographical distribution of animals*. 1110 pp. London: Macmillan.

White, M. J. D. 1968. Models of speciation. *Science* **159**: 1065–1070.

White, M. J. D. 1974. Speciation in the Australian morabine grasshoppers – the cytogenetic evidence. pp. 57–68 *in* White, M. J. D. (Ed.), *Genetic mechanisms of speciation in insects*. Sydney: Australia & New Zealand Book Co.

White, M. J. D. 1978. *Modes of speciation*. 455 pp. San Francisco: Freeman.

Wilson, E. O. & Brown, W. L. 1953. The subspecies concept. *Systematic Zoology* **2**: 97–111.

Young, F. N. 1960. Regional melanism in aquatic beetles. *Evolution* **14**: 277–283.

CHAPTER 2

Semispecies, sibling species and superspecies

G. B. White

In the context of population genetics, it can be shown that reproductive isolation is an important evolutionary force. As long as cross-breeding occurs between populations, gene flow causes their gene pools to contain mostly the same alleles, albeit at differential frequencies determined by natural selection. After genetic exchange has ceased, new mutations and coadapted gene arrangements are restricted to the populations in which they arise. Thus, as soon as populations become separated in contrasted habitats that exert differential selection pressures, their phenotypic divergences accelerate. Any mutant or genotype giving characteristics that inhibit cross-breeding will tend to promote speciation. Studies concerned with the evolution of isolating mechanisms are thus an important part of research into the origin of species.

The differences accumulated by incipient species involve various attributes – anatomical, ecological, ethological, seasonal, physiological and so on. Species remaining allopatric have no opportunities for cross-breeding, whereas those which attain overlapping geographical distributions (sympatry) depend on having developed some kind of reproductive isolating mechanism to prevent interspecific hybridisation. Without such mechanisms, sympatric species might interbreed regularly and lose their separate status. Lewis & Bloom (1973) reported the loss of *Clarkia* species in this way. In most cases, the primary isolating mechanism is some sort of specific 'sex appeal', usually backed up by hybrid zygotic failure. These phenomena are not necessarily consistent in all the members of any particular species-group. For example, after test-crossing 28 species of Hawaiian Drosophilidae, Carson *et al.* (1970) reported six cases of hybrid fertility, although most of the crosses were incompatible or unproductive. They concluded 'that sexual isolation is an essentially fortuitous accompaniment of speciation.'

Case studies on many representative groups of organisms have revealed the following categories of isolating mechanisms causing reproductive barriers between species (Dobzhansky, 1970):

Pre-mating or pre-zygotic
Geographical separation – distributions of the species do not overlap.
Ecological separation – species occupy different niches, biotopes or habitats in the same territory.
Temporal separation – species flower or reach sexual maturity at different times of day or year.
Ethological separation – males and females are not attracted to, nor are they receptive to, the opposite sex of other species.
Mechanical separation – structural differences of flowers or genitalia prevent cross-pollination or insemination of individuals of one species by those of another.
Gametic separation – male and female gametes from individuals of different species fail to reach and fertilise each other.
Post-mating or post-zygotic
Hybrid inviability – eliminates many or all hybrids through mortality before they reach sexual maturity.
Hybrid sterility – disrupts gametogenesis so that hybrids produce no functional gametes
Hybrid breakdown – reduces the viability or fertility of hybrid progeny (as for hybrid inviability and hybrid sterility) affecting both backcrosses and successive filial generations.

In each species group, several different isolating mechanisms tend to reinforce each other in preventing cross-breeding between even the closest related species. Between two sympatric mosquito species of *Aedes* subgenus *Stegomyia*, for instance, Leahy & Craig (1967) identified no less than six intrinsic isolating mechanisms. Since this closely related pair of species was probably allopatric until recently, these reproductive barriers would not have been needed to stop them interbreeding.

Generally, we may regard the first sort of isolating mechanism to arise and prevent populations from interbreeding as the primary source of speciation in each case. Supplementary reproductive barriers are also likely to develop and would become effective if the primary barrier breaks down at any time, artificially or by mistake. This may happen, for example, if a male inadvertently copulates with a female from another species; of course she may reject him but, if cross-insemination occurs, hybrid sterility may result. The latter isolating mechanism would not have existed before the populations speciated. Thus, we can conclude that pre-mating barriers are more likely to be primary means of speciation, while post-mating barriers negate the effects of attempted cross-mating or accidental cross-pollination. The essential point is that species must breed true in order to maintain their integrity.

Modes of speciation

A recurring theme in this book, and throughout evolutionary biology, is the circumstances leading to species formation. What are the forces that cause parental populations to segregate and differentiate into derived species, and under what conditions do these forces operate? Major reviews by Bush (1975) and by M. J. D. White (1978) have helped to clarify current thinking about the various ways that speciation might occur. Although complete geographical separation of sub-populations is the most obvious and arguably the most frequent scenario for speciation, there is mounting evidence that reproductive barriers might arise between populations that are not entirely allopatric.

Each sort of isolating mechanism listed above, except the first mentioned could, in principle, cause speciation under sympatric circumstances. Much controversy over the likelihood of this stems from the impossibility of knowing whether currently sympatric species evolved sympatrically in the first place. Further confusion arises from semantics. Species that might be regarded as partially or completely sympatric, depending on how much their geographical ranges overlap, can seldom be also regarded as syntopic (filling the same niche or biotope). Three species of human lice are undoubtedly sympatric, infesting *Homo sapiens* in all parts of the world, but their topographical distribution on the human body is decidedly allotopic: *Phthirus pubis* in the axillae, *Pediculus humanus* on the body and mainly the clothing, *P. capitis* exclusively on the head hairs (Busvine, 1978). Thus it is possible to explain many alleged cases of sympatric speciation by adopting stricter definitions for the spatial distribution of populations. The idea that sympatric populations are those within 'cruising' range of each other whilst being reproductively active (Mayr, 1963) can hardly be applied to plant populations; also, for invertebrate animals the concept of sympatry is irrelevant when populations are effectively allopatric or allotopic on a miniature scale, due to divergences of habitat, biotope and ecological niche specialisation.

Between the broad extremes of allopatry versus sympatry comes the possibility of what is usually termed parapatric speciation. This system involves polarisation of the ancestral population by development of atypical breeding characteristics by individuals on the periphery of its range. Separate species may result in different geographical or ecological situations through the operation of disruptive selection (Murray, 1972). When there is evidence that a derived species has budded off from one point at the edge of, or even within, the range of another, it has been termed stasipatric speciation (White, M. J. D., 1978).

Theoretically at least, viable mutations might be able to promote sudden speciation anywhere. Goldschmidt (1940, p. 390) coined the expression 'hopeful monster' for any such deviant individuals as may produce a population of descendants that are reproductively isolated from the parental stock, within the range of which they must first occur.

It is implicit that these and other possible modes of speciation, through parthenogenesis (Chapter 7) or polyploidy (Chapter 8) for instance, will involve biological adaptations and divergences that lead to the exploitation of new resources and the occupation of new niches and biotopes by the freshly forming species. And yet, during the early stages of their divergent evolution, closely related species will retain a majority of shared characteristics, making it easy to classify them together as species-groups and superspecies.

Superspecies as phylogenetic units

According to the outmoded dogma that allopatric speciation is the invariable rule, the superspecies was defined as 'a monophyletic group of entirely or essentially allopatric species that are too distinct to be included in a single species' (Mayr, 1963, p. 672). Clearly, this left it open to taxonomists to make subjective judgements on ranking geographical populations as species within superspecies, or merely as subspecies within species. Unfortunately the biological species concept is sometimes difficult to apply to the components of superspecies since, like subspecies, they may lack reproductive isolating mechanisms other than simple geographical separation. Indeed, the functions of intrinsic reproductive barriers only begin to operate when incipient species become sympatric. Because of the problems encountered in deciding whether to treat some populations as species or as subspecies, the category of semispecies was devised for borderline cases. We shall comment on this category later. Here we should clarify some of the terminology used in phylogenetics.

Bearing in mind that evolutionary inferences are not proven facts, we can never be sure of knowing any phylogeny, such as the pattern of evolutionary relationships interpreted from fossil evidence and the characteristics of contemporary species. However, we do strive to classify organisms phylogenetically; that is, on the basis of common ancestry. The only practical alternative is to opt for a system of phenetic classification, whereby associations and groupings are based on overall similarities. The latter approach is useful for microorganisms and other taxa with features we cannot judge as being either ancestral or derived. In evolutionarily puzzling groups, such as some kinds of bacteria and protozoa, the most straightforward classification is a phenetic arrangement, based on features such as size, shape, numbers of nuclei, cytoplasmic organelles. For larger and better known plants and animals, however, we prefer to employ characters that are assumed to reflect real evolutionary relationships for the purpose of plotting the best possible approximation to a truly phylogenetic classification. In any such scheme, monophyletic units are composed of all the descendants of an hypothetical ancestor. Pairs of equivalent monophyletic units are termed sister-groups, and the closest related pairs of species are sister-species. Where it is possible to interpret, with reasonable certainty, that several still similar species had a common ancestor, this evidently monophyletic species group is treated as a superspecies. The constituents of most recognised superspecies display allopatric, often also contiguous, specific distributions. And yet, if non-allopatric modes of speciation are to be currently accepted, then there is no reason why superspecies should not sometimes include sympatric species.

Semispecies as the components of superspecies

Under whatever circumstances each act of speciation happens, panmictic populations normally seem to pass through gradual stages of differentiation, from the localised deme to the race and subspecies, before becoming fully distinct biological species. It follows, therefore, that incipient species spend some time and generations in a transitional state on the gradient from subspecific to specific status. Populations of this kind, as they acquire effective isolating mechanisms, can be categorised as semispecies (Lorković, 1958). Because the components of superspecies sometimes cross-breed where they come together, being relatively recently separated taxa with weak reproductive barriers, some taxonomists (particularly ornithologists and dipterists) prefer to treat all such partially interbreeding populations as semispecies.

Now we have come to appreciate more of the importance of interspecific biological divergences, it is easy to see how species, which are differentially adapted but closely related, may tend to merge in zones where their habitats intergrade. One of the earliest studied cases is that of the European Hooded and Carrion Crows (*Corvus cornix* and *C. corone*), largely allopatric taxa with a narrow (up to 80 km) hybrid belt stretching across Scotland and down through Europe (Fig. 2.1 A) having remained in apparent equilibrium for more than a century. *C. cornix* and *C. corone* are unmistakably distinct taxa, with essentially different breeding grounds, but authorities disagree on whether to rank them as subspecies, semispecies or species. As work on other groups reveals more subtle population divergences, it becomes more acceptable to regard these crows as a superspecies comprising two semispecies.

The mouse *Mus musculus*, usually considered to be a single species, presents a curious parallel with the superspecies *C. cornix/corone*. For many years *M. musculus* has been subdivided taxonomically into four European subspecies, two of which need not concern us here. The boundary between the light-bellied *musculus* and the dark-bellied *domesticus* populations runs down much the same line, from Jutland to the Alps, that divides *C. cornix* in the east from *C. corone* in the west. A narrow hybrid zone (Fig. 2.1 B) also exists between eastern *musculus* and western *domesticus*. Recently obtained karyological and isoenzyme information on these two taxa of mice has led M. J. D. White (1978) to consider them as semispecies, since they have more genetical differences than ordinary geographical subspecies. [It should be noted, however, that M. J. D. White refers to them as if they were species since he uses the names *M. musculus* and *M. domesticus*.]

By these standards, then, the category of semispecies might be set aside in favour of species themselves. After all, semispecies are species in the making and should be commonplace among all groups of organisms actively involved in allopatric speciation. Only where intermediate phenotypes cause taxonomic confusion in the hybrid zones does it seem advantageous to invoke the conept of semispecies for analysing superspecies composition. But we can hardly dismiss the term semispecies without mentioning its extensive usage in relation to flowering plants and to fruitflies.

Among the many cases of botanical semispecies, as reviewed by Grant (1971), *Quercus* oak trees provide the best evidence of two conditions not normally encountered among animals. Firstly, species such as *Q. douglasii*, *Q. dumosa* and *Q. turbinella* in California or *Q. petraea* and *Q. robur* in Britain maintain their integrity over extensive areas of sympatry, but form localised 'hybrid swarms' in disturbed or newly colonised habitats. Secondly, evidence from Miocene fossil leaf impressions suggests that these taxa became differentiated no less than 10 million years ago, so they may be regarded as stable semispecies rather than incipient species. Perhaps the explanation lies in temporary regional variations of the pollen transfer mechanism.

Finally, it is necessary to mention the superspecies *Drosophila paulistorum*, one of four main taxa making up the *D. willistoni* species group of fruitflies in central and South America. *D. paulistorum* itself was reported by Dobzhansky & Spassky (1959, p. 419) to be 'a cluster of species *in statu nascendi*'. Further work led to the view that 'ethological isolation among the semispecies of *D. paulistorum* is strong enough that two and sometimes three semispecies exist sympatrically in many localities' (Richmond & Dobzhansky, 1976, p. 746). Altogether, six so-called semispecies have been detected. They are morphologically indistinguishable but produce sterile male hybrid progeny when artificially crossed. Female hybrid progeny are usually fertile, and the semispecies share sufficient of their polytene chromosome banding patterns and polymorphisms to make identification difficult

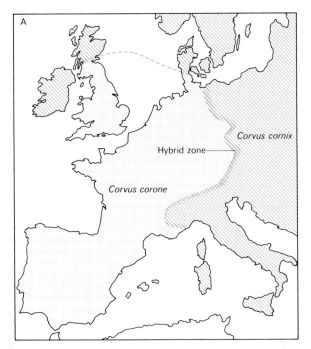
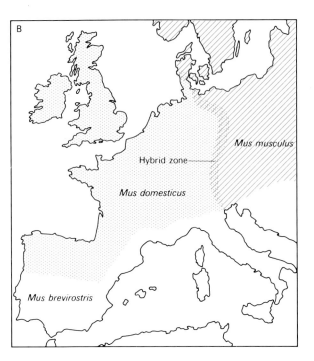

Figure 2.1 Distributions in Europe of two semispecies pairs showing congruent hybrid zones: A, the crows *Corvus cornix* and *C. corone* and B, the mice *Mus domesticus* and *M. musculus*.

or impossible from cytotaxonomic evidence (Dobzhansky & Powell, 1975). Comparable situations have since been detected in other Drosophilidae and among further families of dipterous insects on which studies of evolutionary genetics have been intense. Such a cluster of cryptic taxa would nowadays be called a complex of sibling species, some allopatric and some sympatric, with adequate reproductive barriers to prevent them cross-breeding. A point of interest about *D. paulistorum*, however, is that Dobzhansky chose to regard it as a superspecies consisting of sympatric semispecies, although most workers treat the superspecies as a category for allopatric arrays of species.

Sibling species complexes

By using cross-breeding tests for detection of post-mating isolating mechanisms, and by looking for ethological and other signs of pre-mating isolating mechanisms, the existence of many groups of anatomically very similar species has been demonstrated in recent years. Morphologically indistinguishable or 'identical' species are known as 'sibling species', although Steyskal (1972) proposed the term 'aphanic species' as an alternative to avoid possible ambiguity of the word sibling. Indications that a complex of sibling species exists in nature may come from biological variation in populations of what, at face value, seems to be a single species. Detailed observations and experimental investigations on representative specimens may show that

a number of reproductively isolated populations fulfil the qualifications of biological species, without gene flow between them, despite their lack of distinguishing morphological characteristics. It requires only a simple but effective reproductive barrier to maintain the independence of sibling species. For example the lacewings *Chrysopa carnea* and *C. downesi* differ by specific alleles causing asynchronous seasonal reproductive cycles that prevent them from interbreeding (Tauber *et al.*, 1977).

Probably the first recognised cases of sibling species were three leaf warblers found in Britain and Europe: the Chiffchaff (*Phylloscopus collybita*), the Willow Warbler (*P. trochilus*) and the Wood Warbler (*P. sibilatrix*). Songs of these birds were studied by Gilbert White (1768) of Selborne who wrote 'I have now, past dispute, made out three distinct species...which constantly and invariably use distinct notes'. His comparative observations were facilitated by the way that these three species of birds occur sympatrically in the woodlands of southern England, where they form breeding pairs according to their specific songs which thus act as highly effective ethological isolating mechanisms: cross-breeding never occurs in nature. By careful examination of specimens identified from their songs before capture, Gilbert White found that *P. sibilatrix* averages about 20 per cent larger than the others, while some slight colour contrasts of beak, legs and plumage aid the identification of specimens in the hand. Moreover, these *Phylloscopus* spp. have quite distinctive nests and habits, and their eggs bear specifically different markings.

The latter point raises the question of whether these warblers should actually be regarded as sibling species, since some small but reliable morphological distinctions are exhibited. It seems advantageous to employ the term sibling species more or less strictly according to the circumstances in any major group of organisms. As there are few better examples among birds, these leaf warblers may as well be treated as sibling species. The numerous cases of autopolyploidy among morphologically indistinguishable plants are also eligible (see Chapter 8), although botanists seldom adopt the term sibling species (Lewis, 1973).

Invertebrate animal species often have only a few, minor micromorphological features to distinguish them, and the structural minutiae of male external genitalia are sometimes the only means of specific identification. Adult Dark Dagger and Grey Dagger Moths (*Acronycta psi* and *A. tridens*) provide a familiar example of species with identical outward appearances. Their larvae differ strikingly, however, and identification can be based on the male genitalia of the adults. Rather limited but still reliable differences of this degree are too frequently encountered among arthropods for them to be rated at the sibling species level.

The discovery of hybrid sterility between morphologically very similar populations has repeatedly been the clue to recognition of sibling species. Yet investigators tend to hesitate over accepting that reproductive incompatibility is a sufficient indicator of specific difference. Much cross-breeding work with mosquitoes (Culicidae) has shown that routine testing for hybrid sterility is a practical and realistic way to demonstrate sibling species distinctions between biologically heterogeneous but morphologically uniform populations (Davidson, 1977). Among protozoans, *Paramecium aurelia* appears to be a complex of at least 16 sibling 'species' that produce inviable hybrids when caused to cross, although those who first detected these reproductive barriers thought they warranted only the status of varieties (Sonneborn, 1957).

The idea that cross-sterility can occur between what were thought to be mere races or varieties has been a conceptual obstacle to the realisation that non-interbreeding populations must be distinct species. Dobzhansky & Epling (1944) created a precedent over the case of *Drosophila pseudoobscura* races A and B when they described the latter as *D. persimilis*, because it is a reproductively isolated and chromosomally unique sibling species. *D. pseudoobscura* × *persimilis* crosses give fertile hybrid females but sterile hybrid male progeny. Male hybrid sterility is often more pronounced than female hybrid sterility and has been explained by Haldane's Rule which states that it is the heterogametic sex (that bearing the XY or XO sex chromosomes) that is most likely to be severely affected among hybrids.

Likewise, various kinds of hybrid failure come from crosses among the nine sibling species comprising the European *Anopheles maculipennis* complex of mosquitoes. So-called races or varieties of this complex were first

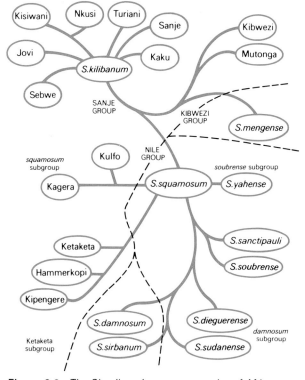

Figure 2.2 The *Simulium damnosum* complex of African blackflies consists of sibling species differing from each other by chromosomal mutations (e.g. inversions) that can be interpreted precisely from changes in the banding sequences on the polytene chromosomes. This diagram shows relationships between 24 species, based on their chromosomal similarities and differences, step by step. Ten of the species have been given formal Latin names, while the others have provisional names based on the localities where they were first found. At present, cytotaxonomy is the only practical means of distinguishing these sibling species. Based on World Health Organisation (1977) and Rothfels (1979).

named in 1926 when character variations of egg patterns and wing lengths were correlated with the way that some populations transmit human malaria while other 'races' are harmless because they prefer to bite animals other than man. Despite ample proof of both pre-mating and post-mating reproductive barriers between members of the *A. maculipennis* complex, as many as five of which may co-exist sympatrically, it was not until recently that their status as distinct sibling species has become generally accepted (White, G. B., 1978).

As will be apparent from examples already given, by far the most elaborate sibling species complexes have been found in several families of flies (Diptera). This order of insects is almost unique in the possession of giant polytene chromosomes in nuclei of some tissues (for example, larval salivary gland and ovarian nurse cells). Specific banding patterns on the polytene chromosomes provide a good guide to karyotype rearrangements between species, and it is found that changes in the banding sequence nearly always accompany speciation. In addition, there are chro-

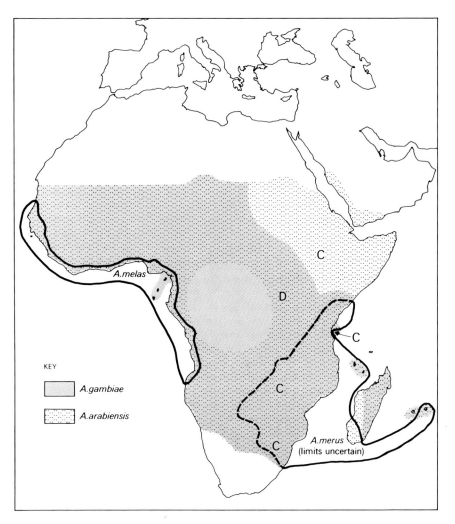

Figure 2.3 Distributions in the Afrotropical Region of six sibling species comprising the *Anopheles gambiae* complex of mosquitoes. The true *A. gambiae* is widely sympatric with *A. arabiensis*, breeding in freshwater habitats; both of these man-biting mosquitoes are principal vectors of human malaria and other parasitic diseases. *A. melas* and *A. merus* are western and eastern saltwater-breeding species, respectively, which also transmit diseases and are sometimes sympatric with *A. gambiae* and *A. arabiensis*. Two other species, shown here as C and D, have restricted distributions. These sibling species rarely interbreed, but are morphologically so similar that their specific identification is based on cytotaxonomy; that is, recognition of specific banding patterns on the polytene chromosomes.

mosomal inversion polymorphisms in many Diptera, and these can help to characterise species and populations. Since shared chromosomal inversions are usually clues to phylogenetic affinity, it has been possible to trace the evolutionary relationships within whole groups of Diptera, such as the Hawaiian Drosophilidae (Carson *et al.*, 1970), Holarctic Simuliidae (Rothfels, 1979) and global Chironomidae (Martin, 1979). Among these Diptera, therefore, it seems that the total numbers of biologically distinct sibling species are several-fold greater than the numbers of morphological species recognised by conventional taxonomy. Such proliferation of species with minimal morphological differentiation is apparent in a few other insect orders, notably the Orthoptera, and in certain families of Acarina, Crustacea, Nematoda and doubtless many more invertebrates. Sibling species complexes can thus be expected in almost any genus or group that exploits marginally different habitats, biotopes or niches, for which ecological and behavioural specialisations are clearly more important than morphological divergences would be. From this point of view, we may assume that sibling species remain morphologically similar for indefinite periods of time, and there is no overall justification in assuming that sibling species have necessarily arisen recently.

Two final examples of intensively studied dipteran complexes should serve to emphasise the significance of sibling species in understanding ecological (and epidemiological) problems. During the last decade, international agencies have launched a widespread control campaign using insecticides against the Afrotropical blackfly *Simulium damnosum* (Le Berre *et al.*, 1978). These Simuliidae are of medical importance in Africa because the adult female blackflies are pests which transmit filarial parasites responsible for the human disease onchocerciasis, or river blindness. Blackfly control measures are based on larvicidal treatment of the breeding sites in rivers and streams; currently this is being undertaken on a vast scale throughout seven West African countries. Due to their specialised habitat requirements in fast flowing water-courses, it is difficult to study *Simulium* populations under natural conditions, or to breed them in the laboratory, so comparatively little is known of their autecology and population genetics. By making cytogenetic studies on many samples from representative habitats, however, it has been revealed that *S. damnosum* is actually a complex of no less than 24 sibling species (Fig. 2.2), eight of which are present in the areas currently being subjected to control measures (Vajime & Dunbar, 1975). Several of these sibling species tend to be sympatric in each river, but there are indications that the larvae and pupae may have slightly different niche adaptations. Certain members of this complex are restricted to forest zones, others to savannah, and there are signs that the adults of some sibling species are more vagile than others, giving different degrees of dispersal and powers of reinvasion. Members of the complex are not equally involved in transmission of onchocerciasis, because the adult females of some species are more strongly anthropophilic (man-biting) than others; this cannot be evaluated without more accurate methods of specimen identification. There is an urgent need to fully analyse and investigate the *S. damnosum* complex, in order that understanding more about these sibling species can facilitate their improved control. So far it seems impossible to distinguish between all the species by any means other than examination of the specific banding sequences on polytene chromosomes prepared from the larval salivary glands. The material on which the newly described members of the *S. damnosum* complex is based consists of permanent chromosome preparations which have been deposited in the British Museum (Natural History).

Like Drosophilidae, some Culicidae are easy to keep and breed under laboratory conditions, and this has permitted cross-breeding studies on various populations, species and suspected species. Perhaps the most interesting discovery among mosquitoes is the *Anopheles gambiae* complex, comprising a well-defined set of six sibling species from the Afrotropical Region (White, 1974). Their collective importance lies in the long established view that *A. gambiae* is the principal vector of malaria in Africa. Different sibling species of *A. gambiae* were detected by chance 25 years ago (Davidson, 1956) and it has since been shown that hybrid male sterility results from all but one of the 30 possible interspecific crosses within the complex. In addition to these post-mating isolating mechanisms, strong ethological barriers exist between the sibling species, since wild hybridisation rates in sympatric populations do not surpass 0·02 per cent.

Specimens can best be identified by recognition of specific polytene chromosome characteristics, both in larvae and adult females (Coluzzi *et al.*, 1979), and it is also possible to separate the species on the basis of biochemical characteristics (Miles, 1979). Distribution surveys have often employed hybridisation tests for definitive identification of broods from wild females, by crossing them with a series of reference strains. Comparative field investigations, mainly using the cytotaxonomic identification method, have revealed many biological contrasts between members of the *A. gambiae* complex. Two sibling species, *A. melas* and *A. merus*, show slight morphological points of distinction and breed in brackish water habitats of West and East Africa, respectively; a third species is restricted to the vicinity of its breeding-sites in geothermal springs in Uganda; the remaining three sibling species are more widespread and breed in freshwater (Fig. 2.3). Of these, *A. quadriannulatus* has a restricted distribution in eastern Africa, while *A. arabiensis* and *A. gambiae* (*sensu stricto*) are largely sympatric throughout Africa. However, *A. arabiensis* spreads further into arid zones, including southern Arabia, while *A. gambiae* is more prevalent in humid parts of Africa. The latter is the most efficient malaria vector, having greater man-biting tendencies and average longevity. All the sibling species except *A. quadriannulatus* are apparently involved to some extent in transmission of malaria and other human diseases such as elephantiasis. The reason why *A. quadriannulatus* is not a vector lies in its strong proclivity to bite animals other than man. Because the five vector sibling species do not shelter indoors to equivalent degrees in different areas, they are not equally well controlled by the standard antimalarial measures of house-spraying with residual insecticides.

A remarkable number of additional species complexes of this sort, comprising arrays of sympatric and allopatric sibling species, are being found among mosquitoes, black-flies, sandflies and other medically important insects, where the applied incentive is greatest to understand sibling species problems. Parasitic protozoa and helminths are similarly the focus of much experimental taxonomy, revealing comparable amounts of what may be interpreted as sibling speciation (Muller & Taylor, 1979). This is a salutory indication that taxonomic methods based purely on the morphological approach are bound to overlook the reproductively distinct sibling species that are so numerous and significant in some groups.

References

Bush, G. L. 1975. Modes of animal speciation. *Annual Review of Ecology and Systematics* **6**: 339–364.

Busvine, J. R. 1978. Evidence from double infestations for the specific status of human head lice and body lice (Anoplura). *Systematic Entomology* **3**: 1–8.

Carson, H. L., Hardy, D. E., Speith, H. T. & Stone, W. S. 1970. The evolutionary biology of the Hawaiian Drosophilidae, pp. 437–543, *in* Hecht, M. K. & Steere, W. C. (Eds) *Essays*

in evolution and genetics in honor of Theodosius Dobzhansky, New York: Appleton-Century-Crofts.

Coluzzi, M., Sabatini, A., Petrarca, V. & Di Deco, M. A. 1979. Chromosomal differentiation and adaptation to human environments in the *Anopheles gambiae* complex. *Transactions of the Royal Society of Tropical Medicine and Hygiene* **73**: 483–497.

Davidson, G. 1956. Insecticide resistance in *Anopheles gambiae*

Giles: a case of simple Mendelian inheritance. *Nature* **178**: 861–863.

Davidson, G. 1977. Anopheline species complexes, pp. 254–271, in Gear, J. H. S. (Ed.) *Medicine in a tropical environment*. Cape Town: Balkema.

Dobzhansky, T. 1970. *Genetics of the evolutionary process*. 505 pp. New York: Columbia University Press.

Dobzhansky, T. & Epling, C. 1944. Contributions to the genetics, taxonomy, and ecology of *Drosophila pseudoobscura* and its relatives. *Publications of the Carnegie Institution of Washington* **554**: 1–183.

Dobzhansky, T. & Powell, J. R. 1975. The *willistoni* group of sibling species of *Drosophila*, pp. 589–622, in King, R. C. (Ed.) *Handbook of genetics, vol. 3, Invertebrates of genetic interest*, New York: Plenum Press.

Dobzhansky, T. & Spassky, B. 1959. *Drosophila paulistorum*, a cluster of species *in statu nascendi*. *Proceedings of the National Academy of Sciences of the United States of America* **45**: 419–428.

Goldschmidt, R. B. 1940. *The material basis of evolution*. 436 pp. New Haven: Yale University Press.

Grant, V. 1971. *Plant speciation*. 435 pp. New York: Columbia University Press.

Leahy, M. G. & Craig, G. B. 1967. Barriers to hybridization between *Aedes aegypti* and *Aedes albopictus* (Diptera: Culicidae). *Evolution* **21**: 41–58.

Le Berre, R., Walsh, J. F., Davies, J. B., Philippon, B. & Garms, R. 1978. Control of onchocerciasis: medical entomology – a necessary pre-requisite to socio-economic development, pp. 70–75, in Willmott, S. (Ed.) *Medical Entomology Centenary Symposium Proceedings*. London: Royal Society of Tropical Medicine and Hygiene.

Lewis, H. 1973. The origin of diploid neospecies in *Clarkia*. *American Naturalist* **107**: 161–170.

Lewis, H. & Bloom, W. L. 1972. The loss of a species through breakdown of a chromosomal barrier. *Symposia Biologica Hungarica Research* **12**: 61–64.

Lorković, Z. 1958. Die Merkmale der unvollständigen Speziationsstufe und die Frage der Einführung der Semispezies in die Systematik. *Uppsala Universitets Arsskrift* **1958**: 159–168.

Martin, J. 1979. Chromosomes as tools in taxonomy and phylogeny of Chironomidae (Diptera). *Entomologica Scandinavica* (Supplement) **10**: 67–74.

Mayr, E. 1963. *Animal species and evolution*. 797 pp. Cambridge, Massachusetts: Belknap Press.

Miles, S. J. 1979. A biochemical key to adult members of the *Anopheles gambiae* group of species. *Journal of Medical Entomology*, Honolulu **15**: 297–299.

Muller, R. & Taylor, A. E. R. (Eds) 1979. Problems in the identification of parasites and their vectors. *Symposia of the British Society for Parasitology* **17**: 1–221.

Murray, J. 1972. *Genetic diversity and natural selection*. 128 pp. Edinburgh: Oliver & Boyd.

Richmond, R. C. & Dobzhansky, T. 1976. Genetic differentiation within the Andean semispecies of *Drosophila paulistorum*. *Evolution* **30**: 746–756.

Rothfels, K. H. 1979. Cytotaxonomy of blackflies. *Annual Review of Entomology* **24**: 507–539.

Sonneborn, T. M. 1957. Breeding systems, reproductive methods, and species problems in *Protozoa*, pp. 155–324, in Mayr, E. (Ed.) *The species problem*, Publications of the American Association for the Advancement of Science.

Steyskal, G. C. 1972. The meaning of the term 'sibling species'. *Systematic Zoology* **21**: 446.

Tauber, C. A., Tauber, M. J. & Nechols, J. R. 1977. Two genes control seasonal isolation in sibling species. *Science* **197**: 592–593.

Vajime, C. G. & Dunbar, R. W. 1975. Chromosomal identification of eight species of the subgenus *Edwardsellum* near and including *Simulium* (*Edwardsellum*) *damnosum* Theobald (Diptera: Simuliidae). *Tropenmedizin und Parasitologie* **25**: 111–140.

White, Gilbert. Letter dated 17 August 1768, in *The Natural History of Selborne in the County of Southampton*, revised edition, 1937. 300 pp. The World's Classics No. 22. London: Oxford University Press.

White, G. B. 1974. *Anopheles gambiae* complex and disease transmission in Africa. *Transactions of the Royal Society of Tropical Medicine and Hygiene* **68**: 278–301.

White, G. B. 1978. Systematic reappraisal of the *Anopheles maculipennis* complex. *Mosquito Systematics* **10**: 13–44.

White, M. J. D. 1978. *Modes of speciation*. 455 pp. San Francisco: W. H. Freeman.

World Health Organisation 1977. *Species complexes in insect vectors of disease (blackflies, mosquitoes, tsetse flies)*. Document WHO/VBC/77.656, 56 pp. Geneva: World Health Organisation.

CHAPTER 3

The allopatric model of speciation with special reference to birds

D. W. Snow

It has often been pointed out that Charles Darwin's great work, *The Origin of Species*, was misnamed, as one important aspect of evolution that it did not deal with at all satisfactorily was, precisely, the origin of new species. The criticism is justified, if by the origin of species we mean the multiplication of species, that is, the splitting of a parent species to form two or more daughter species. Darwin was concerned mainly with evolutionary change and the processes by which such change takes place. Evolutionary change may take place without any multiplication of species. For example, it is probable that the flightless Dodo of Mauritius evolved gradually over millions of years from a flying ancestor which colonised the island from overseas, and that there never was at any one time more than a single species. If one could somehow follow all the stages in this long process of gradual change, one would have no doubt that the last stage of all, the Dodo as it existed at the end of the seventeenth century, should be considered a different species from the original colonising population. But how many intervening species should be recognised would be very largely arbitrary, depending on one's subjective judgment of the degree of morphological difference required between distinct species.

The modification of a single lineage in time has been termed 'phyletic evolution'. The evolution and diversification of life on earth, however, is the outcome not simply of phyletic evolution but of phyletic evolution and the multiplication of species, or speciation. If this is visualised in the conventional way, as a branching tree, phyletic evolution corresponds to the changes occurring as one moves up one of the ascending branches, and speciation refers to what happens at the point where one branch splits into two. It is the second of these two processes that is the subject of the present chapter.

In essence, the process of speciation as we now understand it is very simple – so simple, in fact, that it may be wondered why Darwin, so astonishingly perspicacious in most matters, did not see it more clearly. There are several reasons why he did not do so, foremost amongst which was the lack of a satisfactory concept of the species. The history of our understanding of speciation, from Darwin up to more recent times, is a fascinating chapter in the history of biology, but is beyond the scope of this chapter. It has been well dealt with by Mayr (1963, pp. 482–488).

The main stages of the currently favoured theory of speciation are shown diagrammatically in Figure 3.1. First, an originally continuous population is split into two (or more) spatially (geographically) isolated populations, either by some sort of barrier or by the colonisation of an isolated outlying area. Under the influence of the different environments, organic and inorganic, to which they are exposed, genetic differences then begin to accumulate in the isolated populations and they diverge (see Chapter 1). When they have diverged to a certain degree there comes a point at which members of the two populations can no longer interbreed with one another or, if they still can, will produce offspring with reduced viability. If the barrier dividing the populations then breaks down, they may extend their ranges so as to meet one another. Providing that they are distinct enough ecologically not to compete severely, the final outcome will be that two species coexist where there was only one before.

As mentioned above, the isolation of sections of an originally continuous population – the first step in the speciation process – may be produced in two different ways. The population may be split by the development of a barrier within its range (a desert area or a sea, for instance), or an existing barrier may be crossed by a few pioneering individuals or groups, who then found a new population in a suitable area beyond. The distinction, shown diagrammatically in Figure 3.2, is potentially important because, for reasons discussed later, small pioneering populations are likely to evolve more rapidly than larger populations, and their relationship to the stock from which they were derived may be obscured. In particular cases it is, of course, not always possible to tell in which way the isolation came about.

Even from this very bald statement of the stages of speciation it will be obvious that geographical isolation is the first essential, without which none of the ensuing stages is possible. This is reflected in the two terms that are commonly used for this mode of species-formation, 'geographical speciation' or 'allopatric speciation'. It is now widely accepted that in most animal groups this is the main, or indeed the only, way in which speciation takes place (but see Chapter 4, where models of speciation without prior geographical isolation are discussed). In what follows, we consider the evidence for the allopatric model of speciation, then explore more thoroughly the details of the process, and finally discuss how it affects the taxonomist, whose aim is to classify his animals in an orderly and objective manner that reflects, as far as possible, their phylogeny.

Inferred examples of allopatric speciation

Allopatric speciation is far too slow a process to be observable directly within the time span of modern biology; no one has ever seen a species arising. The evidence for allopatric speciation must therefore be indirect. It is probably fair to say that the zoologists who readily accepted the hypothesis, at the end of the nineteenth and beginning of the twentieth centuries, were largely influenced by *a priori* considerations, chiefly the argument that it is difficult to see how reproductive isolation can originate and be maintained between two sections of a geographically continuous population. The observed facts of geographical distribution and variation (Chapter 1) gave massive support to the hypothesis once it had been formulated, but probably were insufficient in themselves to suggest it. 'Evolution may go on from age to age, but without isolation no new species can be produced. However much a species can be changed by Evolution, "the swamping effects of intercrossing" prevent there being at any given time more than one species' (Seebohm, 1887, pp. 21–22). In any case, whatever the historical development of the idea it is most convenient now to treat allopatric speciation as a model and see how far the predictions derived from it are consistent with observations.

According to the model one would always expect to find the closest relatives of any population of animals in a neighbouring area, and not within its own range. (It is, perhaps, a measure of our general acceptance of allopatric speciation that this prediction seems almost too obvious to be worth making.) Furthermore, where there are several allopatric and more or less closely related populations one would expect to find that the differences are greatest between those forms that have been isolated from one another for longest. Usually, these predictions are fulfilled as closely as can reasonably be hoped for in the present state of knowledge.

The Coal Tit (*Parus ater*) provides a good example. This species has a vast range in the Palaearctic region, with

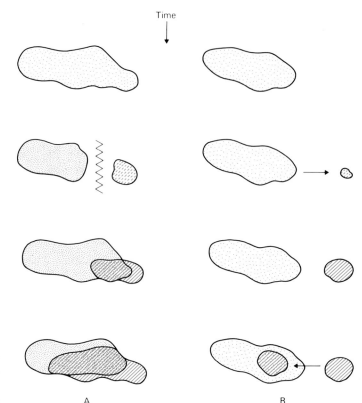

Time

A   B

Figure 3.1 Schematic representation of two hypothetical cases of allopatric speciation as a result of (A) development and break-down of a barrier in a continental population, and (B) colonisation of an outlying area, followed by recolonisation of original area by modified isolate. Different shadings represent degrees of genetic difference between populations in the time sequence; thus in A both parts of the divided population develop genetic differences from one another and from the parental stock after they are isolated from one another. In the final stage, the different ranges occupied by the two species reflect differences in ecological tolerance; thus in A the species that had been isolated in the west remains better adapted to conditions in the north and west than the species that had been isolated in the east.

numerous distinct populations. Almost without exception, the degree of difference between any two populations is related to what we know or suspect to be the degree of isolation between them, isolation being defined in terms of the time since gene-flow between them was interrupted. Thus, taking the central European populations as a standard for comparison, other northern and eastern European and Siberian populations (which are continuous with the central European population) are very similar or identical, the English and (at least partly isolated) southwestern European populations are slightly different, the Irish population a little more different, and the northwest African populations even more different, while the populations of Iran, the Himalayas, southern China and Formosa are strikingly distinct. Indeed many of the outlying populations are so different that they have been treated

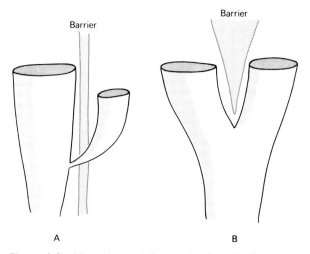

Barrier

Barrier

A   B

Figure 3.2 Allopatric speciation can be thought of as either of two responses to a geographical barrier. A – Barrier always in existence but at a single point in time a fraction of the original population manages to cross the barrier and diverge in isolation. B – Barrier newly formed and is the cause of the fragmentation of the original population. These differing assumptions of historical events have implications for biogeographic hypotheses (see Chapters 20 and 21).

as separate species, a point to which we shall return below.

As a corollary to the first prediction, discussed above, if a species is found to overlap geographically with the species to which it seems most closely related, it will be expected that examination of their ranges and probable past histories will show that they have differentiated in isolation from one another and that the overlap in range is a later development. Again, the evidence is generally fully consistent with this. Two examples will show the kinds of situation that are found.

The first example concerns two treecreepers, *Certhia familiaris* (Common Treecreeper) and *C. brachydactyla* (Short-toed Treecreeper) which form a pair of species so similar to one another that they are difficult to distinguish, even in the hand (though their songs are more distinct). These overlap widely in central Europe but their overall ranges, habits and habitat preferences strongly suggest that the overlap is recent, dating from the end of one of the later glaciations. *C. familiaris*, a cold-adapted species that prefers conifer woodland over most of its range and in the south is found mainly in mountains, has a very wide range in Eurasia as well as in North America; while *C. brachydactyla*, a less hardy species inhabiting mainly broad-leaved woodland, has an essentially southern and southwestern distribution in Europe. It is hypothesised that during one of the glacial periods the population that gave rise to *brachydactyla* was isolated from the parental species (*familiaris*) in a comparatively mild southwestern refuge, most likely in the Iberian peninsula. When the climate ameliorated both were able to spread with the spreading forests, so that they now overlap widely. Being

adapted to different habitats within the area of overlap, they are not ecological competitors and so can coexist. It is of interest to British ornithologists that *familiaris* apparently spread more rapidly into northwestern Europe from its nearest forest refuge, presumably somewhere to the southeast, than did *brachydactyla* from the southwest, so that only *familiaris* had reached Britain when it was isolated from the continent by the cutting of the English Channel. *C. familiaris* in Britain has now become adapted to broad-leaved habitats of a kind that, on the continent, would be more suitable for *brachydactyla*.

A second example of suspected prior isolation concerns the grebe *Tachybaptus rufolavatus* of Madagascar. This species is most closely related to the widespread Little Grebe or Dabchick *T. ruficollis*, which occurs not only over most of Eurasia and Africa but also in Madagascar. It seems certain that *rufolavatus* represents an earlier colonisation of Madagascar by *ruficollis* stock, which differentiated in isolation from the rest of the species, so that when a second colonisation occurred, probably fairly recently from Africa, the earlier colonists had become specifically distinct. This is an especially interesting case which deserves further investigation, because there is evidence that reproductive isolation between the two species is not complete. A few hybrids have been reported. It would be very interesting to know whether the hybrids are at a disadvantage compared with individuals of the two parent species. If they are, selection should act against the hybrids and reproductive isolation between the two species should improve and finally become complete. If, on the other hand, the hybrids are not at a disadvantage there may ultimately be a fusion of the two parental stocks (though this seems to be a very rare occurrence, at least in birds). A third possibility is that *ruficollis* may prove to be competitively superior and so may cause the extinction of *rufolavatus*.

Islands obviously provide particularly good opportunities for speciation, at least for speciation of terrestrial organisms. It is for this reason that species that have colonised isolated archipelagoes are sometimes able to multiply by a process of repeated divergence (in isolation on separate islands) followed by rejunction (when the isolated populations come together again as a result of movement between islands). This is impossible for species that colonise isolated single islands. For them, repeated colonisations from the mainland may be the only way in which speciation can take place, as in the case of the grebes of Madagascar; and such cases are rare.

The Galapagos and Hawaiian islands are the most famous examples of archipelagoes in which striking allopatric speciation, especially of birds and insects, has resulted in a rich and diverse fauna from a comparatively small number of original immigrants. Thus nine or ten species of Darwin's finches coexist on the larger Galapagos islands, all descended with little doubt from one common ancestral immigrant. It is hardly necessary, after the evidence and arguments presented above, to comment on

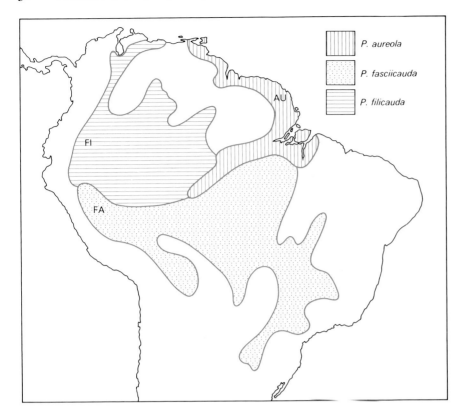

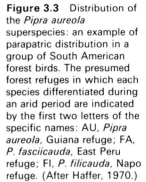

Figure 3.3 Distribution of the *Pipra aureola* superspecies: an example of parapatric distribution in a group of South American forest birds. The presumed forest refuges in which each species differentiated during an arid period are indicated by the first two letters of the specific names: AU, *Pipra aureola*, Guiana refuge; FA, *P. fasciicauda*, East Peru refuge; FI, *P. filicauda*, Napo refuge. (After Haffer, 1970.)

the significance of the fact that the very isolated Cocos Island, half way between the Galapagos and Central America, was colonised by a finch apparently of the same stock as Darwin's finches, and that this finch, though much modified, remains a single species.

It is ironical that, although the finches of the Galapagos Islands (with the extinct mammals of South America) were acknowledged by Darwin as 'the origin of all my views', he did not see that they provided not only the strongest evidence for evolutionary adaptation and radiation but also the key to speciation. Indeed, their full significance for the understanding of speciation was not clearly recognised until the publication of David Lack's now classic *Darwin's Finches* (1947).

Divergence in isolation

It is an essential part of the model of allopatric speciation that geographically isolated populations should be expected to diverge from one another. Divergence in genetic constitution or in mating behaviour (or in both) is necessary if the two populations are to be reproductively isolated from one another when they come together again; additionally ecological divergence is necessary if they are to be able to coexist.

Divergence may not take place, or may do so slowly, that there is little opportunity for speciation. This may be the case if a widespread species with a very stable, balanced

gene pool is split into fairly large sections (Mayr, 1963). If we assume that genetic stability is reflected in morphological stability then there are examples of lack of differentiation within widespread and disjunct bird species, but they are not common. One example is the Fulvous Whistling Duck (*Dendrocygna bicolor*) which occurs in tropical America, Africa, southeast Asia and Australia, and shows little, if any, morphological variation throughout its range. Much more often, however, isolated populations begin to diverge, and it is not difficult to see why they should do so. The factors promoting divergence are of two kinds – genetic and environmental (Chapter 1).

Widespread animal populations are not genetically homogeneous. Clinal variation, often affecting a number of characters that vary independently of one another, is the general rule. When such variation is analysed it usually reflects adaptation to geographically changing environmental factors such as climate. Hence, if an originally widespread population is split by the development of some kind of ecological barrier, for example a forest species by a tract of arid, open country, the two sections are likely to diverge genetically because there will be few matings between members of the two populations.

If one of the populations resulting from a geographic split is very small, the genetic differences between it and the parent population are likely to have far-reaching effects. Thus, if an island or some other isolated outlying area, is colonised by a small number of pioneers from a

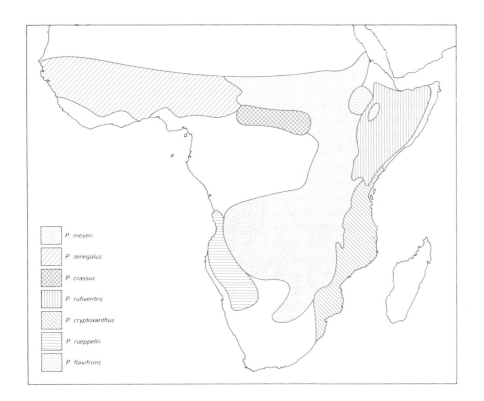

P. meyeri

P. senegalus

P. crassus

P. rufiventris

P. cryptoxanthus

P. rueppelli

P. flavifrons

Figure 3.4 Distribution of the *Poicephalus meyeri* superspecies: an example of parapatric distribution in Africa.

continental population or segregated from the main population by geological events, then the newly established population will contain only a fraction of the total gene pool of the species. Mayr (1963) has shown how this is likely to result in a 'genetic revolution', leading to a new and different genetic constitution that will be reflected in morphological and physiological divergence from the parent population. A most important aspect of such genetic revolutions is that they may, so to speak, prepare the ground for major adaptive changes in the newly established population. This idea corresponds, in a general way, to Simpson's (1961) idea of quantum evolution.

When an originally continuous population is split, whether by the development of ecological barriers or by the founding of a new population in an isolated outlying area, environmental conditions in the two parts of the range are certain to be different. Thus, on top of whatever genetic differences there may be between the populations, divergence will be further enhanced by selection for adaptation to different environmental conditions.

Hence divergence is to be expected as a consequence of geographical isolation. If this involves mainly fine adaptation to differences in climate, habitat and other aspects of the organic and inorganic environment of the two areas, and if the two species so formed later become sympatric, one will expect the outcome to be a subdivision of the original ecological niche. The result of repetition of the process will be a close 'packing' of species within essen-

tially the same adaptive zone. This 'packing' is indeed what we regularly find from the analysis of any rich fauna. But occasionally the effect of isolation may be to enable a species to enter an entirely new adaptive zone; and it may indeed be easier for geographical isolates, which have undergone a 'genetic revolution', to do this than for widespread continental species with very stable, buffered genetic constitutions.

Rejunction after isolation

After the break-down of a barrier that has separated two diverging populations, both reproductive isolation and ecological segregation are necessary if they are to coexist as separate species. It is convenient to treat these two prerequisites for speciation separately, though they are not in fact independent of one another.

The evolution of reproductive isolating mechanisms has been studied experimentally in some invertebrates, but in vertebrates we are dependent on inference from what can be observed in natural populations. Conclusions are tentative and must be stated in rather general terms. First, it seems likely that a moderate to high degree of reproductive incompatibility vis-à-vis the parent population is a probable by-product of the genetic differences accumulated by any isolated population that is sufficiently distinct to be a 'candidate' for speciation. This will especially be true of any isolated population that has undergone a 'genetic revolution', as discussed in the previous section. Thus, if the isolate and parent population meet again, hybridisation either will not occur or, if it does, the hybrid individuals will be less well adapted than either of the parent populations. If such hybrids are occasionally produced, natural selection will tend to strengthen and perfect the isolating mechanisms. The nature of the isolating mechanism will differ in different groups of animals. In birds, it may involve genetic incompatibility (causing infertility, or reduced viability of hybrid off-spring) or it may be primarily behavioural (as in many ducks, which can interbreed with close relatives, producing fully fertile hybrids, but are prevented from doing so in the wild by their species-specific courtship displays). Ecological differences, for example different habitat preferences and different breeding seasons, may strengthen other isolating mechanisms, but are probably insufficient to be effective by themselves. These remarks on hybrids must be qualified by making reference to hybrids involving polyploidy, which may be very important in plant speciation (see Chapter 8).

The degree of ecological segregation necessary for coexistence is more amenable to direct investigation. Charles Darwin recognised the general principle that competition will be expected to be more severe between closely related species, since their requirements are likely to be similar, than between more distantly related species. The validity of this generalisation has been established in numerous instances. The minimum degree of ecological difference necessary to allow coexistence is not, however, easy to state quantitatively, mainly because of the extreme complexity of the ecological relationships in which all species are involved. Indirectly, one can sometimes arrive at a measure based on size differences between species. This is especially applicable to predatory species that feed on the same general kind of prey and which 'partition' between themselves the range of available prey. The four species of forest kingfishers comprising the neotropical genus *Chloroceryle* are a good example. The two species *Chloroceryle amazona* and *C. americana* coexist extensively in the same kinds of habitat, where they feed mainly on freshwater fishes. They are very similar, certainly more closely related to one another than to other kingfishers, but *amazona* is considerably larger than *americana*, approximately in the linear proportion 1·5:1. The two remaining species, *C. inda* and *C. aenea*, are also a very closely related pair, differing mainly in size approximately in the linear proportion 1·7:1. The size differences, in this instance, between the members of these two pairs are perhaps larger than is usual in such cases. A number of cases in mammals and birds have been studied and it has been found that the mean ratio in linear measurements of 'trophic structures' (skull in mammals, bill in birds) was about 1·3:1. Obviously, one will not necessarily expect the same ratio to obtain in all instances, especially in view of the different spectrum and size range of potential prey species.

If two closely related species, presumed to have diver-

ged in isolation, show only a limited amount of geographical overlap, it might be expected that they should be ecologically most distinct, and perhaps as a consequence should be most different in overall size or in size of feeding apparatus, in the area of overlap. Vaurie (1951) described such a case in two species of nuthatches (*Sitta neumayer* and *S. tephronota*) which overlap in western Asia, and this example has been widely quoted; other examples have been adduced, and the term 'character displacement' has been introduced and generally accepted for this observation, that differences are accentuated in areas where populations of two species overlap. A closer analysis of the nuthatches has, however, shown that the situation is complex and open to various interpretations (Grant, 1975). Thus there is no evidence that bill size in either species has undergone character displacement in the area of overlap, but there is some evidence that body size may have done so. Character displacement remains a reasonable hypothesis for explaining certain size relationships in sympatric species, but each case needs detailed analysis and all other sources of variation must be taken into account (a more lengthy discussion of character displacement is presented in Chapter 17).

There are many examples of closely related species now occupying the same geographic area that had presumably developed in allopatry. The potential to live together is conditional on the establishment of both reproductive and ecological isolation (see also Chapter 6).

But there is another type of distribution pattern, particularly common in birds, which needs comment in any discussion of speciation. Many continental bird species, especially in Africa and South America, show a parapatric distribution. Species are said to be parapatric if their ranges adjoin one another without any overlap.

A parapatric distribution may be explained by one of two speciation models. It might represent the result of allopatric speciation in which the two species having attained reproductive isolation in separate areas are not ecologically isolated and so are unable to penetrate each other's ranges. This is the explanation favoured here, at least for birds, for it is assumed in this chapter that geographic isolation is necessary for the establishment of reproductive isolation. The other explanation invokes an hypothesis of parapatric speciation in which it is suggested that species can arise by a parent population splitting into daughter species while they remain side by side and in contact with one another. Endler (1977) has recently argued that this form of speciation may be important in birds. A fuller discussion of parapatric speciation is given by White (1978) and Hammond (Chapter 4); all that need be stressed here is that parapatric speciation (an hypothesis) and parapatric distribution (an observation) should be clearly distinguished.

In the most striking cases of parapatric bird distributions mosaics of distinct species are formed where the ranges adjoin one another more or less exactly.

Figure 3.3 shows a typical example of a group of

parapatric bird species in South America. In several areas the ranges of the three species abut one another, but apart from evidence of occasional hybridisation between *Pipra filicauda* and *P. fasciicauda* where they meet one another in eastern Peru, they behave as good species. It is noteworthy that neither the Amazon nor any of the other major rivers separates the range of one species from another, but they abut in areas where there is no obvious ecological barrier. Haffer (1970, 1974), who has examined many such cases of parapatric distribution in Amazonian forest birds, explains them as the result of alternating humid and arid periods in the Pleistocene. According to his argument, which is supported by palaeoclimatic and palaeobotanical data, the South American forests have altered greatly in their extent, probably several times, during the course of the Pleistocene. During arid periods (which may have coincided with glacial periods in the north, though this is not certain) the forests contracted into a number of relatively small pockets or refuges, the positions of which are probably much the same as the areas that today are especially humid. The ranges of forest birds similarly contracted and were split up, enabling them to differentiate in isolation. During humid periods they spread again with the spreading forests, eventually meeting their counterparts which had differentiated in other forest refuges. It seems that in many cases these isolates had differentiated sufficiently to be reproductively incompatible but that at the same time they were too similar ecologically to be able to coexist. Hence they could not penetrate one another's ranges, and the result was a mosaic of ranges occupied by closely related species.

Haffer identifies as many as 16 forest refuges on the basis of avian distribution and present rainfall patterns. His model provides a plausible and consistent explanation of present-day bird distribution patterns (even if there is an element of circular reasoning) and especially of the many cases of parapatric species. At the same time the model does not enable one to give a precise date to the differentiation of any species produced as a result of fragmentation during arid periods, since there is no means of telling whether their differentiation took place in the most recent or in some earlier phase of aridity.

Parapatric distributions in Africa may well have the same kind of origin as those found in South America; that is, they may have resulted from the contraction and expansion of vegetation belts caused by alternating wet and dry periods. In general, however, it seems that they are of more ancient origin than those that are such a prominent feature of the Amazonian avifauna. Figure 3.4 shows a striking example in the parrot genus *Poicephalus*, seven species of which adjoin one another over almost the whole of Africa south of the Sahara, except for evergreen forest areas, semi-desert and desert. They are all of much the same size and are ecologically similar, except for *P. flavifrons* of the Ethiopian highlands (which may be less closely related to other members of the group than they are to one another), but in plumage they are strikingly

different, so much so that it seems unlikely that they are the product of recent isolation. Though they are to some extent adapted to different habitats (for instance, *P. rufiventris* and *P. rueppellii* to more arid habitats than the other species) it seems that where the different species abut their ranges are limited by the presence of the neighbouring species.

Parapatric bird species need further investigation, especially by means of thorough statistical analyses. They seem to be particularly common in primarily frugivorous families, possibly because partitioning of food niches is less easy for frugivorous than for insectivorous birds (since insects constitute a far more complex and diverse food resource), thus rendering it more difficult for closely related species to coexist (Chapter 13). Parapatric distributions also seem to be especially common in birds with well-developed display plumage in the males, perhaps because behavioural isolating mechanisms, based on visual signals, are rather rapidly evolved in such birds and so tend to be developed before the evolution of ecological divergence sufficient to permit coexistence.

We have seen that the indirect evidence for the allopatric speciation model includes geographic variants and this has presented certain problems for the recognition of species which, for the vast majority of organisms, are based on morphological criteria. Taxonomists are all too familiar with one of the consequences of allopatric speciation, in so far as it affects their attempts to make orderly and consistent classifications of the animals that they study. The difficulties encountered are, in fact, exactly what would be predicted from the model; though they may be annoying to a tidy-minded taxonomist, for the evolutionary biologist they provide satisfactory corroboration for the allopatric speciation model. Thus, when two forms have come together after differentiation in isolation and are sympatric, as in the case of the treecreepers discussed earlier, there is usually no doubt about their specific distinctness. Hence naturalists who are concerned with reasonably well-known local faunas usually have no difficulty in knowing how many species they are dealing with. But when two forms have diverged in isolation and have remained isolated, it may be very difficult or even impossible (without elaborate experiments, which are usually impracticable) to decide whether they should be treated as separate species or as geographical races (subspecies) of the same species. The degree of morphological difference is not necessarily a sufficient guide. This is exactly what would be expected; divergence is gradual, and one will expect to find examples of every stage, depending on the length of the period of isolation and the rate of divergence.

It is this difficulty that is chiefly responsible for the variation in the number of species recognised by different authorities in groups of animals that show well-marked geographical variation, even by authors who have studied the same material and have much the same taxonomic philosophy. Thus the genus *Penelope* in South America, a group of arboreal galliform birds (curassows and guans), consists of a number of strictly allopatric populations showing various degrees of morphological differentiation. In three recent reviews, Vuilleumier (1965) recognises 6 species, Vaurie (1968) 13 species, and Delacour & Amadon (1973) 15 species.

It is of no special significance that in this example the more recent authors recognise more species; in fact, the long-term trend has been towards a reduction in the number of species recognised. In the late nineteenth century, when the world's avifauna was fairly thoroughly known but before the general adoption of the concept of the subspecies, all well-marked geographical races were treated as separate species and the total number of bird species recognised was much higher than it is today. In Sharpe's *Hand-list of the Genera and Species of Birds* (1899–1909), nearly 19000 species of living birds are listed; by the middle of the twentieth century the number had been reduced to about 8600 (Mayr & Amadon, 1951). Groups of forms such as the coal tits, with many isolated and distinct outlying populations, suffered the greatest reduction. Thus in the *Hand-list* 13 species of coal tits are listed; by 1951 this had been reduced to two – *Parus ater* and *P. melanolophus* – the latter being a little known species with very dark plumage, and confined to part of the western Himalayas. More recently, with the discovery that *Parus melanolophus* is not only very like *P. ater* in behaviour but intergrades with it in a part of Nepal that had not previously been explored ornithologically, it has become apparent that only a single species should be recognised.

It is important to realise that these problems of practical taxonomy – of the formal naming of animal populations –

are not only to be expected, they are inevitable consequences of the process of allopatric speciation. They can only be resolved, if at all, by a fundamental change in our system of nomenclature; and no fundamental change is likely to be adopted, as the present system has great advantages as well as short-comings. Refinements in the system of nomenclature allowing for subdivision of categories do not seem to provide the answer. A good many of these have been advocated (for instance, semispecies, megasubspecies, microsubspecies), but there always comes a point where a more or less arbitrary decision must be made, and different taxonomists will decide in different ways.

The only way in which the specific status of allopatric populations that are on the border-line of speciation can be tested is by experiment, by bringing them together under natural conditions and seeing whether they are reproductively isolated. For small animals that can be maintained in the laboratory, artificial experiments may be made to approximate to natural conditions and give a more or less reliable answer; but for vertebrates this is usually impracticable or prohibitively expensive. For this reason, we must admit that many border-line cases will continue to be debated, and that an agreed taxonomy is an ideal rather than an achievable objective. Certainly in birds, the most significant advances in our understanding of speciation are likely to come, not from vain attempts to fit the infinitely variable facts into a formal system, but from thorough, long-term studies of 'natural experiments' that are in progress today, such as those involving the grebes of Madagascar and the other cases where closely related forms, on the threshold of speciation, are in contact with one another.

References

Delacour, J. & Amadon, D. 1973. *Curassows and related birds.* 247 pp. New York: American Museum of Natural History.

Endler, J. A. 1977. *Geographic variation, speciation, and clines.* 246 pp. Princeton: Princeton University Press.

Grant, P. R. 1975. The classical case of character displacement, pp. 237–337 in Dobzhansky, D., Hecht, M. K. & Steere, W. C. (Eds) *Evolutionary Biology* 8. New York: Plenum Press.

Haffer, J. 1970. Art-Entstehung bei einigen Waldvögeln Amazoniens. *Journal für Ornithologie* 111: 285–331.

Haffer, J. 1974. *Avian speciation in tropical South America.* 390 pp. Cambridge, Massachusetts: Nuttall Ornithological Club.

Lack, D. 1947. *Darwin's finches.* 208 pp. Cambridge: Cambridge University Press.

Mayr, E. 1963. *Animal species and evolution.* 797 pp. Cambridge, Massachusetts: Belknap Press.

Mayr, E. & Amadon, D. 1951. A classification of recent birds. *American Museum Novitates* 1496: 1–42.

Seebohm, H. 1887. *The geographical distribution of the family Charadriidae.* 524 pp. London: Henry Sotheran.

Simpson, G. G. 1961. *Principles of animal taxonomy.* 247 pp. New York: Columbia University Press.

Vaurie, C. 1951. Adaptive differences between two sympatric species of nuthatches (*Sitta*). *Proceedings of the tenth International Ornithological Congress*: 163–6.

Vaurie, C. 1968. Taxonomy of the Cracidae (Aves). *Bulletin of the American Museum of Natural History* 138: 131–260.

Vuilleumier, F. 1965. Relationships and evolution within the Cracidae (Aves, Galliformes). *Bulletin of the Museum of Comparative Zoology, Harvard College* 134: 1–27.

White, M. J. D. 1978. *Modes of speciation.* 455 pp. San Francisco: Freeman.

CHAPTER 4

Speciation in the face of gene flow–sympatric–parapatric speciation

P. M. Hammond

The intention of this chapter is to discuss models of species multiplication that do not conform to the 'classical' allopatric or geographical model (see Chapter 3). It may be said at the outset that my interest in this topic, and therefore the way in which it will be approached, is that of the insect taxonomist. With insects it is clear that abundant speciation has played an integral part in their various successful adaptive radiations. Considering the numbers of extant species, and what is known of their likely phyletic relationships, it is recognised that many millions of speciation events have been involved. Furthermore, as insects are primarily sexually reproducing and biparental organisms one may ask: was each of these many millions of events instigated by a sudden interruption of gene flow of the kind most usually associated with the classical allopatric speciation model? Or: do other models provide a real challenge to the accepted idea that new species of sexually reproducing animals arise only after a period of complete geographic isolation and gradual genetic change?

A taxonomist concerned with the origins of diversity may wish to test the allopatric model by using the results of his own attempts to discover the phylogenetic relationships between species. He will examine the assemblages for which he has provided phylogenetic hypotheses, paying special attention to existing pairs of sister species. Palaeogeographical and palaeoecological information will be of interest, in so far as this enables some light to be shed on former distributions. The distribution and variation of Recent species will be of importance (see Chapter 1). For instance, it will be of relevance to know if a morphologically uniform species is distributed in two or more widely separated areas (disjunct distribution), or if a species varies over a continuous geographic area (as found in clines). Of special interest will be the coexistence of related species, and the structure of communities and guilds in which they operate. In the case of closely related coexisting species the means by which reproductive isolation is maintained will be of particular interest and any available information concerning cytological differences within and between

species will be examined carefully. Against a background of what information he has on the general biology and evolutionary history of the group of organisms, the taxonomist will review his data in the hope of identifying items having some direct bearing on the dynamics of past speciation events. Do any of his data provide a test for the allopatric or other speciation models? Finally, he will wish to learn what other workers, such as population geneticists have to offer in terms of testing either model.

The substance of this chapter then, derives largely from biogeographical, ecological and phylogenetic interests and the hypotheses which stem from a taxonomist's work. I shall begin by briefly reviewing the circumstances under which, until recently, the validity of the allopatric speciation model was little questioned. Alternative models will then be discussed, paying particular attention to the ways these may be categorised. The role of geographical distribution, including disjunctions of range (see also Chapter 1), chromosomal rearrangements and other factors possibly initiating divergence will then be considered. It will then be pertinent to ask: in what circumstances does divergence lead to the completion of speciation? By focusing attention on ways in which a speciation process may be completed, a strict antithesis of allopatric and non-allopatric models will be abandoned. Here, emphasis will be on the ways in which reproductive isolation may develop in diverging populations which are conjunct (parapatric) or overlapping (sympatric). In other words, how may species status be achieved in the face of gene flow? From the outset it will be clear that views on modes of speciation will be influenced by our ideas of what constitutes a species. Much has been written on this and the reader is referred to Sokal & Crovello (1970), Scudder (1974) and White (1978, pp. 1–5) for contrasting views. For the purposes of this essay the concept of the biological species is accepted.

Speciation models – the problems

In general, 'the process of speciation has to be reconstructed by inference' (Mayr, 1979). Observations on extant species have been the principal source of hypotheses concerning the process of speciation. The fossil record, unfortunately, has proved of little direct value in testing hypotheses. Population genetics also, perhaps surprisingly, appears to have contributed relatively little to our understanding of speciation. The study of speciation remains very much an '*ad hoc* science' (Bush, 1975).

Recent controversy has centred on the question of whether the allopatric model is still acceptable as the virtually universal model for speciation. The work of Mayr (1942, 1963, 1970) emphasises the allopatric model, almost to the exclusion of other modes of speciation, and has undoubtedly been of great influence. To quote White (1978, p. 10): 'Mayr's views on speciation have been presented with so much supporting evidence and developed with so much erudition that for many biologists they undoubtedly appear to be the ultimate truth on the matter'.

The allopatric model states that speciation occurs by an original population becoming split into two or more fractions which thus occupy different geographic areas (allopatry). The fractions undergo genetic divergence such that when, and if, they meet again they are no longer able to interbreed. That is, species status is attained. During the last ten or fifteen years the allopatric model has come increasingly under fire. A range of feasible alternative models is now available. Many of these are elaborated and discussed in recent works (Murray, 1972; Bush, 1975; Endler, 1977). In particular, the reader is referred to White's (1978) comprehensive review of the majority of currently proposed models of speciation.

Alternative speciation models may provide answers to either of two questions: are models containing a non-allopatric element feasible? And are models totally lacking an allopatric element feasible?

Discussion of these questions, unfortunately, is bedevilled by much confusion of terminology. For instance, there may be little agreement concerning what is meant by the speciation process itself; at what point speciation may be regarded as completed and what is considered to be the crucial step in the process. A static view of what is patently a dynamic process may also lead to difficulties. It needs to be stressed that differences between closely related species may not necessarily be associated with the attainment of reproductive isolation – speciation. Characteristics exhibited by species today, whether of geographical range, karyotype, behaviour or morphology may have existed before or have been acquired after speciation.

Mayr's view is that virtually all speciation involves a geographical barrier which enables two separated gene pools to be reorganised (Mayr, 1942, 1963, 1970, 1979). Certainly, proponents of the allopatric model '...are impressed by the role of heterozygosity, and by the difficulty of getting from one peak of a well integrated, well coadapted gene complex, through a valley of disintegration to another peak of coadaptedness' (Mayr, 1979, p. 479).

However, a general problem that arises in comparing the allopatric model with alternatives is to decide just what should be regarded as a geographical barrier. Some formulations of the allopatric model embrace as geographical any barrier, including a difference of host in the case of parasites, which is sufficient to protect diverging gene pools from the 'disruptive influence of gene flow.' Such models, however legitimate, are clearly no longer identical with what may be regarded as the classical allopatric one, in which a more restricted definition of geographical barriers, such as mountain ranges and seas, applies. It would seem preferable to reserve the term *allopatric* for speciation models of the latter type, in which *geographical disjunction*, in the biogeographical sense, is a factor.

Speciation models of supposedly wide applicability are sometimes erected on the basis of inferences drawn from a rather restricted group of organisms. For example, much of the supporting evidence for the allopatric model is taken from the study of birds, in particular from consideration of the ranges occupied by what are considered to be incipient and recently formed species. A problem often met in discussions of speciation models is the extent to which the general feasibility of a model is confused with the likelihood of its broad applicability. It should be remembered that groups of organisms, such as insects, in which most speciation is, in fact, likely to have taken place received little attention in speciation studies. Most organisms are small and, to the insect taxonomist, an insistence on extrapolating from species-poor groups of large vertebrates may appear almost perverse.

Comparison of alternatives is also hampered by disagreements on which events form part of the speciation process. Sometimes such differences stem from a genuine divergence of views on how the speciation process proceeds, and as such will be reflected in the salient features of different models. In other cases apparently important differences between models may represent little more than disagreement on when the process may be considered to have been completed or begun, not on the actual method of speciation.

Several people have pointed out that it is only in retrospect that the initial stages of differentiation or the establishment of a partial or potentially reversible barrier to reproduction may be identified as part of a *particular* speciation event. However, it would seem legitimate to acknowledge these components in the construction of speciation models. This, of course, should not be taken to suggest, for example, that all geographically disjunct segregates or adjacently distributed chromosomal races are necessarily incipient species. It is clear that not all differentiation, especially that of populations of widely distributed organisms, is associated with discrete and readily identifiable barriers. Nevertheless, differentiation of this type is incorporated in a variety of speciation models. The precise point at which speciation is to be regarded as having begun will, in practical terms, be somewhat arbitrary.

The problem of assessing when a particular speciation event may be regarded as completed stems from difficulties in applying the notion of 'biological species' in those instances where gene pools are completely isolated by an extrinsic barrier. Here, the question as to whether reproductive isolation, as measured by genetic contact, is sufficient for the two segregates to be regarded as 'good' biological species remains unresolved unless secondary contact (natural or artificial) intervenes to provide a test. Differences of view regarding the extent of gene flow (if any at all) that may be allowed for in applying the biological species concept gives rise to a more general problem. There is, perhaps, much to be said for the position adopted by Key (1974), that the simplest course is to recognise biological species only when their reproductive separation from related species is complete. Even here, it must be conceded that, in practice, there are difficulties. The problem may be more apparent in the case of parapatric (species living adjacent) rather than sympatric species. Certainly, some populations which appear to have achieved full reproductive isolation are distributed parapatrically, although care must be taken to distinguish parapatric distributions, in this connection, from those which are, in fact partially sympatric.

Thus, biological species *may* exclude each other geographically, but it does not necessarily follow that parapatric distributions are exhibited by populations between which gene flow is negligible or non-existent. Indeed, many parapatric distribution patterns are likely to be maintained by the operation of intrinsic but partial reproductive barriers. The occurrence of such barriers is insufficient in itself for regarding the segregates as biological species. However, the actual status to be accorded parapatric populations that exhibit less than complete reproductive isolation is bound to remain contentious. Populations of differing karyotype that lack detectable differences of morphology would seem to be best regarded informally as 'chromosomal races'. Where morphological, behavioural or other differences are found it would seem appropriate to treat the populations in the same way as those which are disjunct, and refer to them as subspecies.

The classification of speciation models

For the purposes of discussing the speciation models that have been proposed as alternatives to the classical allopatric model it will be useful to group them in some manner. White (1978) has noted that three main sets of variables are involved in speciation. These are underlying genetic mechanisms, genetic isolating mechanisms and geographical factors. Other variables, such as degree of vagility of the organisms, are also of significance. A complete categorisation of these models on purely genetic criteria is, as yet, scarcely feasible. However, White's (1978) classification, although based primarily on the geographical component, also involves population structure and genetic mechanisms. Bush (1975), employs three ostensibly geographic

categories – allopatric, parapatric and sympatric – but also defines these in other than geographical terms. The attempt to directly equate geographic and other components in this case results in a classification that is not entirely satisfactory in purely geographical terms. For example, Bush's 'sympatric' category is virtually limited to parasites and parasitoids in which premating reproductive barriers are considered to develop before the shift of one population to a new niche. However, other types of organism and other types of relationship between the production of reproductive barriers and niche shift may be considered feasible and are found in models which are nonetheless sympatric in the strictly geographical sense. It would seem that any complete congruence of speciation model categories based on different components of the process is scarcely possible.

Despite the fact that several variables have been used in the classification of speciation models, the geographic component would appear to retain some usefulness. Indeed, in discussing explicit alternatives to allopatric models, based on geographical criteria, the use of a geographic component is scarcely avoidable. However, two possible sources of confusion should be noted. The use of the term *allopatric* for models involving any extended spatial separation has been mentioned above. To avoid ambiguity the term is used here in its usual purely geographic sense. However, geography also involves questions of scale. The distinction between small-scale geographic separation ('micro-allopatry') and separation which is more clearly 'intermeshing', as in ecological separation based on choice of host, may not always be easy to make in practice.

A second possible source of confusion stems from the fact that different stages of the speciation process may take place in different geographic settings. When methods of speciation are inferred from existing distributions of species this confusion may be compounded. A firm distinction should be drawn between the geographical context in which, according to various speciation models, principal stages of the process were thought to have been undertaken, and post-speciation geographical range. Possible patterns of geographic distribution in relation to speciation are presented in tabular form in Table 1. All patterns which, at least by implication, are features of what may be regarded as the principal feasible speciation models are included in Table 1; others are theoretically possible.

It will be clear from Table 1 that the current distribution of an organism is unlikely to be a reliable guide, in geographical terms, to the mode of speciation. Although secondary allopatry may not be common it is certainly always possible. This often appears to be overlooked by those who are most ready to identify sympatric and parapatric distributions as secondary.

In purely geographical terms, four modes of speciation can be recognised – sympatric, parapatric, allo-parapatric and allopatric. Here, I have followed in essence, the classification of Endler (1977). The principal features of these categories are:

Table 1 Speciation models and the geographic component

Initial differentiation	Completion of reproductive isolation	Subsequent distribution	Mode of speciation
1 Sympatric	Sympatric	Sympatric	Sympatric
2 Sympatric	Sympatric	Allopatric	Sympatric
3 Parapatric	Parapatric	Parapatric	Parapatric
4 Parapatric	Parapatric	Sympatric	Parapatric
5 Parapatric	Parapatric	Allopatric	Parapatric
6 Allopatric	Parapatric	Parapatric	Allo-parapatric
7 Allopatric	Parapatric	Sympatric	Allo-parapatric
8 Allopatric	Parapatric	Allopatric	Allo-parapatric
9 Allopatric	Allopatric	Allopatric	Allopatric
10 Allopatric	Allopatric	Parapatric	Allopatric
11 Allopatric	Allopatric	Sympatric	Allopatric
12 Parapatric	Allopatric	Allopatric	Allopatric
13 Parapatric	Allopatric	Sympatric	Allopatric

Ranges of incipient species are compared with those occupied subsequent to completion of speciation. Ranges may be disjunct (allopatric), conjunct (parapatric) or overlapping (sympatric).

In truly *sympatric* models there is no spatial segregation of diverging populations, in the sense that members of both remain within 'cruising' range of each other throughout the process.

In *parapatric* speciation, spatial segregation (but not disjunction) and differentiation are followed by the eventual attainment of reproductive isolation while the populations remain contiguous.

In *allo-parapatric* speciation, initial segregation and differentiation takes place in disjunction (allopatry), while complete reproductive isolation is achieved in parapatry.

In allopatric speciation the entire process, or virtually the entire process, takes place in disjunction. Secondary contact plays no part in the process and, if subsequently achieved, is immaterial to the independent future of the species which have already achieved complete, or virtually complete, reproductive isolation.

A modified version of Endler's (1977) scheme, employing the geographic component to demonstrate various pathways which may be involved is presented in Figure 4.1.

Apart from the category of 'instantaneous' speciation none of the pathways involving differentiation includes a guarantee that complete reproductive isolation will eventually be achieved. It also will be clear again that the ranges finally occupied by a species pair provide no indication, in themselves, of the speciation mode involved.

The feasibility of non-allopatric speciation

Characteristics exhibited by what are thought to be 'recently' evolved species pairs or by possible 'incipient' species, remain the principal source of evidence for assessing the merits of various speciation models. However, it should be made clear that few of the hypotheses incorporated in speciation models are actually *testable* on this basis. Apart from establishing that available evidence

is consistent with a variety of models, the most that we may expect is the demonstration that a particular model is not, or is very unlikely to be, applicable to a given case. This will generally have no direct bearing on the validity of the model in other circumstances.

A particular difficulty in the case of extant species pairs, even those possibly 'newly minted', is recognising which of their features are relevant to the speciation event by which they arose. The acquisition not only of geographical range (Table 1) but of other characteristics may be post-speciation phenomena. Evidence from extant populations which have achieved less than full reproductive isolation may also be difficult to evaluate. For instance, we often recognise races, geographically separated or otherwise, and these races may provide evidence for a number of speciation theories in which a gradation from a uniform population to distinct species is predicted. However, it is the *model* that enables one to 'recognise' the population as a possibly incipient species; the attributes of diverging populations, in themselves, entail no guarantee of future progress to speciation.

Before examining in more detail the evidence that recently evolved or possibly incipient species may provide, the general feasibility of at least some of the alternative (non-allopatric) models needs to be established. Here, the speciation process will be regarded as being composed of two principal components – differentiation, and reproductive isolation (the irrevocable severance of all, or virtually all, genetic interchange). It should not be inferred that speciation is, or is regarded here, as simply a two-stage process. The two components may be inextricably linked. Differentiation may be initiated by the imposition of reproductive separation through an extrinsic barrier or through a variety of intrinsic factors (Chapter 5). It continues, of course, as 'phyletic evolution' after reproductive isolation is achieved and speciation is complete.

Differentiation and its initiation

Barriers to gene flow created by geographical disjunction, if persistent, may form the basis for marked differentiation of the isolated populations. Complete subdivision of a gene pool by other than strictly geographical barriers, such as those implicit in some models of host-race formation may similarly lead to differentiation. Initial subdivision of a gene pool is discussed more fully in connection with the allopatric speciation model (Chapter 3) and in Chapter 1. However, just how frequently such differentiation is likely to be ultimately divisive (by leading to the development, in isolation, of intrinsic reproductive barriers) is another question.

The belief that complete subdivision of a gene pool by geographical disjunction, is a prerequisite for intense differentiation may be seen as the principal basis for the widespread acceptance of the allopatric model. It seems, that this view has stemmed, at least in part, from underestimates of the extent of genetic diversity within a single

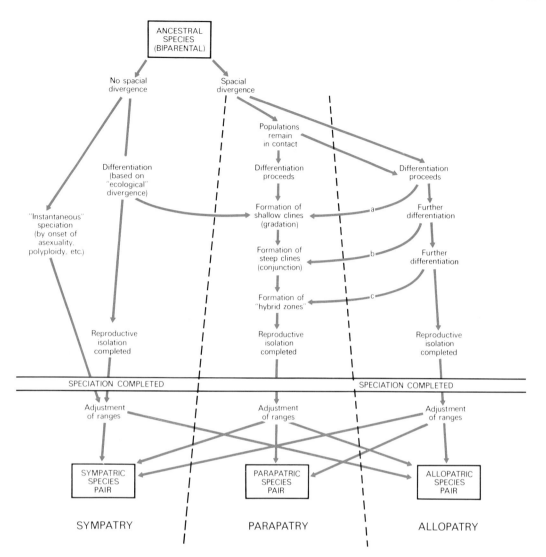

Figure 4.1 The geographic component in speciation models (modified from Endler, 1977). The three columns separated by broken lines indicate the three geographic situations of sympatry, parapatry and allopatry. Sympatric, parapatric and allopatric speciation models follow pathways which, above the speciation completed level, are generally confined to their respective columns. The pathway for allo-parapatric speciation follows the right-hand column and then shifts to the middle column at 'a', 'b' or 'c'. Not all possible shifts from sympatry to parapatry, or parapatry to allopatry, above the speciation completed level, are included in the diagram.

gene pool. With the introduction of new techniques for demonstrating genetic variability the existence of high levels of genetic polymorphism and of heterozygosity within populations of sexually reproducing species has become increasingly apparent. The amount of genetic variability in natural populations is reviewed at length by Lewontin (1974) who concludes that 'the overwhelming preponderance of genetic differences between closely related species is latent in the polymorphisms existing within species'. It is of some interest that this variability is found to differ markedly from group to group; for example, levels of both polymorphism and genic heterozygosity are generally much higher in invertebrate than vertebrate species (see White, 1978, table 2.2).

Recent studies have also established the extent to which gene frequencies are heterogeneous within the geographical range occupied by a species. Such geographic or spatial variation is a feature of both genetically continuous series of populations and those which are disjunct or separated by other barriers to gene flow. Can such variation then evolve in the absence of fully effective barriers to gene flow?

It has rarely been suggested that *all* geographic variation stems from the effects of disjunctions, but what of the more intense differentiation exhibited at steps in clines and across hybrid zones? These questions are specifically addressed in the recent works of Murray (1972) and Endler (1977) (see also Chapter 1). Endler states (1977) that repeated fragmentation and concomitant interruption or reduction in gene flow may accelerate the differentiation process but is not necessary for population differentiation and speciation. I will not attempt a summary of the mathematical modelling on which this conclusion rests, but will stress that two points are at issue: the extent of gene flow in natural populations and the influence of gene flow as a cohesive force. In practice, particularly in organisms with patchy microdistributions, the extent of gene flow is likely to be quite restricted. Generally, gene flow will be spatially restricted compared with migration and dispersal of individuals, from both of which it should be carefully distinguished. The finding that dispersal distances most commonly follow a leptokurtic pattern is of particular relevance to gene flow; successful establishment of a migrant gene over a long distance is unlikely when dispersal over such a distance is infrequent and by few individuals. Moreover, experimental work on *Drosophila* fruitfly populations (see Endler, 1977) provides evidence that, even where levels of gene flow are high, marked differentiation may evolve rapidly. The properties of naturally occurring and experimental clines enabled Endler to conclude (1977) that gene flow is not as strong a dedifferentiating factor as has often been supposed. His models demonstrate not only the feasibility of a modicum of spatial differentiation in the face of gene flow, but also how shallow clines so produced may steepen, leading to the eventual formation of 'hybrid zones'.

If marked differentiation may originate in either a genetically continuous series of populations or in geographically disjunct populations, which of these two sets of circumstances is likely to be of greater significance? In any given instance, as with species, the distributions of races (disjunct or not) provide no test of the geographic context in which differentiation took place. However, in so much as the geographic relationships of races may be considered more likely to be primary than secondary phenomena, existing patterns of disjunction and conjunction may be instructive. We find that disjunction unaccompanied by readily observable differentiation is, perhaps surprisingly, common, while true monomorphism in widely distributed but spatially continuous populations is relatively rare. Distance, just as much as disjunction, may be responsible for the geographic differentiation observed in natural populations.

What then of differentiation without geographical segregation? Maynard Smith (1966) and others have explored, by means of mathematical models, the conditions under which disruptive selection in a heterogeneous environment may lead to the development of a stable genetic polymorphism. The type of multi-niche polymorphism which Maynard Smith's investigations suggest may arise in sympatry is found among natural populations (for instance, that of a species of spittle-bug described by Halkka (1978)). Frequency-dependent selection will be responsible for maintaining such polymorphisms. The correlation of morphs with different ecological niches will naturally result in the establishment of quite different allele frequencies in various, geographically distinct sub-populations, albeit that the morphs have arisen in sympatry.

Another type of divergence for which there is good evidence of sympatric origin is that exhibited by host-specific or other 'biological' races which occur sympatrically. Host-races which apparently arose rapidly in recent historical times are known in several insect groups. How the genetics of host-race formation may be compatible with a *sympatric* shift of host has been shown by Bush (1975). He points out that the genetic variation needed to establish a new host-race is likely to be present in the parent population even before a new host appears on the scene. In the tephritid fruitflies studied by Bush there is good evidence that both host and mate selection are closely linked and genetically controlled, mating taking place on the fruits of the host-plant. In those instances where mating is not so closely linked with the host, Bush concludes that a difference in the timing of mating, in association with a change of host, may also form the basis of sympatric divergence. An important feature of the model for sympatric host-race formation is that only a few changes at key gene loci are required to enable a switch of host (for a full account see Huettel & Bush, 1972). The process is not initiated by a 'genetic revolution'. It should be stressed that in Bush's model a switch of host is the *first* step; differentiation associated with adjustment to a new ecological niche follows. If the initial switch involves virtually complete reproductive separation of the new host-race from the parent population then the differentiation of host-races may be akin to that initiated by geographical disjunction.

The sympatric origin of host-races in tephritid fruitflies seems rather well established and is even accepted as likely by proponents of allopatric speciation. It is probable that a generally similar model of sympatric divergence is applicable to many other parasitic organisms. However, the view that the model is necessarily relevant to the great majority of parasites should be treated with some caution. It is not at all clear that the majority of parasites are host-specific, as appears to be suggested by Bush (1975), although this is likely to be true for very many members of one enormous insect group – the parasitic Hymenoptera. It must also be remembered that a close association of host and mating site may be the exception (for instance, pollen-feeding beetles of the family Nitidulidae) in a number of groups in which host-specificity predominates.

Differentiation in genetically continuous series of populations is a feature of chromosomal models of speciation. These are reviewed at length by White (1978) who stresses, in particular, the role that chromosomal rearrangements may play in initiating divergence. White's own 'stasipatric' model (see Fig. 4.2) provides a convincing explanation of how parapatrically distributed races may arise. This model was initially developed through intensive cytological and taxonomic work on a group of Australian grasshoppers (Morabinae), and is likely to be applicable to other groups of organisms of low vagility and similar genetical population structure, such as lizards and rodents.

Morabine grasshoppers are flightless and exhibit great variation of karyotype even within species. In addition to chromosomal rearrangements that are able to exist in a polymorphic state, other rearrangements, of a type likely to produce inferior heterozygotes, are also found. These rearrangements, mostly fusions and fissions (sometimes called dissociations), are characteristic not only of most closely related species but also of parapatrically distributed chromosomal races. In all, some 61 fusions and fissions have been identified in the approximately 200 morabine species which have been examined cytologically.

White's stasipatric model involves the generation of such chromosomal rearrangements, giving adaptively superior homozygotes but inferior heterozygotes, 'within the range of a widespread species'. Populations in which a rearrangement has become established adopt a parapatric distribution with respect to the parent population. The narrow zone of hybridisation separating the chromosomal races acts as a selective boundary. Only the chromosome, or chromosomal segment responsible for heterozygote inferiority, along with any genes which are closely linked, will necessarily be blocked from crossing this boundary. Nevertheless, progressive accumulation of genetic differences about the zone is likely. The precise circumstances in which the chromosomal rearrangement arises and initially spreads is not pertinent to what I regard as the model's principal feature – that differentiation is *initiated* by the establishment of partial postmating incompatability.

Chromosomal races, clines of various types and multiniche polymorphisms are all commonly observable in natural populations. It may reasonably be concluded therefore that there is considerable circumstantial, and even experimental evidence (Endler, 1977) for differentiation being initiated and proceeding in a variety of geographical settings. Strict allopatry is not a prerequisite.

Equally, allopatry does not necessarily lead to marked differentiation. Although barriers to gene flow created by geographic disjunction are bound to play an important role in generating divergence, other barriers or impediments to gene flow may also have a vital part to play. In some instances, for example the development of balanced multiniche polymorphism, no discrete barrier is responsible for initiating differentiation. In other cases the imposition of an irreversible barrier to gene flow (for instance, a change of ploidy) initiates differentiation in populations that may be regarded as having already achieved species status.

Barriers may be intrinsic or extrinsic, partial or complete. Such impediments to gene flow that may play a role in generating divergence are of four principal types:

1. Complete and irreversible subdivision of a gene pool, (e.g. change of ploidy).
2. Complete but always potentially reversible subdivision of a gene pool (e.g. by geographic disjunction).
3. Partial but generally irreversible barriers to gene flow (for instance, chromosomal rearrangements).
4. Partial and potentially reversible impediments to gene flow (for instance, distance, as in a cline).

Populations in which differentiation is proceeding have an uncertain future. Many are doomed to extinction before any extensive divergence has taken place, while reversal of partial or temporary barriers to gene flow will reunite many diverging segregates with their parent population. Differentiation in itself is not necessarily ultimately divisive.

Reproductive isolation and its attainment

Just what comprises the final stage of a speciation process is, theoretically at least, unequivocal: it is signalled by the attainment of reproductive isolation, that is, an effectively complete and irreversible rupture of genetic continuity. It allows no or only a negligible amount of exchange of genetic material and the nature of reproductive barriers ensures the future integrity of the species. This view of the completion of a speciation process, of course, is applicable only to those organisms (including the great majority of insects) in which species conform fairly rigorously to the biological species concept.

Reproductive isolation may be achieved at a stroke. For example, polyploids are, from the moment of their origin, isolated from their diploid progenitors. The gradual attainment of full reproductive isolation in allopatry is clearly another way in which a speciation process may be completed. Similarly, the reproductive barrier between host races may, in the absence of any extensive gene flow, provide the basis for the development of full reproductive isolation, in which coevolution of host and parasite may play a part.

Leaving aside those cases where the final stage of speciation takes place in virtual immunity from gene flow the possibility of reproductive isolation being attained in the face of extensive gene flow is to be considered. Is the gradual development of an effective reproductive barrier between populations which retain a measure, although gradually diminishing measure, of genetic contact feasible?

To deal with the question of reproductive isolation we return to patterns of differentiation observed in series of populations which remain in genetic contact. Such patterns will be either parapatric or, in broad terms, sympatric. Parapatric distributions, which may be primary or secondary, are exhibited not only by populations in genetic contact but also by those which may be regarded as distinct species. Where genetic contact is nil, geographic exclusion may be the result of competition, although this

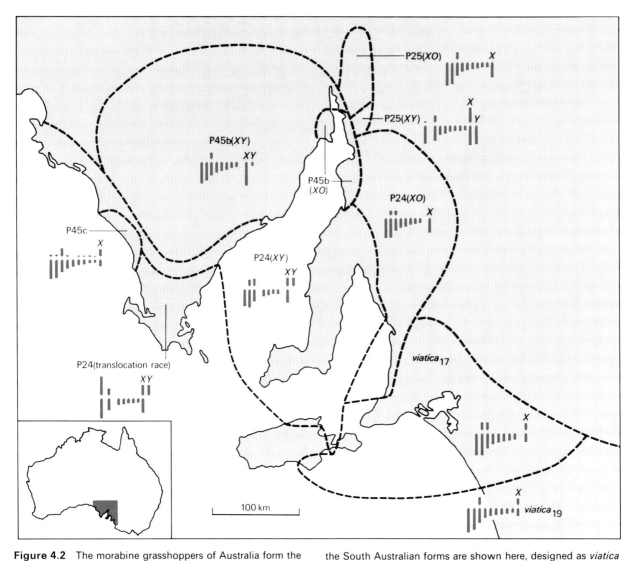

Figure 4.2 The morabine grasshoppers of Australia form the basis of M. J. D. White's theory of stasipatric speciation which proposes that it is possible that chromosomal rearrangements (e.g. inversions, deletions, fusions or translocations) may initiate differentiation within genetically continuous populations. Within the *Vandiemenella viatica* group there are many chromosomal races which differ in karyotype; some of the South Australian forms are shown here, designed as *viatica* 17 etc. or as P_{24} etc. and the chromosome patterns are shown as ideograms. Some of the chromosomal races have males in which there are two sex chromosomes (XY), others have just one (XO). The race P_{456} is, karyologically, indistinguishable from $P_{25(XO)}$. Redrawn and simplified from White (1973).

might be expected to produce a characteristically inter-digitating interface between species. Perhaps more often parapatric distributions are maintained by a '*lethal confrontation*' which may arise when populations unable to produce viable hybrids but with no effective premating reproductive barrier meet.

Whether or not there is some exchange of genetic material through a 'hybrid zone', it is also clear that parapatric distributions may be extremely durable. The Pleistocene fossil record of insects, in particular, testifies to this, although the precise boundary separating the parapatric populations may wander geographically over time. What bearing then, does this observed stability of parapatric distributions have on the likelihood of a parapatric completion of speciation? Allo-parapatric and purely parapatric models of raciation (for instance, the stasipatric and clinal models referred to above), provide convincing explanations of the way in which parapatrically distributed races may arise, but how relevant are these models to the attainment of full reproductive isolation in parapatry? How often are partial postmating reproductive barriers, such as those stemming from chromosomal rearrangements in the stasipatric model, ultimately divisive? It seems very likely that eventual rupture of genetic continuity between chromosomal races, as postulated by the stasipatric model, does in fact take place in groups such as the Morabinae. The group includes parapatrically distributed populations separated by apparently completely effective postmating barriers, as well as parapatric races where barriers are only partial. However, not all closely related species of Morabinae differ in karyotype, and it is clear that the majority of speciation events in this group have not been accompanied by major chromosome rearrangements of the type exhibited by many chromosomal races. There are approximately 240 extant species of Morabinae. Even if each of the 80 to 100 chromosome rearrangements estimated by White (1974) to have undergone fixation in this group could be associated with a speciation event, and some are not as yet, speciation of this type is evidently far from the general rule in terms of the entire history of the Morabinae.

In the Morabinae, closely related species with markedly different karyotypes appear never to be sympatric. Continuing geographical exclusion is unlikely to be conducive to survival of each member of a group of parapatric species in times of environmental change. It would seem possible that many parapatrically distributed species, as well as chromosomal races, are doomed to extinction before potential for extending their geographic range is realised by the development of appropriate premating reproductive barriers. If this is the case, there may be some justification for regarding the reproductive isolation attained by parapatric species as often relatively 'trivial'. The use of the term 'semispecies' to embrace such instances, even though genetic contact between populations is nil, may have some value.

Indeed, in the Morabinae as in other groups, there would appear to be little evidence for the incipient development of premating reproductive barriers between chromosomal races (or 'semispecies') or that such barriers have arisen in this way in the past. In the case of groups such as the European mole-crickets, some parapatric taxa differ in calling song, but it is not at all clear that these songs are effective premating barriers. If speciation of the existing parapatrically distributed mole-crickets has been stasipatric, and if the acquisition of premating reproductive barriers is a likely eventual outcome of such a process, then some sympatric taxa might be expected to occur in this group. The various rounds of stasipatric race-formation which might be assumed to have preceded that generating the existing parapatric races (semispecies) should have led to this conclusion. However, sympatric mole-cricket species, even distantly related ones, are not to be found.

White found that even closely related species of morabine grasshoppers often differ in karyotype and this led him to believe that chromosomal rearrangements are particularly important in the attainment of reproductive isolation – speciation. But just how general and how significant chromosomal rearrangements are considered to be, will depend on interpretation of available cytological evidence, and also on which stage of the speciation process is regarded as being most crucial. White (1974) is quite explicit on this matter stating that 'it seems legitimate for

the time being to regard the origin of genetic isolating mechanisms rather than subdivision of the gene pool as the prime cause of speciation'. However, I have suggested in the preceding discussion that creating a partial postmating barrier by no means inevitably lead to speciation. Parapatric differentiation may often be initiated in this way, but intense differentiation and eventual reproductive isolation may be no more likely than if differentiation is triggered by a partial extrinsic barrier.

The available cytological evidence certainly supports White's (1978) contention that karyotypic differences between (but also within) species are common. However, unlike the major chromosomal rearrangements of morabine grasshoppers, the role of most other karyotypic differences in speciation is difficult to assess. In the majority of cases we cannot be sure that differences are not post-speciation phenomena. We must also consider the nature of the sample of organisms for which cytological evidence is available. Even beta karyological data (chromosome numbers and lengths of chromosome arms; sex chromosomes) are on record for only a small minority of known animal species. Data concerning the most species-rich groups are especially poor. In more than 60 pages of discussion of chromosomal differences between species White (1978) quotes no examples from the largest insect groups – Coleoptera and Hymenoptera – which together may represent half of extant animal species! Whether the majority of sister-species pairs truly differ in visible metaphase karyotypes remains to be established. Homosequential species-complexes (those in which chromosome banding patterns and hence the gene arrangements are considered to be identical) are known to occur only in *Drosophila*. Available evidence thus supports the view that such complexes represent a 'very exceptional situation' (White, 1978). We must, however, be cautious when assigning chromosomal rearrangements a key role in the process of speciation; cytological data for the many insect groups most similar in general biology and reproductive behaviour to drosophilids are almost totally lacking.

The observed stability and durability of balanced multiniche polymorphisms and of parapatrically distributed races might suggest that these usually are the final outcome of differentiation within genetically continuous series of populations. In fact, probable primary parapatric distributions are also exhibited by species (or semispecies) no longer in genetic contact. However, here also parapatric stability may be seen to originate from the operation of postmating reproductive barriers (inviability or sterility of hybrids). Development of premating barriers avoiding these lethal effects would appear to offer a way out of this impasse. Where populations retain genetic contact are such developments, step by step, feasible?

The idea that reproductive barriers may gradually be developed and improved by selection is an old one, stemming originally from A. R. Wallace. The essential premise is a simple one. If there is more than one optimal phenotype favoured by selection, intermediates having

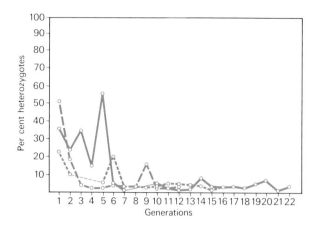

Figure 4.3 The progress of artificial selection for reproductive isolation between *Drosophila pseudoobscura* and *D. persimilis*. The decrease in the proportion of hybrids is shown for three separate laboratory populations. (After Koopman, 1950)

lowered fitness, then natural selection will tend to eliminate 'hybrids'. Any postmating disadvantage of 'hybrids' should thus generally favour the development of barriers to heterogametic matings. More specifically, if heterozygotes are at a disadvantage compared to homozygotes then any genes causing a reduction of heterozygote frequency will be favoured over other genes that have no effect on mating preference. Development of reproductive barriers in this manner has been termed the Wallace Effect. An important point to make is that selection of this type is not selection *for* the development of reproductive isolation as such. It is heterogametic matings which are selected *against*. Postmating barriers with no direct bearing on mating preference will not be involved in the Wallace Effect, but the possibility of hybrid inviability arising as a by-product of selection has been discussed by Coyne (1974).

Although their relevance to natural conditions remains to be clarified, several experimental studies (listed by Endler, 1977, p. 145) have successfully employed artificial selection to produce assortative mating (individuals select preferred mates). Under intense selection, premating barriers may be produced very rapidly. However equivocal the results of such experiments are considered to be they permit one clear conclusion: genetic resources to explain the Wallace Effect do exist. Based on postmating disadvantage of hybrids, the mathematical models of Endler (1977) demonstrate how in certain circumstances, a gene causing an increase in assortative mating may spread to fixation. The same models show how the stability of a 'hybrid zone', such as that of the oft-quoted hooded and carrion crows in Europe (see Fig. 2.1A), may be maintained as long as the effects of fitness modifiers do not give a net heterozygous disadvantage. Endler also discusses the conditions for a limited spread of assortative mating genes, so that their effects remain largely confined to the area in which contact is made. Such cases, where premat-

ing reproductive barriers are well developed only in zones of overlap, are known in nature (see Chapter 5).

A variety of mathematical models has been produced to attempt to explain the origin and maintenance of polymorphism in a heterogeneous environment. That of Maynard-Smith has already been mentioned. A more intricate mathematical treatment of the effects of disruptive selection in an heterogeneous environment has been presented by Dickinson and Antonovics (1973). Their model, although containing many assumptions (concerning population size and its constancy, constancy of gene flow, migration of individuals, and so on), clearly demonstrates the far-reaching effects that environmental heterogeneity has on the genetic structure of a population. They show that in a two-niche situation, and employing either single gene or polygenic characters, a polymorphism may be established over a wide range of conditions. Disruptive selection may rapidly extend the reproductive separation of the morphs. For a single gene A (not dominant) controlling a character undergoing disruptive selection, with selection pressure on genotypes AA, Aa and aa $= 0.8$, gene flow 0.3 and degree of assortative mating 0.9, complete reproductive separation of morphs arises in as little as 18 generations. Despite the admitted assumptions that their model contains, Dickinson and Antonovics conclude that it provides an unequivocal demonstration that sympatric speciation is feasible.

Returning to the formulation of the Wallace Effect, it will be seen that continuing selection against more or less heterozygous individuals of low fitness was considered to favour the spread of assortative mating genes. But it is pertinent to ask the question: can we, in practice, distinguish this from possible effects of selection against heterogametic mating *per se*? Selection of this type, if it takes place, may be assumed to be based on a 'loss of reproductive potential' associated with successful heterogametic matings. The distinction is of some significance as loss of potential is also a feasible basis for selection against indiscriminate mating between populations which have effectively severed genetic contact. Some artificial selection experiments suggest that assortative mating may develop between populations which have (artificially) severed genetic links. For example, by removing hybrids between each experimental generation, Koopman (1950) conferred effective lethality on hybrids appearing in mass cultures of *Drosophila pseudoobscura* and *D. persimilis*, fruitfly species which cross fairly readily under laboratory conditions. The experimental populations for each generation contained virgin males and females, equally distributed by species and sex, and it was found that the proportion of hybrids among the offspring decreased rapidly, within a few generations. However, whether selection against indiscriminate mating based on 'gamete wastage' or 'loss of reproductive potential' *alone* may lead to the development of completely effective premating reproductive barriers is less clear.

Selection favouring assortative mating as such is not the

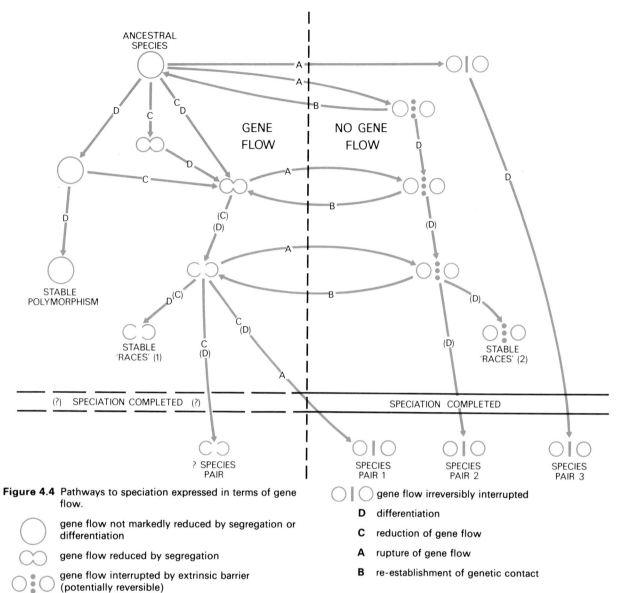

Figure 4.4 Pathways to speciation expressed in terms of gene flow.

○ gene flow not markedly reduced by segregation or differentiation

∞ gene flow reduced by segregation

○⋮○ gene flow interrupted by extrinsic barrier (potentially reversible)

○│○ gene flow irreversibly interrupted

D differentiation

C reduction of gene flow

A rupture of gene flow

B re-establishment of genetic contact

only way in which premating reproductive barriers may arise in a series of genetically continuous populations. In theoretical discussions of sympatric divergence and speciation (Maynard-Smith, 1966) changes in mating frequency resulting from 'habitat selection' alone are also envisaged as possible means by which reproductive barriers may be developed. Habitat selection, in this context, is the preferential occurrence of a genotype in the habitat or niche for which it is best fitted. Where there is a tendency for mating to take place within a 'niche', disruptive selection favouring an increase in 'habitat selection' is likely to produce a decrease in heterogametic matings. The more discrete are the niches of the genotypes (spatially or temporally) the greater will be the tendency for premating reproductive separation to develop. Such a process has much in common with the model proposed by Bush for

host-race formation and eventual speciation, referred to above. However, here the emphasis is on a *gradual* development of premating reproductive barriers as a corollary of increasing habitat selection (not necessarily a different host).

Conclusion

In this chapter we started with the classical postulates that competing hypotheses of speciation are based on the geographical setting in which one or other methods of speciation may have taken place. Throughout this essay I have tried to develop the idea that, because it is sometimes very difficult to establish *when* a speciation event may have taken place, it may often be more useful to regard speciation strictly in terms of gene flow and to speculate on what might cause its interruption. The establishment of premating reproductive barriers is undoubtedly crucial to the speciation process.

A distinction has already been made between instances where speciation is completed in isolation from gene flow and others where at least a measure of genetic contact persists until the moment when speciation may be said to be complete. The same distinction will be found useful in dealing with premating reproductive barriers. Adopting genetic continuity or discontinuity as the basic component, and emphasising the final stage of the speciation process, speciation models may be categorised in a relatively straightforward manner (Fig. 4.4). This greatly simplified representation of possible pathways to speciation is essentially a re-statement, in terms of gene flow rather than geographical circumstances, of the diagram employed earlier in this chapter (Fig. 4.1). In Figure 4.4 populations to the left of the main vertical division maintain genetic contact with each other. Genetic discontinuity, indicated by a position to the right of the vertical division, may be of guaranteed durability or may be temporary. Despite a variety of possible pathways, some of them crossing from one side of the diagram to the other, the final step to unequivocal species status is by three possible routes, leading to species pairs which are labelled 1, 2 and 3.

Speciation mode 1: following a period of differentiation (some of which *may* take place in immunity from gene flow) reproductive isolation is finally attained by the perfection of reproductive barriers in the face of gene flow.

Speciation mode 2: following a period of differentiation, reproductive isolation is attained in immunity from gene flow.

Speciation mode 3: irreversible rupture of gene flow produces 'instant' reproductive isolation. This is theoretically synonymous with speciation, but new species will generally only be recognised if they survive long enough to exhibit a degree of differentiation.

Many, but not all, models of sympatric, parapatric and allo-parapatric speciation clearly conform to Mode 1, while allopatric speciation in the strict sense conforms to Mode 2. Speciation models that involve host or habitat selection may be more difficult to refer to one of these modes. If the initial switch of niche is envisaged as providing immunity from gene flow (but no guarantee that genetic contact will not be re-established) then speciation is by mode 2. However, host or habitat selection may also form the basis for progression to speciation by mode 1.

Despite possible ambiguities and a lack of any reference to the genetic mechanisms underlying the speciation process, the categorisation of speciation models in gene flow terms may be of some value in the way in which it could reflect differences of interrelationship between the sister species resulting from a speciation event. This could prove more useful to the taxonomist than a classification based solely on the geographical component (Fig. 4.1).

It is important to realise that speciation consists of two identifiable yet interdependent processes – differentiation and reproductive isolation. Several models (some purely mathematical) have been proposed in which both processes could take place without the intervention of a geographical barrier. It remains for the experimentalist to test the validity and general applicability of these non-allopatric models.

References

Bush, G. L. 1975. Modes of animal speciation. *Annual Review of Ecology and Systematics* 6: 339–364.

Coyne, J. A. 1974. The evolutionary origin of hybrid invariability. *Evolution* 28: 505–506.

Dickinson, H. & Antonovics, J. 1973. Theoretical considerations of sympatric divergence. *American Naturalist* 107: 256–274.

Endler, J. A. 1977. *Geographic variation, speciation, and clines.* 246 pp. Princeton: Princeton University Press.

Halkka, O. 1978. Influence of spatial and host-plant isolation on polymorphism in *Philaenus spumarius*, in *Diversity of insect faunas*, Mound, L. A. & Waloff, N. (Eds) *Symposia of the Royal Entomological Society of London* 9: 41–55.

Huettel, M. D. & Bush, G. L. 1972. The genetics of host selection and its bearing on sympatric speciation in *Procecidochares* (Diptera: Tephritidae). *Entomologia Experimentalis et Applicata* 15: 465–480.

Key, K. H. L. 1974. Speciation in the Australian morabine grasshoppers: Taxonomy and ecology, pp. 43–56, in White, M. J. D. (Ed.) *Genetic mechanisms of speciation of insects.* Sydney: Australia and New Zealand Book Co.

Koopman, K. F. 1950. Natural selection for reproductive isolation between *Drosophila pseudoobscura* and *Drosophila persimilis*. *Evolution* 12: 135–148.

Lewontin, R. C. 1974. *The genetic basis of evolutionary change.* 346 pp. New York: Columbia University Press.

Maynard Smith, J. 1966. Sympatric speciation. *American Naturalist* 100: 637–650.

Mayr, E. 1942. *Systematics and the origin of species.* 334 pp. New York: Columbia University Press.

Mayr, E. 1963. *Animal species and evolution.* 797 pp. Cambridge: Harvard University Press.

Mayr, E. 1970. *Populations, species and evolution.* 453 pp. Cambridge: Harvard University Press.

Mayr, E. 1979. Reviews. Modes of speciation. – Michael J. D. White. 1978. *Systematic Zoology* 27: 478–482.

Murray, J. 1972. *Genetic diversity and natural selection.* 128 pp. Edinburgh: Oliver & Boyd.

Scudder, G. G. E. 1974. Species concepts and speciation. *Canadian Journal of Zoology* 52: 1121–1134.

Sokal, R. R. & Crovello, T. J. 1970. The biological species concept: a critical evaluation. *American Naturalist* 104: 127–153.

White, M. J. D. 1974. Speciation in the Australian morabine grasshoppers: The cytogenetic evidence, pp. 57–68, in White, M. J. D. (Ed.) *Genetic mechanisms of speciation in insects.* Sydney: Australia and New Zealand Book Co.

White, M. J. D. 1978. *Modes of speciation.* 455 pp. San Francisco: W. H. Freeman & Co.

CHAPTER 5

The origin and development of reproductive barriers

P. M. Hammond

The circumstances that conspire to impede or prevent successful crosses between populations of different species manifest themselves to evolutionary biologists in a number of ways. In the case of coexisting species these circumstances may bring themselves to the attention of ecologists in terms of differences in niche. Inasmuch as factors responsible for the reproductive separation of closely related species are reflected in morphological or other readily accessible data these will form the basis of taxonomists' attempts to delimit biospecies. Just how these circumstances arose may appear of less crucial concern to the taxonomist and ecologist than the manner in which they now serve to maintain the reproductive separation of populations. However, one aim of the present chapter is to explore the possible implications which the mode of origin of reproductive barriers may have for work in both ecology and taxonomy.

For those concerned with the dynamics of speciation the origin of reproductive barriers is more evidently of interest. Extrinsic factors, such as geographical disjunction, play an indisputable role in initiating differentiation of populations and, at least in some cases, permitting divergence of these populations to the point of intersterility. There is less agreement, however, concerning the origin and significance of differences in behaviour and compatibility within a single gene pool.

In the present chapter the origin and development of reproductive barriers that serve to separate sympatric species are discussed. In particular, the theme that barriers to reproduction may arise in the face of gene flow (a topic introduced in Chapter 4) is developed. In Chapter 4 Wallace's view that reproductive barriers may gradually be developed and improved by selection was briefly discussed in relation to possible modes of speciation that do not conform to the classical allopatric model. Reference was made to mathematical modelling and artificial selection experiments that lend some support to the view that reproductive barriers may arise as a direct result of selection for assortative mating. Unfortunately, such models and experiments cannot incorporate the complexities of any given natural situation. They can do no more than establish the possibility of ways in which premating reproductive barriers may develop. In this chapter the pattern of occurrence of reproductive barriers in nature will be examined for evidence that the Wallace Effect is instrumental in the development of intrinsic premating barriers to reproduction.

The term 'isolating mechanism', coined by Dobzhansky (1937), has frequently been used with reference to factors involved in the reproductive separation of populations. The term *reproductive barrier* is preferred here (and in Chapter 4), partly because reproductive *isolation* has been reserved for the situation *resulting* from the imposition of a completely effective and irreversible barrier, and partly because of the teleological overtones of 'mechanism'. Even so, the alternative term is not without its own suggestion of something 'purpose-built'. This, however, is not what is implied in the use of 'barrier' in the present discussion; reproductive barrier is intended merely as a convenient means of referring to the set of conditions involved in reproductive separation.

A categorisation of reproductive barriers

It is possible to categorise reproductive barriers in a number of different ways. The classifications most widely employed, especially in discussing speciation models, probably are those of Mayr (1970). My own classification, erected for the purposes of the present discussion, is outlined in Table 1. Certain difficulties in precisely delimiting some of these categories, notably the various types of premating (more strictly prezygotic) barriers will be considered in more detail below. Of course, reproductive barriers may be either partial or complete and may occur in a variety of combinations. In certain circumstances a potential barrier may exist but be redundant. For example, any intrinsic barrier will remain inoperative in the presence of a completely effective extrinsic barrier. Similarly, where an intrinsic prezygotic barrier is completely effective then postzygotic barriers, although present, will not be called into play.

The role of extrinsic barriers to reproduction (such as geographical disjunction) in speciation is explicit in the allopatric and other speciation models. The immunity from gene flow provided by an extrinsic barrier clearly provides a possible basis for the development of intrinsic barriers, as a by-product of differentiation. Completely effective postzygotic barriers which, generally speaking, may be regarded as non-adaptive *per se*, would seem especially likely to develop in this way. If the allopatric speciation model is of universal applicability intrinsic premating barriers will also be dismissed as incidental products of differentiation, or their production will be regarded as a post-speciation phenomenon.

However, the development of intrinsic premating barriers may be directly adaptive, through the Wallace Effect, discussed in Chapter 4. Although *some* premating reproductive barriers are likely to result from general divergence in allopatry, particularly any marked divergence of niche, others may result from selection favouring assortative mating between populations in genetic contact. The question is: may such cases, if they exist, be recognised? Each of the three categories of prezygotic barrier listed in Table 1 could conceivably be involved, although some types, if selected for, will be more obvious. It is these barriers which are deserving of particularly close scrutiny in any attempt to establish the general validity of the proposition that selection for assortative mating does, in fact, take place. Before examining the proposition more closely the classification of prezygotic barriers to successful reproduction outlined in Table 1 requires a little comment and elaboration.

Mate 'avoidance'

Factors that decrease opportunities for meetings between potential mates may be temporal or spatial, both factors sometimes operating together. A difference of niche, especially for host-specific species or races, may necessarily involve separation of potential mates. In other cases, where habitats are broadly overlapping, a difference of mating site within the habitat may separate potential mates, although this may be a corollary of site specificity for other activities to take place (for instance, feeding or pupation in the case of insects). Temporal factors may be cyclical, seasonal, or involve daily rhythms. An obvious but important point concerning mate 'avoidance' is that it is the behaviour of individuals with readiness to mate which is in question; the pattern of occurrence of sexually immature or other individuals is immaterial. A second point which should not be overlooked concerns the nature of cues which enable potential mates to meet at specific sites or times. Although these may often be extrinsic (for instance, host odour), intrinsic attractants – either specific or non-specific – may also be involved.

Table 1 Factors involved in the reproductive separation of populations

Type	Effect on 'potential' mates
I. EXTRINSIC: may or may not have a genetical basis	
A. Geographic separation	Do not meet
B. Other spatial separation (e.g. certain parasites)	Do not meet
II. INTRINSIC: has a genetical basis	
A. Asexuality, thelytoky, etc.	No potential mates
B. Postzygotic factors (e.g. hybrid inviability or sterility, etc.)	Potential mates may meet and mating may occur (see also Chapter 8)
C. Prezygotic factors	
1. Mate 'avoidance' (spatial or temporal barriers)	Do not meet
2. Mate 'rejection' (largely, the 'ethological isolation' of authors)	Potential mates may meet but do not mate
3. Mate incompatibility (mechanical, physiological or chemical impediments to successful fertilisation)	Potential mates may meet and may mate or attempt to mate

Mate 'rejection'

The factors in this category are those involving specific signals and response (or lack of response) between potential mates. Signal reception may involve any of the sensory modes – visual, auditory, olfactory, gustatory or tactile. The overall shape of complex courtship behaviour may be specific or, at the other extreme, a single specific signal may suffice for mate recognition. The essential 'coyness' of one partner, most usually the female, is a striking feature of much behaviour involving potential mates.

Although the distinction between mate 'avoidance' and mate 'rejection' factors is not always easy to make, it is of some significance. Both *may* involve an exchange of signals but to refer all such signals to a single type of reproductive barrier – as 'ethological isolation' – is likely to be unhelpful.

Mate incompatibility

This category is intended to include all non-behavioural impediments to fertilisation that come into play after mating or attempted mating has begun. Amongst insects a variety of structural incompatibilities may occur. There may be difficulties in coupling stemming from form or size of claspers and other structures, and copulation may be hindered by the size and form of the intromissive organ itself. Appropriate deposition of sperm may be impeded by the shape of the male endophallus and female bursa copulatrix, by the length and form of male flagellar structures, the position of the opening to and structure of the female spermatheca (sperm-store), or by spermatophore structure. Whether or not such factors as pollen–stigma interactions in plants should be referred to this or the previous category will not be pursued here. An absolutely strict distinction between mate incompatibility and mate 'rejection' may be impossible.

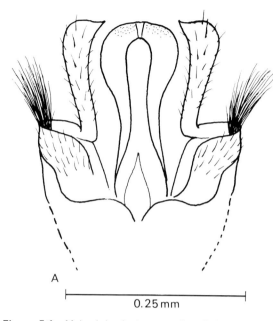
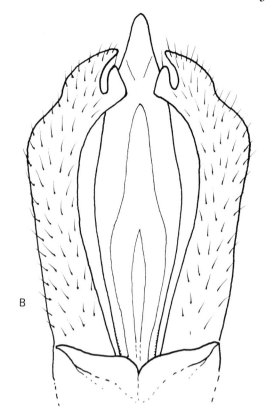

Figure 5.1 Male abdominal apex and genitalia of the woodworm beetle – *Anobium punctatum* (A) and its sympatric close relative *Anobium inexspectatum* (B). Females of the two species cannot be reliably distinguished.

Clearly, a set of conditions in which partial or complete reproductive separation of two populations by intrinsic prezygotic factors is achieved may arise in many different ways. In pursuing the question whether selection for assortative mating may not be one of these ways, it would be well to avoid any tendency to prejudge the issue. We must be careful not to imply that intrinsic prezygotic barriers are 'designed' specifically for the purpose of reproductive separation. Explanations for the widespread occurrence of intrinsic prezygotic barriers without selection against heterogametic matings should also be examined carefully.

The view that the development of such barriers is always incidental, in the sense that they are simply the outcome of selection favouring homogametic mating success has some adherents. Let us briefly explore the implications of this view. Firstly, it is clear that many factors impeding heterogametic matings also favour the chances of homogametic matings being achieved. This is true of specificity of mating site or time of mating, in both daily and seasonal terms, swarming behaviour of insects and so on. If populations differ in any of these respects, the differences may be a necessary corollary of divergent niches. For this reason mate 'avoidance' factors will generally be a rather poor source of evidence that selection for assortative mating does take place. From this, of course, it should not be inferred that a divergence of mating site or time of mating is *never* the results of selection against heterogametic matings.

A marked divergence of mating time or site between closely related sympatric species may also arise as a result of a general divergence of niche. Where the range of such species overlap, competition for resources may generate what is generally termed character displacement (see Chapter 17). Divergent character displacement may involve a single niche dimension, such as seasonal utilisation of a resource. In this case allochrony of breeding season may well develop and function as an effective reproductive barrier, even though the two species are already reproductively isolated. In some instances divergence may be greater in areas of sympatry than in areas of allopatry, so that reproductive separation by mate 'avoidance' factors is seen to be effective only in the area of overlap. This sort of pattern might be expected where selection is thought to operate directly against heterogametic matings. The fact that an alternative explanation – character displacement based on divergence of niche – may be applicable to many cases should be borne in mind.

Unfortunately, the term character displacement has also been employed with reference to divergent developments unrelated to questions of niche. As noted by White (1978) divergence of behaviour or morphology resulting from selection against heterogametic matings (for instance, signal patterns of butterfly wings) is a different phenomenon from divergence stemming from competition (for example, bill widths of birds). Although not all divergence in sympatry may be definitely referred to one or other of these categories (see discussion of 'interference' effects on attractant or other signals below) it would seem preferable to mark this distinction by restricting the usage of character displacement to the competitive situation.

To the extent that they are directly associated with specificity of mating site or time, signal systems are also likely to be involved in ensuring that potential mates of the same population meet. While mate recognition involving touch or taste is restricted to instances where mates are already in 'contact', other types of signal may play a part in behaviour where mate 'avoidance' is the result. Pheromones, songs and, in some instances, visual displays, may all act purely as attractants. Where the effect of such signals is to bring a potential mate to a specific mating site its function is similar to that of any extrinsic cue associated with the site. Attractants which are employed to bring potential mates into close proximity are by no means always specific (for example, many pheromones). However, specificity of attractants is also common, although this too, in some cases, may simply play a role in ensuring the effectiveness of the signal, thus favouring homogametic mating success. In environments otherwise virtually free from competing signals the effectiveness of an attractant may be diluted by 'interference' from identical or sufficiently similar attractants emanating from a different source, often from a related species. Such interference, if marked, might form a basis for the development of divergent signals. Long distance movements in response to scents are, perhaps, particularly likely to be affected in this way, although songs that draw responses over some distance and isolated visual signals (such as patterns of flashing lights) may have much in common with scents in this respect.

Attractants then may play an important role in providing greater opportunities for homogametic matings but, to the extent that they are specific, may also be involved in the reproductive separation of populations. Generally speaking specific attractant signals may be regarded as contributing to reproductive barriers which are of a mate 'avoidance' type. But, as many such signals are essentially similar to those which may be employed after the onset of courtship, a clear distinction between mate 'avoidance' and mate 'rejection' is not always possible. An apparent lack of awareness that attractants may (or may not) involve specific signals, but that not all signals are attractants, seems to underly at least some disagreements over the role that species-specific signals play in reproductive separation. The fact that some attractants are not specific has been used to suggest that signal systems are generally not involved in reproductive barriers, or to contend that the role of specific signals has been much exaggerated. Such reasoning is clearly faulty. The primary function of attractants is to bring potential conspecific mates together. Specificity of signals involved in attraction is one way in which reproductive separation of populations may be achieved. However, where attractants are employed and lack specificity a variety of other reproductive barriers may come into play. It may also be noted that non-specific

attractants may play an incidental part in reproductive separation by mate 'avoidance', for example in closely related species of moth with similar attractants but a difference of daily flight and mating time. More often, non-specific attractants are only one component of a series of signals involved in courtship. For example, many bark-beetles assemble at mating sites in response to non-specific pheromones. Further signals, generally acoustic ones, which are species-specific, are then exchanged, the reproductive barrier operating finally being of a mate 'rejection' type. A lack of specificity in attractant signals in such situations has no particular bearing on the general significance which species-specific signals may have for reproductive separation of species.

Reproductive barriers involving mate 'avoidance' factors, in sum, are unlikely to prove the best source of evidence that selection for assortative mating is of common occurrence. Much of the available evidence will be generally consistent with the view that differences between species in place and time of mating are the outcome of developments simply favouring homogametic mating success. Divergence of niche, whatever the origin of the divergence in geographical or other terms, will tend to result in divergent behaviour entailing a difference of mating time or place. There is no need, in many instances, to invoke 'disruptive' selection of any type operating directly on this behaviour. The fact that other explanations are frequently available does not, of course, demonstrate that selection against heterogametic matings is never a factor involved in the divergence of time or place of mating. Indeed, selection for assortative mating may quite frequently be responsible for such divergence, although evidence that this is the case will commonly be equivocal.

For a more rigorous examination of the view that all species-specific elements of mating behaviour are explicable as devices furthering the efficiency and economy of homogametic matings we must turn to prezygotic reproductive barriers of a different type. Reproductive barriers involving mate 'rejection' factors are a more likely source of evidence that selection for assortative mating does play a role in the development of divergent behaviour.

Specific mate recognition systems

Species-specific behaviour is shown by many kinds of animal immediately prior to or during the process of courtship. The sometimes bewildering complexities of courtship, including the exchange of signals once potential mates are 'in contact', are scarcely interpretable as means of achieving rapid and purposeful homogametic matings. Although intra-population epigamic selection is clearly involved in the elaboration of much courtship behaviour, the precise development of complexity as such in courtship is a topic which cannot be pursued here. It is the *pattern* of inter-population variation with respect to courtship behaviour that is of concern in the present discussion. If we restrict attention to signals that are exchanged at close quarters, it would seem legitimate to assume that marked inter-population differences are not an automatic corollary of a shift of niche. Where signals are not also attractants and where they are employed only after courtship may be said to have begun, their divergence would not appear to favour opportunities for homogametic matings *per se*.

A degree of intra-population variation, both geographical and individual, in courtship elements is well documented, and differences in courtship may also be detected among groups of species with entirely allopatric distributions. However, closely related sympatric species which differ markedly in one or more courtship components have been noted in a great variety of animal groups. Examples of groups in which the sympatric species exhibit exaggerated differences in signals exchanged at close quarters, while allopatric species exhibit no more than slight differences, are probably familiar to most entomologists. A vast literature on courtship behaviour exists, much of it dealing with songs, pheromones and colour-patterns. Unfortunately, most of the more detailed treatments do not include any comparison of closely related species. Another evident disadvantage of the available evidence is the extent to which signal systems studied (for example, those in many birds, Anura, Lepidoptera, Orthoptera) are concerned with *attraction* of potential mates. Studies of courtship at close quarters, especially in the exceptionally species-rich groups where this would seem most likely to be of importance for discrimination, are relatively few. However, studies of drosophilid fruitflies have been intensive, although even here the signals, including visual, acoustic and chemical stimuli are little understood. Nevertheless, components of *Drosophila* signal systems are known to be heritable. That signal differences between closely related species actually form the basis for discrimination is not always easy to document, but there is sufficient experimental evidence (work on *Drosophila*) and demonstration of selective response by females (to calls in frogs) to suggest that marked signal differences are employed as recognition features.

Reproductive separation of closely related animal species by means of specific mate recognition appears to be common. Indeed, the generalisation that in the many instances where prezygotic barriers of other types (see Table 1) are not completely effective then mate discrimination comes into play, probably has some validity, at least for insects. Exceptions, where postzygotic barriers alone ensure the complete reproductive separation of closely related sympatric species, appear to be rare. The nature of signal differences and the circumstances in which markedly divergent signals between closely related species are found may be expected to provide some evidence bearing on the origin of signal divergence. Does this, at least in some cases, result from selection for assortative mating? If so, is such selection likely to have been involved in a speciation process or to be a post-speciation phenomenon?

In surveying mate recognition systems for evidence concerning the ways in which divergent developments may have arisen it is clear that, depending on the complexity of the signals and on the sensory mechanisms involved, information concerning signal differences is of widely varying accessibility. It is not surprising that relatively few courtship rituals of a complex type, especially those involving more than one sensory mechanism, have been fully analysed and compared with those of related species. The composition and mode of employment of pheromones used at close quarters in courtship are also rather poorly documented. Acoustic signals, on the other hand, are not especially difficult to record and analyse. These and the simpler kinds of visual signals, especially those involving display of features of characteristic form or colour-pattern, are now relatively extensively described in the literature. Another eminently accessible but largely untapped body of evidence is that to be derived from morphological features which may be used in tactile discrimination among potential mates.

A brief digression concerning these various signalling systems, with particular reference to arthropods, is in order here.

Mate recognition systems involving tactile discrimination have received little attention. It is, of course, widely accepted that much specific courtship behaviour, especially that of insects, involves tactile signals. However, it is probably fair to say that specificity of such signals has been interpreted most generally as originating from the characteristics of movements or sequence of movements made, usually by a male. Tactile stimulation by tapping, drumming, stroking, and so on is commonplace in the courtship of many animal groups. Although skin texture is known to be involved in discrimination among potential mates in some Anura, the idea that tactile recognition of form or texture plays a part in insect courtship has received little consideration. Where tactile discrimination, like much visual recognition of form or colour-pattern, is based largely on differences of form alone, the advantages for investigations of specific mate recognition systems are several. There is no possibility that tactile signals fulfil an attractant function, and there is little likelihood that divergent developments result from 'interference' of the type (see above) which may affect, for example, mating calls. If the specificity of a tactile signal (or series of signals) is reflected in structural specificity there is the added advantage that a good deal may be learned from the study of preserved specimens.

The idea that male secondary sexual structures of insects may be directly involved in premating reproductive barriers has been the subject of much discussion since the days of Léon Dufour who, in 1844, suggested that the observed variation in genital structures of male insects could be explained on the basis of 'lock and key' adaptations of female and male. The demonstration that specificity of male genital structures is most commonly not matched by any equivalent variation in females – that is the 'keys' vary but the 'locks' are generally similar – led to the rapid demise of Dufour's theory. In fact, both partial and complete physical incompatability of genital structures between males and females of closely related species is now known to occur in a variety of insect groups. However, the greater part of the array of variation which attracted Dufour's attention is clearly not of this type. It should be noted, first of all, that the majority of male secondary sexual structures, even those associated closely with the genitalia, are not involved in actual intromission. Many also have no, or no recognised mechanical function, such as clasping the female, during mating. Indeed, a lack of any obvious function has resulted in a tendency for many male secondary sexual structures, especially those which exhibit marked differences between closely related species, to be dismissed by some entomologists as merely bizarre or relegated simply to the category of 'function unknown'. However, a particularly thorough survey of variation in male secondary sexual characteristics associated with the genitalia of beetles led Jeannel (1955) to the conclusion that many of these features must, somehow, be involved in the prezygotic reproductive separation of species. Assuming that it was the male which discriminated among potential mates, Jeannel was, nevertheless, at a loss to explain how variation in *male* form was related to *reception* of variable *female* signals. Today, a role of many male secondary sexual structures in signal transmission is more widely recognised, although signal differences likely to stem directly from the form of these structures have received relatively little mention.

Of course, some male secondary sexual features (for example, elaborate developments of insect antennae) *are* involved in reception of female signals. Other features are involved in the production of chemical or acoustic signals, and some with a variety of mechanical functions in mating. Others still are employed in activities involving intrasexual selection, or derive from a division of labour between the sexes. Tactile signals, which in no way depend on specificity of form but involve specificity of male behaviour, may also be conveyed by means of sexually dimorphic features. It is clear that the pattern of variation exhibited by secondary sexual structures of insects and other organisms does not stem from adaptation to serve a single function. It is not a phenomenon for which a single explanation can be found. Unfortunately, the lack of recognition that secondary sexual features fulfil a variety of different functions has been responsible for at least some of the confusion to be found in arguments both for and against ethological isolation developing in the face of gene flow. Evidence for the view that secondary sexual features which have no mechanical function in mating are, in a number of instances, involved in the transmission of tactile signals can not be fully reviewed here. Some evidence that the specific form of certain structures in male beetles is employed in mate recognition is discussed by Hammond (1972). Further work on this topic is in progress. The feasibility of signal transmission from male to female may be established by morphological studies and observations of mating behaviour. This idea prompts an obvious question: are female receptors appropriately sited and likely to be stimulated at an appropriate stage in courtship?

Putative tactile recognition features may be recognised in sympatric species of a variety of animal groups, but are especially frequent in the major species-rich groups of Arthropoda. They are most common in those groups where reproductive barriers of the 'mate avoidance' type and signal systems of other types (songs, visual displays) are of least apparent significance.

Differences between closely related species with respect to male secondary sexual structures likely to be implicated in tactile signalling during courtship are, of course, not restricted to those which are sympatric. However, it is in coexisting species that differences are, if not always dramatic, generally most clear-cut. Putative tactile recognition features are most often associated with the abdominal apex, genitalia (see Fig. 5.1), or legs. Especially characteristic are prongs, hooks, divided lobes, slots, angled projections, angled incisions, combs, bundles of bristles and textured patches, these occurring in a variety of combinations. Differences between closely related sympatric species are generally of a type which would be readily detected by simple mechanical sensors. In some instances the differences are analogous to those of patterns designed for ready discrimination (tactile or visual) by humans – the formulations of braille letters, the patterns of playing cards or dice, or the recognition marks of tactile mazes used in psychological experimentation. They should prove amenable to statistical investigation.

The origin of intrinsic prezygotic reproductive barriers

Employing data concerning the occurrence of putative tactile recognition features (Hammond, unpublished work), but making reference also to specific mate recognition systems of other types, let us return to a discussion of the manner in which specificity of recognition signals may develop. First, are we able to determine whether such signals are generally an effective means not only of facilitating homogametic matings but also of preventing successful crosses between species? Mayr (1963) has expressed the belief that single factors are rarely responsible for 'all or none' prezygotic reproductive separation of species. This is thought to be achieved by a combination of partially effective barriers, generally including postzygotic factors. The results of laboratory hybridisation attempts demonstrate clearly that postzygotic barriers are rather frequently ineffective in preventing successful crosses between closely related species. That this is also true of some prezygotic barriers is shown by the occurrence of hybrids under 'natural' conditions. However, naturally occurring hybrids between sympatric insect species may be largely confined to groups in which species pairs are separated, not always entirely effectively, by reproductive barriers of particular types. A survey of the literature

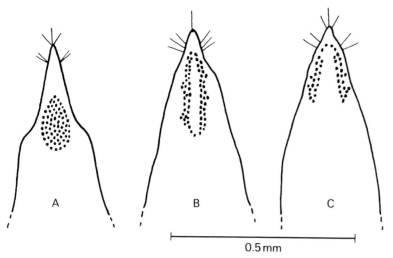

Figure 5.2 Parameres of male staphylinid beetles of the genus *Quedius* in North America. The ranges of all three members of the '*brunnipennis*' species-group, *Q. brunnipennis* (A), *Q. densiventris* (B) and *Q. breviceps* (C) overlap. Females of *Q. densiventris* and *Q. breviceps* cannot be reliably distinguished.

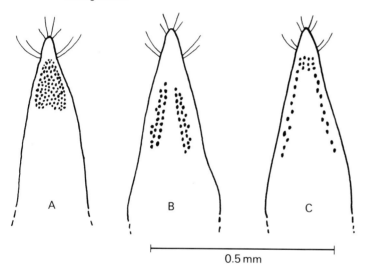

Figure 5.3 Parameres of male staphylinid beetles of the genus *Quedius* in North America. The three native North American species of the '*molochinus*' species-group are: *Q. neomolochinus* (A), *Q. strenuus* (B) and *Q. labradorensis* (C). The range of *Q. neomolochinus* overlaps those of both the other species. Females of *Q. strenuus* and *Q. neomolochinus* cannot be reliably distinguished.

pertaining to cross-matings and successful hybridisation in beetles reveals *no* examples where distinctive specific recognition signals are known to be employed during courtship. The great majority of hybridisation records, including those under laboratory conditions, relate to species which (generally) achieve reproductive separation by 'mate avoidance' – allochrony of mating, host specificity or habitat selectivity. The numerous records of naturally occurring hybrids between butterfly species also generally relate to instances where reproductive separa-

tion is achieved essentially by 'mate avoidance', although attractant signals may be of importance in courtship.

The indications would seem to be, then, that specific mate recognition signals may frequently be completely effective in ensuring premating reproductive separation of closely related sympatric species. That this effectiveness may be maintained even under laboratory conditions is demonstrated by the results of experiments with a variety of animals. Finally, it should be noted that although a battery of signal systems may apparently be employed in courtship by any one species, the marked divergence characteristically found between closely related sympatric species is generally restricted to one type of signal. If this is a valid generalisation, we may conclude that, despite the variety of prezygotic factors that is likely to have some bearing on the reproductive separation of any given pair of closely related species, completely effective separation may be ensured by a single factor.

Second, the extent and nature of recognition signal differences between closely related species require scrutiny. Although a generally poor level of understanding of phyletic relationships precludes confident generalisation, the observed pattern of variation in signals exchanged at close quarters would seem to support the view that exaggerated differences between closely related species are common only in those which coexist. Such a pattern of variation (see Figs 5.2–5.4) could be explained by invoking selection for assortative mating. However, it has been suggested (H. E. Paterson, pers. comm.) that such an explanation is merely teleological. Where effective intrinsic prezygotic reproductive barriers are a prerequisite for coexistence, adherents of the allopatric speciation model may reasonably suggest that sympatry (secondary in their view) is achieved, naturally enough, only by those species already so equipped. Certainly, the immunity from gene flow provided by an extrinsic barrier creates a possible basis for the development of potential intrinsic barriers. Populations separated by extrinsic barriers are likely to acquire a variety of characteristics that would contribute to reproductive separation on secondary contact. This will undoubtedly be true of prezygotic as well as postzygotic factors. Here, the role of intra-population epigamic selection should not be ignored. Heavy sexual selection of this type is likely to have an unpredictable outcome in a small isolated population. Where exaggerated signal characters occur, as in some groups of birds, their continuing development might fairly rapidly take a new direction. However, it is difficult to envisage the acquisition in allopatry of 'ready-made' signal system differences of certain types which are exhibited by coexisting species. If such developments are at all common, we should expect to find markedly divergent tactile, visual and other recognition signals in at least some allopatric isolates which are separated by particularly durable extrinsic barriers. Examples are difficult to find. Indeed, available evidence suggests that intrinsic barriers to mating of all types may generally be slow to develop in allopatry. When the reproductive separation of closely related species which, in nature, are separated by extrinsic barriers is put to the test under laboratory conditions, intrinsic premating barriers are frequently found to be ineffective. The thesis that divergence of specific mate recognition systems in disjunct populations is a likely result of selection to fit a new environment (Paterson, 1978) is, no doubt, applicable to the development of small differences in a variety of signal systems. In some cases a marked divergence may have an environmental basis: for example, louder or more piercing songs, or more elaborate visual displays may be more effective in one setting than another. The effect of 'interference' from signals of related species (see above) will also vary from place to place. But, differences of environment, however extreme, would seem unlikely to be responsible for marked divergence of signals which are exchanged at close quarters during courtship.

Support for a sympatric origin of divergent recognition signals has often been adduced from the absence of signal characters in species which are geographically isolated from their near relatives by barriers that may be assumed to have some durability. A well-known example of this type concerns ducks of the genus *Anas*, where sympatry of closely related species is the general rule. However, a few species of *Anas* are restricted to individual oceanic islands. These are monomorphic, males lacking the distinctive plumage of that sex found in almost all continental species (Sibley, 1957). Similarly, structures likely to be involved in tactile recognition are poorly developed or lacking in males of several staphylinid and hydraenid beetle species which are restricted to individual subantarctic islands (Hammond, unpublished work); among their relatives extensive sympatry of closely related species and specific secondary sexual features of a striking type are again the general rule. Examples of a similar nature relating to beetle species from isolated cave systems and isolated mountain peaks in East Africa and Mexico are provided by Jeannel (1955). However, in such extreme cases factors other than the absence of coexisting congeners are likely to have been of at least some influence in relation to the poor development of signal-generating features. More significant in relation to the present discussion is the very general lack of divergence in mate recognition signals of the kind exchanged at close quarters in less dramatically isolated species. Poorly developed signal systems or, more usually, weak specificity of such signals appear to be especially frequent in insect species which are reproductively separated from their close relatives by disjunctions of range, allochrony of mating, specificity of mating site or other circumstances in which potential mates of differing species do not meet.

A tendency for 'ethological speciationists' to assume an adaptive significance in reproductive separation of species for each and every characteristic exhibited by specific mate recognition systems has attracted deserved criticism. The majority of design features of mate recognition systems are no doubt related simply to the furtherance of homogametic

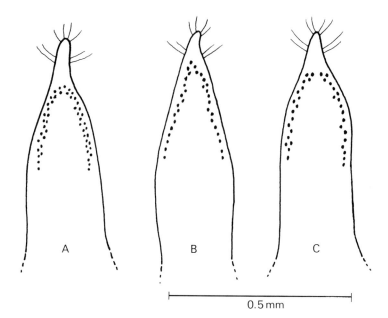

Figure 5.4 Parameres of male staphylinid beetles of the genus *Quedius* in North America. The three North American species of the '*horni*' species-group are: *Q. molochinoides* (A), *Q. lanei* (B) and *Q. horni* (C). The ranges of the three species, although closely approaching each other, apparently do not overlap.

matings, or are the legacy of intra-population competition. Complex and specific signal systems that show little variability between individuals of a species may be explained as devices to achieve one hundred per cent conspecific matings. What is more interesting is the origin of *marked divergence* of mate recognition signals in closely related species inhabiting the same area. Divergence of this type is characterised by its nature as much as by its extent. Differences in recognition signals employed by sympatric species are typically simple, yet clear-cut, with similar patterns of difference often repeated in a number of species-groups (for instance, patches of stumpy rounded setae in males of two different species-groups illustrated in Figs 5.2 & 5.3). In sum, a prevalent characteristic of the signals in pairs or groups of sympatric species is that ease of discrimination between them is the most striking feature of their design. The overwhelming impression to be gained from their pattern of occurrence is that selection (not necessarily that for assortative mating) has favoured signal divergence *per se*. A possible role of such selection in the elaboration of courtship behaviour is not difficult to envisage. For example, a study of the bizarre secondary sexual features exhibited by members of such groups as the pselaphid beetles suggests that their complexity derives from progressive acquisition of new and distinctive features. Repeated rounds of selection for the development of distinctive signal characters may thus be ultimately responsible for this elaboration.

Another source of evidence that divergence in mate recognition signals may be a response to selection for assortative mating derives from the somewhat controversial

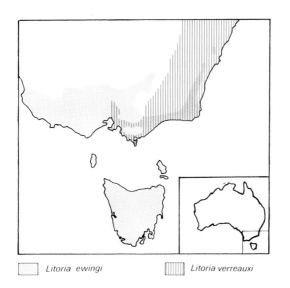

Litoria ewingi   Litoria verreauxi

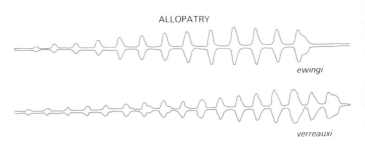

ALLOPATRY

ewingi

verreauxi

SYMPATRY (VERREAUXI EXTENSION)

ewingi

verreauxi

SYMPATRY (EWINGI EXTENSION)

ewingi

verreauxi

Figure 5.5 Mating calls of the related frog species *Litoria ewingi* and *Litoria verreauxi* from areas of allopatry and sympatry (after Littlejohn, 1965).

subject of what has been termed 'reinforcement' of reproductive barriers. Where closely related species are partially sympatric, differences in signals involved in reproductive separation are sometimes greater in the areas of overlap than elsewhere. These have been particularly well investigated in the case of frog and toad mating calls. Instances of 'reinforcement' in North American Anura are reviewed by Blair (1964). An oft-quoted example concerns two closely related species of frog – *Litoria ewingi* and *L. verreauxi* in south-eastern Australia. In areas of allopatry male mating calls are very similar, while in two principal areas of range overlap the calls are much more distinct (Fig. 5.5). Behavioural responses of females have been studied and shown to follow closely the characteristics of male calls. Although insect calls have generally not been investigated so intensively, variations of 'dialect' reported in species of Plecoptera and in corixid bugs with partially overlapping ranges are also suggestive of 'reinforcement'. Of course, calls, like pheromones, may be subject to displacement by interference from other signals, and the possibility of pseudo-adaptive patterns of 'reinforcement' must not be overlooked. However, that selection for assortative mating is responsible for the greater divergence of calls in areas of overlap is the explanation which would seem best to fit at least some of the cases. The term 'reinforcement' itself unfortunately suggests that signal divergence derives from secondary contact, although Endler (1977) and others (see Chapter 4) have discussed how the same pattern of geographically variable signals might stem from a primary 'hybrid zone'.

Prezygotic reproductive barriers – always a post-speciation development?

Where prezygotic reproductive barriers between sympatric species bear the apparent stamp of selection against heterogametic matings, are we entitled to conclude that selection has operated on populations which, while diverging, retain at least a measure of genetic contact? Or, are there grounds for concluding that the development of such barriers is generally a post-speciation phenomenon? In the latter case the basis for selection against indiscriminate mating would be a simple loss of 'reproductive potential', through wastage of valuable gametes or a reduction of homogametic mating success. It is, of course, only prezygotic (in fact mostly premating) barriers which will reduce such wastage where it occurs. Just how significant any loss of reproductive potential through heterogametic mating is likely to be, and whether it forms the basis for strong selection may be assumed to vary considerably according to the reproductive biology of particular species.

In many soldier-beetles of the family Cantharidae, for example, males are notoriously indiscriminate in their choice of potential mates. Mating attempts by cantharid males on individuals, not always female, of a wide range of other beetle families have been recorded. These soldier-beetles are to be found commonly on the inflorescences of various plants, where *in copula* pairs are sometimes more numerous than unpaired individuals. It may be assumed that conspecific mates in this situation are not difficult to find; expenditure of time or gametes by some males in a non-productive fashion would seem to involve little disadvantage. Closely related species of soldier-beetle are found together but there appear to be no reliable records of females found *in copula* with other than conspecific mates. Indeed, observations suggest that specific male signals are necessary to overcome female coyness. The effectiveness of postzygotic barriers to reproduction, if any, between closely related species in this group is unknown. A loss of 'reproductive potential', at least on the part of males, is clearly not the basis on which apparently effective prezygotic barriers have developed here.

The development of *ad hoc* premating reproductive barriers, operating through pollinating insects, where certain sympatric plant species are already more or less effectively isolated by postmating barriers (Levin & Kerster, 1967), certainly suggests that loss of reproductive potential may form the basis for selection against heterogametic matings. Although the full range of intrinsic prezygotic barriers may be subject to 'secondary' development in sympatry between species already isolated by postzygotic barriers, it may be supposed that those involving innovations of some complexity (in genetic terms) are especially likely to have developed in this way. For example, the partial or complete mechanical incompatability characteristic of the genitalia of some closely related insect species (for example, certain carabid beetles) may involve a variety of structures that do not generally exhibit any great lability. Such mate incompatability factors are no doubt often incidental products of the differentiation which accompanies and follows speciation, but unusual divergent developments are to be found in some sympatric species pairs. If these are the result of selection against heterogametic matings it would seem more probable that this has taken place in gene pools already separated by the establishment of an effective postzygotic barrier.

The crucial point in demonstrating that the development of prezygotic barriers is not always a post-speciation phenomenon is that they are by no means always supplemented by fully effective post-mating barriers. Sympatric species-pairs which do not (or very rarely) hybridise in nature, but which can be crossed successfully when prezygotic barriers break down under laboratory conditions are not uncommon. Mecham (1961) contends that postmating barriers between sympatric frog species of the same group are more often ineffective than effective. Examples of a similar type for a variety of animal groups may be found, although the belief that it seems to be generally true that diverging populations of animals usually acquire mechanisms of reproductive isolation before they have diverged genetically to the point of intersterility is likely to be an overstatement.

Conclusion

It has not been the purpose of this discussion to claim that the development of every distinctive mate recognition system is associated intimately with a speciation event (speciation mode 1 of Chapter 4, p. 48). However, available evidence does appear to suggest that intrinsic prezygotic barriers, exhibiting signs of specific adaptiveness to that end are frequently responsible for the reproductive separation of closely related sympatric species. The development of some barriers of this type (especially those involving 'mate avoidance') may be explained as incidental correlates of niche displacement in secondary sympatry. A marked shift of a specific mate recognition system may also result from various types of interference, such as that between competing attractants. However, some divergent developments of recognition signals in pairs or groups of sympatric species would seem best explained as the direct results of selection against the deleterious effects of heterogametic matings. Such selection may act to minimise loss of reproductive potential in species already separated by effective postzygotic barriers which have arisen in immunity from gene-flow (speciation mode 2, see Chapter 4, p. 48). In a number of cases, however, the indication appears to be that the 'Wallace Effect' (see Chapter 4) has been at work. Selection for assortative mating in the face of gene flow has been responsible for the gradual development of reproductive barriers between diverging populations. The best evidence for this mode of origin of prezygotic barriers to reproduction is provided by mate recognition signals exchanged at close quarters. However, because these provide the least ambiguous evidence it does not follow that other types of mating barrier may not arise in the same way.

The type of barrier, whatever its mode of origin, that operates to impede or prevent inter-mating of closely related species inhabiting the same area will depend not only on extrinsic environmental characteristics, but also on dynamic factors intrinsic to particular species. As well as the heterogeneity of habitat occupied and the full range of the species' niche dimensions, additional variables – the genetic and breeding structure of populations, vagility, existing courtship and mating behaviour, the potential flexibility of barrier-inducing variables such as mating season and site, and the existing variability and lability of signal-producing structures – may all have their effect. If reproductive barriers developed in the face of gene flow play an integral part in some speciation events we may expect the pre-adaptations of particular organisms for producing one or other type of barrier to have considerable impact on their patterns of speciation.

For example, Murray (1972) concluded that the existence of old, stable hybrid zones constitutes a notable obstacle to acceptance of the Wallace Effect as an important phenomenon. However, such zones are not especially common and seem likely to be exhibited largely by members of only a few groups. In parapatrically distributed species (or races), even where hybrid infertility or invia-bility is complete, the lethality of their confrontation is seen to exclude the possibility of sympatric overlap. The development of prezygotic barriers to reproduction pro-vides a potential escape from this impasse. However, for some pairs of differentially adapted populations separated by hybrid zones the development of 'mate avoidance' barriers entailing a shift of mating time or place may be effectively precluded by their biology, or at least involve a net disadvantage. If preadaptations for the development of other mating barriers (such as recognition signals) are lacking, inferior heterozygotes may simply continue to be selected against on a narrow front. Is the stability of narrow hybrid zones sometimes due to a poor or inap-propriate repertoire of potential intrinsic mating barriers in the organisms concerned?

The different ways in which reproductive barriers arise and serve to separate closely related species is of concern also in the study of communities. It will readily be appreciated that the various types of reproductive barrier differ in the minimal constraints which they place upon the overall behaviour of populations involved. 'Mate avoidance' barriers (see Table 1), involving segregation in space of potential mates, may entail a relatively pronounced divergence of niche. On the other hand, reproductive separation by means of 'mate rejection' or mate incom-patibility factors (see Table 1) in itself will have little if any effect on the extent of niche overlap. In certain habitats and in certain groups of organisms the number of closely related species which cohabit is much greater than in others. Is the occurrence of many closely related species in a given area related in a simple fashion to environmental characteristics such as heterogeneity? In heterogeneous habitats where resources are patchily distributed the extent to which niches are more nearly coextensive and the tightness of species-packing may be determined as much by the nature of reproductive barriers between closely related species as by the amount and nature of resources. Do the more deterministic models of local diversity take enough account of the varying pre-adaptations of species that make up different communities? Niches, as phenotypic attributes of populations of conspecific individuals, are most usefully defined in terms of the niche occupant, while resource characteristics (such as their distribution or harvestable productivity) are increasingly assessed by ecologists in terms of the particular organisms which experience them. However, is it enough to regard the size of organisms and other intrinsic characteristics that influence the grain-size of their habitat as the principal intrinsic determinants of comparative species-richness? It would seem likely that the presence (or absence) of pre-adaptations for the development of reproductive barriers allowing extensive reciprocal niche overlap will also have a role to play.

In as much as reproductive barriers separating similar and related species find expression in those data most readily accessible to him they will be of particular interest to the taxonomist. A crucial concern will be the identifi-cation of structures that may result from selection for assortative mating. Where divergence of mate recognition systems finds expression in differences, sometimes dram-atic, of morphology the experienced taxonomist with some knowledge of the courtship behaviour and repro-ductive biology of the organisms with which he works will be well equipped to predict the limits of biospecies with some confidence. Indeed, the 'false aura of respectability' ascribed by Scudder (1974) to 'almost all' species now being described on morphological grounds, may not be true of quite so many species as he suggests. Recognition of divergent developments resulting from selection for assortative mating is also of significance for studies of relationships. Only when these are recognised for what they are is it possible to appropriately weight and to recognise likely convergences among divergent signal characters.

For zoogeographical studies the mode of origin of reproductive barriers between closely related species is of some considerable significance. The possibility that such barriers may arise in the face of gene flow will be overlooked by zoogeographers who apply the allopatric model of speciation in its classical form to each and every case studied. Palaeogeographical reconstructions based on the assumption of a past vicariance event that may never have occurred should be examined critically.

An open mind concerning ways in which reproductive barriers arise is likely to be helpful in tackling a variety of evolutionary problems. For example, the most parsi-monious explanation for the presence of flocks of closely related species of cichlid fish in some East African lakes (see Chapter 6) is that intra-lacustrine speciation has taken place. Do the types of prezygotic reproductive barrier separating closely related species of these fish support a view that they may have developed as a response to selection for assortative mating? Are sister-species separated by fully effective postzygotic barriers?

The emphasis in this and a previous chapter (Chapter 4) has been not so much on the origin and nature of differentiation between populations as on the situations and ways in which a single gene pool may be irrevocably split. It is concluded that the manner in which this may take place is likely to vary, and does not always accord with the classical allopatric speciation model. It should be remembered that most animals are, in fact, small. As noted by Hubbell (1954 p. 114) the diversity of 'factors that produce isolates within species' is probably much greater in small animals, such as insects, than in large ones, such as vertebrates.

60 SPECIES AND SPECIATION

References

Blair, W. F. 1964. Isolating mechanisms and interspecies interactions in anuran amphibians. *Quarterly Review of Biology* **39**: 334–344.

Dobzhansky, T. 1937. *Genetics and the origin of species.* 1st edition, 364 pp. New York: Columbia University Press.

Endler, J. A. 1977. *Geographic variation, speciation, and clines.* 246 pp. Princeton: Princeton University Press.

Hammond, P. M. 1972. The micro-structure, distribution and possible function of peg-like setae in male Coleoptera. *Entomologica Scandinavica* **3**: 40–54.

Hubbell, T. H. 1954. The naming of geographically variant populations. *Systematic Zoology* **3**: 113–121.

Jeannel, R. 1955. L'édéage. Initiation aux recherches sur la systématique des Coléoptères. *Publications du Muséum National d'Histoire Naturelle, Paris* **16**: 1–155.

Levin, D. A. & Kerster, H. W. 1967. Natural selection for reproductive isolation in *Phlox*. *Evolution* **21**: 679–687.

Littlejohn, M. J. 1965. Premating isolation in the *Hyla ewingi* complex (Anura: Hylidae). *Evolution* **19**: 234–243.

Mayr, E. 1963. *Animal species and evolution.* 797 pp. Cambridge, Massachusetts: Harvard University Press.

Mayr, E. 1970. *Populations, species, and evolution.* 453 pp. Cambridge, Massachusetts: Harvard University Press.

Mecham, J. S. 1961. Isolating mechanisms in anuran amphibians, pp. 24–61 *in* Blair, W. F. (Ed.) *Vertebrate speciation.* Austin: University of Texas Press.

Murray, J. 1972. *Genetic diversity and natural selection* 128 pp. Edinburgh: Oliver & Boyd.

Paterson, H. E. 1978. More evidence against speciation by reinforcement. *South African Journal of Science* **74**: 369–371.

Sibley, C. G. 1957. The evolutionary and taxonomic significance of sexual dimorphism and hybridization in birds. *Condor* **59**: 166–191.

Scudder, G. G. E. 1974. Species concepts and speciation. *Canadian Journal of Zoology* **52**: 1121–1134.

White, M. J. D. 1978. *Modes of speciation.* 455 pp. San Francisco: W. H. Freeman & Co.

CHAPTER 6

Species-flocks and explosive evolution

P. H. Greenwood

The species-rich, narrowly endemic, and ecologically diverse flocks of cichlid fishes from certain east and central African lakes are sometimes described as the products of 'explosive speciation' and 'explosive evolution' (Mayr, 1976, pp. 168–170; also Fryer & Iles, 1972 and Greenwood, 1974).

Such graphic imagery is perhaps justified when one considers the relatively short time-scales involved (750 000 years to 1·5 or 2 million years), the wide range of morphological and ecological diversity produced, and the fact that each species flock has evolved within one lake basin (see Fig. 6.1).

That explosive evolution in these lakes is confined virtually to one family, the perch-like Cichlidae, seems further to highlight its unusual nature. Are there, then, unusual evolutionary processes involved (as has been suggested), and are there special characteristics of cichlid fishes not shared by other families with which they coexist in each lake? There is also the question of whether explosive speciation could have played a significant role in evolution generally.

These are some of the problems that have intrigued students of evolution and ichthyology for almost seventy years.

Equally intriguing are the ecological consequences of lacustrine explosive evolution. Frequently this has led to the coexistence in one niche of two or more species (often close relatives) with what appear to be identical demands on that niche. In other words, this appears to be an apparent negation of the competitive exclusion principle (see Chapter 10). Such close coexistence may, of course, be more apparent than real, the consequence of insufficiently detailed research. Be that as it may, even apparent total interspecific overlap in such fundamental requirements as food, breeding times, and spawning sites raises the question of whether interspecific competition is inevitably, or even commonly, a major factor in evolution.

Cichlid fishes, the subjects used in this essay to investigate certain questions relating to species-flocks, are perch-like teleosts. These may be familiar to readers since many species are kept in aquaria. Cichlids are widely distributed in the freshwaters of Central and South America, Africa, Syria, Madagascar and in certain brackish waters of the Indian subcontinent. There are at least 1000 species but well over half of these are found in Africa and especially in the Great Lakes where endemism reaches a very high level.

The Great Lakes of tropical Africa, in which the cichlid species flocks occur, lie in or between the eastern and western Rift Valleys, and owe their existence to earth movements and surface changes associated with rift formation. The actual ages of the lakes are still uncertain, but Lake Tanganyika is probably the oldest (c. 1·8 million years BP) and Lake Victoria the youngest (c. 750 000 years BP). Since the various lakes differ in age and developmental history, and because most were populated from different river systems, it is not surprising that each has its distinctive hydrological and geomorphological features, and a characteristic fish fauna.

But, despite these differences all the larger lakes, with the two exceptions of Lakes Turkana and Albert, share one prominent feature: the taxonomic and ecological dominance of their fish-faunas by numerous species belonging to one family – the Cichlidae. The cichlids have entered into and exploited most of the many habitats and ecological niches available in the lakes, sometimes coexisting with species from other families, sometimes as the sole inhabitants. Even in Lakes Turkana and Albert, where the predominating fishes belong to other families, small cichlid species flocks have evolved. These too show a surprising range of ecological diversity (Greenwood, 1979a).

The larger flocks inhabiting lakes stand in marked contrast to the relatively few cichlid species inhabiting the rivers of Africa. Although there is a fairly high level of endemism for a particular river system, there has been far less speciation (Fig. 6.1) and adaptive radiation amongst the fluviatile species. Clearly, environmental conditions within a lake, coupled with the developmental history of

61

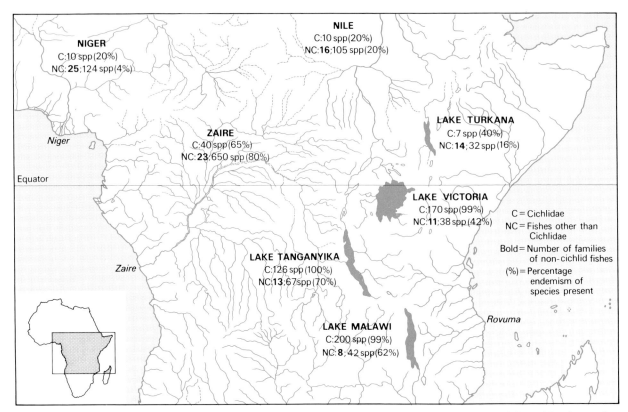

Figure 6.1 Map showing the number of cichlid and non-cichlid fishes in the major lakes and rivers of Africa. Abbreviations: C = number of cichlid species present, NC = non-cichlid fishes; the figure in **bold** type is the number of non-cichlid families present, and is followed by the number of species in those families. The percentage of endemic species in a lake or river is given in parenthesis.

the lake, provide a more suitable background for cichlid radiation than does a riverine environment in the course of its history. By contrast, fishes from other families show a somewhat higher level of major anatomical diversity, and a much higher level of speciation, in rivers than in lakes (see Fig. 6.1).

When comparing the number of endemic cichlid species in the three major centres of explosive speciation (Lakes Malawi, Tanganyika and Victoria) one finds only a slight difference in the total number of species for each lake, and no great difference in their degree of endemicity. In contrast, many more endemic genera have been described from Lakes Tanganyika and Malawi than in Lake Victoria, a fact which reflects the greater morphological differentiation within the cichlids of the two former lakes. In Lake Victoria most of the endemic species have been referred to a single, non-endemic genus, *Haplochromis*, a genus that also looms large in the cichlid fauna of Lake Malawi. Recent research (Greenwood, 1979*a*, 1980), however, indicates that a more precise phylogenetic picture is presented if the 'genus' *Haplochromis* is divided into a number of distinct lineages (or genera). It is also clear that there is little close phyletic relationship between the so-called *Haplochromis* species of Lake Malawi and those of Lake Victoria (Greenwood, 1979*a*). That until now

more genera have been described from Lakes Tanganyika and Malawi is, nevertheless, a valid reflection of the relatively muted morphological diversity seen amongst the cichlids of Lake Victoria. Why there should be this pronounced difference is still a debated question (see Fryer & Iles, 1972).

Lake Victoria has about 170 known *Haplochromis*-like species but only 38 species of non-cichlid fishes and can be taken as an example of a cichlid dominated lake. As compared with Lakes Tanganyika and Malawi it is shallow (100 metres, *cf.* 1470 and 704 metres) and has a much greater surface area (69 000 km², *cf.* 34 000 and 29 604 km²); and its indented shoreline, shallow offshore regions and deep central region provide a variety of macro- and microhabitats, all of which are inhabited by cichlid and non-cichlid species. The Lake's relative youth (750 000 years, compared with 1.5 to 2 million years for Malawi and Tanganyika), the moderate level of morphological differentiation shown by its species flock, and our knowledge of its geological history, all make Lake Victoria a particularly informative example of lacustrine explosive evolution (but still an imperfectly studied one; see Greenwood, 1974, 1980).

Despite the low level of superficial diversity among the Lake Victoria cichlids (Fig. 6.2), there is a wide spectrum

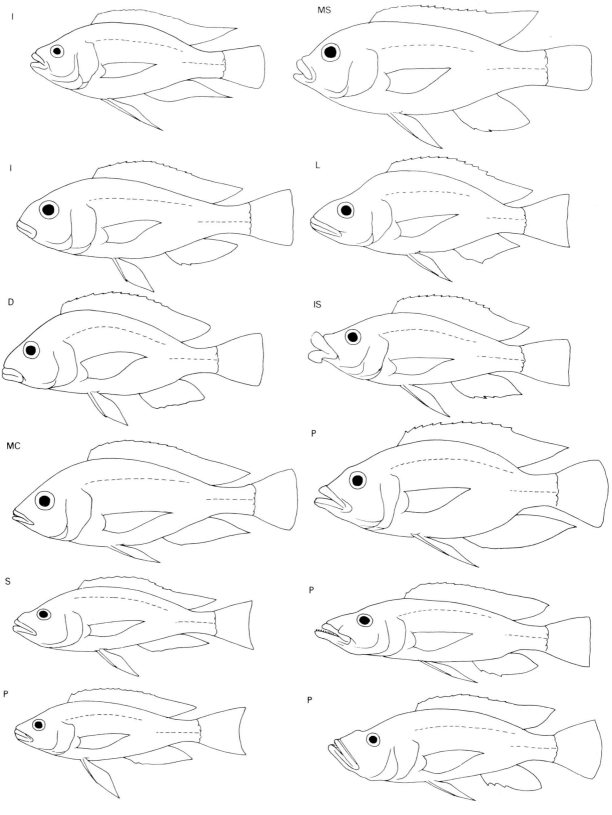

Figure 6.2 The range of body form found amongst *Haplochromis*-group cichlids in Lake Victoria. The feeding groups of the species illustrated are: D–Detritus eater, I–Insectivore, IS–Specialised insectivore (removes larvae and pupae from burrows), L–Paedophage, MC–Mollusc eater (pharyngeal crusher), MS–Mollusc eater (oral sheller), P–Piscivore, S–Scale eater. The drawings are not to the same scale.

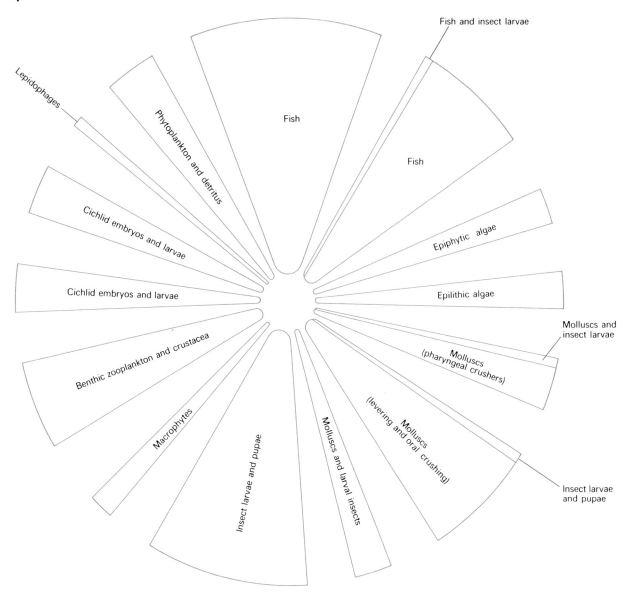

Figure 6.3 Diagram showing the variety of feeding habits amongst species of the *Haplochromis*-group from Lake Victoria.

The area of each segment is proportional to the number of species utilising a particular food source.

of smaller differences, especially in dental characters and in cranial anatomy. These differences, brought together in various combinations (and doubtless with certain physiological specialisations too) have produced a species flock that differs little from those of Malawi and Tanganyika in its range of trophic specialisations and habitat exploitation.

The evolution of many different trophic specialisations is apparently a key element in the ecological success of the lacustrine cichlids, and Lake Victoria is no exception in this respect. As in the other lakes, the cichlids of Victoria

encompass the entire range of feeding habits (Fig. 6.3) practised by species in other families, and in addition they show specialisations not represented among those fishes. The term specialisation, it must be emphasised, does not mean that a species feeds exclusively on one food; it is used in this context to indicate its usual diet.

Adults of most Lake Victoria *Haplochromis* species are rarely more than 100–130 mm long, but some piscivorous species reach lengths of between 150–250 mm. There is a marked sexual colour difference in all species (but virtually no other dimorphism), with males brightly coloured and

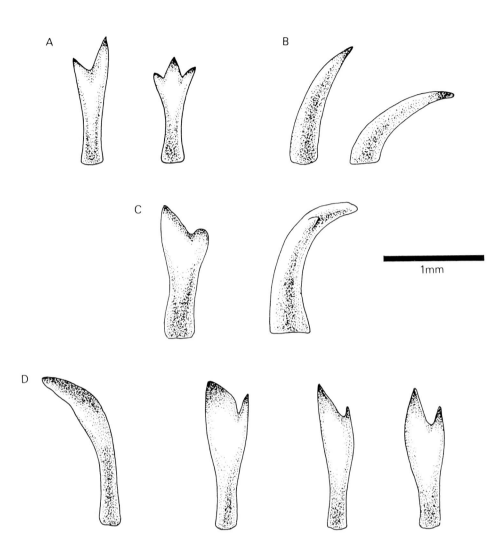

Figure 6.4 Tooth form in various Lake Victoria *Haplochromis*-group species. A–Outer (left) and inner (right) teeth of a generalised type; taken from an insectivore. B–Unicuspid, canine-like teeth, seen in side view to show variation in the degree of curvature; taken from piscivorous species. C–Stout bicuspid form (frontal view left, side view right) from a scale eating species. D–Teeth from four different species to show the gradual change in cusp form; on the right is a generalised bicuspid type, and on the left an extreme obliquely cuspid type associated with algal scraping. The second and third teeth in the row are also from algal scraping species.

females drably so; no species are known to have identical breeding liveries.

Figure 6.2 shows the relatively narrow range of variation in body form encountered amongst the species so far discovered, a uniformity that is even more obvious when compared with the diversity found in Lake Tanganyika (see Fryer & Iles, 1972). The most strikingly different body shape is that of certain piscivores (Fig. 6.2), but even in this group many species have what may be described as the modal form for the flock.

It is only when the anatomy of the head is examined closely, in particular skull and jaw morphology and the shape and arrangement of the teeth (Fig. 6.4 & 6.5), that one encounters any real diversity. Important elements in the cranial anatomy are the upper and lower pharyngeal bones and teeth. These modified segments of the gill arches (Fig. 6.5) provide, in effect, a second pair of jaws situated immediately in front of the oesophagus.

The basic oral dentition, found in over 50 Lake Victoria species, comprises an outer row of moderately stout bicuspid teeth, backed by two or three rows of small, tricuspid teeth. Both cusps of the outer teeth are acutely

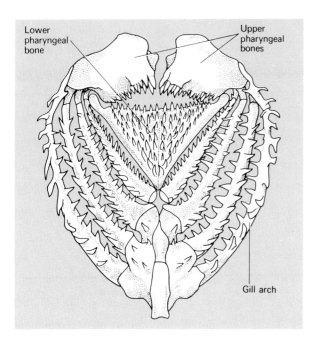

Figure 6.5 'Prey's-eye' view of the gill arches and the pharyngeal bones in a *Haplochromis*-group fish.

pointed, the anterior one distinctly larger than the posterior one (Fig. 6.4). Individuals in most species having outer teeth like these rarely exceed an adult size of more than 100 mm; specimens from the upper end of this size range generally have some unicuspid teeth interspersed amongst the bicuspids. In other words there is a change in the type of replacement teeth correlated with body size, a feature of some importance in the evolution of certain trophic types.

The basic pharyngeal dentition (shown on the lower pharyngeal bone in Fig. 6.5) is also composed of bicuspid teeth, with those in the two median rows slightly stouter than the lateral teeth.

Species in Lake Victoria with this type of oral and pharyngeal dentition are mostly insectivorous, feeding chiefly on the benthic larvae of dipterous insects. They closely resemble many of the *Haplochromis* that inhabit the rivers of East Africa. Since the developing Rift Valley lakes were first populated from these rivers, it seems reasonable to assume that it was from such fishes that the more specialised species of the lake evolved.

Slight departures from the generalised type are seen in five species that feed on phytoplankton and on organic bottom ooze (mostly moribund phytoplankton). Jaw and pharyngeal teeth are finer and more numerous in these species, and the intestine is two or three times longer than in the insectivores. The two species that browse on macrophytic vegetation also have a lengthened gut but otherwise barely depart from the modal type.

Somewhat greater dental deviation occurs in a group of at least six species which graze algae from rocks and rooted plants. Here the teeth are longer, finer, and are moveably

attached to the jaw bones. There is also some change in cusp form, with the crowns having an oblique rather than an acute cutting edge; in a few species the posterior cusp is almost as large as the anterior one. The inner teeth are arranged in several rows (5–8, compared with 2 or 3 in the generalised type), but retain a tricuspid crown.

The skull departs from the generalised shape in having the snout region noticeably downcurved (Fig. 6.7). The lower jaw is short and stout, relative to the modal type, and the bones of the upper jaw are also strengthened.

A second group of algal grazing species retains a generalised type of skull and jaw shape, but shows greater departure in tooth form (Fig. 6.4). The anterior cusp is drawn out and the posterior cusp is much reduced or even suppressed completely.

Unfortunately the significance of the dental and cranial differences between the two lineages is not really understood. Apparently the same algal species are consumed, but some field observations suggest that each group of algal grazers has different feeding methods. The species with stout jaws, decurved snouts and teeth with cusps of subequal size seem to graze algae directly from a hard, unyielding substrate. Members of the other group take a leaf in the mouth and, holding it loosely, then swim along its length, scraping off the epiphytes as the leaf passes between the several rows of teeth in each jaw. The shape of the teeth would thus seem to be associated with the method of scraping employed.

Of particular interest is the fact that when the five species of the second group are compared, they can be arranged in a graded series according to cusp shape in the outer teeth. At the lower end of the series is a species in which the cusp is close to the generalised type (but with some protraction of the major cusp and its oblique cutting edge already apparent). The series terminates with a species in which the minor cusp is suppressed and the major cusp greatly protracted (Fig. 6.4).

A third type of scraping dentition has evolved in a species feeding on scales rasped from the tail fin of other *Haplochromis*. As in the algal grazers there are broad bands of inner teeth but, unlike those species, the lepidophage has stout, strongly recurved and acutely bicuspid teeth in the outer rows. This species also differs in being more slender and streamlined, in this way closely resembling many of the piscivorous predators. Presumably its feeding habits require a greater turn of speed when approaching and leaving its prey than is demanded of the algal grazers.

Molluscs, both gastropods and bivalves, form an important or even exclusive element in the food of at least twenty Lake Victoria *Haplochromis* species. Two quite distinct methods of handling this hard-shelled prey have been evolved. One method involves crushing the entire mollusc between the upper and lower pharyngeal 'jaws'. The other involves removing the mollusc from its shell, either by crushing the shell between the fish's jaws or by actually levering out the body and leaving the shell intact.

In the pharyngeal crushers the oral dentition is of a

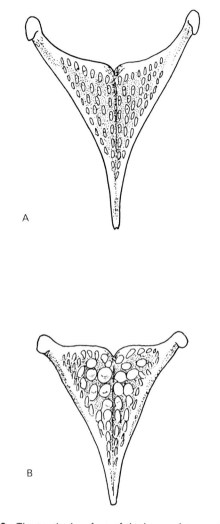
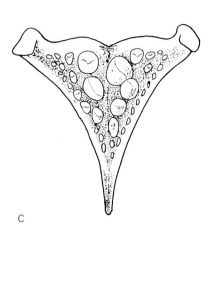
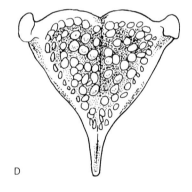

Figure 6.6 The toothed surface of the lower pharyngeal bones from four *Haplochromis*-group species in Lake Victoria. A–Detritus eater; B–D–Mollusc crushing species. In B the diet is a mixed mollusc-insect one, in C small gastropods and bivalves, and in D large, hard-shelled gastropods. Drawn to different scales, C greatly enlarged.

generalised type but the pharyngeal bones, especially the lower one, are much enlarged and some or almost all the teeth are strong and molar-like (Fig. 6.6). Skull shape in these species hardly differs from the generalised type.

Within the species of pharyngeal crushers one can observe a series of intergrading steps involving increasing hypertrophy of the pharyngeal bones and a correlated molarisation of their dentition (see Fig. 6.6). Species with slightly enlarged and molarised pharyngeals tend to have a mixed mollusc-insect diet, those with hypertrophied pharyngeal mills feed, at least when adult, almost exclusively on snails.

Those species that lever out the snail's body, or crush molluscs orally, show, as one might expect, modifications to the jaws and oral dentition. In several species there is also some departure from the generalised skull type, involving a strong downward flexure of the snout region. The latter modification allows the upper jaw to be protracted straight downwards, rather than forward and downward as in the generalised skull, thus enabling the fish to grab its prey from above rather than having to scoop it into the mouth from below.

The jaws are stout and the tooth-bearing surfaces are broad, markedly so in a few species. The teeth are strong, sharply pointed, and have a pronounced curvature towards the buccal cavity.

Functionally, the teeth, their disposition in the jaws and the shape of the jaws, all combine to form a strong, vice-like

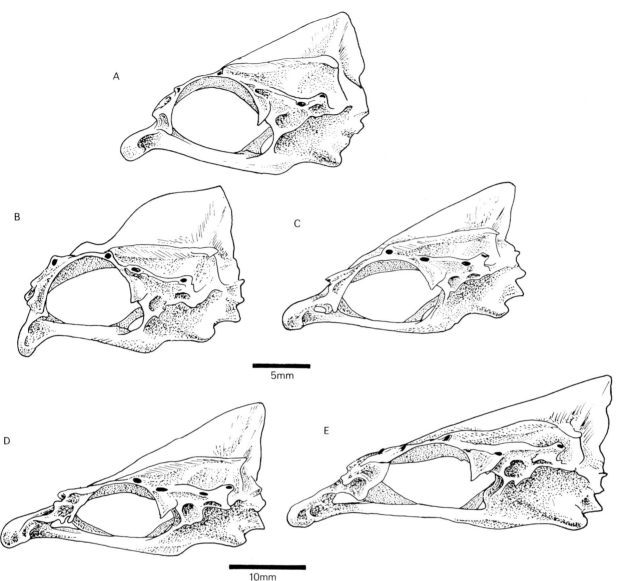

5mm

10mm

Figure 6.7 Skull shape in some Lake Victoria *Haplochromis*-group species. A–The generalised type (an insectivore). B–Skull with strongly decurved snout region (from an algal grazer). C, D and E–Skulls from piscivorous predators. Note the gradual change in proportions, particularly the relative elongation of the snout region and the overall increase in streamline form of the skull; compare with Figure 6.2.

mechanism. With this vice, the body of the snail is held whilst the fish, with lever-like movements of its body, wrenches the snail from its shell. Alternatively, the shell is crushed free from the body (the usual process when bivalves are eaten). In contrast to the jaws, neither the pharyngeal bones nor their dentition is modified.

As with the algal grazers and the pharyngeal snail crushers, one can observe intergrading changes in tooth form, tooth pattern and skull type amongst the 'winkle-picking' and oral crushing species. These changes involve various combinations of characters which, starting with a species that differs only slightly from the generalised type, end with species so modified that systematists have placed them in genera distinct from one another and from other members of the trophic group.

A similar sequence in morphology can be seen in a lineage of specialised insectivores which are able, with their forceps-like dentition, to remove the larvae of burrowing insects from rocks and wood.

All species in this group have the lower jaw foreshortened and robust. The upper jaws are strengthened and capable of being protruded downwards, and there are correlated changes in the preorbital skull region like those seen in the mollusc eaters described before. The strong jaw teeth are reduced in number, uni- or weakly bicuspid, have sharply pointed crowns and protrude forwards. The unusual angle of tooth implantation, combined with an anteriorly narrowed dental arcade, imparts to the biting action of the jaws a forceps-like action as well. Because the teeth protrude, the fishes are able to insert what is virtually an extension of the jaws into the larval burrows.

Finally, we must consider what is probably the largest trophic group in Lake Victoria, the piscivorous predators, to which about 30–40 per cent of the known *Haplochromis* species of that lake belong. The principal food of these species is other *Haplochromis*, although a few non-cichlid fishes are also eaten.

Predatory *Haplochromis*, unlike their non-cichlid counterparts in the lake, do not bolt their prey whole, but macerate it between the upper and lower pharyngeal teeth. Thus the jaws and oral teeth serve to capture the prey and then to hold it during the lengthy process of pharyngeal mastication.

In the light of what we know about other trophic groups, it is not surprising to find an almost complete morphological sequence in adaptive characters amongst the piscivores (of which there are at least two phylogenetically distinct lineages). Most obvious are an increase in size, improved streamlining of the body, lengthening and increased protrusibility of the jaws, and a change from bicuspid to sharply pointed, unicuspid jaw teeth. Many predatory species also show a deepening of the pharyngeal and buccal cavities, and show a well-marked trend towards an increase in skull length coupled with a decrease in its height (Fig. 6.7). The decrease in skull height (affecting as it does that part of the skull over which the toothed part of the upper jaw slides during its protrusion), correlated with a deepening of the pharyngeal and buccal cavities, are important factors in developing an efficient prey catching device.

Modifications to the pharyngeal dentition are relatively slight, incorporating some reduction in tooth number, coarsening of the teeth, and the tendency for the crowns to develop a strong cutting edge.

The mouth-brooding habits of female cichlids have been exploited as a source of food by at least eight piscivorous species, the so-called paedophages, whose diet is principally the embryos and larvae obtained from the mouths of brooding *Haplochromis* females.

Two distinct lineages have adopted this bizarre feeding strategy. Both lines share certain oral modifications, namely a widely distensible and protrusible mouth, a reduction in the number of teeth and an hypertrophy of the oral mucosa so that the teeth are deeply embedded in soft tissue. Aquarium observations suggest that these modifications are associated with the paedophages' feeding methods, that is, engulfing the snout of the brooding female and forcing her to jettison the brood into the predator's mouth.

Although trends of increasing specialisation are seen within both paedophage lineages, these species do seem to differ from other specialist feeders in showing a more trenchant morphological gap between their least derived members and those of other trophic groups (including the true piscivores with which they share certain features).

There are a few lineages amongst the Lake Victoria *Haplochromis* (for example the small, benthic feeding zooplankton eaters) in which these seemingly orthoselective trends (Grant, 1977) cannot be detected because the member species all show a comparable level of specialisation. Within the flock as a whole, however, lineages having a distinct trend in morphological specialisation are commoner than those which do not. Nevertheless, it must be stressed that even in lineages showing clear-cut morphological sequences one invariably finds two or more species at any one level of specialisation.

This latter phenomenon is particularly obvious amongst the two major piscivore lines (but not the paedophages), the mollusc crushers (both oral and pharyngeal types), the specialist insectivores, and the algal grazers. Thus, from an anatomical point of view, it seems that two kinds of speciation may be recognised: an apomorphic kind in which one or more anatomical characters show further development in daughter taxa, and a stasimorphic one in which there is only a multiplication of species without further anatomical differentiation. These latter, then, may display characteristics seen in sibling species-groups (see Chapter 2). The adjectival terms apomorphic and stasimorphic refer to extreme points in a continuum of the changes in external characteristics that are associated with speciation. Apomorphic is that which shows greater deviation from what is presumed to be the primitive condition.

Such a brief summary of morphological features in a trophically multiradiate species flock, perforce, be an oversimplification, but it does serve to underline certain aspects which are of importance in attempting to understand this example of explosive evolution.

On analysis it seems that the anatomical features that underlie the trophic specialisations originated and developed through simple morphological transformations. The skeletal transformations (especially in the head) involved only slight alterations in the differential growth patterns of the various skull regions, coupled with similar changes in the jaws and the suspensorium. Likewise, taking the generalised bicuspid tooth as a starting point, the observed dental changes seem to stem from altered differential growth rates, this time in cusp sizes, curvature of the tooth, its length and robustness. Other dental changes involved an increase, or decrease, in the number of teeth or tooth rows. Changes in pharyngeal bone shape and in the form of the pharyngeal teeth also seem attributable to the same processes.

Finally, there are some modifications, especially important among piscivorous lineages, linked with an increase in adult size and possibly an enhanced growth rate. Particularly obvious are shape and proportional changes associated with allometric growth patterns, and the ontogenetic changes from bi- to unicuspid teeth with body growth beyond a certain size.

The simplicity of these structural changes, individually

or in combination, may help to explain not only the wide range of trophic specialisations that were produced but also the speed with which this happened (that is, in about 750000 years).

It would seem unnecessary to postulate a genetical revolution to produce morphological changes of this magnitude, and that too may account for the rapidity of change.

Regrettably, nothing is known about the genetical basis for the various transformations. Their phenotypic manifestation, however, suggests that genes with a regulatory or epigenetic function could have been more important than those with structural effects (see Stansfield, 1977, p. 331 for discussion of epigenesis).

In many respects (and with particular regard to the way in which speciation probably occurred in Lake Victoria) the origin of this flock seems to accord well with Grant's concept of quantum speciation, except that the genetical changes may not be of the revolutionary kind which he associates with that phenomenon (see Grant, 1977; see also Chapter 3).

Parenthetically, it should be stressed that such seemingly simple, genetical and anatomical transformations are not confined to the species flock in Lake Victoria. They would seem equally applicable to the superficially more complex cichlid flocks of Tanganyika and Malawi.

That it is the cichlid species and not members of other families which have undergone explosive evolution and adaptive radiation probably is due in large measure to the suitability of the cichlid 'bauplan' as a substrate for the type of anatomical and dental specialisations discussed above. Most cichlids, unlike members of other families in the lakes, have a 'bauplan' that is neither too generalised nor too specialised for such simple changes to be effective. Perhaps, over a greater time-span, species from other families could have produced the necessary modifications to rival those of the cichlids. From their performance elsewhere the Cyprinidae (minnows) and Characidae might well have succeeded in so doing, particularly if the African characid species were of a dentally less specialised kind. To produce the range of dental and cranial specialisations seen in the Characidae as a whole (the only basis for comparison with a lake cichlid flock) from even the most generalised African species would require very considerable remodelling of the skull and dentition. It could not be achieved, as apparently it was in the cichlids, by simple changes in differential growth patterns. The cyprinids lack jaw teeth, and have only the lower pharyngeal bones toothed, two inhibiting factors that are compensated for only by the development of extensive modifications to mouth and lip form, and to the morphology and pattern of the pharyngeal teeth. Thus in this family too it would seem reasonable to hypothesise a greater level of genetical reorganisation than was required in the cichlids. The cichlids have two further advantages over the non-cichlids. Firstly, their ability to breed throughout the year and, secondly, the fact that most species have short generation

times. Thus, compared with the annually breeding, slower maturing non-cichlids, the cichlids have far greater opportunity for genetical reshuffling, so providing more raw material for the production of novel morphotypes.

No matter what epigenetic mechanisms control the morphological differences underlying the trophic specialisations in a cichlid species flock, each flock owes its origin to repeated acts of speciation, many of which still persist in the contemporary lake. In other words, the Lake Victoria species flock, in its origins, consequent results and the evolutionary potential contained in those results, would seem to fit closely Grant's concept of speciational evolution, a thesis further developed in Gould & Eldredge's model of punctuational evolution (see Grant, 1977; Gould & Eldredge, 1977; Stanley, 1975).

Gould & Eldredge believe that the history of life has been dominated by concentrated outbursts of rapid speciation (followed by the differential success of certain species), rather than by slow directional transformations within a lineage (that is, evolution by accelerated cladogenesis rather than through phyletic gradualism).

Admittedly, within lineages of the Lake Victoria flock one can see many examples of gradual change in a particular character or suite of characters. But, each point in the grade is a species, and the different species are contemporaneous, not successive elements in a temporal and phyletic continuum as they would be in a case of true phyletic gradualism. Perhaps one should call this phenomenon 'cladistic gradualism'?

Unless truly quantum jumps occur, cladistic gradualism must be the norm in punctuational evolutionary histories, but it is doubtless a difficult feature to recognise when only fossils are available.

The concept of cladistic gradualism raises some interesting questions, especially when, as in Lake Victoria, it invokes characters associated with trophic specialisations. Take, for example, the phyletic lineages whose members are, respectively, mollusc eaters, piscivores, or algal grazers. From the data currently available there seem to be no qualitative differences between the diet of most species in a lineage (nor between members of different lineages with similar feeding habits). What factors, one asks, could have initiated development towards further levels of character derivation when, apparently, the 'lower' levels provide an equally effective degree of trophic specialisation? And, how is it that the latter species still remain in successful coexistence with their derived relatives if the concept of competitive exclusion has general applicability?

Paradoxically, within a moderately wide range of anatomical specialisation a trophically specialised species can also be a more generalised one. That is, it retains the ability to utilise the food sources tapped by its ancestors, and also has the capabilities to exploit sources not open to the ancestors because these lacked the dental and other necessary specialisations. Thus, in Lake Victoria today specialised algal grazers can, and do, feed on insects, but the insectivores are incapable of grazing algae.

Questions relating to the origin and further differentiation of derived features are relevant to the continuing debate on the role of natural selection in creating evolutionary novelties (see review by Rosen, 1978, pp. 371–373).

The slight and seemingly orthoselective differences between some species in a lineage might be interpreted as the sort of 'fine tuning' (Avise, 1977) that could be produced by natural selection (that is, the differential survival and ultimate fixation of certain alleles within a genotype). Such an interpretation would imply that the species are competing for a limited food resource, and that their morphological differences bring about more effective resource partitioning. Data currently available would not appear to corroborate the 'selection' hypothesis since there is apparently complete interspecific overlap in environmental requirement. However, more refined analyses are needed before these ideas can be tested fully.

Also uncorroborated is an alternative hypothesis, that such slight interspecific differences have no 'selective' value, and are stochastic in nature, the consequence of genetic sampling inevitably associated with a speciation event (that is, the so-called founder effect of Mayr). The high level of intraspecific variability in the characters under discussion, and the interspecific overlap in ecological requirements, however, seem to add support to the 'non-selectionist' viewpoint.

The major morphological differences, the obvious evolutionary novelties, which characterise each of the various lineages appear to be changes associated with altered growth patterns. These features, at least superficially, seem to be of greater magnitude than the intralineage differences discussed above. They certainly exceed the scale of change that could be labelled 'fine tuning'. The origin of such features is impossible to explain realistically on the basis of either true Darwinian or neo-Darwinian selection. The alternative is to hypothesise an origin stemming from chance mutations established in a population at the time of the speciation event. The products of such changes would be, effectively, Goldschmidtian 'hopeful monsters' (although their origin, in the case of Lake Victoria cichlids, would not seem to involve the large-scale systemic mutations postulated by Goldschmidt; for further discussion of Goldschmidt's ideas, the reader is referred to Stansfield [1977]). In other words, the major lineage traits could have evolved in essentially the same non-selective way as the minor, intralineage traits.

Whatever evolutionary factors are involved, most lineages in the contemporary cichlid species flock of Lake Victoria clearly display the phenomenon of cladistic gradualism in several morphological traits. In this respect they retain a more complete account of their history than do the flocks of Lakes Tanganyika and Malawi, especially the former. The reasons for these differences are difficult to determine. The relative youth of Victoria, and its mode of origin, could both be contributory factors, as could differences in the diversity of the riverine forms that first colonised the developing lakes.

One can virtually be certain that cichlids were not the only fishes to populate a developing lake. If the riverine fish faunas of east Africa during the Pleistocene were like those of today, then both in the number of their species and in their trophic diversity, the cichlids were very much a minority group; the little fossil evidence available supports this conjecture. Thus, any suggested origin of a cichlid flock through multiple invasions of the lake can be discounted as an important element in its history, but the early non-cichlid invaders could well have been important in shaping its early development.

If multiple invasions are ruled out, then speciation within the lake basin, at all stages of its ontogeny, would seem to be the only way the cichlid flocks could have evolved.

Much debate has centered around the actual way in which speciation has taken place in the African lakes, in particular whether or not the cichlid flocks provide a *prima facie* case for sympatric as opposed to allopatric speciation (see Fryer & Iles, 1972; Greenwood, 1974 for references).

Supporters of sympatric speciation in Lake Victoria have generally suggested habitudinal segregation as the means whereby the evolving species are isolated from one another. It is difficult to conceive how such segregation alone could have led to the origin of 170 species, especially when the existing species are segregated more by differences of feeding habits than by differences in habitat, and when there is no or little appreciable spatial segregation of preferred food organisms within a habitat. Also it could be argued that if the original cichlid invaders did, by some means or other, become habitudinally segregated, then any subsequent speciation would, in fact, conform to the allopatric model. Indeed, it is difficult to know at what spatial cut-off point one should draw the line between so-called allo- and sympatric speciation (see discussion in White, 1978, p. 146). Perhaps only stasipatric speciation, speciation through disruptive selection and speciation through polyploidy in plants (see Chapter 8) should qualify as truly sympatric models.

As we have no genetical or karyotypic information for the Lake Victoria cichlids (and very little for other cichlids either) one cannot usefully speculate on the possibility of stasipatric speciation playing some part in the origin of the flock.

Evidence from an American cichlid does, however, suggest a way in which disruptive selection could have been involved (Sage & Selander, 1975). In a Mexican lake studied by these authors, the supposedly single species present exists in a state of balanced polymorphism. The three morphs show distinct anatomical differences associated with the oral and pharyngeal dentition, and there are correlated differences in feeding habits (algae and detritus, molluscs, and fishes respectively). If the genes controlling these features were to become linked with genes affecting male breeding coloration (or other reproductive features), then, through the effects of assortative mating, the morphs could become true species (a state which the Mexican species seemingly has not yet reached). In other words, this model proposes that the primary isolating mechanism is selective mating, effected through differences in breeding livery. Unfortunately, there is no biological information to support Sage & Selander's suggestion that balanced polymorphism could explain the situation in the African lake flocks (see Greenwood, 1974), although it might have played a part in their origins.

The geomorphological and hydrographical history of Lake Victoria seems to suggest that allopatric speciation, through actual geographical isolation, played the prime role in the establishment and diversification of its species flock.

When the lake first formed (during the mid-Pleistocene *c.* 750000 years BP) its future basin was crossed by four or more westward flowing rivers. These drained the eastern highlands of present-day Kenya and emptied into the Zaire river system. A gradual but large-scale surface warping in the west led to a reversal of river flow and a consequent backponding in the western reaches of the rivers. As the shallow western river valleys filled with water each became an expansive, dendritic lake. Eventually these lakes overflowed, joining their neighbours to form a large but shallow water body occupying an area considerably greater than that of the present lake. Later, the lake basin was subject to further periods of tectonic instability which considerably modified the margins of the lake. Such changes would, at times, result in the isolation of water bodies on the periphery, and at other times their reunion with the main lake. Still later, and when the lake had assumed its present form, local climatic changes led to alterations in the lake level. These changes again resulted in the formation of small peripheral lakes and their ultimate reunion with the main lake (see Fryer & Iles, 1972; Greenwood, 1974 for references).

In effect then, present-day Lake Victoria can be considered the product of repeated fractioning and reunion; that is, an amalgam of several lakes which developed in a closed drainage basin.

Initial differentiation, through acts of speciation, of the main phyletic lines within the cichlid species flock, and the concomitant origin of what are now seen as the various trophic specialisations, must have taken place in the shallow lakes formed when the rivers were first reversed and ponded back. Potentially each lake could be the cradle of a species. Later, as the number of species was increased, each peripheral isolate could serve as the cradle for as many new species as there were existing species cut off in it, a kind of exponential species growth.

Lake Nabugabo, a small (30 km²) lake now separated from Victoria by a narrow sand bar, is a good example of this process, and also of the rate at which certain cichlids can speciate (Greenwood, 1965). Nabugabo was cut off about 4000 years ago; yet five of its seven *Haplochromis* species are endemic. Each endemic species closely resembles a species in Lake Victoria, but differs both in male

coloration and, less markedly, in some morphometric characters (an example of stasimorphic speciation).

A model of alternating allopatry and sympatry may also help to explain a notable feature of the Lake Victoria flock. That is, the occurrence in many lineages of more than one species at a particular level in a trend of trophic specialisation. That is to say, these species are the products of speciation events that did not involve the evolution of any change in anatomical or physiological characters associated with feeding habits, but simply the evolution, in different peripheral lakes, of only specific reproductive barriers between daughter species derived from segments of the same parental species.

Although, ultimately, the cichlid flock in Lake Victoria came to resemble those of Lakes Tanganyika and Malawi, the history of the lakes, and the phylogeny of their cichlid flocks, differ in several respects (see Fryer & Iles, 1972; Greenwood, 1974, 1979a). Each lake, in its own way, contributes to our understanding of explosive evolution, a multifaceted phenomenon.

In this context it is relevant to compare the cichlid species flocks with certain other, and better publicised, examples of adaptively multiradiate and geographically restricted species groups, the Galapagos finches, the Hawaiian honeycreepers and the Hawaiian fruitflies (see Grant, 1977; White, 1978; Dobzhansky et al., 1977). As in the lakes, these island faunas are, so to speak, dominated by one group of related animals, a feature further emphasised in the islands by the absence of many components of the mainland faunas from which they were first populated.

There are many broad similarities in the ways in which these otherwise dissimilar organisms have differentiated. In all three island flocks (as in the lake cichlids) differentiation has been primarily towards trophic specialisation. All share, as contributory features in their evolution, the historical elements of geographical (or at least spatial) isolation and the availability of new habitats and niches for exploitation. A high level of endemicity is a further common feature, as is the fact that the island birds and fruitflies, like the lake faunas, show a greater range of morphological diversity than is encountered in their close relatives from other regions. Except for the fruitflies, however, far fewer species are involved in the islands flocks (for example 13 finches and 22 honeycreepers, but as many as 160 species of fruitflies).

The manner in which the spatial isolation necessary for speciation was achieved may differ in the different island flocks (as indeed it probably did in the African lakes). For example, the pattern amongst the Galapagos finches, of evolution in isolation followed by inter-island migration and the ultimate occurrence of several species coexisting on any one island, closely parallels the pattern proposed in the Lake Victoria fishes. There, the first speciation events occurred in isolated lakes, whilst the later union of the separate small lakes would be equivalent to the inter-island migration phases of evolution in the Galapagos flock. Isolation through the development of ecological barriers within an island (for example, lava flows surrounding patches of vegetation) is thought to be the principal factor in speciation amongst Hawaiian fruitflies. The parallel here would be with the way in which part of the Lake Malawi cichlid flock probably evolved, with sandy beaches and rocky outcrops providing the ecological barriers (see discussion in Fryer, 1977; Fryer & Iles, 1972).

The fish, bird and insect flocks are similar, too, in that their adaptational successes have been effected through simple anatomical changes. The complexity, or otherwise, of the underlying genetical changes involved are unknown for the birds and the fishes. Those concerned in the fruit-fly radiation would not appear to be revolutionary ones (see White, 1978).

It is generally believed that one factor stimulating the evolution of trophic adaptations in the Galapagos finches and the Hawaiian honeycreepers was an absence of competitors on the newly created islands. To a large extent this hypothesis seems inapplicable to the cichlid fishes. The original cichlid invaders would be in the company of other species from other families which, in total, would seem capable of occupying all the trophic and ecological niches provided by the developing lake (and, from their present-day habits, those of the mature lake as well). These interfaunal reactions (especially in the exploitation of micro-niches), rather than their absence, may have played an important role in the early shaping of the cichlid flocks. Nevertheless, the creation of increasing numbers of niches by members of the flock itself could well have been an important influence on later stages of differentiation within the species flocks of the island birds and the fruitflies, as it undoubtedly was amongst the lake fishes.

In general then, there would seem to be greater levels of similarity than dissimilarity in the way these various radiations have evolved. White (1978, p. 232), on the contrary, believes that '...there is probably not much resemblance between the patterns and modes of speciation

in ancient lakes and on oceanic islands'. His views are based essentially on supposedly differing degrees of isolation and ecological diversity obtaining in the two situations, a viewpoint that is not substantiated by information from the African lakes, nor from Lake Baikal (see Greenwood, 1974; Fryer & Iles, 1972; Kozkov, 1963). White (1978) would, however, concede a greater level of similarity between the situation in Lake Victoria and that in the Galapagos and the Hawaiian islands.

The species flocks and other localised, sometimes explosive, radiations discussed above are by no means the only ones known amongst animals and plants (see White, 1978). From what we can reconstruct of their histories, and learn from the living animals, it seems unnecessary to invoke any special kind of evolutionary phenomenon to explain their origin and development. Rather, each appears to represent a combination, and temporal condensation, of many factors thought to have been generally operative in the evolution of bisexual organisms. In particular, these flocks focus attention on the primary factor in evolution – the act of speciation (Stanley, 1975; Greenwood, 1979b). It is the elements of temporal and spatial condensation, coupled with the persistence of all or nearly all stages in the development of different anatomical specialisations, that make species flocks and explosive evolution so conspicuous; on studying a species flock one experiences the feeling of looking into a factory where prototypes are still in production alongside the latest models.

The question of whether or not explosive speciation has contributed to the origin of major animal and plant groups is unlikely to be answered by direct evidence and certainly not by fossil evidence. But, since the origin of every major group lies in a single speciation event, the origin of numerous and diverse species in a circumscribed area, and over a geologically short time-span, could well be of importance in providing the raw materials for further evolutionary development and diversification (see also Greenwood, 1979b; Gould & Eldredge, 1977; Rensch, 1959 for further discussion).

References

Avise, J. C. 1977. Is evolution gradual or rectangular? Evidence from living fishes. *Proceedings of the National Academy of Sciences of the United States of America* 74: 5083–5087.

Dobzhansky, T., Ayala, F. J., Stebbins, G. G. & Valentine, J. W. 1977. *Evolution*, 572 pp. San Francisco: Freeman.

Fryer, G. 1977. Evolution of species flocks of cichlid fishes in African Lakes. *Zeitschrift für Zoologische Systematik und Evolutionsforschung* 15: 141–165.

Fryer, G. & Iles, T. D. 1972. *The cichlid fishes of the Great Lakes of Africa. Their biology and evolution.* 641 pp. Edinburgh: Oliver & Boyd.

Gould, S. J. & Eldredge, N. 1977. Punctuated equilibria: the tempo and mode of evolution reconsidered. *Paleobiology* 3: 115–151.

Grant, V. 1977. *Organismic evolution.* 418 pp. San Francisco: Freeman.

Greenwood, P. H. 1965. The cichlid fishes of Lake Nabugabo, Uganda. *Bulletin of the British Museum (Natural History), Zoology* 12: 315–357.

Greenwood, P. H. 1974. Cichlid fishes of Lake Victoria, east Africa: the biology and evolution of a species flock. *Bulletin of the British Museum (Natural History), Zoology* Supplement 6: 1–134.

Greenwood, P. H. 1979a. Towards a phyletic classification of the 'genus' Haplochromis and related taxa. Part I. *Bulletin of the British Museum (Natural History), Zoology* 35: 265–322.

Greenwood, P. H. 1979b. Macroevolution – myth or reality? *Biological Journal of the Linnean Society of London* 12: 293–304.

Greenwood, P. H. 1980. Towards a phyletic classification of the 'genus' *Haplochromis* and related taxa. Part II. *Bulletin of the British Museum (Natural History), Zoology* **39**: 1–101.

Kozkov, M. 1963. Lake Baikal and its life. *Monographiae Biologicae* **11**: 1–344.

Mayr, E. 1976. *Evolution and the diversity of life*, 721 pp. Harvard: Belknap Press.

Rensch, B. 1959. *Evolution above the species level*, 419 pp. London: Methuen.

Rosen, D. E. 1978. Darwin's demon (A review of *Introduction to natural selection*, by C. Johnson [1976]), *Systematic Zoology* **27**: 370–373.

Sage, R. D. & Selander, R. K. 1975. Trophic radiation through polymorphism in cichlid fishes. *Proceedings of the National Academy of Sciences of the United States of America* **72**: 4669–73.

Stanley, S. M. 1975. A theory of evolution above the species level. *Proceedings of the National Academy of Sciences of the United States of America* **72**: 646–650.

Stansfield, W. D. 1977. *The science of evolution.* 614 pp. London: Collier Macmillan.

White, M. J. D. 1978. *Modes of speciation.* 455 pp. San Francisco: Freeman.

CHAPTER 7

Species, sex and parthenogenesis in aphids

R. L. Blackman

Biological species, the discrete packages into which life is organised, and the process of speciation by which they are formed, seem at first sight to be inextricably linked to sexual reproduction. Each species has its own pool of genes and each individual is the result of a separate dip into that gene pool. Sexual reproduction not only provides the mixing process that ensures that two individuals never have exactly the same genetic composition, but it also provides the vehicle through which isolating mechanisms act to prevent the gene pools of different species becoming mixed. The result is that one rabbit, for example, although individually unique, will always have more genes in common with another rabbit than either will have with a hare. The existence and integrity of species seem to be completely dependent on sex.

If life manifests itself in the form of species, and species exist only because of sexual reproduction, is sex then essential to life itself? Amazingly, this question, which is perhaps the most fundamental that could be asked in biology, still awaits a proper answer. Certainly the occurrence of sex in living organisms is almost universal, although it was not until the late seventeenth century that people recognised the sexuality of plants; the conjugation of protozoa was not regarded as a sexual process until late in the nineteenth century, and the recombination processes that represent sex in bacteria and viruses have been recognised only in the last thirty years.

However, together with the gradual recognition of the ubiquity of sex in the living world, came the demonstration that some organisms have dispensed with sex entirely without suffering any obvious ill-effects. There are relatively few such parthenogenetic organisms, probably less than 0·1 per cent of animals, but rather more plants. They tend to be scattered randomly through the plant and animal kingdoms, and usually have close relatives that still reproduce sexually. If present-day life is anything to go by, this suggests that parthenogenetic reproduction (or apomixis, the term generally applied to parthenogenesis in plants) has cropped up again and again in the evolution of both plants and animals. Since no major group of animals or plants is completely parthenogenetic, it would appear that, for some reason, parthenogenesis cannot stand the test of time as an alternative strategy to sex. Why should this be?

It is fairly easy to see the long-term advantages of sex. Continual stirring of the gene pool produces an almost infinite array of new genetic combinations on which natural selection can act, and favourable mutations occurring in two different individuals can be combined in their descendants. It seems intuitive that such processes should hasten evolutionary change, so that in the course of time sexual organisms should prove more capable than asexual ones of evolving new ways of coping with a constantly changing environments. But there are problems, and they are substantial ones. These problems immediately become apparent if we stop asking the question 'Is sex really necessary?' and start asking 'Why does sex survive at all?'.

The trouble is that natural selection is an opportunistic process. It acts on each and every generation, and it cannot be expected to take into account the future course of evolution. The survival of one individual's genes into the next generation depends on its fitness as measured by reproductive success. Those individuals that are fittest leave the largest number of descendants. But sexual reproduction is definitely not the way to leave the largest number of descendants. Think, for a moment, what it entails. The sexual female must await fertilisation by a male, which may itself be a risky process. Then usually half its offspring, the males, will be infecund. How can such an individual compete with a parthenogenetic female that dispenses with males and lays unfertilised eggs, all of which develop into more parthenogenetic females? With such advantages the obvious question is: why does parthenogenesis not replace sexual reproduction in all living organisms? This question still worries evolutionary biologists, and has led in recent years to books such as those of Ghiselin (1974), Williams (1975) and Maynard Smith (1978).

Cyclical parthenogenesis

There is probably no better way to compare the advantages and disadvantages of sexual and parthenogenetic reproduction than to look at organisms that have achieved a mastery of both. Several animal groups have independently evolved the strategy known as cyclical parthenogenesis, in which sexual and parthenogenetic reproduction both play a very effective part in the life-cycle of a single species. Cyclical parthenogenesis is found typically in rotifers, water fleas (Cladocera), cynipid gall wasps and aphids. Sexual reproduction occurs, often as a regular part of the life-cycle, but it alternates with one or more generations in which males are absent and the females produce their offspring parthenogenetically.

An alternation of sexual and asexual phases is, of course, nothing special in itself. All higher plants base their reproductive cycle on an alternation of gamete-producing and spore-producing generations. Indeed, if we take a broad view, all multicellular organisms can be said to have a phase of sexual reproduction, which results in the formation of the fertilised egg-cell, and a phase of asexual reproduction, when that egg-cell divides and develops into a new, sexually mature adult animal or plant. Cyclical parthenogenesis is special mainly because it involves differences between individuals, so that we can compare the effects of parthenogenesis and sexuality on the structure of populations and in so doing we can also examine more critically what we really mean by terms such as the *individual*, the *population*, and the *species*. I wish to structure the discussion around aphids, which are well-known insects that have perhaps gone furthest in their exploitation of the advantages of cyclical parthenogenesis.

Aphids have complex life-cycles, but there is no need to become too involved with this complexity. Let us start in winter. In temperate climates most aphids pass the winter in the egg stage. Their shiny black, cold-resistant eggs are produced by sexual reproduction, so each one has a unique combination of genes. The aphids that hatch from these eggs are all parthenogenetic females, and each one has the potential to produce a large number of offspring (usually 50–100). All the offspring will again be female, all parthenogenetic, and all with the same genetic constitution as the individual that hatched from the over-wintering egg.

Parthenogenetic female aphids do not lay eggs. Instead, they give birth rapidly to live young, often starting within a day of becoming adult. There is no need to wait around or adopt elaborate procedures to find a mate, and more importantly, there is no need for the egg-cells within parthenogenetic aphids to be fertilised before they start to divide. In fact the embryos of many species start to develop before their mothers are born, and even before their grandmothers are adult (Fig. 7.1). So, throughout the time an aphid is completing its own embryonic development, hatching, growing and moulting through its successive larval stages until it finally becomes adult, the next one, and eventually two, generations will also be developing inside it. The timing is such that only a day after becoming adult it will usually be giving birth to live young. Moreover the newly-born aphids are not only completely independent of their parent and immediately ready to start feeding and growing, but they are also already themselves confirmed as mothers-to-be. This precocious development of embryos in parthenogenetic female aphids must have evolved secondarily after a parthenogenetic phase had become established as part of the aphid life cycle, but now it overshadows the more immediate demographic advantage of dispensing with males.

The parthenogenetic generations of aphids are thus effectively telescoped into one another, and the generation time, that is, the time between one aphid giving birth and its offspring themselves giving birth, is very short indeed. As Figure 7.2 shows this is a remarkably effective way of achieving a high potential rate of population increase. In this diagram, keeping as many factors constant as possible, we compare a parthenogenetic aphid with a sexually reproducing insect. In each example we start with a single female egg, make the embryonic, larval and adult stages occupy the same periods of time, and assume that the fecundity of both females is 50 progeny. The numbers of the aphid reach astronomical proportions while the numbers of the sexually reproducing insect are still relatively low. The assumption that half the individuals produced by the sexually reproducing species are males is included in Figure 7.2, and also contributes to the contrast between the rates of increase in the two populations.

The natural enemies of aphids, mainly insects that prey upon or parasitise them, reproduce sexually and can hardly be expected to keep pace with parthenogenetic aphid

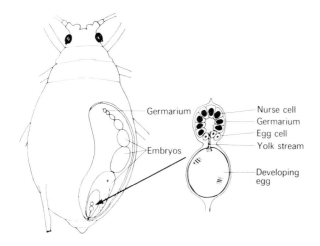

Figure 7.1 A diagram of an immature parthenogenetic female aphid just prior to becoming adult, showing one of its 12 ovarioles, with a string of developing embryos. The largest embryo has about one day to go before birth and has eggs already developing in its ovarioles; the structural details of one ovariole are shown in the larger-scale diagram.

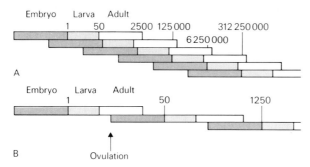

Figure 7.2 Diagram comparing the powers of multiplication of A – an aphid, and B – a sexually-reproducing organism, assuming in each case that both produce 50 progeny per female and that embryonic, larval and adult stages occupy the same periods of time. The figures show the potential number of descendants in successive generations starting from a single female individual.

populations. Such a remarkable potential for rapid increase in population size is a potential problem that has to be controlled by the aphid itself, otherwise aphids would rapidly overrun and destroy their food-plants. Parthenogenesis makes such control possible. A colony of aphids consists, at least initially, of a clone of genetically identical individuals. Individual survival does not matter, it is only the survival of the clone, the genotype, that is important. Selection will favour traits that make the survival of the clone more likely, even if such traits may be detrimental to individual aphids. It is usual in the early development of an aphid colony for each individual to stay close to its parent and sisters, so that a compact family crowd is formed, occupying only a small part of the host-plant. The members of the crowd continuously receive tactile stimuli from each other, and there may also be chemical communication between them by means of pheromones. Thus, as a result of this self-induced crowding, the individuals of a clone interact with one another and 'compete' for food long before there is any real danger of the food supply being exhausted. As the size of the colony increases, so does the degree of interaction, with the result that smaller adults are produced with reduced fecundity, and a gradual brake is applied to the multiplication rate.

In most aphid species the females that develop initially in a colony have no wings, their energies being entirely devoted to producing as many offspring in as short a time as possible. But interactions within an aphid colony have another important effect. As a colony becomes more crowded, an increasing number of individuals develop wings. Eventually all the offspring produced in the colony develop wings and fly away to found new colonies elsewhere. This migration to a fresh host-plant is an extremely hazardous undertaking for the individual aphid, but not for the clone as a whole. Because it produces so many winged migrants, all bearing the same genotype, by sheer numbers alone the clone is certain to have a good chance of survival.

Winged and wingless parthenogenetic females are one

example of how a single aphid genotype can manifest itself in a variety of guises. Presumably, systems of genes are switched on or off according to which particular set of environmental cues are provided, and development proceeds along one of several alternative pathways. In this way individuals that are suited to particular environmental circumstances are produced as a direct response to the situation, but without any change in the genetic constitution of the population. Such environmentally-cued polymorphism can operate far more easily in a parthenogenetic system, where selection is acting not on the individual, but on the clone as a whole. In a sense, the aphid clone *is* the individual. The different morphs such as winged and wingless females are different functional components of the clone in much the same way as the heart and the liver, for example, are functional components of the individual human body. All the constituent parts of the clone-body must work effectively together to ensure its survival.

From clones as individuals we now turn to populations. Aphid populations can consist of any number of clones. In spring, each aphid hatching from a fertilised egg has the potential to found a clone, so the spring generations of aphids should initially consist of numerous clones. As the season progresses, certain clones will prove fitter than others, either in direct competition or in their relative ability to cope with changes in the environment. More and more clones are likely to be eliminated, some through their lack of fitness and others solely by chance through some catastrophic event, so that in the autumn only relatively few genotypes contribute to the gene pool from which the following year's aphids emerge (Fig. 7.3).

This, of course, is exactly the outcome we would expect to find in most organisms that reproduce sexually. It is the basis of natural selection; many genotypes are produced, but only the fittest survive and reproduce successfully. Aphid populations differ only in that during the spring and summer each genotype is represented not just by one individual, but by all the members of a clone, which may amount to many thousands or millions of individuals.

We still know very little about the way in which selective forces operate in such clonally-structured populations. Genetic diversity must be at a maximum in the egg stage and at a minimum in the autumn just prior to sexual reproduction, but whether selection is any more or less intense than in conventional sexually-reproducing species is still an open question. An aphid clone has as many 'lives' as it has individual aphids, so it is unlikely to be eliminated by natural enemies such as parasites and predators, especially if it disperses itself widely over the habitat. On the other hand, we would expect competition between clones for the same food resources to be intense, and this could be the principal force selecting the fittest genotypes.

Males and sexual, egg-laying females are usually produced in autumn, again in response to environmental cues; a changing photoperiod is frequently used, perhaps because it is the most reliable indicator of season. Males

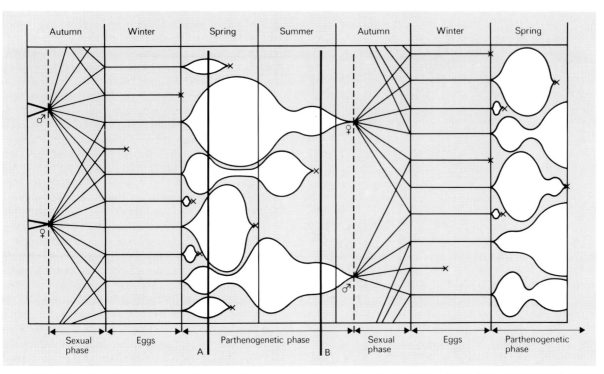

Figure 7.3 Alternation of sexual and parthenogenetic phases in the aphid life-cycle over two years. The width of each shaded area represents the number of individuals of one genotype. Sexual reproduction in the autumn generates a diversity of genotypes that spend the winter as eggs. In spring and summer many of these genotypes are eliminated (X); others build up large clones of genetically identical individuals. Some of these clones survive until autumn to contribute the sperm and eggs that will produce the following year's aphids. The genetic structure of a population at A is obviously very different from that of a population at B.

are produced by a cytological mechanism that has remained latent during the parthenogenetic phase. Since males have the XO type of sex determination, eggs destined to produce males must lose half their sex chromatin during the maturation process. The X chromosomes of such eggs pair and undergo a sort of meiosis all on their own, while the autosomes divide as in a normal mitosis. How this process is triggered by an environmental stimulus such as a change in photoperiod is still a mystery.

In many aphids sexual reproduction takes place on a particular species of tree or shrub, the parthenogenetic generations having been spent on a quite unrelated herbaceous plant. This means that there must be an autumn migration, and the arrival on one plant of aphids from diverse sources should lead to plenty of outbreeding. In the major subfamily Aphidinae, the sexes migrate independently to the overwintering host-plant, a particularly effective way of maximising recombination between diverse genotypes. This is somewhat comparable to many hermaphroditic invertebrates where male and female gametes from the same individual are shed at different times, or in monoecious angiosperms where pollen and ova mature at different times.

In the subfamily Pemphiginae, something happens that

emphasises the different roles of parthenogenetic and sexual reproduction in the aphid life-cycle. In this group of aphids sexual females each lay only a single egg. In fact, in one pemphigine tribe (the Fordini), the female does not even get as far as laying her one egg, but dies with it inside her, where it remains protected by the body of the dead mother aphid. The net result of sexual reproduction in this group is thus a halving of the number of individuals, which would be rather unsatisfactory, to say the least, if it were not for the preceding and following parthenogenetic generations.

The typical aphid life system seems to be a very effective way of exploiting fully the various advantages of parthenogenesis, while retaining a single, annual sexual generation. But why retain sex at all? Perhaps the best clue to the answer lies in the time of year at which sex occurs in the life-cycle, namely just prior to the winter or the harshest period. The new genotypes enter a resting stage, and when eventually they emerge it is to face a whole new year, a new season, a new succession of host-plants, natural enemies, and vagaries of the weather. Conditions will be very different from those of the previous autumn, and genotypes that were most successful then may not be so well suited to the new year, or to the next. Recombination ensures the genetic diversity needed to cope with the environmental pressures of a new season.

In fact, not just aphids but most animals and plants, as a general rule, accompany their recombination processes with either a resting stage, or a phase of dispersion, or both, with the result that the new genotypes appear at a time and place where they are most likely to encounter fresh conditions. This phenomenon is discussed by Bonner (1958). In many aphids, notably those that change host plants in the autumn, there is a dispersive phase immediately before sexual reproduction, as well as the resting stage afterwards.

The loss of the sexual phase

So far we have been concerned only with those aphids, the majority, that have an obligatory alternation of parthenogenetic and sexual generations. Genetic recombination happens only once each year as a regular event, preserving the integrity of the species' gene pool in exactly the same way as in other animals and plants with annual sexual reproduction. But many species, having demonstrated what can be achieved by parthenogenetic reproduction during spring and summer, have gone at least part of the way towards dispensing with the sexual portion of their life-cycles. In what circumstances is the sexual phase lost? What happens to a species that has gone along the road of permanent parthenogenesis?

Before looking further into these questions, there are one or two points about aphid biology that need to be remembered as they tend to complicate the issue. First, loss of the sexual phase in an aphid is not the same as the acquisition *de novo* of parthenogenesis by most other organisms, because aphids already have parthenogenesis as an evolved adaptation. No new cytological or reproductive mechanism is required, all that is needed is for parthenogenetic females to continue giving birth to parthenogenetic females at a time when they would normally start to produce sexual morphs. Judging from the number of different aphid species in which it has occurred, that is a fairly easy adjustment to make. However, this leads to the second consideration: an aphid that loses its sexual phase also loses its means of getting through the winter in a resting, cold-resistant egg stage. Aphids predominate as a group in the northern temperate parts of the world, where most plant-feeding insects pass the winter in a resting condition. Survival through the winter without an egg stage would be an impossible proposition, for example, in the many aphids that feed on the leaves of deciduous trees. Even if not deprived of their food source, over-wintering parthenogenetic females have to overcome the problems of feeding and survival at low temperatures. It does not seem surprising, therefore, that aphids without a sexual phase tend to occur in places that have mild winters, or no winters at all. The principal aphid pests of tropical and subtropical crops all go on from year to year without sexual reproduction, as do several aphid species that have specialised in feeding on plants grown in glasshouses.

It is tempting to look at this problem in another way and conclude that if it were not for the need to produce eggs as a means of surviving cold winters or other adverse periods, aphids might dispense with sex altogether. But such a conclusion cannot be justified. In the first place, there is no obvious reason why aphids need a sexual phase in order to produce cold-resistant eggs. It might be a bigger evolutionary step to produce such eggs without mating, but if the advantage were there, it would probably have happened. Furthermore, there are plenty of aphid species, notably those of Greenideinae and Hormaphidinae occurring in south-east Asia and the Fordini of the Mediterranean region, that live in climates where there is no need for a cold-resistant egg stage, yet sexual morphs and eggs are still an integral part of their life-cycle. In fact the Fordini have a two-year cycle, spending alternate winters in the egg stage and as young parthenogenetic females. This effectively nullifies the argument that sex is retained merely as a necessary adjunct to winter survival.

Can we then, find an explanation for the pattern of sexual and parthenogenetic reproduction in aphids throughout the world that applies more generally? Let us look again at the sort of situations where aphids have abandoned the sexual part of their life-cycle. We mentioned aphids that are pests on tropical and subtropical crops, and in glasshouses, but in fact most pest aphids show some tendency to abandon their sexual phase, even in temperate climates. A pest species, newly introduced to another continent, is particularly liable to dispense with sexual reproduction, at least until it is well-established in its new environment. Crop environments are, from the aphid's viewpoint, all disturbed habitats, very different from the original environment in which each species evolved. Above all, they have far fewer interacting biological components, and it is this difference that may be the key to the problem, the reason why sex is no longer such an essential part of the life system.

Evolutionary biologists, G. C. Williams in particular, have tried to demonstrate how changes in the physical environment from one generation to the next can be significant enough to necessitate the continuous production of genetic diversity, but they have failed to find a satisfactory, generally applicable model. Others, however, have looked upon biotic rather than abiotic factors in the environment as the chief contributors to environmental uncertainty (Levin, 1975, Glesener & Tilman, 1978). According to this line of thinking, life for most organisms is unpredictable mainly because it is dependent on what their 'opponents' do next. In a natural situation, host plant, natural enemies and competitors are all generating their own variability, producing new permutations which test their opposition, so that a continuous cycle of moves and countermoves is played out like a multi-dimensional game of chess. Only species with sexual reproduction can produce the continuous genetic diversity required to participate successfully in this evolutionary 'chess game'.

So, an aphid living in a crop environment has escaped the complexities of the 'chess game' taking place in its original habitat. Most natural enemies and competitors are left behind and even the variability of the host-plant is suppressed. The aphid is thus presented with a much more constant, predictable habitat, and the need for continual adjustments to the genotype vanishes. This is especially the case in glasshouses, and also on tropical crops, where the aphid is far removed from its original, temperate environment. But it would be wrong to conclude that when an aphid abandons sexual reproduction it also throws away the ability to respond to any changes in its environment. In fact, there are certain types of change to which a parthenogenetic aphid can adapt more readily than a sexual one.

The case of the Spotted Alfalfa Aphid

We can take as an example the story of one aphid's invasion of North America. The Spotted Alfalfa Aphid, *Therioaphis trifolii* forma *maculata*, first appeared on alfalfa (lucerne) in North America in 1953–4, probably starting in New Mexico, and by 1956 it had spread through most of the southern United States. It was quite evidently a species introduced from the Old World, but there were several peculiarities in the New World form. First, *T. trifolii* which, in the Old World, ranges through Europe, south-west Asia and North Africa, has a very variable morphology, whereas the form introduced into the United States had a much narrower range of variation, bearing closest resemblance to populations in the Mediterranean area. Second, in Europe and Asia this species colonises a range of leguminous plants in several genera, whereas the North American introduction restricted its feeding almost entirely to alfalfa. Third, for several years after its introduction, the North American form apparently reproduced solely by parthenogenesis. Sexual females were occasionally produced but males were very rare and no eggs were ever observed to hatch.

Thus, all the evidence suggested strongly that here was a single clone, perhaps resulting from the introduction of just one female with a genetic constitution quite untypical of the species from which it originated.

Lack of genetic recombination, however, did not prevent the Spotted Alfalfa Aphid from overcoming some of the more feeble attempts by man to oppose its progress. Two years after the initial appearance of the Spotted Alfalfa Aphid in North America, some populations in California had developed resistance to organophosphorus insecticides, and this resistance soon spread to many areas. In 1958, a variety of alfalfa that had proved resistant to infestation through three years of exposure to the Spotted Alfalfa Aphid suddenly started to show symptoms of attack, indicating that the aphid had developed a way of breaking down its resistance.

Until 1960, the Spotted Alfalfa Aphid could only overwinter in the southern United States, because the parthenogenetic females could not survive the cold winters of the north. Alfalfa fields in some more northerly states such as Nebraska were invaded annually each summer by aphids migrating from the south. Then, in the autumn of 1960, a population in the centre of Nebraska suddenly started to produce viable sexual morphs, and passed the winter successfully in the egg stage. By 1964 this new variant had spread from its point of origin to cover the whole of Nebraska, and had extended the range of the species northward into parts of South Dakota not reached by the previous annual migration from the south. By this time also a new pocket of sexual reproduction had developed in another northern state, Wisconsin. Today the Spotted Alfalfa Aphid is found throughout the United States, and sexual reproduction and overwintering as eggs occur over most of the northern part of its range.

These are only some of the many genetic changes that have occurred in the Spotted Alfalfa Aphid since its introduction, perhaps as just one female, into New Mexico only a little more than a quarter of a century ago. New variants capable of colonising previously resistant varieties of alfalfa appear regularly, and are a particular problem in the south-western states even though sexual reproduction still does not occur there. Mutations are commonly regarded as rare events. They usually occur at rates of between 10^{-5} and 10^{-9} per gene per generation. To assess the likelihood that any given mutation will occur in a population, we have to take the size of that population into account. Dickson (1962) estimated that in the two years from the first appearance of Spotted Alfalfa Aphid in California to its development of resistance to organophosphorus insecticides there had been $1·7 \times 10^{11}$ aphids in one valley alone, so that even a rare mutation would be likely to have occurred at least 170 000 times. Mutations, provided that they occur at all, become matters of certainty rather than chance if the population size is large enough. Moreover, in a parthenogenetic population, a single individual with an advantageous mutation passes on the advantage to all its descendants, and we have seen that the multiplicative powers of aphids are such that in next to no time the population could be taken over by the new mutant.

We are, of course, discussing a highly artificial situation; a species colonising virgin territory and responding to coarse selection pressures such as insecticidal treatments imposed by man. In nature, species only rarely encounter such circumstances. Normally the genetic changes that are required of them are far more subtle, minor adjustments to much more complex situations in which any mutation with a significant effect would almost always be deleterious. But by its very extremeness, the case of the Spotted Alfalfa Aphid serves to emphasise the advantages and limitations of life with and without sexual reproduction.

There is another facet to this story. *T. trifolii* was in fact first described in the eastern United States in 1882, over

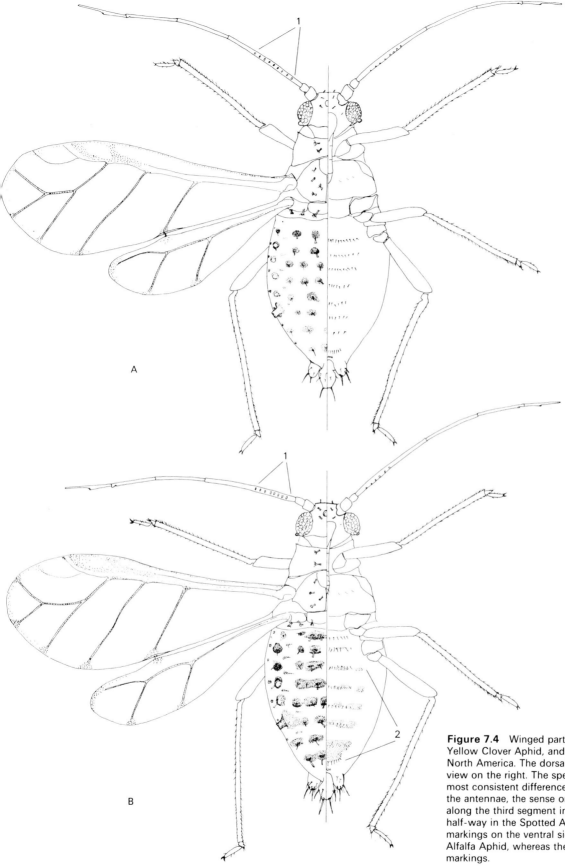

Figure 7.4 Winged parthenogenetic females of A – the
Yellow Clover Aphid, and B – the Spotted Alfalfa Aphid from
North America. The dorsal view is on the left and the ventral
view on the right. The specimens illustrated are typical, but the
most consistent differences between the two forms are: 1 – on
the antennae, the sense organs extend more than half-way
along the third segment in the Yellow Clover Aphid, less than
half-way in the Spotted Alfalfa Aphid; 2 – there are dark
markings on the ventral side of the abdomen in the Spotted
Alfalfa Aphid, whereas the Yellow Clover aphid has no ventral
markings.

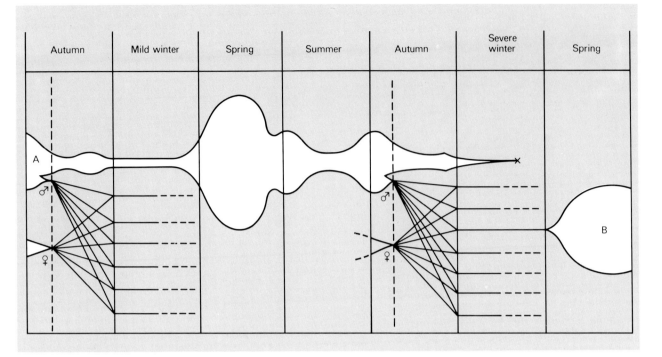

Figure 7.5 Maintenance of life-cycle variability in the Peach-potato Aphid, *Myzus persicae*. Clone A continues to reproduce parthenogenetically in the autumn and can survive in this way through a mild winter, but not through a severe one. However, it produces some males in the autumn that mate with sexual genotypes, and some of the progeny (at B, for example) inherit the ability to carry on through the winter parthenogenetically.

70 years before the Spotted Alfalfa Aphid first appeared in North America. It became known as the Yellow Clover Aphid, as it lives primarily on species of *Trifolium*. When the Spotted Alfalfa Aphid appeared in the west it was at first thought that the Yellow Clover Aphid had simply extended its range. Then it was realised that the two aphids could be readily distinguished not only by their different host plant ranges, but also by various morphological differences, some of which are illustrated in Figure 7.4. These differences still remain quite constant although the Spotted Alfalfa Aphid has now spread into many areas where the Yellow Clover Aphid has been established for years. Laboratory tests have shown that hybridisation might occur were it not for the strict host plant preferences which effectively isolate the two forms in the field.

In the Old World, *T. trifolii* feeds on many legumes including both alfalfa and clover, and its morphology varies considerably throughout the distributional range of the species, encompassing forms with the characters of both the Spotted Alfalfa Aphid and the Yellow Clover Aphid of North America, but with all manner of intermediates. It is quite clear that the two North American pests do not exist as separate entities in the Old World. They seem to represent separate introductions, at least 70 years apart, of two rather extreme genotypes drawn from the gene pool of Old World *T. trifolii*.

There remains the problem of how this situation should be treated taxonomically. The North American aphid taxonomist is naturally inclined to the viewpoint that the forms are two distinct species, whereas to the European worker they are merely two variants of the Old World *T. trifolii*. Clearly we have a case of incipient speciation, and given time the North American aphids are likely to diverge further from each other, and from the parent species. We can at least understand what has happened, which is perhaps all that really matters.

In 1977, the Spotted Alfalfa Aphid was accidentally introduced from North America into Australia, and rapidly spread through alfalfa fields in Victoria and Queensland. It will be interesting to see whether history repeats itself on a new continent. One thing, at least, is different: there is no Yellow Clover Aphid on the clover in Australia. Given time, will the Spotted Alfalfa Aphid be able to move over onto clover?

The case of the Peach-potato Aphid

The Spotted Alfalfa Aphid has shown that suppression of the sexual phase in an aphid is not necessarily an irreversible process, as long as it does not go too far. The form originally introduced into North America was sexually sterile, yet a variant appeared, presumably by back-mutation, with the ability to undergo successful sexual reproduction. Now these two coexist in North America, the original asexual form in the south where it can reproduce all the year around by parthenogenesis, the derived, sexual form replacing it further north where the winters can only be survived in the cold-resistant egg stage. This arrangement may meet the needs of the present situation, but in another species, the Peach-potato Aphid *Myzus persicae*, there is evidence of a more subtle system for balancing the complementary benefits of parthenogenesis and sexual reproduction (Blackman, 1974).

M. persicae is a world-wide pest of many different agricultural crops. Its origins can probably be traced to south-east Asia, since its sexual phase is associated with the peach tree, *Prunus persica*, which is a native of China. Where winters are severe, as in central Europe, this aphid produces sexual morphs in the autumn, and the sexual females lay their eggs on peach trees. In tropical climates it reproduces continuously by parthenogenesis. Between the two extremes, in regions where the winters are usually mild, *M. persicae* may have a sexual phase but can also pass the winter parthenogenetically. However, in many areas with such mild temperate climates, including the British Isles, the occasional severe winter may wipe out the overwintering parthenogenetic females.

Let us suppose that the Peach-potato Aphid, like the Spotted Alfalfa Aphid in North America, had just two forms, one with continuous parthenogenesis and the other with an annual sexual phase. The relative success of these two forms would depend largely on winter conditions. Mild winters would favour the continuously parthenogenetic form, which could build up its numbers more rapidly the following spring, but a severe winter might eliminate overwintering parthenogenetic females completely. There would be no way, short of recolonisation of the area by the continuously parthenogenetic form after each severe winter, of keeping the options open from year to year.

In *M. persicae*, however, there is a crucial difference. In autumn, those genotypes that are otherwise continuously parthenogenetic produce a few males that can mate with sexual females from other clones. Laboratory breeding experiments have shown that these males are able to transmit the odd but very useful trait – 'continuous parthenogenesis with production of a few males in autumn' – through the sexual phase. Because of this, the options are kept open; even after a severe winter some of the next season's aphids still have the potential to reproduce parthenogenetically through the following winter (Fig. 7.5). This is the immediate benefit, but there is also a significant long-term consideration; those few males produced by otherwise parthenogenetic genotypes retain an all-important link with the gene pool maintained by sexual reproduction, thus preserving the integrity of *M. persicae* as a species. For the Peach-potato Aphid, continuous parthenogenesis is not so much a step into the dark; it is just another way of exploiting the environment.

Permanent parthenogenesis

Now it is time to consider what happens when the link with the sexual phase is irretrievably lost.

In glasshouses in various parts of the world there is an aphid that looks very similar to *M. persicae*. However, this aphid occurs only on carnations (*Dianthus*), and its saliva is toxic causing spotting of the leaves, whereas when *M. persicae* feeds on carnations the leaves do not become spotted. This carnation aphid has been given the specific name *Myzus dianthicola*, and it is permanently parthenogenetic.

Examination of its chromosomes shows that it is most unlikely ever to regain the ability to reproduce sexually. Whereas *M. persicae* has a diploid set of 12 chromosomes (six pairs), *M. dianthicola* has 14 chromosomes, seven of them long and seven short, which cannot be matched into homologous pairs. The chromosome set of *M. dianthicola* is never required to undergo the pairing process of meiosis, and consequently it is no longer diploid in character. Since parthenogenesis became permanent there must have been a series of chromosome breaks and unequal exchanges of parts between chromosomes to give the present arrangement. We do not know the time scale of these changes, or whether *M. dianthicola* was permanently parthenogenetic on carnations before they were first cultivated, probably over 2000 years ago.

There are other permanently parthenogenetic aphids that seem to be quite distinct species, yet whose origins are even more of a mystery. The shallot aphid, *Myzus ascalonicus*, suddenly appeared for the first time in 1941 on some bulbs in storage in Lincolnshire, eastern England, and has since spread rapidly as a pest to all parts of the world. One can be reasonably certain that *M. ascalonicus*, a rather distinctive species feeding on a wide variety of plants, had not been previously overlooked, and in fact did not exist anywhere else in the world much before that date. No sexual stages are known. Many cases of permanent parthenogenesis in other organisms are thought to be due to hybridisations between bisexual species, and one can only surmise that *M. ascalonicus* may have arisen by the chance cross-breeding of two other species, whose identity remains unknown. The chromosomes of *M. ascalonicus* hold no clues as to its origin, and look like a normal diploid set, but then if its existence as a parthenogenetic species dates only from about 1941 we would not expect many chromosomal rearrangements to have become incorporated in that time!

The Fordini were mentioned earlier in this chapter because of their two-year cycle in the Mediterranean region, where they spend every other winter as eggs on *Pistacia* species, living parthenogenetically at other times on the roots of other plants, especially grasses. Many Fordini have abandoned their sexual phase and spread to parts of the world where there are no native *Pistacia*. Some are pests on the roots of crop plants, many live permanently on the roots of grasses. In areas such as north-western Europe, heavily forested until recently, grasses are sub-climax vegetation and may be regarded as a disturbed habitat. In some genera, *Forda* and *Geoica* for example, permanently parthenogenetic populations have undergone chromosomal changes which indicate that the link with the sexual part of the life-cycle has been broken, never to be regained. If *Pistacia* were to become extinct, *Forda* and its relatives would then exist only as parthenogenetic forms on the roots of grasses.

There is another group of aphids in a different subfamily from the Fordini, which is already in this situation. The Tramini are only known from the roots of herbaceous plants, mainly Compositae. Although there are 32 described species, which seem to group naturally into three distinct genera, sexual morphs have only been recorded twice from the group as a whole, and not at all from the principal genus *Trama*. Again, as evidence of a rather long history of parthenogenesis, the chromosomes of *Trama* have lost all semblance to a normal, diploid set. Moreover they also contain a large quantity of heterochromatic, genetically-inactive material.

How do we explain the present state of the Tramini, where in spite of the virtual absence of sexual reproduction throughout the group there are still clearly recognisable genera and species? We can conjecture that most of the speciation events, which gave rise to the Tramini as they exist today, occurred while there was still a sexual phase in some part of the world, perhaps on some group of woody host plants that had a restricted distribution for some time before finally dwindling to extinction. By becoming permanently parthenogenetic on the roots of Compositae many species were able to spread away from the original habitat, much as the Fordini do on grass roots today. There is, however, a problem with this idea: if the ancestors of modern Tramini migrated regularly between a woody and a herbaceous host plant, it is rather odd that none of the existing species in the sub-family to which the Tramini belong (the Lachninae) has such an alternation of host plants. The groups most closely related to the Tramini live on a variety of woody hosts, and none has any association at all with herbaceous plants, so they provide no clues as to how and when the Tramini arrived on the roots of Compositae.

Perhaps it is significant that the parthenogenetic populations of Fordini and Tramini have something else in common. They have developed intimate associations with ants. Ants look after aphids, protect them from natural enemies, and generally provide them with a more amenable environment in which to live. Could it be that ants can provide some aphids with a sufficiently predictable habitat so that it is no longer necessary for them to generate continuous genetic diversity by sexual reproduction?

The Spotted Alfalfa Aphid story showed how a parthenogenetic aphid can adapt by mutation and selection to certain changes, given a fairly simple environment. But the environment cannot be expected to stay simple forever. Certainly, man-made environments are so recent and transient as hardly to count in evolutionary terms. Permanently parthenogenetic aphids may also encounter another basic genetic problem – the acquisition in time of an unacceptable burden of mutations.

Organisms have of necessity evolved genetic systems that are finely tuned to their needs. So most mutational changes, if they affect the function of genes at all, will be deleterious to some extent. One of the advantages of the diploid system is that with two chromosome sets each gene is present as at least two alleles. A deleterious mutation, even one that produces a completely inactive allele, may have very little effect on the phenotype because a fully-functional form of the gene is still present and can normally be made to compensate to a large degree for the mutant allele; that is, the mutation is generally recessive to the pre-existing allele. Sexual reproduction provides an efficient method of culling out genotypes with deleterious mutations, even when they are completely recessive, because it tests them in homozygous condition; offspring that inherit a deleterious or non-functional allele of the same gene from each of their parents are likely to be eliminated. Also, parental genotypes, each of which have a different deleterious mutation, can give rise to offspring free from either mutant allele.

In a parthenogenetic line, however, deleterious or non-functional alleles will tend to accumulate, by what H. J. Muller has described as a kind of ratchet mechanism (Muller, 1964). A permanently parthenogenetic aphid can always acquire additional mutations; each aphid will pass on to its offspring all the mutations it inherited from its parent, plus any new ones it has acquired itself. Individuals with heavy loads will be selected against, but there is no way in which the overall load can be lightened, because even the least loaded lines must gradually increase their load. Eventually the accumulation of mutant alleles, especially non-functional ones, may severely limit the possibilities of further genetic change. This may be the real reason why parthenogenetic organisms cannot stand the test of time. To make a simple analogy, it is a bit like starting out on a journey of indefinite length in a vehicle that has every control system duplicated; if one part fails we can use a back-up system and still keep going. But, without being able to pick up any spares along the route, one of the back-up systems will eventually fail and we will come to a halt even while some parts of the vehicle are still fully functional.

In conclusion, aphids demonstrate in a unique way the advantages that can accrue from parthenogenesis. At the same time they show that loss of the sexual phase is only possible under certain circumstances, generally those that seem to involve the dispersal of the species from its original habitat, where it has evolved in complex association with other organisms, into newer and simpler environments. Even in such situations some link is often retained with the sexual phase; a link that may have adaptive significance, as in the case of the Peach-potato Aphid, but which will also have the effect of preserving the integrity of the species. When sexual reproduction is abandoned, a species starts down a road of no return, and although it may be able to respond adequately to environmental changes in the short term, its long-term evolutionary prospects are bleak.

References

Blackman, R. L. 1974. Life-cycle variation of *Myzus persicae* (Sulz.) (Hom., Aphididae) in different parts of the world, in relation to genotype and environment. *Bulletin of Entomological Research* **63**: 595–607.

Bonner, J. T. 1958. The relation of spore formation to recombination. *American Naturalist* **92**: 193–200.

Dickson, R. C. 1962. Development of the spotted alfalfa aphid population in North America. *Internationaler Kongress für Entomologie (Wien 1960)* **2**: 26–28.

Ghiselin, M. T. 1974. *The economy of nature and the evolution of sex.* 346 pp. Berkeley: California University Press.

Glesener, R. R. & Tilman, D. 1978. Sexuality and the components of environmental uncertainty: clues from geographic parthenogenesis in terrestrial animals. *American Naturalist* **112**: 659–671.

Levin, D. A. 1975. Pest pressure and recombination systems in plants. *American Naturalist* **109**: 437–451.

Maynard-Smith, J. 1978. *The evolution of sex.* 222 pp. Cambridge: Cambridge University Press.

Muller, H. J. 1964. The relation of recombination to mutational advance. *Mutation Research* **1**: 2–9.

Williams, G. C. 1975. *Sex and evolution.* 201 pp. Princeton, New Jersey: The University Press.

CHAPTER 8

Polyploidy and its evolutionary significance

M. Gibby

A polyploid is an organism (or cell, or tissue) possessing more than two complete sets of chromosomes. Normally there are two sets in each cell; cells in this state are called diploid. A polyploid which has three chromosome sets is called a triploid, one with four sets is a tetraploid, one with five a pentaploid and so on. Polyploidy can be restricted to certain cells or tissues of a diploid organism, and this condition is referred to as endo-polyploidy. This chapter is confined to a discussion of polyploidy of the whole organism, and the role that polyploidy has played in evolution. A polyploid organism arises from one or more diploid progenitors by some failure during the process of somatic cell division (mitosis) or during the cell division that leads to the formation of gametes (meiosis). The newly arisen polyploid is often genetically isolated from its diploid parents, since backcrossing can result in the formation of sterile hybrids. Polyploidy can, therefore, be regarded as a form of speciation.

Polyploidy is very common in the plant kingdom, particularly where it is combined with hybridisation; in certain groups, like the grasses and ferns, at least half the species are thought to be polyploids. In the animal kingdom polyploidy is less common, but it can be found, for example, in some oligochaete worms, amphibians and teleost fishes. Where polyploidy does occur in animals it is often found to be associated with parthenogenetic (asexual) reproduction (for example, in lizards, blackflies (Simuliidae) and oligochaete worms). Stebbins (1950; 1971) has made an extensive study of polyploidy in plants, and the subject has been reviewed more recently by White (1973; 1978), with particular reference to animals, and by Jackson (1976). Together these authors provide an extensive list of publications on the subject.

Historical review

The term polyploidy was first coined by Winkler (1916), who observed the phenomenon in Tomato (*Lycopersicon esculentum*) and Black Nightshade (*Solanum nigrum*).

Callus tissue that forms at the cut surface of a stem can produce a new shoot and Winkler found that some shoots formed in this way had twice as many chromosomes in the cells as the original plant, and he called these polyploids. The diploid tomato had 24 chromosomes ($2n = 24$, where n equals the gametic number of chromosomes) whereas the new polyploid shoot had 48 chromosomes. Winkler observed that the polyploids could be distinguished from their progenitors by their increased size. The mutant *gigas* of *Oenothera lamarkiana*, the Evening Primrose, described by de Vries, showed a similar effect in having larger petals, anthers and seeds; Gates (1909) demonstrated that *O. gigas* had a chromosome number of $2n = 28$, whereas *O. lamarkiana* had $2n = 14$.

Tahara (1915) published chromosome counts for *Chrysanthemum* species of $2n = 18, 36, 54, 72$ and 90. Winge (1917) noted that these numbers, and those from other genera like *Chenopodium* ($2n = 18, 36$), formed a simple arithmetic series. He suggested that a tetraploid could develop by hybridisation between two diploid species forming a sterile hybrid, followed by chromosome doubling which would restore fertility (Fig. 8.1). He believed that sterility of the diploid hybrid would result from failure of the two sets of chromosomes to form the normal bivalents or pairs of chromosomes which are seen during meiosis in a diploid species. Winge suggested that chromosome doubling in the hybrid to give a tetraploid would provide each chromosome with a homologue, a similar partner with which it could form a bivalent, and the tetraploid would therefore be fertile. Similarly a hexaploid, with six sets of chromosomes, could arise by hybridisation between a diploid and a tetraploid species, with subsequent chromosome doubling.

Winge's hypothesis for the formation of polyploids was subsequently demonstrated experimentally in *Nicotiana* and in *Raphanobrassica*. Karpechenko (1927; 1928) produced hybrids between two diploid species, *Raphanus sativus* – the Radish ($2n = 18$), and *Brassica oleracea* – the Cabbage ($2n = 18$). Spontaneous chromosome doubling

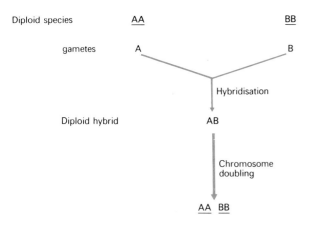

Diploid species   AA   BB

gametes   A   B

Hybridisation

Diploid hybrid   AB

Chromosome
doubling

AA BB

Figure 8.1 The formation of a polyploid by hybridisation and chromosome doubling. A and B represent non-homologous chromosome sets or genomes. Underlining represents chromosome pairing at meiosis between homologous genomes.

in the sterile diploid hybrids resulted in the formation of the tetraploid *Raphanobrassica* ($2n = 36$), which was fertile. Work with *Nicotiana* by Clausen & Goodspeed (1925) gave similar results, but in this case a diploid, *N. glutinosa* ($2n = 24$) was crossed with tetraploid tobacco, *N. tabacum* ($2n = 48$), to give a fertile hexaploid ($2n = 72$).

This early work on polyploidy soon distinguished two types of polyploid: an autopolyploid which is derived from a single species; chromosome doubling in a diploid AA (where A represents one chromosome set) results in the formation of an autotetraploid AAAA. The four chromosome sets in this tetraploid are identical, and during meiosis quadrivalents – associations of four chromosomes – may be present. An allopolyploid is derived by hybridisation between two different species followed by chromosome doubling (Fig. 8.1); the chromosomes from one diploid parent, AA, are unable to pair with those from the other diploid parent, BB, and only bivalents form in meiosis in the allopolyploid. Winkler's tetraploid tomatoes were autopolyploids, whereas *Raphanobrassica* is a good example of an allotetraploid.

Origins of polyploidy

Chromosome doubling results from some breakdown in the process of cell division, either during mitosis or meiosis. For example, the tetraploid *Primula kewensis* ($2n = 36$) developed by somatic chromosome duplication in the diploid hybrid *P. floribunda × verticillata* ($2n = 18$). A plant of the normally sterile hybrid produced a branch bearing fertile flowers which set good seed. Newton & Pellew (1929) were able to demonstrate that the vegetative tissue of the fertile stem was tetraploid; a breakdown of mitosis must have occurred early in the development of the stem, resulting in chromosome doubling. The seed from the fertile flowers was sown, and developed into

fertile tetraploid plants of *P. kewensis*. Winkler's polyploid tomato and Black Nightshade developed in a similar fashion, by somatic chromosome duplication in callus tissue. In somatic chromosome doubling mitosis probably begins normally but the chromatids fail to separate during anaphase, and all become incorporated in one nucleus.

Breakdown during meiosis, leading to chromosome doubling in the formation of unreduced gametes, is found at low frequencies in many organisms, including Man. Four per cent of spontaneous human abortions have been found to be triploid, being the product of fusion between an unreduced gamete and a normal haploid gamete. Fusion of two unreduced gametes would give rise to a tetraploid. *Raphanobrassica* resulted from the fusion of two unreduced gametes. Karpechenko examined meiosis in the sterile diploid hybrid, *Raphanus sativus × Brassica oleracea*. This showed 18 univalents at first metaphase. The chromosomes segregated unevenly during anaphase 1, giving gametes with different chromosome numbers, or with unbalanced complements, that were inviable. In some cells the univalents failed to separate at anaphase 1, and formed one large restitution nucleus. At the second division of meiosis the chromatids separated to form two nuclei and these developed into gametes with 18 chromosomes. Fusion of two such unreduced gametes resulted in the synthesis of the tetraploid *Raphanobrassica*, with 36 chromosomes. The normal process of meiosis is illustrated in Figure 8.2, together with meiosis in a hybrid, illustrating the mechanisms of uneven segregation and restitution.

In the Potato, *Solanum tuberosum*, Ramanna (1974) has shown that unreduced gametes are produced as a result of fusion of the two spindles at second metaphase of meiosis, producing two nuclei at telophase 2 instead of four, each with the doubled chromosome number. Hybridisation between two species of sugarcane, *Saccharum officinarum* ($2n = 80$), and *S. spontaneum* ($2n = 64$) usually results in the formation of hybrids with 112 chromosomes, instead of the expected $2n = 72$ (Roach, 1968). These polyploid hybrids with $2n = 112$ have arisen by fusion of an unreduced female gamete of *S. officinarum* (with 80 chromosomes) with a normal haploid male gamete of *S. spontaneum* (with 32 chromosomes). This is an example of an allopolyploid being produced without the intermediate formation of a sterile hybrid; Harlan & deWet (1975) have suggested that this is a common phenomenon.

Triploids can give rise to gametes with a range of chromosome numbers. These may be diploid, triploid or higher ploidy levels, or aneuploids (with addition or loss of individual chromosomes). Fusions involving combinations of such gametes, or backcrossing to progenitors of the triploid, can give rise to further polyploid offspring. This can be illustrated by the Banana, where triploids give tetraploid offspring when pollinated with normal pollen from diploids, as the triploids produce triploid female gametes (Simmonds, 1966). Heptaploids and various aneuploids can also develop from this cross, when the triploid has produced hexaploid or aneuploid gametes.

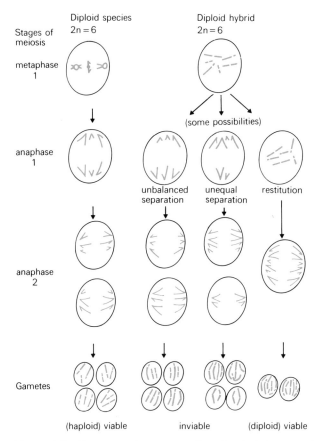

Stages of meiosis

Diploid species
2n = 6

Diploid hybrid
2n = 6

metaphase 1

anaphase 1

(some possibilities)

unbalanced separation

unequal separation

restitution

anaphase 2

Gametes

(haploid) viable   inviable   (diploid) viable

Figure 8.2 Diagrammatic representation of meiosis in a diploid species and in a diploid hybrid showing the origin of unbalanced and unreduced gametes.

The recognition of polyploids

Winge postulated that since the chromosome numbers of a polyploid series are related in an arithmetic fashion, for example *Chrysanthemum* ($2n = 18, 36, 54...$), the lowest number found, 18, representing the diploid number, is two sets of nine chromosomes. Nine is said to be the *basic number* of this polyploid series, and is represented as $x = 9$. The diploid is $2x$, with 18 chromosomes ($2n = 18$) and the tetraploid $4x$, with $2n = 36$. The basic number is not always so clearly recognisable as that in *Chrysanthemum*. For example, *Brassica napus*, the Swede has $2n = 38$. From a simple inspection of its chromosome number it could be inferred that it is a diploid species with $x = 19$. In fact, it is a tetraploid, derived from two diploids with different basic numbers, *B. campestris* – the Turnip ($2n = 2x = 20$), and *B. oleracea* – the Cabbage ($2n = 2x = 18$); the Swede is referred to as a dibasic tetraploid (Stebbins, 1971).

In some genera chromosome numbers can be very high. The fern *Ophioglossum costatum* has a sporophytic ($2n$) number of 240, but a chromosome number of approximately 1320 has been observed in an American population of *O.*

vulgare (Löve, Löve & Pichi Sermolli, 1977). *O. costatum* has the lowest known number in the genus, and if it were diploid, the base number would be 120; *O. vulgare* would then be 11-ploid. Several workers have suggested that $x = 120$ is too high to be a basic number, and that *O. costatum* itself must be an ancient polyploid. Löve *et al.* (1977) think the base number for the genus is $x = 15$, the lowest number that cannot be halved to give a non-fractional number. This is hardly a scientific approach! They would regard *O. vulgare* as 88-ploid.

Tatuno & Kawakami (1969) made detailed studies of the chromosome morphology of three species of the fern genus *Asplenium*, all with $2n = 72$, this being the lowest chromosome number that is found in the genus. They found that the chromosome complements could be divided into six groups; they suggested that the primary base number for the genus is $x = 12$, and that these species with $2n = 72$ are hexaploids.

In attempting to determine the degree of polyploidy in angiosperms, Stebbins (1971) defined as polyploid any species with a chromosome number of $2n = 24$ or higher. Similarly, Grant (1971) classified as polyploid species with $2n = 28$ or greater, and he estimates that 47 per cent of angiosperms are polyploids. Clearly, this cannot give an accurate picture since some polyploids are known that have fewer than 24 chromosomes, and there is no firm evidence to support the view that species with numbers greater than $2n = 24$ are in fact ancient polyploids. However, it may be useful to *suspect* chromosome complements greater than $2n = 24$ as polyploid.

Matthey (1945) showed that the number of major chromosome arms can be of greater evolutionary significance than chromosome number, since chromosome number can be altered by Robertsonian changes; these involve the fusion of two one-armed telocentric or acrocentric chromosomes (in which the centromere is situated at or near one end of the chromosome) to give a two-armed or metacentric chromosome (in which the centromere is located centrally), or the dissociation of a metacentric to give two telocentrics. Robertsonian rearrangement and polyploidy are both involved in speciation within the genus *Cymbispatha* (Jones, 1978). *C. geniculata* and some populations of *C. commelinoides* have a somatic number of $2n = 14$. In the former the chromosomes are all acrocentrics, but the latter show 14 metacentrics. *C. geniculata* is bivalent forming at meiosis, but *C. commelinoides* shows up to three quadrivalents and it appears to have had a polyploid origin. It has probably evolved from a diploid with 14 acrocentrics by chromosome doubling and progressive chromosome fusions.

Comparison of DNA amounts per cell reveals that, as expected, many tetraploid species show a doubling of the amount of DNA compared with their related diploids (Van't Hof, 1974). However, such differences in DNA amount do not always reflect polyploidy, since similar differences can be found between closely related diploids. For example, Rees & Hazarika (1969) found a three-fold difference in DNA amount when comparing diploid species of *Lathyrus*. In the 18 species examined the chromosome number was consistent ($2n = 14$) but there was considerable variation in chromosome size between different species. Increase (or decrease) in the amount of DNA may not be uniform throughout the chromosome complement, but may be restricted to certain fractions of the DNA.

Effects of polyploidy

In trying to assess the evolutionary significance of polyploidy it is clearly necessary to try to understand the effects on the organism of this increase in chromosome complement. Chromosome duplication means that the polyploid will have more DNA per cell than its diploid progenitors. This in itself may have some effect on the organism. Cell size may be increased, and this in turn could change growth patterns and may result in altered size, shape and functioning of various tissues and organs. The increased number of chromosomes may affect cell division, in particular the first division of meiosis, where chromosome associations greater than bivalents may form. This can have consequences for the fertility of the organism. Duplication of chromosomes results in large-scale gene duplication that can have wide-ranging effects on the inheritance of genetic characters. Polyploidy may also affect sex determination mechanisms and the breeding system of some plants.

Increase in the amount of DNA per cell

It has been shown that there is a relationship between DNA content, nuclear volume and the mitotic cycle time in diploid species (see review by Van't Hof, 1974) and, in general, both nuclear volume and mitotic cycle time increase with increasing DNA content. Similarly, the duration of meiosis lengthens with increasing DNA content of diploid nuclei. However, there appears to be no general increase in cell cycle time with polyploidy; the mitotic cycle may be longer or shorter in polyploids than in their related diploids. Colchicine-induced autotetraploid cells of *Vicia faba* (Broad Bean) have the same mitotic cycle time as diploid cells in the same tissue. Polyploid cereals show a shorter duration of meiosis than do their related diploid species. These results suggest that cell cycle time is correlated with chromosome size, rather than the actual DNA amount.

Cell size

An increase in cell size with polyploidy was recognised by Winkler and other early cytogeneticists. In tomato pollen grains, stomata and palisade parenchyma cells are larger in the tetraploid than in the diploid. *Oenothera gigas* showed increase in size of the sepals, petals and seeds – the so-called 'gigas' effect, which tends to be restricted to organs of determinate growth form. Often the individual organs remain the same size in polyploids as in diploids, but are composed of larger, and therefore fewer cells. The larger cell size may result in thickening of leaves or other appendages (Stebbins, 1971).

Physiological changes may result from an increase in cell size. Hall (1972) studied the oxygen requirement of diploid and autotetraploid rye, and found not only that the tetraploid seedlings required higher oxygen concentrations at the root surface, but also they were less tolerant to soil temperature. Since respiration rates of the diploid and tetraploid were similar, he believed that the changes in the tetraploid were caused by its larger root meristem volume. This had a 16 per cent longer radius than the root meristem of the diploid, and the increase appeared to be a result of increased cell size.

Chromosome pairing

Polyploidy results in an increase in the number of chromosome sets. In an autotetraploid the four chromosome sets are all similar and during meiosis may form quadrivalents, associations of four chromosomes. The basic number in the genus *Lycopersicon* is $x = 12$; autotetraploid plants of Tomato, *L. esculentum*, have $2n = 4x = 48$. In theory the maximum number of quadrivalents forming at meiosis in the tetraploid should be 12. Upcott (1935) analysed 50 pollen mother cells, and found variation from one to eight quadrivalents per cell, with a mean value of $3 \cdot 1$. That the maximum number of 12 is not reached does not imply that the four chromosome sets in this autotetraploid are not fully homologous, since multivalent formation can be restricted in ways other than from lack of homology. These may be chromosome size, chiasma frequency or position or genetic regulation. For example, chromosome association may be restricted to bivalents where each chromosome pair is held together by a single chiasma. To form a trivalent, an association of three chromosomes, at least one of the chromosomes must be involved in two chiasmata. An organism with small chromosomes that show only one chiasma per bivalent will be unable to form multivalents.

In selecting for improved fertility in autotetraploid rye, Hazarika & Rees (1967) found that an increase in quadrivalent frequency was correlated with an increase in the chiasma frequency.

Chiasma position may also influence quadrivalent frequency. *Allium porrum*, the Leek, an autotetraploid with $2n = 32$, shows proximally localised chiasmata (that is, the chiasmata occur close to the centrometres). Early in meiosis, associations of four chromosomes form, but later, at first metaphase quadrivalents are rare and bivalents more common. Other autotetraploid *Allium* species with chiasmata that are more distally located show a higher frequency of quadrivalents at first metaphase (Levan, 1940).

Genetic control of chromosome pairing in a polyploid has been clearly demonstrated in Breadwheat, *Triticum aestivum* (Riley & Chapman, 1958; Sears & Okamoto, 1958). This is a hexaploid species ($2n = 42$), and the chromosomes are assigned to three groups, AA, BB and

DD, the groups being slightly different in morphology. Each group has been derived from a different diploid, each with $2n = 14$. *T. aestivum* normally forms bivalents, but in the absence of the long arm of one chromosome (5B) multivalent formation occurs. Although the AA, BB and DD chromosome sets show differences in morphology, chiasmata can form between them, giving multivalents, and the chromosome sets are said to be homoeologous. In the presence of chromosome 5B pairing is restricted to full homologues and so only bivalents form.

Fertility of polyploids

Allopolyploids are usually bivalent forming and fully fertile. Chromosome doubling to give an allopolyploid can be regarded as a means by which a sterile hybrid can gain fertility. This is seen in the formation of the fertile tetraploid *Primula kewensis* from the sterile *P. floribunda × verticillata*. It is often implied that multivalent forming polyploids are sterile, and that for fertility to be restored, chromosome association must be limited to bivalents by genetic control or other means. However, multivalents in themselves do not lead to infertility, for so long as segregation of the chromosomes during anaphase is balanced, then multivalent forming polyploids are potentially fertile. The presence of trivalents and to a lesser extent chain quadrivalents may result in infertility following mis-orientation at metaphase 1, but ring quadrivalents can give numerically equal segregation. It is the presence of univalents that leads to chromosome imbalance and infertility. In autotetraploid rye, which shows the presence of trivalents and univalents as well as quadrivalents and bivalents, selection for improved fertility results in a decrease in the number of univalents and an increase in quadrivalent frequency (Hazarika & Rees, 1967). A similar picture is seen in autotetraploid *Lolium perenne*, but here the higher quadrivalent frequency is achieved by a redistribution of chiasmata (Crowley & Rees, 1968), rather than by an increase in chiasma frequency.

Whilst even-numbered polyploids are generally fertile, the fertility of odd-numbered polyploids is usually reduced as a result of the presence of univalents or odd-numbered multivalents. Some triploids are fertile, but the gametes that they produce are diploid, triploid etc. or aneuploid, and these give rise to progeny with a variety of chromosome numbers. The only way in which an odd-numbered polyploid can consistently produce offspring of similar cytogenetic structure is by some form of asexual reproduction (apomixis or parthenogenesis).

Gene duplication and increase in genetic heterozygosity

The duplication of a chromosome set involves duplication of genes. Somatic doubling of chromosomes to give an autopolyploid adds no new genetic information. However, gene duplication in itself can have significant effects. A diploid with two chromosome sets has two copies of each gene; the tetraploid has four. In a diploid the two alleles, A and a, may be present in the homozygous state, AA or aa, or in the heterozygous state, Aa. The tetraploid also shows two homozygotes, AAAA and aaaa, but there are more possible heterozygotes, AAAa, AAaa and Aaaa, and the frequency of heterozygotes in a tetraploid population is, therefore, likely to be higher than in a diploid population. A hybrid is likely to be heterozygous at several loci, and chromosome duplication will produce an allotetraploid with fixed heterozygosity at these loci.

The presence of duplicated genes can buffer the effects of deleterious recessive mutations, and a polyploid will have the potential to accumulate a large number of new recessive mutations. Ohno (1970) has stressed the importance of gene duplication for evolution. He suggests that gene duplication allows redundancy at some of the loci, and that mutations that accumulate at these points may emerge ultimately as gene loci with new functions. He believes that polyploidy occurred during the development of fishes, and that this large scale gene duplication allowed for mutations to accumulate, leading eventually to the evolution of mammals.

Enzyme systems have been studied in some tetraploids and their related diploids to discover whether levels of heterozygosity are increased with polyploidy. Enzymes are particularly useful in this type of analysis. Different enzymes can be identified since each digests a specific substrate. They can be separated by electrophoresis, since each enzyme has a particular charge and so a change in the amino acid sequence may be detected if it alters the charge of the enzyme. This means that changes in the coding sequence in DNA can be recognised.

Ralin & Selander (1979) have investigated enzyme variation in the diploid frog, *Hyla chrysoscelis*, and its probable autotetraploid derivative, *H. versicolor*. The frequency of polymorphic enzymes in *H. chrysoscelis* is similar to that found in other diploid sexual species. The polymorphic loci of *H. versicolor* show band patterns that are consistent with an allele dosage effect expected in a tetraploid, where each locus is represented by four alleles. The average heterozygosity (average frequency of heterozygosity per locus) of the diploid was determined as $\bar{H} = 0.068$, whereas in the tetraploid this was much greater, $\bar{H} = 0.322$. Populations of *H. chrysoscelis* from different areas showed some differences in their genetic composition. Although *H. versicolor* carries most of the alleles represented in *H. chrysoscelis*, none of the populations of the diploid had the enzyme patterns expected for a directly ancestral population of the tetraploid. Where the two species are sympatric, they show the greatest similarity. This could be the result of introgression via triploid hybrids, although this seems unlikely, since triploids produced in the laboratory show reduced viability, and there is heavy mortality in backcross progeny. It is possible that the diploid population from which the tetraploid arose has not yet been discovered, or it may have diverged subsequently. A further possibility is that hybridisation between populations of *H. chrysoscelis* with differing enzyme patterns may have occurred prior to the formation of *H. versicolor*. This may account for some of the increase in heterozygosity. The greater individual genetic variability in the tetraploid appears to be attributable in part to the larger number of alleles per locus. However, *H. versicolor* also shows the presence of unique alleles at several loci, alleles that have not been found in any of the diploid populations. These may be the result of new mutations, or novel intra-genic recombinations.

Gottlieb (1976) has studied enzyme variation in two recently evolved allotetraploid species of composite angiosperms, *Tragopogon mirus* and *T. miscellus*, species that have evolved from diploids as a result of Man-made disturbances (Ownbey, 1950). The diploids from which the tetraploids have evolved are quite distinct morphologically and possess different alleles at a high proportion of their loci. Hybridisation between the diploids, and subsequent chromosome doubling results in tetraploids that combine the different alleles from the diploids; as a result both tetraploids show substantial heterozygosity. In addition, some new enzymes are found in both tetraploids, which are intermediate in mobility, and probably reflect interaction between enzymes formed by the diploid genomes.

Biochemical change associated with polyploidy

Chromosome duplication has been found to alter flavonoid synthesis in a grass species, *Briza media*. Flavonoids occur in vascular plants and are water soluble secondary compounds that are usually found in cell vacuoles; Murray & Williams (1976) found that the leaves of the diploid form of this species accumulate apigenin 8-C-galactoside and an unidentified acyl derivative, whereas the major leaf constituent of the autotetraploid form is a different chemical, probably luteolin 8-C-galactoside. By producing aneuploids, they were able to show that the change in flavonoid synthesis was related to the duplication of two small acrocentric chromosomes.

Polyploidy and sex determination

Muller (1925) suggested that the reason for the rarity of polyploid animals was that an XX, XY or XX, XO sex-determining mechanism was incompatible with polyploidy. There are two genetic types of XY mechanism, one depends on 'genic balance', and the other on the 'dominant Y' principle. Muller believed that the genic balance mechanism, as found in the fruitfly *Drosophila*, was universal. In this case an organism with equal numbers of X and Y chromosomes is male, one without Y chromosomes is female. An XXYY tetraploid will be male and XXXX will be female, but XXXY or XYYY will be intersexes. A similar type of sex-determining mechanism can be found in plants. The moss, *Pogonatum microstomum* ($n = 7$; the major part of the life-cycle of a moss is spent in the haploid or gametophytic state) is an example. Populations with $n = 14$ consist of monoecious (male and female on one plant), all male and all female plants, coresponding possibly to XY, YY and XX complements (Sharma, 1963). Many

animals, including mammals, show the dominant Y method of sex determination, where the Y chromosome is male-determining (or dominant W, where the female is the heterogametic sex). *Silene alba*, the White Campion, has this type of sex determination. Tetraploids with XYYY, XXYY or XXXY are all males. From studies of the phenotypes of plants where portions of the Y chromosome are missing, it has been shown that genes that suppress the development of female organs are tightly linked to those that determine another development, and both sets of genes are located on the Y chromosome (Stebbins, 1971). Species of *Rumex* subgenus *Acetosella* (Polygonaceae, the dock family) have a strong male-determining Y chromosome; even for artificial dodecaploids ($2n = 12x$) with eleven X chromosomes and one Y chromosome the plants are still male (Löve, 1969).

Polyploidy and breeding systems

Some monoecious plants are obligate outbreeders and self sterile. Self sterility is caused after pollination by an incompatibility reaction between the pollen and the style. There are two types of breeding system, gametophytic and sporophytic. In a gametophytic system the incompatibility reaction is controlled by the pollen genotype (which is haploid) and the diploid genotype of the style. This system is found in the Rosaceae, Leguminosae and Onagraceae. The sporophytic incompatibility system, found in the Cruciferae and Compositae, involves reaction between the style and the pollen coat; the pollen coat is laid down by the cells of the tapetum, and reflects the diploid genotype of the male parent.

Polyploidy has been found to cause breakdown of the gametophytic incompatibility system in the pear, *Pyrus communis*, a member of the Rosaceae, which is normally diploid. A branch bearing self-fertile flowers was found on one particular tree, and cytological examination revealed that it was tetraploid (see Lewis, 1979). *Oenothera organensis*, also with a gametophytic incompatibility system, shows self-fertility in polyploid plants.

Diploid plants that are heterozygous for incompatibility, with the alleles S_aS_b, produce two genetically different haploid pollen – S_a and S_b. The heterozygous tetraploid, $S_aS_aS_bS_b$, produces three types of diploid pollen – S_aS_a, S_bS_b and S_aS_b. The heterozygous pollen, S_aS_b, was the only type found to be compatible on its own style. The mechanism of pollen genotype recognition by the style is unknown, as is the reason for the compatibility of the heterozygous pollen in the tetraploids. Lewis has suggested that it may involve a competitive reaction, with the different alleles in the diploid pollen grain competing for a common substrate; this could result in the lack of S-allele products that are needed for the incompatibility reaction.

The establishment of polyploids

In natural habitats polyploids arise at low frequency and may first appear as isolated individuals. A new tetraploid can arise from within a population of diploids and it will immediately be isolated genetically from its diploid parents, because backcrossing usually results in the formation of sterile triploid hybrids. The establishment of a tetraploid with the production of viable offspring can give rise to a new species – thus polyploidy can be considered to be a form of speciation, it will usually be sympatric and occur very rapidly, appearing within just one or two generations. Initially the new polyploid, often an isolated individual, will be in competition with its parents and so how can it become established?

Morphological and biochemical changes

The initial growth and development of a new polyploid may be influenced by changes that result from chromosome duplication, such as increase in cell size, in the size of organs and tissues, or changes in the physiology and biochemistry of the organism. Some polyploids may have an advantage over diploids during early development in the larger size of the seed, in the case of plants (Stebbins, 1971), or the yolk sac in animals (Jackson, 1976), as these may contain greater amounts of stored products. This may be outweighed by the slower growth rate found in some polyploid plants (Stebbins, 1971), with later flowering and fruiting. Oxygen requirements in diploids and tetraploids may differ (Hall, 1972), and biochemical synthesis may be altered (Murray & Williams, 1976). The survival of the new polyploid depends on whether these changes are advantageous.

Breeding system, duration and fertility

Once the polyploid has reached maturity, survival becomes dependent on the production of viable offspring, and this is affected by the breeding system, duration and fertility of the organism. A polyploid individual derived from dioecious (with separate male and female individuals) or outcrossing diploid species will be unlikely to survive. It may be swamped with haploid gametes from the surrounding diploid population, resulting in the production of sterile triploid hybrids. In addition, the low level of production of natural polyploids, makes it unlikely that both a male and a female individual of a dioecious polyploid will evolve at the same time and so sexual reproduction will be prevented. This probably accounts for the rarity of polyploids amongst animals. Outcrossing plants will be similarly affected, although occasional self-fertilisation may be possible through the action of heterozygous diploid pollen grains, as in *Pyrus communis*. Backcrossing to parental diploids by outcrossing polyploids can be prevented by physical and physiological changes, and by behavioural differences in animals. The diploid and autotetraploid tree frogs, *Hyla chrysoscelis* and *H. versicolor*, are sympatric over parts of their ranges and practically indistinguishable. The tetraploid has a fast-trilling call whilst the diploid is slow-trilling (Wasserman, 1970), and this difference may be important in preventing hybridisation in the wild.

A newly arisen polyploid that is self-fertilising will have greater opportunity to produce offspring than an outcrossing or dioecious polyploid. Similarly, a long-lived polyploid that has several cycles of sexual reproduction is more likely to become established than an annual or biennial species, where sexual reproduction is limited to one season. Stebbins (1971) found that in flowering plants there is a higher frequency of polyploids amongst perennial herbs, than in annuals. In polyploid perennial herbs outcrossing is as common as self-fertilisation, but polyploid annuals are predominantly self-fertilising. Isolated individuals of short-lived outcrossing polyploids will have little opportunity for fertilisation during their single season of sexual reproduction. Longevity of a sterile diploid hybrid will increase the possibility for chromosome doubling to occur, resulting in the formation of an allotetraploid.

Bivalent-forming allopolyploids are immediately capable of producing balanced gametes, and are usually fully fertile. Fertility in an autopolyploid may be low when it first arises. This can result from the presence of univalents at metaphase 1, but fertility may be improved by artificial selection, for example by increasing the frequency of multivalents, or through restriction of chromosome pairing to bivalents by genetic control. In an autopolyploid pairing may occur between homologous chromosomes that are derived from different parents and are slightly differentiated; crossing over between such chromosomes could result in genetically deficient gametes. The initial establishment of an autopolyploid may, therefore, be less successful than an allopolyploid.

Asexual reproduction

Many polyploids reproduce asexually. Asexual reproduction in plants is termed apoximis, and this may be vegetative reproduction through stolons and rhizomes, or agamospermy, the production of seeds without fertilisation. In animals the term parthenogenesis is used to describe the development of an individual from an egg without fertilisation.

All these forms of asexual reproduction result in the formation of offspring that are genetically identical with each other, and with the parent, except in a few of the species where meiosis is retained.

Polyploidy is more frequent in those perennial herbs that are able to reproduce vegetatively. Stebbins (1971) shows that in the grass genera *Agropyron* and *Elymus*, the known diploids are annual or caespitose (of tufted growth), whereas all the rhizomatous species are polyploids. Of animal species known to be polyploid, the majority reproduce asexually.

In some organisms asexual reproduction may alternate with sexual reproduction. This is well illustrated in aphids, and an account and discussion of their life-cycle is given by Blackman in the previous Chapter. Organisms that are exclusively parthenogenetic or apomictic have often been found to be of hybrid or of polyploid origin. For example, an Australian grasshopper, *Warramaba virgo*, a diploid all-female parthenogenetic species, has been shown to have arisen by hybridisation between two sexually reproducing diploids (White, 1978). There are several parthenogenetic lizards in the genus *Cnemidophorus*. *C. tesselatus* consists of diploid and triploid forms. Enzyme studies suggest that some of the triploids are tri-hybrid in origin, combining the genomes of three diploid sexual species (Neaves, 1969).

Irregular meiosis is usually a barrier to reproduction in that it leads to the production of inviable gametes but by parthenogeneis hybrids and odd-numbered polyploids can reproduce. *Dryopteris pseudomas* is an apomictic fern, and both diploid ($2n = 82$) and triploid ($2n = 123$) forms are known. These produce unreduced spores with the sporophytic ($2n$) number of chromosomes; the new sporophyte grows directly from the tissue of the prothallus, without the formation of gametes and fertilisation. Sixteen spore mother cells develop in each sporangium of sexual species of *Dryopteris*, and meiosis results in the formation of 64 spores, each with the gametophytic (n) number of chromosomes. In *D. pseudomas* restitution occurs in the mitotic division that precedes meiosis, with the result that only eight spore mother cells are formed, but each has twice the sporophytic number of chromosomes. Meiosis is regular, bivalents form at metaphase 1, with pairing occurring probably between exact homologues; 32 unreduced spores develop.

Sometimes the mitotic divisions that precede meiosis in *D. pseudomas* are normal and 16 spore mother cells each with the sporophytic number of chromosomes form. Where this occurs meiosis is always disturbed. In the diploid, univalents are seen; some triploids show univalents, others have bivalents and univalents, which suggests that there was more than one origin for the triploid *D. pseudomas*. Both diploid and triploid forms appear to have had a hybrid origin (Manton, 1950). These disturbed meioses give rise to abortive spores; this means that *D. pseudomas* can only reproduce asexually. Although the prothalli of this species do not produce archegonia, antheridia do develop, and the male gametes can hybridise with the related tetraploid sexual species, *D. filix-mas*. The resulting tetraploid and pentaploid hybrids of *D. pseudomas* × *D. filix-mas* can reproduce apomictically, but the majority of sporangia are the 16-celled type, with relatively few eight-celled sporangia; this results in a much lower viable spore output in these polyploid hybrids than that found in *D. pseudomas* (Manton, 1950).

In some asexually reproducing organisms the first stage of meiosis is normal but is followed by fusion between second division nuclei, or an equivalent process. The parthenogenetic enchytraeid worm *Cognettia glandulosa* is a hexaploid, and shows 54 bivalents at metaphase 1. In the second division of meiosis in the egg nucleus the chromosomes divide into chromatids but these fail to separate, giving a nucleus with the sporophytic number of chromosomes, $2n = 108$ (See White, 1973). In this organism the sperm nucleus penetrates the egg, but

degenerates without fusion of nuclei occurring. With this type of parthenogenesis segregation at the first division of meiosis must be regular; the presence of univalents would give rise to inviable eggs.

Asexual reproduction provides a means by which many hybrids and polyploids can reproduce. Some apomictic and parthenogenetic polyploids have widespread distributions; apomictic plants may be successful weeds. Most of the races of Dandelion, *Taraxacum officinale*, in north-western Europe are apomictic, many being triploids with inviable pollen (Stebbins, 1971). However, the offspring of an asexually reproducing organism are genetically identical in most cases; they represent a large number of clones of a single organism, rather than individual members of a population. Any change in the ecosystem could result in the destruction of an entire asexual population. Genetic diversity generated by sexual reproduction buffers the population against environmental change.

Heterozygosity and hybridity

Polyploidy is often accompanied by an increase in the level of individual genetic heterozygosity. This results in part from gene duplication, which will allow mutations to accumulate in the polyploid, and from the change from disomic to polysomic inheritance. As mentioned previously, Ralin & Selander's (1979) study of diploid and autotetraploid tree frogs revealed a large increase in average heterozygosity with polyploidy.

Increase in genetic diversity may be accompanied by an increase in stress tolerance, and this may increase the competitive ability of the polyploid. An allopolyploid, combining the genetic material of two or more species, may become more successful than its parents. Allopolyploidy with subsequent recombination and modification of the constituent diploid genomes has been described in the grass genus *Aegilops* (Zohary & Feldman, 1962). The genus includes diploid, tetraploid and hexaploid species. The diploids can be separated into distinct species or species groups fairly easily using morphological characters, and each shows a rather narrow stress tolerance. The diploid species are also separated by sterility barriers. The polyploids, however, are all extremely variable, with the range in morphology of one species overlapping with that of another. The overlap in morphology is usually found where the polyploids share a common genome, and these polyploids can interbreed. The seven tetraploids belonging to the *Pleinathera* section of the genus all share the common genome C^u, derived from the diploid *A. umbellulata* (Table 1). The second genome in each of these tetraploids is derived from a different diploid; however, in only one variety, *A. triuncialis* var. *persica*, is the second genome identical with a known diploid. In the other tetraploids it shows some similarity with a known diploid, but with some modification (Table 1).

Zohary & Feldman have found that, although the tetraploids are primarily inbreeders, hybridisation does occur, particularly in disturbed habitats. The hybrids are

Table 1 Some diploid and tetraploid *Aegilops* species and their genomic formulae

	Species	Genomic formula
Diploids	A. bicornis (Forsk.) Jaub. et Sp.	S^b
	A. sharonensis Eig	S^l
	A. longissima Schweinf. et Musch.	
	A. ligustica Coss.	S
	A. speltoides Tausch	
	A. caudata L.	C
	A. comosa Sibth. et Sm.	M
	A. uniaristata Vis.	M^u
	A. umbellulata Zhuk.	C^u
Tetraploids	A. variabilis Eig	$C^u \underline{S^v}$
	A. kotschyi Boiss.	
	A. triuncialis L.	$C^u \underline{C}$
	A. columnaris Zhuk.	$C^u \underline{M^e}$
	A. biuncialis Vis	$C^u \underline{M^b}$
	A. triaristata Willd.	$C^u \underline{M^t}$
	A. ovata L.	$C^u \underline{M^o}$

Genomes which have been modified are underlined.
(From Zohary and Feldman, 1962)

completely pollen sterile, but can set seed from pollination by the parental species, producing fertile segregating progeny (Feldman, 1965). The fertility of these offspring allows inter-specific gene flow. The C^u genome that is shared by the tetraploids remains unaltered by hybridisation, but chromosomal exchanges can occur between the other two chromosome sets in a hybrid thus modifying these genomes. Because of the inbreeding habit of the species the new genotypes arising after hybridisation can be fixed rapidly, and if suitably adapted may spread widely. Both the morphological and ecological ranges of the tetraploids are increased, with the result that they are often more successful than their diploid progenitors.

Polyploidy and speciation

Polyploid cells probably arise spontaneously in most organisms, but subsequent development to give rise to polyploid species is less widespread. Polyploids are rare in the animal kingdom, in fungi and gymnosperms, and common only in ferns and flowering plants. The distribution of polyploidy throughout these groups appears to be associated with problems of establishment. Polyploidy may be limited in many animals by the need for cross fertilisation; the majority of animals that are polyploid reproduce parthenogenetically. Sexually reproducing polyploids are represented only by a few species of fishes, amphibians, molluscs and earthworms (White, 1973). These latter two groups are hermaphroditic, and the polyploids may be self-fertilising, although there is little information on the relative importance of self- or cross-fertilisation in these groups. No mammals are known to be polyploid although Ohno (1970) has suggested that polyploidy played an important role in their evolution.

Polyploidy is widespread in plants and the success of polyploidy as a method of speciation may be attributable in part to the changes resulting from chromosome doubling – increase in cell size, in the size of organs, changes in physiology and biochemistry, and increase in heterozygosity. However, the majority of polyploid plants appear to be allopolyploids, and the success of polyploidy seems to be closely associated with hybridisation. Hybridisation is widespread in many plant genera. Hybrids combine the genomes of two or more species, and their greater genetic diversity may enable them to compete successfully with their parental species. Sterility in such hybrids may be overcome through polyploidy, chromosome doubling providing each chromosome with a homologue so that pairing at meiosis will be regular, and anaphase separation balanced. Polyploidy has played an important role in providing sterile hybrids with fertility.

References

Clausen, R. E. & Goodspeed, T. H. 1925. Interspecific hybridization in *Nicotiana* II. A tetraploid *glutinosa-tabacum* hybrid, an experimental verification of Winge's hypothesis. *Genetics* **10**: 279–284.

Crowley, J. G. & Rees, H. 1968. Fertility and selection in tetraploid *Lolium*. *Chromosoma* **24**: 300–308.

Feldman, M. 1965. Fertility of interspecific F₁ hybrids and hybrid derivatives involving tetraploid species of *Aegilops* section *Pleinathera*. *Evolution* **19**: 556–562.

Gates, R. R. 1909. The stature and chromosomes of *Oenothera gigas*, de Vries. *Archiv für Zellforschung* **3**: 525–552.

Gottlieb, L. D. 1976. Biochemical consequences of speciation in plants, pp. 123–140, *in* Ayala, F. J. (Ed.) *Molecular evolution*. Sunderland, Massachusetts: Sinauer Associates Inc.

Grant, V. G. 1971. *Plant speciation*. 435 pp. New York: Columbia University Press.

Hall, O. 1972. Oxygen requirements of root meristems in diploid and autotetraploid rye. *Hereditas* **70**: 69–74.

Harlan, J. R. & deWet, J. M. J. 1975. On Ö. Winge and a prayer: the origins of polyploidy. *Botanical Review* **41**: 361–390.

Hazarika, M. H. & Rees, H. 1967. Genotypic control of chromosome behaviour in rye. X. Chromosome pairing and fertility in autotetraploids. *Heredity* **22**: 317–332.

Jackson, R. C. 1976. Evolution and systematic significance of polyploidy. *Annual Review of Ecology and Systematics* **7**: 209–234.

Jones, K. 1978. Aspects of chromosome evolution in higher plants. *Advances in botanical research* **6**: 119–194.

Karpechenko, G. D. 1927. The production of polyploid gametes in hybrids. *Hereditas* **9**: 349–368.

Karpechenko, G. D. 1928. Polyploid hybrids of *Raphanus*

sativus L. × *Brassica oleracea* L. *Zeitschrift für induktive Abstammungs-Vererbungslehre* **48**: 1–85.

Levan, A. 1940. Meiosis of *Allium porrum*, a tetraploid species with chiasma localisation. *Hereditas* **26**: 454–462.

Lewis, D. 1979. *Sexual incompatibility in plants.* 60 pp. London: Edward Arnold.

Löve, A. 1969. Conservative sex chromosomes in *Acetosa*, pp. 166–171. *In* Darlington, C. D. & Lewis, K. R. (Eds) *Chromosomes Today*, 2. Edinburgh: Oliver Boyd.

Löve, A., Löve, D. & Pichi Sermolli, R. E. G. 1977. *Cytotaxonomical atlas of the Pteridophyta.* 398 pp. Vaduz: Cramer.

Manton, I. 1950. *Problems of cytology and evolution in the Pteridophyta.* 316 pp. Cambridge: University Press.

Matthey, R. 1945. L'évolution de la formule chromosomiale chez les vertébrés. *Experientia* **1**: 50–6 and 78–86.

Muller, H. J. 1925. Why polyploidy is rarer in animals than in plants. *American Naturalist* **59**: 346–353.

Murray, B. G. & Williams, C. A. Chromosome number and flavonoid synthesis in *Briza* L. (Gramineae). *Biochemical Genetics* **14**: 897–904.

Neaves, W. B. 1969. Adenosine deaminase phenotypes among sexual and parthenogenetic lizards in the genus *Cnemidophorus* (*Teiidae*). *Journal of Experimental Zoology* **171**: 175–184.

Newton, W. C. F. & Pellew, C. 1929. *Primula kewensis* and its derivatives. *Journal of Genetics* **20**: 405–467.

Ohno, S. 1970. *Evolution by gene duplication.* 160 pp. London: Allen & Unwin.

Ownbey, M. 1950. Natural hybridization and amphidiploidy in the genus *Tragopogon*. *American Journal of Botany* **37**: 487–499.

Ralin, D. B. & Selander, R. K. 1979. Evolutionary genetics of diploid – tetraploid species of treefrogs of the genus *Hyla*. *Evolution* **33**: 395–608.

Ramanna, M. S. 1974. The origin of unreduced microspores due to aberrant cytokinesis in the meiocytes of potato and its genetic significance. *Euphytica* **23**: 20–30.

Rees, H. & Hazarika, M. H. 1969. Chromosome evolution in *Lathyrus*. pp. 158–165 *in* Darlington, C. D. & Lewis, K. R. (Eds) *Chromosomes Today*, 2. Edinburgh: Oliver & Boyd.

Riley, R. & Chapman, V. 1958. Genetic control of the cytologically diploid behaviour of hexaploid wheat. *Nature* **182**: 713–715.

Roach, B. T. 1968. Cytological studies in *Saccharum*. Chromosome transmission in interspecific and intergeneric crosses. *Proceedings of the International Sugar Cane Technologists* **13**: 901–920.

Sears, E. R. & Okamoto, M. 1958. Intergenomic relationships in hexaploid wheat. *Proceedings of the Tenth International Congress of Genetics* **2**: 258–259.

Sharma, P. D. 1963. Cytology of some Himalayan Polytrichaceae. *Caryologia* **16**: 111–120.

Simmonds, N. W. 1966. *Bananas.* 512 pp. London: Longmans.

Stebbins, G. L. 1950. *Variation and evolution in plants.* 643 pp. New York: Columbia University Press.

Stebbins, G. L. 1971. *Chromosomal evolution in higher plants.* 216 pp. London: Edward Arnold.

Tahara, M. 1915. Cytological studies on *Chrysanthemum*. *Botanical Magazine (Tokyo)* **29**: 48–50.

Tatuno, S. & Kawakami, S. 1969. Karyological studies on Aspleniaceae. 1. Karyotypes of three species of *Asplenium*. *Botanical Magazine, (Tokyo)* **82**: 436–444.

Upcott, M. 1935. The cytology of triploid and tetraploid *Lycopersicum esculentum*. *Journal of Genetics* **31**: 1–19.

Van't Hof, J. 1974. The duration of chromosomal DNA synthesis, of the mitotic cycle, and of meiosis in higher plants, pp. 363–377 *in* King, R. C. (Ed.) *Handbook of genetics*, Volume 2. New York: Plenum Press.

Wasserman, A. O. 1970. Polyploidy in the common tree toad, *Hyla versicolor* Le Conte. *Science* **167**: 385–386.

White, M. J. D. 1973. *Animal cytology and evolution.* 961 pp. Cambridge: Cambridge University Press.

White, M. J. D. *Modes of speciation.* 455 pp. San Francisco: Freeman.

Winge, Ö. 1917. The chromosomes. Their number and general importance. *Compte Rendu des Travaux du Laboratoire de Carlsberg* **13**: 131–275.

Winkler, H. 1916. Über die experimentelle Erzengung von Pflanzen mit abweichenden Chromosomenzahlen. *Zeitschrift für Botanik* **8**: 417–531.

Zohary, D. & Feldman, M. 1962. Hybridization between amphidiploids and the evolution of polyploids in the wheat (*Aegilops–Triticum*) group. *Evolution* **16**: 44–61.

PART II

Coexistence and coevolution

Introduction

G. Evelyn Hutchinson's aphorism 'The ecological theater and the evolutionary play' attractively highlights the dynamism and complexities of the world we study. Co-evolution and coexistence are important subplots in that long-running play without denouement.

Coevolution is perhaps the one word that most succinctly describes the historical background to the complex inter-relationships between living organisms and between them and their abiotic environment. Nothing has evolved in isolation; life itself must have originated in conjunction with the changing abiotic environment of the early Earth.

Coexistence can be considered a major consequence of coevolution when it describes the outcome of those evolutionary processes which enable different organisms to share the resources of a habitat. If competition is an important element in the evolutionary play, however, then coexistence is more apparent than real, an artifact stemming from the short time-span over which we can watch the play. Over a much longer time period, millions rather than scores of years, what now seems to be a static situation could be altered drastically, with one or more of the players either disappearing entirely from the stage, or continuing to act but in different roles.

These remarks apply to the broad canvas of nature. At the other end of the scale the concept of coevolution embraces the development of more intimate relationships, particularly those which have led to the near absolute interdependence of two or more species. On this scale, coexistence, as a correlate and outcome of competition, could be one of the factors involved in shaping population structures within species. The entire spectrum is well reviewed by Futuyama (1979, pp. 51–74).

The chapters in this section of the volume look at coevolution and coexistence from these several different viewpoints, and often from a different philosophical angle from that put forward in this introduction.

Although the results of coevolution are manifest, as Darwin (1859) so frequently noted, in the 'infinitely complex relations of organisms', there is immense difficulty in discovering the causal mechanism producing those relationships. The attractions of nature's infinite complexity hide a conceptual minefield – evolutionary theory.

Evolution can be, and often is looked upon as the result of organisms responding to 'problems' set by the external world. The 'solutions' to these problems are recognised as adaptations, and evolutionary histories are often traced and measured through the origin and spread of adaptations. Well and good, but there is a real danger in this situation, namely that the observer is identifying the 'problem' from what he considers to be the 'solution'. This danger is somewhat reduced when the problem is a man-made one that can be studied from its inception (for example the effects of pollution, the spread of drug resistant organisms, or large-scale environmental changes). But even here the natural response to the new selection pressure is often a complex of factors making it difficult or impossible to work out determinate relationships between 'problem' and 'solution' (see Berry, 1977, pp. 118–120 & 124–130).

Evolutionary change – including coevolution at all levels – and thus the origin of adaptations, is usually interpreted as the outcome of natural selection. In recent years Darwin's theory, and in particular its usual method of application, have come under increasingly critical review (Macbeth, 1971; Bethell, 1976; and particularly Brady, 1979).

First, let us consider its application. The theory of natural selection, or the survival of the fittest, sets out to account for differential mortality. In other words, to explain why certain kinds of individuals survive and by surviving produce more offspring thus perpetuating their kind, whilst others die either before they have bred or early in their reproductive lives. Fitness in the struggle for existence can, however, only be measured by the fact that the fitter types survive (that is, leave more offspring). When the question of why they leave more offspring is asked, one can only answer 'Because they survive' (that is, are the fittest). The theory expressed in that way is tautological, nothing is explained because there is no criterion of fitness independent of the empirical fact of reproductive success and ultimate survival. In the words of Brady (1979, p. 603) this...'amounts to a claim that the facts explain themselves – individuals leave more offspring because they leave more offspring...the tautologous repetition is utterly useless for explanatory purposes'.

However, as Brady has also demonstrated, Darwin's theory is not tautological in its original formulation: 'Owing to the struggle for life, any variation, however slight and from whatever cause proceeding, if it be in any degree profitable to an individual of any species, in its infinitely complex relationships to other organic beings and to external nature, will tend to the preservation of that individual, and will generally be inherited by its offspring. The offspring, also, will thus have a better chance of surviving, for, of the many individuals of any species which are periodically born, but a small number can survive. I have called this principle, by which each slight variation, if useful, is preserved, by the term of Natural Selection, in order to mark its relation to man's power of selection' (Darwin, 1859, p. 61).

Obviously Darwin's intention was to designate a causal agency equivalent to that of human intervention in the widely practised and successful art of artificial selection. His aim was to identify in nature a determinate relationship between environmental factors and particular traits in an organism that is analogous to the relationship between human choice and specific traits in selective breeding.

The real problem, however, is to identify traits representing evolutionary fitness, independent of differential reproduction or mortality, which would explain that mortality or, conversely, that reproductive success.

Even in the few cases where the situation looks relatively simple, as for example in the phenomenon of industrial melanism in moths, and the warfarin resistance in rats, closer analysis reveals difficulties in establishing a one-to-one determinate relationship between any specific adaptive traits and the selective 'hand' of the environment (see Berry, 1977, pp. 116–130). Indeed, it is the very complexities of the organism, the environment, and their joint interactions which render the basic Darwinian concept of natural selection impossible to test fully.

If a theory cannot be tested fully it cannot be refuted, and since a theory can only be corroborated and never proved (because there may always be some set of undiscovered circumstances that will refute it), an untestable theory is more in the nature of a 'belief' than a scientific hypothesis. For a further discussion of this and other aspects of evolutionary philosophy the reader should read the chapter 'Proof and disproof' in Patterson's book *Evolution* (1978).

Another of the difficulties inherent in the use of differential reproductive success alone as a criterion of evolutionary fitness, especially of relative adaptiveness, was demonstrated neatly by Lewontin (1978, pp. 166–167).

Lewontin's mathematical model is based on two, resource-limited populations each consisting of 100 animals; in both populations every animal requires one unit of the limiting source (say, food). In one population a mutation occurs, its effect being to make the mutant individuals twice as fecund as the non-mutants; the resource utilisation effectiveness of neither type is altered. The composition, size and growth rate of the population over a period of time can be calculated; this reveals that the mutant forms will eventually replace the non-mutants. In terms of natural selection then, that form would be considered the 'fittest'. However, nothing apart from the fecundity of the individuals has changed; the adult population size and the growth rate of the population remain constant.

In the second population a different mutation arises whose effect is to double the efficiency of resource utilisation in the mutants (fecundity is not altered). Calculations of changes in population size and growth rate over the same period as used in the first model give very different results. As before, the original type is replaced by the mutant form (that is, the mutant form, in terms of selection theory, must be the fittest). However, in this case the growth rate of the population is increased, as is the size of the population. [Remember, there has been no change in individual fecundity as there was in the first model.] Eventually the growth rate drops to its original level but the now stabilised population size remains double that at the outset.

In both models one form has survived, the other has disappeared, but in terms of the data fed into the models how can we tell that the survivors are better adapted (are fitter in an evolutionary sense) than the original types? As Lewontin says... 'Those (individuals) with higher fecundity would be better buffered against accidents such as sudden changes in temperature since there is a better chance that some of the eggs would survive. On the other hand, their offspring would be more susceptible to the epidemic diseases of immature forms and to predators that concentrate on the more numerous immature forms. Individuals in the second population would be better adapted to temporary resource shortages, but also more susceptible to predators or epidemics that attack adults in a density-dependent manner. Hence there is no way we can predict whether a change due to natural selection will increase or decrease the adaptation in general. Nor can we argue that the population as a whole is better off in one case than in another. Neither population continues to grow or is necessarily less subject to extinction, since the larger number of immature or adult stages presents the same risks for the population as a whole as it does for individual families'.

But all this is speculative prediction as to what might happen in changed circumstances since none of these environmental changes was actually built into the model. A change has taken place, one type is shown to replace the other and is thus by definition, 'fitter' than the other. Yet, we cannot tell why it is 'fitter'. Certainly no causal environmental factor can be implicated. On the basis of Lewontin's model it would seem that 'natural selection' (as recognised from its end results) can take place without involving adaptation, at least if that word is employed in a precise way and not as a vague reference to any change that takes place.

Natural selection, it seems, is a theory that can be used to explain everything and thus, unfortunately, it explains nothing (see especially, Brady, 1979, also Bethell, 1976 and Macbeth, 1971; for a counter-attack see Gould, 1976). That, of course, is not to deny the utility of Darwin's original thesis as a source of stimulation, or as a programme from which further research has been developed.

The concept of adaptation, central to the theory of natural selection, is itself very loosely structured, and essentially employed in a competitive sense. That is, one organism is better than another at solving environmental problems, or that successive generations within populations of a species increase the effectiveness of their solutions. More often that not the word adaptation is used as a synonym for morphological and/or physiological specialisations that enable an organism to use some aspect of the environment unavailable to other organisms, particularly related species.

The competitive implications in the concept of adaptation raise the question 'How do we recognise that one organism is better adapted than another?' One answer, the historical one, 'By the fact that it survives' leads into the same tautological entanglement already associated with natural selection. The alternative answer 'That it can utilise some part of the environment not available to another organism or other organisms', is only another way of saying that it is differently specialised, and tells us nothing about survival potential. Implicit, and often explicit in the second answer is the rider that the differently specialised organism can enter or has entered a new or empty ecological niche, thus avoiding or overcoming competition from its ecological neighbours or phyletic relatives in the old niche.

Here one is bedevilled further by the varied concepts of an ecological niche. At one extreme, a 'niche' is equated with what other authors would designate a habitat, and thus includes a number of environmental factors. At the other extreme it takes into account only one element of the environment, say an organism's feeding requirements. Very rarely in any definition is the organism under discussion included as a component of the 'niche' being described.

Possibly the most comprehensive and realistic definition of a niche is that given by Lewontin (1978, p. 159)... 'a multidimensional description of the total environment and way of life of an organism. Its description includes physical factors, such as temperature and moisture; biological factors, such as the nature and quantity of food sources and of predators, and factors of the behaviour of the organism itself, such as its social organisation, its pattern of movement and its daily and seasonal activity cycles'.

Thus the organism itself is part of the niche, which would therefore not exist without the organism. Put in another way, there is no such thing as an empty or new niche, although there may be, in any existing environment, unexploited resources that could be utilised by an organism with the requisite specialisations. Usually, when a new or an empty niche is described it is the specialisation of the organisms presumed to have occupied that niche which is being recognised and detected.

Quite clearly, numerous specialisations have evolved and have thereby initiated whole new phyletic lineages and ways of life (see Chapter 6 and Mayr, 1963). But once again it is exceedingly difficult to pin-point the causative environmental agent involved in their evolution, or to label them as adaptations in any sense other than a means of prosecuting a particular way of life.

Apart from the problems of defining an ecological niche without also including the organism supposedly entering that niche, one faces another problem; determining whether or not the often implicit competition which led to, or was overcome by entry into the niche, ever existed.

Competition is a word that often crops up in the programme notes to the evolutionary play. Not only is it used to identify the selective forces behind the evolution of qualifications for entry into a 'new' niche, but it is also offered as an explanation for the evolutionary processes producing such phenomena as resource partitioning, character displacement and several other interactions leading to apparent coexistence.

These aspects of competition acting at all levels in evolution, and therefore its effects on coexistence, are discussed in several of the chapters which follow. The influence of competition in shaping large-scale ecological situations and evolutionary trends, as well as its effects on the organisms involved, is reviewed with regard to marine plankton (Chapter 11), predatory snails (Chapter 18), and coral reefs (Chapter 9). A closer look at the ecological, behavioural and morphological results of competition is provided by the essays dealing with coevolution of birds and flowering plants (Chapter 13), insects and plants (Chapter 14) and, in particular, by Chapter 17 where, with lizards as models, most of the competition-based concepts of evolutionary interactions are considered. For a rather different viewpoint on these topics, and one which also differs from the arguments developed in this introduction, the reader is referred to the chapter on coevolution in Futuyama's (1979) recent book.

The dynamics of species interactions are obviously of special relevance in the study of competition and evolutionary change. This aspect of the problem is dealt with in Chapters 10 and 16 which, respectively, are concerned with aquatic microbial communities and with the dynamics of microscopic marine invertebrates; the latter essay carries the story further to include speculations on the origin of metazoan animals. The ease with which micro-organisms can be cultivated and used as experimental models, coupled with their short generation times, make them ideal experimental material for certain evolutionary studies. This is particularly so because we can actually observe the effects of interactions which, in other organisms, would only be apparent in hundreds or even thousands of years. One of the fundamental concepts in coevolutionary thinking, that of competitive exclusion (also called Gause's law or principle) is given particular attention in Chapter 10.

Throughout these chapters one is constantly reminded of the 'infinitely complex relations' of organisms which so stimulated Darwin in his search for an explanatory principle of evolution. One is also reminded that this complexity is a reflection of the infinitely complex ways in which energy resources are utilized and partitioned. Competition may

have played its part in this process, but actually to establish the role of competition in evolution is far more difficult than to use it as an apparent explanation for the origin of the evolutionary endpoints we observe.

Apparent competition between species can be observed in artificial situations (Chapter 10), but unequivocal examples of interspecific competition in nature are rare (see Vuilleumier & Simberloff, 1980). Competition between members of the same species, on the other hand, may be more common but its evolutionary consequences are shown in more subtle ways than those postulated to explain larger evolutionary changes. For example, such intraspecific competition may be involved in the evolution of certain kinds of mimetic resemblances in butterflies (see Chapter 12). It is probably also a component in some of the processes which, cumulatively, are concerned with the distributions of plant and animals, and hence of biogeography (see Chapter 19).

As viewed from our restricted point in geological time the results of competition would be impossible to identify. If the unsuccessful competitor was rendered extinct by its successful rival there is really no means of establishing that competition has taken place; the fossil record can only tell us of extinction and not how it came about, at least for whole species. If, on the other hand, the competition was not lethal there is no way of telling that the erstwhile competitors did indeed once have that relationship, and thus we cannot identify their differences as evolutionary responses to selection acting through competition. The competitive exclusion of one species by another can be contrived in the laboratory (see Chapter 10) but would be difficult to recognise in nature, particularly over the short time scale we have available for observation.

Likewise, one may question the concept of competitive restriction as a causal evolutionary mechanism. That the situation described as competitive restriction is real, cannot be denied, and is well documented by the examples cited in Chapter 17. But the term would seem to be a means of identifying and describing a particular kind of coexistence and resource partitioning rather than an explanation of how the situation arose. The organisms involved could just as well have been equipped with their particular specialisations before they came into contact rather than to have evolved them after contact. In other words, the specialisations were 'preadaptations' that, in changed circumstances, channelled the organisms into their currently observed ways of life. Unfortunately, we have no positive means of testing either hypothesis.

So far our concern has been mainly with the 'adaptation-competition' aspects of coevolution, and the search for a theory explaining the causal mechanism underlying the interrelationships we observe.

There are, of course, many other intriguing aspects of coevolution, particularly those concerned with whole ecosystems. In many respects the coevolution of parasites and their hosts (see Chapter 15) emphasises the close links existing between evolution at the ecosystem level and that at the level of individual organisms. Other examples are the different patterns of species numbers and diversity in different parts of the world (see Chapter 9); the consistent average number (three) of trophic levels in a food web, irrespective of the considerable variability in the amount of energy flow through various ecological systems (Chapters 10, 11 and 16); the prevalence of small over large animals, and the phenomenon of 'island equilibrium' (Chapter 19).

The concept of island equilibrium, first formulated by MacArthur & Wilson (1967) has wide implications in coevolutionary studies. It is concerned with the equilibrium number of species comprising an 'island' community, that number representing a state of ecological saturation and one in which local extinctions are balanced by fresh immigration of the same or other species. The 'island' may be a real one, or a region ecologically isolated from the surrounding terrain (for example mountain tops), or one artificially isolated (for example a nature reserve or game park). Indeed, many ecological and biogeographical situations can be reduced to 'islands' in this sense, and so the theory of 'island equilibrium' has extensive ramifications in biology.

As a rough but consistent rule, the equilibrium number of species on an 'island' is related to the size of the 'island' so that a tenfold size increase results in a two-fold increase in the number of species present, other factors being held constant. Conversely, for an 'island' of any given size the immigration rate will diminish as the distance of that 'island' from a source of immigrant species increases. Thus the equilibrium number of species will be highest on large 'islands' close to a source of migrants, and lowest on small and distant 'islands'. (See Chapter 19, fig. 19).

Many recolonisation experiments, both natural (for example, after the volcanic eruption of Krakatoa) as well

as artificial, confirm the numerical predictions of the equilibrium hypothesis. What cannot be predicted, however, is the actual species returning; in many instances these were different from those originally inhabiting the area. Palaeontological evidence also matches the theory of 'island equilibrium'.

For a very readable account of these multifaceted aspects of ecosystem evolution, as well as reflections on the vexed question of competition, the reader is referred to May (1978).

Although the phenomena observed in the broader, ecosystem view of coevolution differ somewhat from those at the 'adaptation–competition' level, the latter, since they involve individual organisms, are an integral part of the former. In turn, the wider patterns of coevolution and the evolution of ecosystems form part of the biogeographical approach to evolution, the subject matter in the final part of this volume.

Unfortunately, the patterns and reactions of coevolution, the evolutionary play itself, can no more be satisfactorily explained than can the 'adaptations' and interactions seen amongst individual actors in the ecological theatre. As May (1978, p. 133) notes '...the empirical patterns are important, widespread and abundantly documented, but they lack a convincing explanation'...'The task of understanding how ecological systems work is in the middle of its own successional process'.

There certainly is abundant empirical evidence for coevolution (in all the various usages of the word) and of apparent coexistence as well, just as there is empirical evidence for evolution in its most fundamental sense. What still eludes us, however, is a single, scientifically satisfactory hypothesis explaining the causal mechanism underlying what we observe. There is a long and exciting road ahead.

References
Bethell, T. 1976. Darwin's mistake. *Harpers Magazine* **252**: 70–75.
Berry, R. J. 1977. *Inheritance and natural history.* 355 pp. London: Collins.
Brady, R. H. 1979. Natural selection and the criteria by which a theory is judged. *Systematic Zoology* **28**: 600–621.
Darwin, C. 1859. *On the origin of species.* 502 pp. London: John Murray.
Futuyama, D. J. 1979. *Evolutionary biology.* 565 pp. Sunderland: Sinauer.
Gould, S. J. 1976. Darwin's untimely burial. *Natural History* **85**: 24–30 (Reprinted in Gould, S. J., 1977. *Ever since Darwin.* 285 pp. London: Burnett Books & Andre Deutsch).
Hutchinson, G. E. 1965. *The ecological theater and the evolutionary play.* 139 pp. New Haven & London: Yale University Press.

Lewontin, R. C. 1978. Adaptation. *Scientific American* **239**: 157–169.
MacArthur, R. H. & Wilson, E. O. 1967. *The theory of island biogeography.* 203 pp. Princeton: University Press.
Macbeth, N. 1971. *Darwin retried.* 178 pp. New York: Delta.
May, R. M. 1978. The evolution of ecological systems. *Scientific American* **239**: 119–133.
Mayr, E. 1963. *Animal species and evolution.* 797 pp. Cambridge, Massachusetts: Harvard University Press.
Patterson, C. 1978. *Evolution.* 197 pp. London: British Museum (Natural History).
Vuilleumier, F. & Simberloff, D. S. 1980. Ecology versus history as determinates of patchy and insular distributions in High Andean birds. *Evolutionary Biology* **12**: 235–379.

P. H. Greenwood
British Museum (Natural History)

CHAPTER 9

The tropical high diversity enigma – the corals'-eye view

B. R. Rosen

It is well known that the variety of animals and plants is not the same at different places on earth. One of the best known but nevertheless puzzling patterns of this kind is the fact that most groups of animals and plants exhibit their greatest variety in the tropics. 'Variety' in this particular sense is more generally referred to as 'diversity' (or more precisely as 'taxonomic richness'). In other words there are more species, or indeed taxa generally, in the tropics than in higher latitudes. This pattern applies to corals, for which there are reasonably good supporting data, so they are a convenient group for discussing some of the ideas that have been put forward to explain the general phenomenon of tropical high diversity. This chapter attempts to evaluate these ideas by exploring how far they are applicable to corals, and how far corals contribute to a general explanation.

Diversity as I shall discuss it here is essentially a synthetic concept and as such can be measured in many ways. But for the kind of global pattern that I shall be considering it is simplest to use the number of taxa that occurs within a natural geographical region like an archipelago in comparison with another such natural region. No two geographic regions have identical areas or environments and therefore cannot represent equivalent samples. For this reason I have chosen to base my diversity measurements on the number of taxonomic records within particular latitudes or mean prevailing temperature ranges. Diversity can then be plotted against a series of sample latitudes or temperatures to obtain diversity profiles or gradients (Figs 9.2 to 9.4, 9.7 and 9.8). It is also possible to make contour maps of diversity to give an areal picture, by interpolating lines of particular diversity values between the spot values yielded by particular sample regions (Fig. 9.1). These approaches confirm that there are more coral taxa in the tropics than in higher latitudes, and the discussion that follows will be based on the particular plots compiled here.

In general, the broadest diversity patterns are evident at most taxonomic levels, but they are most evident at species level, and become progressively less marked (for corals at least) at higher taxonomic levels. Table 1 for example illustrates that the Indopacific is taxonomically richer in 'zooxanthellate' scleractinian corals than the Atlantic, but this difference is not apparent at the ordinal level. Ideally, diversity should be handled at the biologically fundamental level of species, but the inadequate state of species level taxonomy in many corals frequently makes it necessary to resort to genera, especially in the Indopacific. There is some justification for this approach. Table 1 shows that generic patterns follow species patterns quite closely and in Figures 9.3 and 9.4 similar curves are obtained for both species and genera. Genera therefore still have reasonable resolving power, especially if it is remembered that the aim is to establish the *relative* changes in diversity along latitude gradients. For similar reasons some of the plots (Figs 9.3, 9.4, 9.7 and 9.8) are presented not in absolute diversity values (which are largely of interest only to other coral specialists, and which will be published elsewhere) but re-expressed in terms of percentages of the total number of taxa known from the whole oceanic region. It is, however, important to remember that whatever taxonomic base is used in the data presentation here, the theories to be discussed concern the ecology and evolution of species.

An intuitive explanation of tropical high diversity is to attribute it to an obvious universal feature of the tropics, like warmth. But there is no reason why warmth should generate more species, rather than say, a greater abundance of the same number of (or even of fewer) species than occurs in higher latitudes. Factors like warmth relate most immediately to physiology, but the high diversity question is one of ecology and/or evolution. Establishing causal connections between these three fields is an elaborate and complicated task. The range of the problem is further beset by two more mundane but seriously frustrating difficulties. First, it is difficult to quantify the many kinds of factors that are usually invoked to explain geographical diversity patterns. How, for example, does one usefully

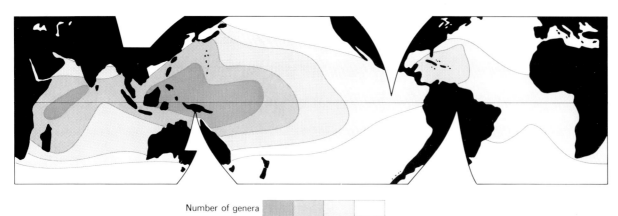

Number of genera
>50 50-30 30-10 10-2

Figure 9.1 Contour map of world distribution of numbers of genera of zooxanthellate corals.

This is based on an earlier data compilation (Rosen 1971 *b*) than Figs 9.2, 9.7 and 9.8, but it serves to show the principle of diversity contours and the high diversity foci. More recent data, used in these figures require that the highest Indopacific diversity areas be latitudinally expanded, resulting in closer spacing of contours at the northern and southern fringes of the distribution areas.

quantify warmth, stress, predation, stability or disturbance in such a way that useful correlations can be made with the diversity patterns? Second, it is difficult to know how far one can successfully switch scales. For instance, an explanation for the high diversity of one small group of organisms in one particular habitat may or may not be reasonably extrapolated to explain high diversity of most organisms in the tropics generally.

The emphasis in science on the simple generality, the universal law and on parsimony makes an explanation like warmth seem plausible and attractive. It also tempts us to debate the merits of one such generality against another on the vague assumption that these generalities are mutually incompatible. This is not parsimony; it is oversimplification. Biology commonly reveals multiple causes for one phenomenon, and single causes of several phenomena. For this reason, I have had to consider a range of different explanations with the aim of finding threads of compatibility between them.

After presenting some basic coral distribution data, I will then briefly review the ecology and biology of corals. This will serve to eliminate, by implication, certain high diversity arguments and to provide the facts and ideas needed for all the sections that follow it. I then give a summary of the *kinds* of arguments that have been offered in order to establish a constructive framework for finding common ground between different theories. Within this framework however, I have chosen to concentrate on a combination of explanations that have been put forward for corals in particular, some general explanations that have attracted recent attention, and, more subjectively, some general explanations that just seem attractive because they can be readily conceived as being applicable to corals. Finally, and with hesitation, I suggest some generalities.

I should like to dedicate this chapter to Professor John W. Wells (Cornell University) for his encouragement and because so many of the ideas expressed here are developed from his own work. I am especially indebted to Dr John Taylor (British Museum (Natural History)) as a constant source of ideas, information and discussion which has influenced my way of thinking beyond what is conventionally conveyed by literature citation. My colleague Jill Darrell has provided assistance throughout preparation of the manuscript and has compiled many of the data used.

Coral distribution patterns

Living scleractinian corals (stony corals) are distributed throughout the oceans in all depths down to about 6000 m and in most latitudes. They occur throughout the usual range of open sea water temperatures and salinities. They are strictly marine and essentially sublittoral, though they do range up into the lowermost limits of the intertidal zone according to local conditions. The diversity patterns in the discussion that follows are based almost entirely on the Atlantic and Indian Oceans, compilation of Pacific Ocean data being beyond present scope.

Figure 9.2, based on distribution data in the northern hemisphere of the Atlantic, shows how the number of scleractinian coral species increases towards the tropics. The increase is almost linear from 65° N to 13° N if intermediate fluctuations are ignored. Unfortunately, the geography of the Atlantic is such that there are few environments suitable for coral growth in latitudes lower than 13° N because of the inhibiting effects of the large river systems debouching from the South American and African coasts, and because there are so few oceanic islands in these latitudes. The profile therefore shows a steep fall-

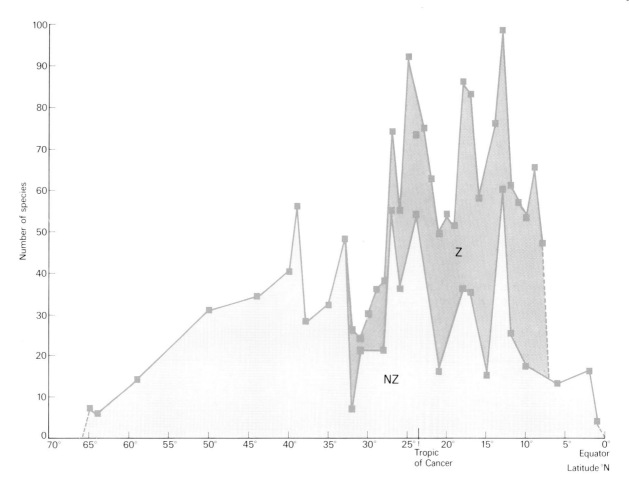

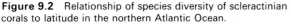

Figure 9.2 Relationship of species diversity of scleractinian corals to latitude in the northern Atlantic Ocean.

Diversity of all scleractinian corals is given by the upper profile. The two areas below the upper profile separate the diversities of zooxanthellate corals (Z) from those of non-zooxanthellate corals (NZ). The upper profile is derived from a summation of NZ and Z information. (Profiles and compilations were prepared jointly with Jill Darrell). Non-zooxanthellate raw data (lower profile) were extracted from Zibrowius (1976) for N.E. Atlantic and from Cairns (1979) for N.W. Atlantic, using the mid-point latitudes of these authors' faunal sample regions. Where several diversity figures were derived for a single latitude, we plotted only the highest figure. Zooxanthellate data are based on unpublished locality information generously provided by Dr J. W. Porter, (University of Georgia). From this we used the latitudes of the localities to obtain the total number of species present within each one-degree span of latitude for the region as a whole (Pielou, 1977).

off in diversity between 13° N and the Equator, though on present evidence alone, there may be factors operating other than these geographical ones. The North Atlantic is, however, the only region for which data based on a reasonable representation of all corals are readily obtainable.

The mean surface water temperatures of the oceans also show a regular increase towards the Equator as shown by Figure 9.5 (Atlantic) and by Figure 9.6 (Indian Ocean). This leads naturally to the relationship between diversity and temperature in Figure 9.3 which is based on Atlantic species and in Figure 9.4, based on a partially global analysis of genera. These demonstrate, not surprisingly, a rise in diversity with temperature, but in contrast to the latitude pattern, there is an almost step-like rise in diversity around a temperature optimum of 17–22 °C. No such step is seen in optimum latitudes (though this should be checked by obtaining profiles for regions additional to the North Atlantic), and there is no step in the temperature–latitude profiles (Figs 9.5 and 9.6). The temperature range over which the steepest diversity rise occurs is not quite the same in the two diversity-temperature profiles, being 19–24 °C in the Atlantic (Fig. 9.3) and 16–19 °C in the partially global profile (Fig. 9.4). It should be mentioned that these plots are not all derived in an identical way, for reasons that will be explained below. The broad patterns just inferred may therefore turn out to be invalid as generalisations. They are all we have at present.

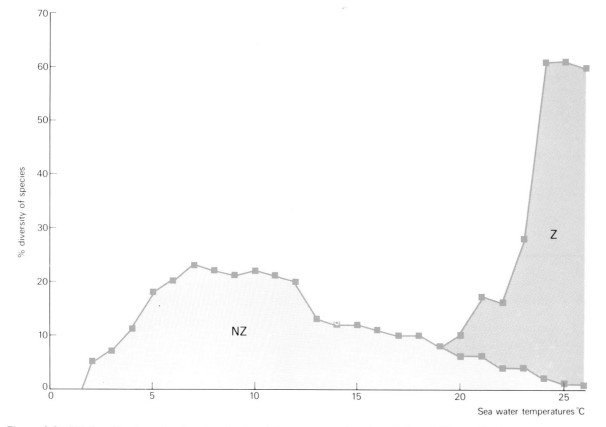

Figure 9.3 Relationship of species diversity of scleractinian corals to sea water temperatures in the Atlantic Ocean.

Diversity of all scleractinian corals is expressed as a percentage of the species total (93) and is given by the upper profile. The two areas below the upper profile separate the percentage diversities of zooxanthellate corals (Z) from those of non-zooxanthellate corals (NZ). The upper profile is derived from a summation of NZ and Z information. (Profiles and compilations were prepared jointly with Jill Darrell). Non-zooxanthellate data (lower profile) are based on Cairns' (1979) temperature ranges for 25 species (that is, not the complete Atlantic fauna). The profile shows the number of non-zooxanthellate species expressed as a percentage of the composite species total of 93, known to occur at each degree of temperature. Zooxanthellate data are based on distributions documented by Dr J. W. Porter (see Fig. 9.2 caption) and by Laborel (1970) combined with minimum average surface sea water temperatures from Sverdrup *et al.* (1946) to give temperature ranges for each of the 68 species. Hence the number of Atlantic zooxanthellates known to occur at each degree of temperature is expressed as a percentage of the composite species total of 93.

It is well known that scleractinian corals fall into two ecological groups. The first are stenothermal (confined to a narrow temperature range) and occur in warm shallow water within the tropics and subtropics. The second are eurythermal species (with a broad temperature range) and have a world-wide distribution. In fact, the general distribution statement for corals as a whole (above) also holds for the eurythermals, while the stenothermals are restricted to just one part of this range. The eurythermals are wide-ranging, however, only when taken as a whole. Individual genera and species within the group may have quite restricted depth and temperature ranges (Wells, 1967). The two groups are distinguished by their physiologies, the stenothermals having a symbiosis with endodermal dinoflagellate algae (zooxanthellae) and eurythermals having no such symbiosis. It will be more convenient and precise from here onwards to refer to the two groups as zooxanthellates and non-zooxanthellates respectively, terms equivalent to the 'hermatypes' (or 'reef-corals') and 'ahermatypes' of most other authors. This is especially intended to avoid ambiguities in a palaeontological context.

Comparison of zooxanthellates and non-zooxanthellates

Since the zooxanthellates have a tropical distribution, it is an obvious matter of interest to break up the diversity profiles derived for corals as a whole (above) into their two component distributions of zooxanthellates and non-zooxanthellates. We need to know whether tropical high diversity in corals is due primarily to zooxanthellates, or non-zooxanthellates, or to a combination of the two. It is first necessary to make some preliminary remarks about the distribution of zooxanthellates.

Although some non-zooxanthellates are regionally

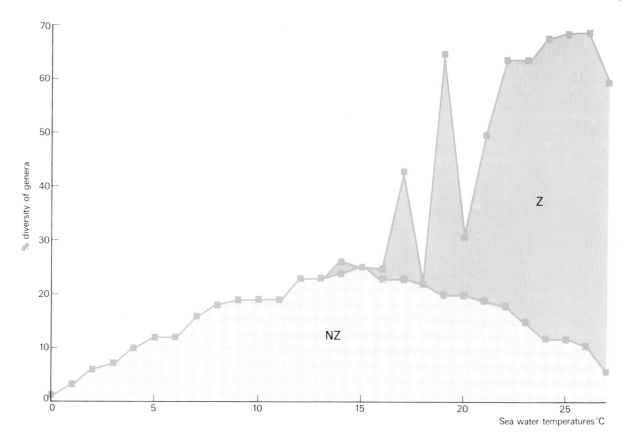

Figure 9.4 Relationship of generic diversity of scleractinian corals to sea water temperatures.

Diversity of all scleractinian corals is expressed as a percentage of the generic total (152) and is given by the upper profile. The two areas below the upper profile separate the percentage diversities of zooxanthellate corals (Z) from those of the non-zooxanthellate corals (NZ). The upper profile is derived from a summation of NZ and Z information. (Profiles and compilations were prepared jointly with Jill Darrell). Non-zooxanthellate data (lower profile) are based on Wells'

(1967) figures for the world-wide temperature ranges of 41 genera. The profile shows the number of non-zooxanthellate genera expressed as a percentage of the composite generic total of 111, known to occur at each degree of temperature. Zooxanthellate data are a combination of Atlantic and Indian Ocean distributions. Atlantic data as in Figs 9.2 and 9.3 captions. Indian Ocean data from unpublished compilation of literature records maintained by Jill Darrell. Profile plotted as for Fig. 9.3.

confined, they are taken as a whole to be cosmopolitan. Zooxanthellates, however, occur in two major realms – the Atlantic and the Indopacific. The geographical distribution of zooxanthellates (Fig. 9.1) shows that their highest latitude limits do not allow migration to take place round the southern tips of Africa and South America. We do not know whether the larvae actually range beyond these adult latitude limits, but if they do, they evidently do not survive as adult colonies in their opposite realm. Nor is there any evidence that corals can spread through the Suez or Panama Canals. It is possible that the physical/chemical obstacles are too great. The only other possible sea connections between the regions lie in Arctic latitudes and must be ruled out as far too cold. The two realms have six to nine common genera (depending on taxonomic points of view), but although certain species are strikingly close morphologically, there are no accepted common species. The generally recognised view is that the two realms

originated by the break-up of a single Tethyan–Indopacific realm in the early Miocene (Chevalier, 1975; Wells, 1956) and that the common genera are relics of a once cosmopolitan distribution pattern rather than the results of recent migration. An important feature of the separation of zooxanthellate distribution into two realms is that there is a great disparity in their diversity (Table 1 and Fig. 9.1).

The above task of splitting coral distribution into its two components is a somewhat artificial one, since the composite (uppermost) profiles of Figure 9.2 to 9.4 are actually constructed by summation of zooxanthellate and non-zooxanthellate data. It is in the nature of the distribution of these two groups that the raw data on which their profiles are based take slightly different forms. For zooxanthellates, one has taxonomic lists for particular places. Since they are largely confined to the uppermost 100 m of water, sea water temperatures throughout their depth range will be very close to surface temperatures.

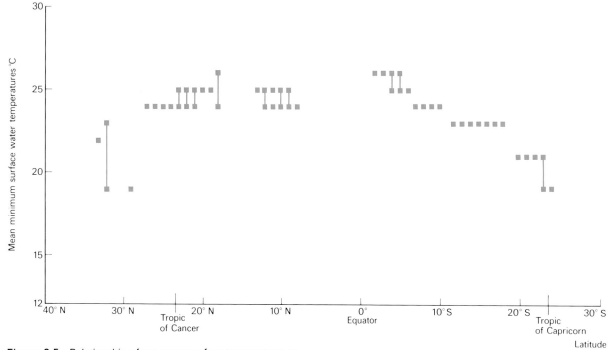

Figure 9.5 Relationship of sea water surface temperatures to latitude for Atlantic Ocean zooxanthellate coral localities.
Mean minimum temperatures (°C) for each of the Atlantic zooxanthellate localities used in Figs 9.2, 9.3 and 9.7, based on Sverdrup *et al.* (1946).

Non-zooxanthellates can be obtained from almost any depths and temperatures and the taxonomic lists in the literature generally give broad regions and temperature ranges rather than localities. The captions to the plots explain in further detail how the raw data have been compiled and processed to produce the profiles.

In the Atlantic patterns of Figure 9.2 the non-zooxanthellates increase in diversity towards the lower latitudes but reach a maximum in the subtropics, thereafter declining towards the Equator. Zooxanthellate corals appear suddenly on the fringe of the tropics coincident with the decrease in non-zooxanthellate diversity towards the Equator. For the probable geographical reasons already mentioned, there are no zooxanthellate records south of latitude 9° N (until 2° S – see Fig. 9.7). The zooxanthellates are evidently more affected by these geographical factors than the non-zooxanthellates (Fig. 9.2). If we now consider temperature instead of latitude (Figs 9.3 and 9.4) we find a similar pattern of differences between the two coral groups, though Atlantic zooxanthellates do not appear to be important until well beyond the temperature optimum of the non-zooxanthellates. In the profiles of Figure 9.4, however, the zooxanthellates' rise in diversity commences just where the non-zooxanthellates begin to decrease.

Figures 9.7 and 9.8 provide further details of the zooxanthellate diversity relationship to latitude by giving complete north-to-south profiles for the Atlantic and Indian Oceans. Neither ocean is an ideal data base. The equatorial data gap in the Atlantic (Fig. 9.7) is once again evident, and the marked asymmetry between the northern and southern sections of this profile reflects the generally recognised barrier (to corals) already mentioned in connection with this data gap. The southern latitude section of this plot is entirely based on Brazilian data, a region regarded by some as an Atlantic subprovince, though it has only one endemic genus and about 10 endemic species (out of Atlantic totals of 25 and 68 respectively). In the Indian Ocean (Fig. 9.8) latitudinal records are truncated north of latitude 30° N, and there are few records north of latitude 20° N. This is simply due to the position of the African and Asian continents. The prominent diversity minimum between 1° S and 4° S may also reflect a shortage of data points, as well perhaps as inadequate sampling on the coastlines of Africa and Western Sumatra at these latitudes. Allowing for these data problems, however, all the zooxanthellate profiles shown here show a steep diversity rise at around 30° N and 30° S. Within this latitude belt there do not appear to be any obvious consistent diversity trends or gradients, apart from a steady diversity increase between 30° N and 8° N in the Indian Ocean (Fig. 9.8). If any other gradients exist they are masked by strong fluctuations due to sampling or other factors. Earlier ideas (Wells, 1954; Rosen, 1971*b*) of steady

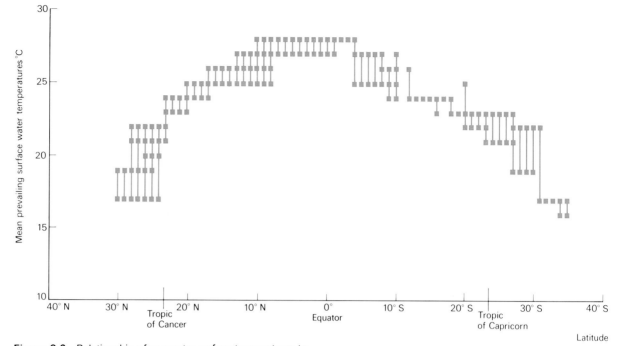

Figure 9.6 Relationship of sea water surface temperatures to latitude for Indian Ocean zooxanthellate coral localities.
 Mean minimum temperatures (°C) for each of the Indian Ocean zooxanthellate localities used in Figs 9.4 and 9.8, based on Sverdrup *et al.* (1946).

equatorial gradients in diversity in both zooxanthellate corals and other organisms do not now seem to hold so markedly for zooxanthellates (though we have seen that they do apply to corals as a whole). A step-like or very steep diversity rise in tropical and subtropical latitudes is, however, known for other organisms; for instance, predatory gastropods, cypraeid gastropods, seaweeds, danaid butterflies, termites, frogs and snakes. Stehli (1968), who has compiled the data for some of these mentioned groups, states that this pattern is frequently found in tropical–subtropical shallow water benthic animals. The diversity–temperature profiles (Figs 9.3 and 9.4) show a similar pattern of zooxanthellate diversity in relation to non-zooxanthellates, with a steep rise in zooxanthellate diversity at an intermediate temperature range.

Distribution has so far been considered largely in terms of changes across latitudes and their equivalent temperatures. Figure 9.1, however, shows that for zooxanthellates at least, diversity can also range from maximum to zero across longitudinal lines especially from west to east in each ocean. Comparison of such patterns with temperature rules out any obvious correlation. The customary explanation is that only a few species can disperse across the large distances in these impoverished easterly regions (Wells, 1969; Rosen, 1971*b*). An alternative view is that zooxanthellate corals are poor dispersers, and that their present-day distribution does not so much reflect their dispersal abilities as the relict effects of previously more

continuous geographical ranges (Heck & McCoy, 1978), as discussed more fully later on.

Conclusions

Two conclusions are suggested by the data presented here. First, since temperature increases regularly with latitude (Figs 9.5 and 9.6), but zooxanthellate diversity increases sharply over a short temperature and latitude range, temperature is probably acting in a threshold fashion, or factors other than temperature alone must be considered to explain this pattern. Second, an explanation of tropical high diversity in corals must concentrate on the biogeography and ecology of zooxanthellate corals.

The natural history of corals

The purpose of this section is to provide an outline of coral biology emphasising the facts and ideas needed for a discussion of diversity. Corals are still poorly understood, even by those who study them. I shall draw on some of the general ideas of evolutionary ecology or niche theory, because this provides a useful framework for organising ecological information.

'Coral' has been applied to a wide variety of organisms, not all of them coelenterates. For the present survey I am referring to the 'stony corals', a group of Anthozoa with rigid calcareous exoskeletons, that is Scleractinia (previously Madreporaria or Hexacorallia). They go back in the

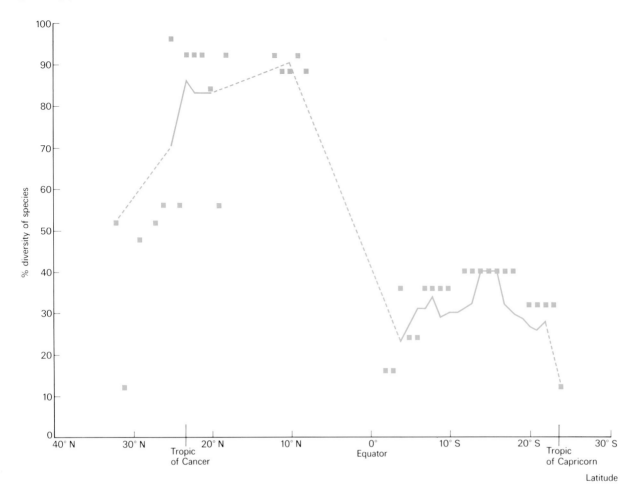

Figure 9.7   Relationship of species diversity of zooxanthellate scleractinian corals to latitude in the Atlantic Ocean.

   Source of zooxanthellate data as in Figs 9.2 and 9.3 and here converted to latitude spans and expressed as a percentage of the Atlantic total of 68 species. Actual plot is based on running mean of 5° latitude spans except for broken line sections. (Profiles and compilations were prepared jointly with Jill Darrell).

geological record to the Triassic (200–230 million years ago). They can be thought of as sea anemones with skeletons. The skeleton confers support and, generally speaking, attachment though many corals do live unattached on the sea floor. About 60 per cent of the 220 living genera are colonial. Unlike most attached invertebrates corals are not true suspension feeders, but are primarily carnivores that catch zooplankton prey with their active tentacles, though they do have other food sources. About 55 per cent of living genera are zooxanthellate, but the full physiological and ecological implications of this symbiosis are still not understood. There is a correlation between zooxanthellates and coloniality, however, since only 25 per cent of living non-zooxanthellates are colonial, compared with 90 per cent for living zooxanthellates. The ecological role of zooxanthellae has long been recognised but this has largely been in a trophic context; that is, the recycling of nutrients both between zooxanthellae and corals, and

within the reef ecosystem as a whole. This correlation between zooxanthellae and coloniality suggests that more attention should be paid to the role of zooxanthellae in the context of functional morphology and the ecological significance of the many colonial forms seen in zooxanthellate corals. Corals are said to reproduce in two ways, sexually and asexually. Sexual reproduction gives rise to free larvae either by internal or external fertilisation. Asexual reproduction only rarely produces detached clonal offspring. Much more commonly, daughter polyps remain attached to the parent and form the basis of colony formation. In this respect it is not true reproduction, but rather an iterative form of growth (Rosen, 1979).

Coral–zooxanthellae symbiosis

This symbiosis has been one of the most studied aspects of coral biology, but there are eight salient points relevant to the question of tropical high diversity: environmental

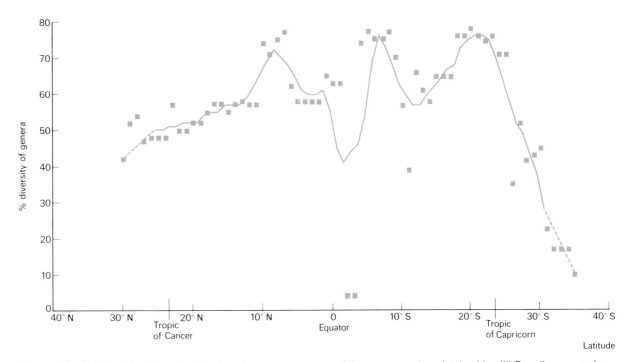

Figure 9.8 Relationship of generic diversity of zooxanthellate scleractinian corals to latitude in the Indian Ocean.
  Zooxanthellate data are based on unpublished compilation of literature records maintained by Jill Darrell, converted to latitude spans and expressed as a percentage of the Indian Ocean total of 93 genera. Actual plot is based on running mean of 10° latitude spans, except for broken line sections.

restriction, trophic generalist character, recycling of nutrients, accommodation to high temperature stress, coloniality, faster calcification rate, plant-like growth habits and geological longevity. These are explained in turn below.

It has already been mentioned that zooxanthellates are restricted to warm shallow water, but it is not shallowness itself that matters so much as a minimum level of illumination for zooxanthellae to photosynthesise (Wells, 1957; Yonge, 1973).

Most corals feed on zooplankton but they also receive their nutritional requirements in other ways such as suspension feeding, bacterial assimilation, uptake of both dissolved organic and inorganic materials and assimilation of products of the zooxanthellae (mainly glycerol together with glucose, alanine and glycolic acid; Muscatine, 1973). This dietary variety makes it difficult to fit zooxanthellate corals into any neat position in the food web, and so makes it harder to speculate on functional morphology in relation to feeding, on limiting resources and on evolutionary issues like selection pressures. Although this variety invites the possibility that coral species might differ from each other in their relative dependence on these categories, and within these categories, there is as yet no direct demonstration of this. I shall have to return to this point when discussing whether or not tropical high diversity in corals is maintained by fine resource partitioning.

The symbiosis seems to allow nitrogen and phosphorus

to be recycled between the zooxanthellae and the corals (and hence, it has been argued, through the whole reef ecosystem). In this way both corals and zooxanthellae overcome the low availability of nitrogen and phosphorous in tropical seas (Lewis, 1973; Muscatine, 1973, 1974). Indeed complementary to this view, is the idea that nitrogen deficiency is a stress factor that might explain the origin of the symbiosis. Low nitrogen levels are also understood to account for low primary productivity in the open tropical surface waters and so might explain the great dietary range of corals, especially the use of zooxanthellar products.

Not only are nitrogen and phosphorus in shorter supply in the tropics, but the higher temperatures raise metabolic rates and hence the demand for nitrogen and phosphorus, so giving the symbiosis additional significance. Higher metabolic rates also strain the respiratory and excretory systems of lower invertebrates like corals, and here the symbiosis is again beneficial to the corals, because the intracellular algae remove waste products (nitrogen, phosphorus and carbon dioxide) more efficiently than the coral would on its own (Beklemishev, 1969; Muscatine, 1974).

Zooxanthellate corals are almost entirely colonial as already mentioned. Moreover, colony structure in zooxanthellates is generally more complex, more highly developed and more closely integrated than in non-zooxanthellates (Coates & Oliver, 1973).

The presence of zooxanthellae enhances calcium deposition rates in zooxanthellates by as much as ten times compared with the same corals growing in periods of darkness (Vandermeulen & Muscatine, 1974). By extrapolation, deposition rates in non-zooxanthellates (assumed to be equivalent to zooxanthellates growing in darkness) will likewise be a tenth of zooxanthellate rates. The precise reason for this is not yet known, but the zooxanthellae are clearly implicated. Zooxanthellates appear to translate this rapid deposition rate in a variety of advantageous ways, the most obvious of which is a more rapid growth of the outward shape of the coral. Zooxanthellate solitaries and colonies (but not necessarily colonial polyps) are also larger than their non-zooxanthellate counterparts, though zooxanthellate colonies have a greater interspecific range of polyp sizes from both larger to smaller extremes than the polyps of non-zooxanthellate colonies.

The plant-like form of many zooxanthellate coral colonies has often been discussed. It is partly inherent to all colonial benthic forms and it is also present in some non-zooxanthellates. I therefore believe that it has been overstressed. The zooxanthellae do, however, need light and the fact that many zooxanthellate species exhibit characteristic colony forms not seen at all in non-zooxanthellates (Wells, 1956; Coates & Oliver, 1973) suggests an evolutionary connection between these forms and the symbiosis. The literal similarity to plants is not so very obvious apart from a light-seeking habit, the frequently bushy shape of colonies and a fanciful floral aspect, but they do share with plants a tendency to maximise surface area. This need not be entirely due to light needs however.

Indirect evidence from the dinoflagellate fossil record, together with stratigraphic extrapolation from growth forms characteristic of living zooxanthellates suggest that the symbiosis is long established in geological time, at least since the mid-Jurassic (170 million years ago) and it was just possibly present in a few Upper Palaeozoic coral groups not directly related to Scleractinia (Rosen, 1977). This, together with the proportion of living corals which are zooxanthellate, points to the success and importance of the symbiosis.

Aspects of niche theory applied to corals

In this review I use niche (Hutchinson, 1965; Valentine, 1973) in its broadest sense to mean the role of an organism in the environment. The role may be measured by an infinite number of parameters such as food requirements, substrate type and amount of light. The niche is thus multidimensional. Niche theory assumes that every species has a unique combination of requirements and adaptations along the different dimensions of a niche and by definition has a unique role to play in any community. The role of the organism or the life-style can be thought of as consisting of groups of traits, like habitat preferences, physical–chemical limiting conditions, diet, styles of searching and obtaining food and life-cycle patterns. Even if two species are identical in one group of traits, there should be differentiation in some other trait. A complete

match would imply that the two supposed species are in reality the same.

In more abstract terms this idea of life styles and traits is formalised as a multidimensional phenomenon in which traits are regarded as axes (Fig. 9.9). This allows information on traits to be organised, especially if it is also quantifiable. Thus various species might be compared along an axis representing the timing and frequency of feeding, or on a prey preference axis. A single axis may need to be broken down in turn into several further parameters. An important subjective element of choice of axes for investigation, is to avoid traits that do not yield recognisable differences between species. This is easy to avoid for readily observed characters; for instance, all corals require an aqueous environment. But for a relatively obscure group of organisms like corals, we do not really know of many axes that are likely to yield useful differences between species.

Differences along axes are conceived as being plotted points or areas spread along the axis, each for a different species. Alternatively they might be overlapping areas whose mid-points are distinct. In addition, we can think of different sizes of areas representing different species, so that an axis might be finely divided or broadly divided. This leads to the idea of 'niche width', an important concept in discussing tropical high diversity, because many authors believe that the tropics are characterised by narrower niches, so enabling more species to be fitted in to any one habitat.

A related further idea of great importance is that many resources important to an organism are available in quantities, or at rates of supply that are limited. Thus the rate of supply of suspended food particles in a particular environment may be sufficient to maintain a hundred suspension feeders but insufficient for two hundred. Different species in the same environment are expected to show different means of obtaining these resources, either directly (for instance diet specialisation) or indirectly, say by living in a habitat that still gives them access to this resource but which is unsuitable to other species. It is hard to judge whether a resource is limiting, however, without reference to the organisms that draw on this resource in a particular environment. The same resource could in theory be present in the same absolute quantities at two different places, but be limiting in one and not limiting in the other because the dependent fauna and biomass were different at each. Nevertheless, tropical high diversity is often discussed in terms of a finer division or partitioning of limiting resources than is found in temperate latitudes. Tropical species are in this way conceived as having narrower or more specialised niches in response to this. Partitioning is also often seen in terms of interspecific competition but more recently there has been interest in subtler possibilities like optimal use of several resources or several strategies, and recognition that if niches overlap, survival is most assured for those species that 'get there first'. It is now appropriate to consider possible significant niche axes for zooxanthellate corals.

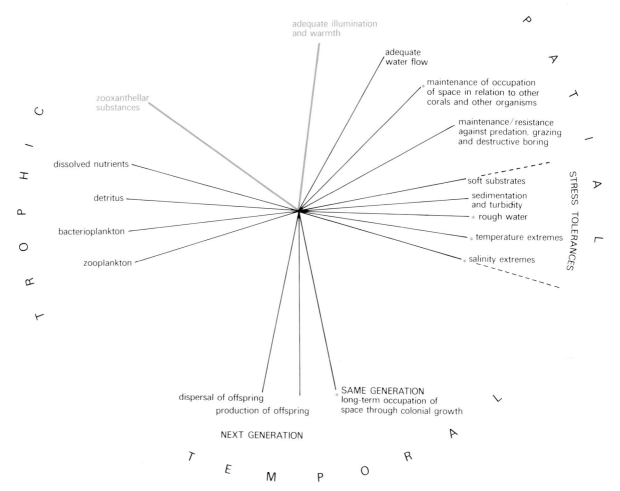

Figure 9.9 Towards a niche concept for corals.
Each radius represents an axis along which coral species appear to be differentiated. Each axis can be thought of as a strategy or trait within which a particular coral species shows some degree of specialisation. Every coral species should have its own combination of specialisations from the sum of the axes.

Bold blue indicates axes for zooxanthellate corals only. Asterisk symbols indicate strategies enhanced by symbiosis with zooxanthellae.

Some niche parameters for corals

Space The two most consistently marked features of corals are that they are often colonial and that they show a striking range of growth habits. 'Growth habits' refers to the outward shape of a coral growth together with the nature of its colony construction (where applicable). Growth habits moreover are very variable within coral species (this being one of the main obstacles to good coral taxonomy). Most coral workers agree that coloniality and growth form are the skeletal expressions of different ways of occupying space. It is not, however, a simple matter of a single type of growth representing an adaptation to a single space or special habitat. The following examples, many of which are based on Jackson (1979), attempt to illustrate this.

A large surface area will increase the chance or rate of intake of dissolved substances, plankton capture, and for zooxanthellate corals, light capture. Highly branched forms will be less susceptible to sediment accumulation. Forms which rapidly encrust over substrate surfaces can exclude other substrate organisms more effectively. Many growth forms have mechanical properties that confer survival advantages in wave or current exposed conditions (Graus *et al.*, 1977). A rapidly growing expanded basal attachment area is also important in these conditions for all forms other than sheets. Rapid growth helps young corals and detached fragments to survive in unstable substrate conditions. It also offsets damaging effects of boring, breakage, grazing and predation, though this can also be achieved by stouter skeletal structure. Strongly upward growing forms like columns and vertical branches give access to food and light at higher levels in the water column and lessen the effects of sediment settling on the colony. Ecological stratification in which some corals have a preference for growing on areas raised up from the neighbouring substrate has a similar advantage (Rosen, 1971a; Crame, 1977). Occupation of space is therefore truly three dimensional in that corals do not just cover substrate areas horizontally. In this way, stalked table forms and shelf-like shapes for example, can overshadow benthic organisms below, and a mass of bushy branches represents a very effective occupation of volume for an optimum amount of coral skeleton.

We lack systematic quantitative observations to support many of these interpretations, but they are plausible and theoretically testable. When it is also realised that the growth form of a particular species can represent an optimum combination of several of these life styles, it is then easy to see how the variety of possible growth forms and life styles in corals might well be comparable with the observed number of species. This is where we should start looking for explanations of coral diversity. I therefore believe that the many ways of occupying space represent a very important starting point for defining niche axes in corals, and that we can expect coral species to be well differentiated along these spatial axes (Fig. 9.9). Two additional points support this view. First, it is coloniality that makes this variety in growth form possible (Wells, 1956; Coates & Oliver, 1973; Rosen, 1979), and this together with rapid growth takes us back to the zooxanthellae. Second, interspecific aggression in corals (Lang, 1973), in which species can 'attack' one another in a consistent pecking order when their growths come in near-contact is another strategy for space occupation. It even appears to be a complementary factor to the growth-form strategies above.

Food For many animals, trophic axes are the most studied and the most finely differentiated between species. We do know that different coral species vary in the manner and timing of their feeding, and that there are some broad differences on emphasis in the principal sources of their food (Muscatine, 1973; Lewis & Price, 1975). Large polyp species are generally active zooplankton feeders, but small polyp species appear to depend more on bacteria, detritus and dissolved substances. Until such time as we have comprehensive systematic data for different species about the relative importance and composition of the different possible dietary components, it will not be possible to say whether every coral species has a truly unique diet. Current information suggests rather the opposite; namely that corals are trophic generalists, or at least exist on a range of options amongst the major possible food sources.

A very special form of dietary distinction between species has, however, been proposed by Porter (1976), who has argued that a finely divided spectrum exists amongst Atlantic zooxanthellates from those corals that are largely

dependent on zooxanthellar products, to those corals that are largely dependent on zooplankton. For reasons that will be given later, I do not believe his observations give strong support to his own claim. His work does show, however, that we should explore the complexity and variety of coral growth forms with regard to their possible trophic significance. If, contrary to Porter's view, corals are really trophic generalists (and if all food sources are indeed limiting), then one might predict that corals partition trophic resources indirectly by species having their own place (and time, or both) for feeding. In this way, I suggest that the idea of spatial partitioning in corals has two component strategies. First, it is habitat specialisation. Second, and especially within the same particular habitat, it is the way in which various adjacent species each ensure their own supply of broadly similar food. We should therefore try initially to interpret particular growth habits in the light of both these strategies. The single example of a colony habit of large surface area in, say, the form of upward growing branches will serve in the context of the previous section to illustrate this point.

Time, growth and reproduction There is a certain amount of partitioning by day and night feeding in zooxanthellate corals, but it does not appear to be a finely differentiated feature, though it may distinguish populations within species.

Production of offspring would appear to be non-seasonal at the species level (Rinkevich & Loya, 1979), which is hardly surprising bearing in mind the relative lack of seasonality of temperatures and of primary productivity in the tropics. However, seasonal breeding does occur within species in relation to their geographic location. Species themselves appear to be very flexible. There is as yet no evidence for the finely controlled breeding times that characterise many insect species for example (Southwood, 1977). This may again relate to the generalist nature of coral diets, but we need much more information on coral breeding before we can make confident statements.

Loya (1976) has argued that certain coral species have a much higher colony turnover than others, but seen as a whole, zooxanthellate corals are more like trees, tortoises and elephants in being remarkable for their longevity. Indeed there is no real information on what constitutes the life span of most coral species, nor even how this concept applies to corals. Growth rates do vary, however, and some species do have an evident colony size limit irrespective of the limiting factors of the habitat.

Corals have generally been regarded as good dispersers but Heck & McCoy (1978) have challenged this, showing that most coral planulae probably settle in less than three weeks. This leads naturally to the question of K and r selection (MacArthur & Wilson, 1967), which has so far been discussed for corals only by Loya (1976). He regards (with more recent reservations) certain coral species as 'r-strategists' in having most of the characteristics associated with high population turnover. His suggestion is misleading if corals are compared with many other organisms, since their longevity and other characteristics point more to corals as a whole having a K-like character. Within their K-like nature, however, this work shows that species can be differentiated along an axis towards the r strategy. If corals do indeed occupy a generally K-like position then one would predict that, compared with other organisms, they devote a correspondingly small proportion of their metabolic energy to production of offspring and that they are not good dispersers. Rather, as already stressed, they have specialised in consolidating their place of attachment, a strategy that would be achieved in part presumably by their devoting most of their energy to attaining a large size.

The K-r spectrum is often considered to be an oversimplification (Grime, 1979; Southwood, 1977). Grime for example has suggested that for plants at least a third strategy should be added, that of stress tolerance, to form a 3-component ecological strategy system. This will be discussed next.

Stress Stress has already been mentioned in connection with the significance of the algal symbiosis. In this context, it applies generally to all corals. Unless Porter (1976) is correct in his inference that corals are differentiated in their nutritional dependence on zooxanthellar products, the kind of stress that appears to underlie the algal symbiosis is not a likely axis for species differentiation.

The problem with the idea of stress is that, even more than the other concepts discussed here so far, it is difficult to define stress absolutely. It may be taken to mean conditions that restrict growth and reproduction. The idea of limiting factors is well known, but stress can be thought of as the gradient between ideal conditions and the ultimate limits of survival. Any given organism will have an ideal set of conditions, generally optimum values of particular environmental parameters like amount of illumination, or rate of sedimentation. Here it is subject to least stress. Passing away from such ideal conditions, the organism becomes increasingly intolerant of the stress created by these less favourable conditions until it reaches the point where it cannot survive at all. Tolerance of stress is species-specific; some species are able to endure wide variations of certain environmental conditions, while others have much narrower ranges of tolerance to the same conditions. A single species moreover may be stress tolerant for one parameter and not for another. There are therefore problems in trying to identify and quantify stress. We may assume that a factor like extreme warmth for example is stressful to other organisms because it is stressful to our own species. Or we may assume that the conditions found at the environmental range extremes of a species, or of a group of organisms, are stressful to them. Thus coarse sandy beaches and heavily exposed rocky shores clearly favour very few intertidal organisms and so we intuitively regard them as stressful.

Grime (1979) turns to the organisms themselves to assess stress. Working with plants he defines stress simply as the external constraints that limit the rate of dry matter production of all or part of the vegetation. This is most generally affected by non-optimal supplies of solar energy, water and nutrients. His hypothesis is that stress tolerance is reflected by the tendency to be small in size and slow-growing. Grime believes his concept is equally applicable to animals. He has incorporated the strategy of stress tolerance within a version of the familiar 'K-r' spectrum of reproductive strategies, modified into a 3-component triangular system (Fig. 9.10, G1). His stress tolerators ('S'-strategists) correspond to the K-strategists of previous authors. His 'ruderal' category ('R'-strategists) corresponds to the r-strategists of previous authors, and his category of 'competitors' ('C'-strategists) represents an intermediate strategy. He is able to assess all three strategies in terms of biomass production and the potential of a plant for lateral spread. Species can then be plotted by triangular ordination (Fig. 9.10, G1) to reveal their particular tendency towards the extremes of being stress tolerant (slow-growing and small), ruderal (fast-growing and small) or competitive (fast-growing and large).

It should be easy to apply Grime's approach to corals. Using his criteria, corals generally seem to be stress tolerant compared with other organisms, with perhaps a greater tendency to be competitive than ruderal (Fig. 9.10, G1). As mentioned already, Loya (1976) has argued that some coral species are r-strategists, that is ruderals in Grime's scheme. It might be better to think of them as showing a ruderal tendency only within the C-S character of corals. I would expect corals as a whole to plot in a similar region of Grime's C-S-R triangle to trees and shrubs in his figure 18d (Fig. 9.10, G1). Within the range of variation of such a plot, individual coral species, as with the plants studied by Grime, should be differentiated from each other. We can then look at how they are differentiated with respect to the stress component in particular, that is, their differentiation along the stress axes of their niches (Fig. 9.9).

In the Indopacific, zooxanthellate corals occur in a sequence of assemblages (groups of coexisting species) that appear to be related to the two gradients of strength of water movement and illumination (Wells, 1954, 1957; Rosen, 1971a, 1975; Faure, 1977). A similar set of assemblages occurs in the Atlantic (Geister, 1977; Faure, 1977). Figure 9.10 shows how I envisage these assemblages plotted against their most likely controlling stress parameters. In that different species appear to occupy their own optimum areas within the scheme, we can suggest that they are differentiated in various degrees of stress tolerance. The kinds and degrees of the particular stresses in Figure 9.10 are associated with particular reef habitats, and so we can also regard this species differentiation as a further manifestation of corals' tendency to specialise in spatial strategies.

Grime's approach provides a basis for testing this scheme in future work.

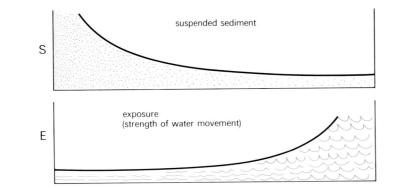

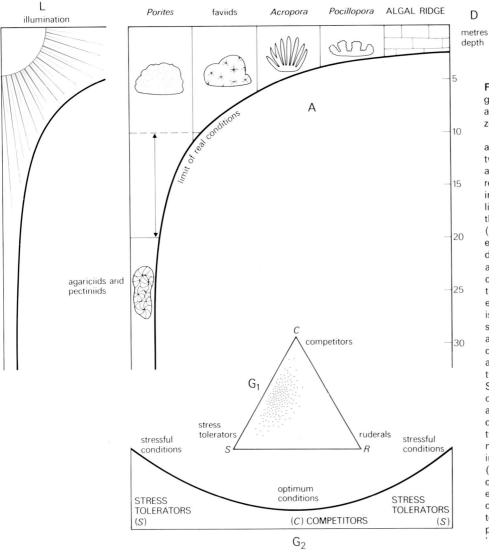

Figure 9.10 Conjectured stress gradient control of ecological assemblages of Indopacific zooxanthellate corals.

In the main central plot (A), coral assemblages are shown in relation to two principal parameters, depth (D) and exposure (E) (see text for references). Since these are not independent of each other, there is a limit of real conditions as shown. Note that depth correlates with illumination (L) and that illumination is really the effective vertical parameter. Precise depths at which illumination-controlled assemblages occur will vary with local conditions and hence in poorly-lit areas they will also encroach upwards on exposure-controlled assemblages. This is indicated by double-headed arrow symbol. On the horizontal axis, the amount of suspended sediment (S) correlates negatively with exposure (E), and these factors together behave as two counteracting stress gradients. Stress tolerant species of corals thus occur at each extreme of the assemblage scheme (G2), while the central area of optimum conditions is typified by dense coverage of numerous species of corals specialised in competing for space. In Grime's (1979) terminology zooxanthellate corals appear to be differentiated into ecological strategies within a broadly competitive-to-stress-tolerant (C–S) tendency, conceived here as plotted points within the coral area of Grime's 'C–S–R' scheme (G1).

Predation, grazing and boring It is widely agreed that predation is intense in the tropics (Vermeij, 1978). Predation on corals is inhibited by their powerful stinging cells, but certain fish, gastropods and echinoderms, including the now famous Crown-of-Thorns starfish have evidently overcome this difficulty. So far as the survival of an individual coral growth is concerned it is immaterial whether the attack on it is for the coral tissues themselves, or as in the case of parrot fish, for the purposes of grazing on the sub-surface filamentous algae (not to be confused with the endodermal zooxanthellae). Once again however we can only speculate. It is easy to imagine that faster growth rate of the skeleton or the more frequent addition of new polyps to the colony might offset the effects of predation, grazing and boring. Spiny skeletal surfaces and extended tube-like corallites may inhibit some predators, but this would depend on the size of the coral ornament with respect to size of the potential predator and its manner of feeding. Most Indopacific corals succumb to the Crown-of-Thorns starfish whose all-consuming capacity is inhibited only by other ecological factors. Although it may prove fruitful to look for defence specialisations in corals as a possible axis for species differentiation, it may be more relevant to consider the general effects of predation on diversity as a disturbance factor (see below).

Summary

I now give five main features of coral biology that bear on coral diversity. These are represented in expanded form in Figure 9.9 as niche axes.

In this synthesis I emphasise corals as specialised space occupiers, a view suggested by the impressive interspecific and intraspecific range of coral growth forms. Growth forms can be interpreted as space occupying strategies. Only a part of this specialisation is represented by fine habitat differentiation between species. Ecological stratification, stress tolerance, interspecific competitive interactions and the intrinsic qualities of many growth forms are also important.

The coral–algal symbiosis is closely associated with a highly developed colonial habit together with the above variability of growth form. It therefore seems that the zooxanthellae help (or cause) corals to specialise in space occupation by enhancing their colonial tendency, by faster calcification rate and by broadening the trophic options available to corals. This last factor effectively eases competition for zooplankton food that is less readily available (more limiting) to the coral than is light for the zooxanthellae.

Zooxanthellate corals in particular are broadly generalist in their diet. Trophic partitioning is achieved not by specialist diet but by acquiring broadly similar food from different places, i.e. spatial partitioning, supported by direct interspecific spatial competition and broad temporal partitioning.

Corals as a whole have considerable longevity and are poor dispersers. They are thus broadly K-like (or C–S in Grime's terms) in character. Temporal partitioning mainly appears to take the form of a tendency for some species to have a more ruderal (r or R) character. These tendencies enable corals to occupy a wider range of habitats by extending the range of space that corals can occupy (Fig. 9.10).

Resistance to predation, boring and grazing may be important by analogy with other organisms, but there is even less information to support this as compared with the above propositions.

Kinds of explanation for tropical high diversity

Although it may seem arbitrary to separate them, there are really two underlying questions to the problem of tropical high diversity. First, what are the factors that might explain why or how more species can live side by side in the tropics? Second, why do the tropics apparently either generate or accumulate more species than the non-tropics? The first question is essentially ecological, and the second essentially evolutionary or historical. The ecological aspects can be usefully divided into environmental factors and biotic factors. In this way we can group ideas about tropical high diversity into three kinds: historical, environmental and biotic. They are not necessarily incompatible alternatives. Almost certainly we must consider all three in a search for a thread of compatibility between them. I will now explain what I mean by these three kinds of ideas.

Historical explanations of tropical high diversity say that evolutionary processes or geological events, or some combination of them both have created or accumulated more species in the tropics; for instance, Wallace suggested that the Pleistocene ice advances drove temperate and boreal species towards the Equator. Only a few survived in or returned to higher latitudes after the ice retreated.

Environmental ideas basically rest on a particular inherent and universal physical or chemical aspect of the tropics which is then thought to favour the existence of more species. The most obvious and popular idea of this kind is that there are more species in the tropics because the tropics are warmer. Warmth is sometimes thought to generate more species because it causes a higher mutation rate. Many environmental 'explanations' are merely observations of the coincidence between a particular aspect of the tropical environment and high diversity, and so are not causal as they stand. These coincidences may be appealing but to investigate them we usually have to turn to physiology and ecology.

Biotic ideas say that organic activities and interactions in the tropics are in some way different from those in other latitudes and as a result of this more species can be accommodated. Examples include more competition, more predation and more specialisation (or more and narrower niches) in the tropics. The extensive literature devoted to these ideas gives the impression that authors believe that these ideas represent self-contained explanations of tropical high diversity. In a sceptic's view they are simply restatements of the original observation. A more positive view is that they are statements about the ecological manner in which high diversity is maintained, but certainly they rarely have much to say about where all the species might have come from in the first place.

Each of the three foregoing kinds of ideas represents a possible starting point for explaining tropical high diversity. We need to start somewhere, and since research emphasis on biotic explanations has possibly been the greatest, this is reason for starting with these.

Biotic explanations

Narrower niches, limiting resources and saturation

If there are more species in the tropics, there must be more niches, according to niche theory. More niches could result from there being, say, more habitats or more resources, but these are strictly environmental arguments. An alternative and much pursued possibility is that, regardless of any differences in amounts of resources or in numbers of habitats that may exist across different latitudes, the tropics are characterised by a finer division of these resources. That is, tropical species are more specialised; they have more narrowly defined niches. These two complementary possibilities are crudely comparable to more cake, and to smaller slices, respectively. There is no way, however, of determining either the absolute size of either the cake or the slices.

A further important idea related to this view is that a particular environment can carry a maximum number of species, and this number represents saturation (for a review of these ideas, see Connell, 1978). The evidence from observations on ecological succession suggests that not all environments have reached saturation point. Succession implies progressive adjustment of an ecosystem towards an equilibrium end point. Saturation is reached at this end point. The maximum number of possible slices has then been cut from the cake. So now we have three possible ways of looking at tropical high diversity, together with various combinations of them, and all without absolute points of reference: the tropical cake is either bigger, or it is more finely sliced, or more of the potential maximum number of slices has been cut from it.

Saturation, equilibrium and succession add the dimension of time to the resource arguments. A growing amount of evidence points to diversity not rising to a maximum when equilibrium is reached but passing through a maximum at some intermediate successional stage, and falling to a low diversity equilibrium (Connell, 1978). It follows that factors that prevent equilibrium from being reached are more important than resources on their own. If this argument is applicable on a global scale then we should be looking for factors that prevent equilibrium from being reached in the tropics as a whole. As far as strictly biotic explanations are concerned, however, we first need to look at possible niche partitioning in corals. These other ideas are best deferred to the section on environmental arguments.

Fundamental to ideas on niche partitioning is the concept of finite or limiting resources (Valentine, 1973). It has been widely argued that competition for a relatively restricted supply of resources like food, nutrients and space leads to finer degrees of sharing. Resources in abundant supply or infinite in amount are not limiting and not finely divided. We must therefore look for niche axes related to limiting resources in order to establish what specialisations might differentiate corals.

Autotrophy, heterotrophy and resource partitioning

The principal resource usually considered in the discussion of partitioning is food. The difficulty in the case of corals seen as a group is that as already explained they are heterotrophic. It is easy to imagine that zooplankton are probably limiting, and possibly also bacteria, detritus and dissolved substances. On the other hand we know that shortage of nitrogen and phosphorus is offset by the recycling relationship between the symbiotic algae and the host corals (zooxanthellates). This, and the use of zooxanthellar substances in coral nutrition, shifts interest to the density of zooxanthellae in coral tissues, and to the light needed for photosynthesis by the zooxanthellae. It is not difficult to think of models by which all these might be limiting, but it is very difficult to think how they might be tested, bearing in mind the numerous trophic options open to corals, and their complex metabolic pathways. It is also difficult to see how these options might be limiting, when taken all together.

Porter (1976) has suggested that the apparent niche overlap (on the feeding axis) that is implied by the heterotrophic (generalist) character of corals is unreal especially in the face of low overall primary productivity in tropical waters. He modifies the idea of niche partitioning by presenting evidence that within the broadly generalist nature of coral diet, zooxanthellate corals are partitioned not by any particular dietary differences, but by a greater or lesser dependence on nutritional substances derived from the zooxanthellae. This is the only published suggestion to date that applies the idea of niche partitioning to corals. If Porter is right, then we do have a purely biotic explanation of how more species of coral coexist in the tropics than in temperate waters.

In place of direct evidence about densities of zooxanthellae, quantities of zooxanthellar products and the relative quantitative importance of different food substances in corals, Porter presents an ingenious indirect approach based on directly observable features of the coral skeleton. He regards heterotrophy as being represented by adaptations for zooplankton capture (that is, in addition to other possible energy sources), for which he takes large polyp diameter to be the simplest indication. Polyp size is equivalent in the skeleton to the more readily measured parameter of corallite size. Autotrophy (that is, dependence on the photosythetic products of the zooxanthellae) is related by Porter to the amount of surface area which a coral presents for light capture. In order to make total surface area independent of arbitrary growth size, Porter uses its ratio to another growth-related feature, volume. In this way he interprets a high surface area to volume ratio as a relatively heavy dependence on zooxanthellae. He then uses his two trophic indicators, polyp size and surface area, as two axes against which he plots a set of Atlantic species and this reveals a striking differentiation between them. Many corals cluster towards having a small polyp diameter, but these small polyp forms are spread out in their surface area ratios.

For the following reasons, Porter's argument seems unconvincing. Active zooplankton capture is not confined to coral species with large polyps (Lewis & Price, 1975). Some small polyp forms appear to be effective predators by virtue of their vigorous and well-developed tentacles. A large surface area moreover would be just as well suited for capturing almost anything that is in the water by way of possible food and nutrients (Jackson, 1979), so unless potential for light capture can be accurately distinguished and demonstrated, surface area alone is not very informative. The ratio of surface area to the volume of the whole coral also seems a strange parameter, since at least 95 per cent of this volume must be skeleton and is therefore no longer an active part of the coral metabolism once it has been deposited. It would be much more useful to measure the surface area of living tissue in relation to volume of living tissue or to find some other measure which relates numbers of zooxanthellae to coral biomass. Porter does not in any case say how he measured the elaborate colony shapes that typify many species, nor does he give the species he used. His plot shows neat points for every species even though each must in reality show a range of variation. In view of these difficulties Porter's work cannot so far be thought of as a convincing demonstration of niche partitioning in corals, with particular regard to food and light dependence. His plots may represent some other form of partitioning, however, and for this we must now turn to Jackson's work.

Jackson (1979) has given surface area to volume ratios considerable theoretical attention, and in general it seems that they are most easily related to space occupying strategies. I have already suggested that spatial strategies are the most attractive axis (or axes) to consider for probable partitioning in corals, and Porter's plot may in fact unintentionally support this better than it does his own argument. The question that now arises is whether space is limiting in corals or whether food is. Porter believes that without the zooxanthellae, both are limiting, but that the symbiosis gives corals access to the virtually non-limiting resource (in shallow water) of light. It is difficult to comprehend why corals should then show his proposed fine gradational dependence on light. Although there is probably a varying dependence on light amongst

zooxanthellate corals, it may be easier to see this in terms of a broadly generalist trait by which corals keep their trophic options open, and to think of space as being a more limiting factor, as an equivalent to trophic partitioning.

The purpose of this section, however, is to consider whether a factor like space is really limiting. Certainly, in many reef environments, there appears to be little room for a greater density of attached benthic organisms than is to be found there already. But corals, by building substantial three-dimensional substrate structures of their own, actually add to habitat complexity and hence can increase the space available to them (and to other benthics). The earlier discussion on stress gradients also suggests that corals have extended their ecological range into otherwise unfavourable environments (Fig. 9.10). If the ability to do this is seen in evolutionary terms then pressure to divide up existing space may be less important than the addition of new habitats to corals' range. In short, how does one conceive of space as finite and limiting in the present global gradient context? The fault may lie not with Porter's or my own efforts to find fine partitioning, but with the very concept of limiting resources.

Predation

Another biotic factor that is frequently mentioned in connection with tropical high diversity is predation. That is, there is more predation in the tropics. This is usually intended to mean that there are more predatory and defensive specialisations, a larger number of predatory species, and a greater overall intensity. This has been convincingly argued by Vermeij (1978). This idea can be interpreted in terms of niche partitioning and limiting resources as above but as already discussed there is as yet no obvious way in which it applies to corals either as predators or prey. The general success and importance of predation in the tropics appears to be related to lack of marked seasons and hence is linked with the environmental factors to be discussed next. There is also the interesting possibility that predation acts as a disturbance factor, but this again is more conveniently considered along with non-biotic disturbance factors in the environmental section that follows.

The simplest counter to predation as an explanation of high diversity is that it is only a restatement of the fact of high diversity. If there are more species in the tropics then there is likely to be a higher absolute number of predators, as well as a higher number of organisms to prey upon.

Summary

The ideas of niche partitioning and limiting resources appear to explain the accommodation aspect of tropical high diversity in a number of other groups of organisms (not reviewed here), but there are serious difficulties in its application to corals. The underlying concepts are disconcerting in their lack of intrinsic absolute reference points. For corals, space is more obviously partitioned than food, but it is far from clear how one might determine the relative niche widths along these two axes, and hence how one can say that either of them might be more finely divided in the tropics than in high latitudes. We are left to note that in the tropics corals are definitely more specialised and that this must relate in some way to their tropical high diversity.

Environmental explanations

So far we have seen that although corals appear to be specialised in particular traits, it is hard to say whether this is a response to limiting resources. We need to know moreover whether this is a cause of high diversity or the consequence of other underlying causes. The approach of this section will be the consideration of various environmental characteristics of the tropics that distinguish them from higher latitudes and to discuss how these characteristics might relate to tropical high diversity. Table 2 lists some possible examples according to whether they are favourable, neutral or adverse factors to population growth of coral species. This is the only immediate way of trying to relate such environmental factors to the biology of organisms and so to gauge which factors may be more important. It is essential, however, to recognise that a factor that is favourable to the population expansion of one particular species or group of species will probably not favour universal high diversity across numerous groups of organisms throughout the tropics. This reservation is derived from the nature of ecological succession, which as already mentioned, points to diversity falling as equilibrium is reached (Connell, 1978). Favourable factors hasten the population growth of the species suited to them so that equilibrium (and a lower diversity state) is reached sooner. Equilibrium is less diverse than intermediate stages which precede it because dominant species emerge at equilibrium at the expense of many early colonisers. Adverse factors like disturbance postpone equilibrium and can therefore act to maintain high diversity. We can draw three possi-

bilities from this to explain tropical high diversity. First, the tropics may be largely in equilibrium, and apparently favourable characteristics of the tropics are offset by there being more space, particularly spatial heterogeneity, than in higher latitudes. Second, it may be adverse factors that cause high diversity, especially those that prevent equilibrium from being reached (e.g. disturbance and predation). Third, it may be favourable factors combined with factors that disturb equilibrium that explain high diversity.

If space is a resource (at least for corals) then its greater availability in the tropics would make it less necessary to invoke finer resource partitioning as an explanation of higher diversity (that is, there is more cake rather than finer slices). It is useful to think of space as having two components – uniform area and heterogeneity. Both may support more species but for different reasons. In the sections that follow, each of the above three possibilities will be considered in turn, with the first divided into considerations of area and heterogeneity.

Greater area

It is a common observation that larger areas generally contain a larger number of species. The Indopacific has a higher number of species than the Atlantic (Table 1). This is often referred to as the area effect. It is not surprising therefore that tropical high diversity has been attributed to the simple geometrical feature of the sphere, by which the tropics have a larger area than any other latitudinal band of the same span. Although this may be a serious part of the explanation of total global diversity, it overlooks the fact that particular environments may be of quite small extent within the tropics, and of smaller extent than in a higher latitude (e.g. continental shelves, and land areas) but their characteristic biota frequently shows the same tropical diversity increase.

Area appears to exert its effect in several ways (MacArthur & Wilson, 1967; Valentine, 1973; Vermeij, 1978). First, a larger area has a higher probability of containing more habitats, and thereby supports a greater number of species. Area must of course be an appropriately inhabitable area for the organisms concerned. Thus the size of the Pacific Ocean is in this respect less significant for corals than the large number of islands that lie within its western tropical region. Second, and theoretically, a very large uniform area may be very great in relation to the areal range and population expansion rate of inhabiting species. The area will have to have been in stable existence a very long time before all its species fully occupy the whole area. Initial colonisers and potential extinctions will take a very long time to reach equilibrium. The larger the area, the greater the probability that all or one part of it will undergo a change in conditions and equilibrium be deferred.

Third, uniform areas may be illusory, and a single large region (e.g. an ocean) may actually be perceived by its inhabiting species as quite differentiated. The problem of

Table 1 Comparison of the number of taxa of zooxanthellate scleractinian corals in the Atlantic and Indopacific Oceans

	Atlantic Ocean	Indopacific Ocean	% difference between Atlantic and Indopacific relative to Indopacific totals
Number of orders	1§	1§	0
Number of suborders	4§	5§	20
Number of families	11§	16§	c. 30
Number of genera/ subgenera	25*	c. 100†	c. 75
Number of species	68*	c. 500‡	c. 85

* From Porter (personal communication) and Laborel (1970).
† Rough estimate based partly on Wells (1956).
‡ From Schuhmacher (1976).
§ Classification according to Wells (1956).

Table 2 Characteristics of the tropical environment and their probable effect on population growth of a single species

Adverse	Adverse/benign	Benign
Low nutrient availability	More radiant energy	?
High frequency and intensity of disturbance (= low interspecific equilibrium), heat stress	Warmth	?
More competition		Seasonal equability
More predation		High primary productivity
		More places to live (= spatial heterogeneity)
		More space to live (= area effect)

the paucity of zooxanthellate corals in the eastern regions of the oceans suggests that better dispersing corals may 'see' the same ocean as being smaller than do poorer dispersers. There is therefore room for several poor dispersers to occur within the single region without mixing and possible extinction, where there is room for only one good disperser. With so little knowledge of coral larval ecology, this possibility cannot be dismissed. In fact, this argument is another way of saying that larger areas may contain more isolating barriers separating populations of different species that might otherwise mix and become selectively extinct. We need to be more flexible, however, about what we consider to be barriers by thinking of them more in terms of filters that vary in how much dispersal they prevent. (The Isthmus of Panama is all but absolute as a barrier to all tropical marine organisms, but cannot be so effective to cosmopolitan marine species.) In this way the apparently uniform region of the Indopacific, as we generally think of it for most tropical marine organisms, may actually consist of a geographical mosaic of species occupying different ranges according to their dispersal capabilities. Closely related to this idea we would also expect to have more species clines across large areas. There are also important implications in this for historical explanations (for speciation). It is interesting that recent advances in coral taxonomy point to regional differences at species level that the earlier state of taxonomy was not able to reveal. For convenience I shall refer to the above group of ideas as the more-area explanation.

Greater spatial heterogeneity

Spatial heterogeneity refers to the variety of habitats present within an area. Unfortunately the only absolute arbiters of the numbers of habitats present are the organisms themselves, and because they appear to overlap in their occupation of particular habitats, there is no direct way of reliably inferring numbers of habitats from numbers of species. On the other hand an environment as physically

varied as a coral reef offers (intuitively) more habitats than a flat floor or sand occupying the same area.

For corals there is greater spatial heterogeneity in the tropics largely because of themselves. First, they have contributed during the Tertiary and Pleistocene (the last 70 million years) to the growth of the limestone structures (reefs) they still occupy. Second, their capacity to transform an initially uniform substrate into geometric complexity by their own colonisation and growth must also be important. The complexity of reef morphology is of course due not just to corals or other organisms, but also to the geological history of these modern reefs. Sea level changes during the Pleistocene exposed the older reef limestones to be cut, carved and dissolved into an intricate surface geometry, now inherited as the foundation and substrate for modern reef organisms (Stoddart, 1973).

There is an implied feedback between diversity and spatial heterogeneity for many attached benthic reef organisms. The greater the diversity of species, the greater the collective range of their morphologies and the greater the heterogeneity they can add to an initial substrate. Spatial heterogeneity is thus increased not only as it affects corals, but also as it affects all other organisms dwelling on them, in them, or otherwise dependent on them in some way. This factor would also apply, for example, to mangroves, marine grasses and many algae in addition to the corals themselves.

Can spatial heterogeneity be applied to tropical high diversity as a whole? There is a greater range of temperatures and temperature regimes in shallow marine and terrestrial environments than in temperate zones (consider the climatic and vegetation range on an equatorial mountain like Mount Kenya), but it is difficult to say more than this without reference to additional factors. The idea remains difficult to test convincingly, and the qualitative support given to it by the feedback effect must be recognised for what it is – the tendency of species, and perhaps of ecosystems, to perpetuate their own occupation and survival, rather than a primary cause of high diversity.

The most likely way in which spatial heterogeneity contributes to high diversity for organisms in general is not so much at the ecosystem level but on the geographical scale (more area), where we can recognise that some regions are geographically more complex than others (Valentine, 1973). This appears to apply to zooxanthellate corals because their highest diversity is in the great archipelagic region of Indonesia and the West Pacific (Fig. 9.1). Geographical complexity provides more inhabitable areas and must also increase the number of dispersal barriers in comparison with a more uniform region, and this would, within such a complex region, enhance the geographical mosaic effect discussed under the more-area explanation, above.

More disturbance and less equilibrium

If we add time to our consideration of space we may find that we do not need to invoke spatial heterogeneity in the rather static manner just discussed. Although time appears to belong more with historical explanations, we can make a rough distinction between ecological time and evolutionary time, though different species must have different time bases. Ecological time is broadly commensurate with species' life spans and includes the cyclical or otherwise repeated events that are a characteristic of an ecosystem, rather than the unique and generally longer term events of speciation and extinction and of major geological changes. Seasons, storms, floods and droughts are within an ecological time-scale even if they might also represent incremental stages in evolutionary time. Disturbances are envisaged as 'intermediate', that is not so intense that whole environments are destroyed or species wiped out (Connell, 1978). They are relatively local. They disrupt progress towards equilibrium within a particular area and delay the emergence of dominant species, so that the higher diversity stages which precede equilibrium prevail more than otherwise. We can regard disturbance in this way as creating or maintaining ecological space and as a factor initially adverse to the welfare of local populations of individual species but favourable to high diversity. On a regional scale, we can think of an ecosystem being at various stages of recovery at different places, so compounding the high diversity effects of single disturbances at single places. For organisms like corals, for which space is such a tangible factor, this is an especially attractive argument, and is supported by observations on reefs (Connell, 1978). Its applicability to more mobile organisms is more abstract because it has to be seen in terms of population densities and distributions.

Can the relatively local concept of disturbance be applied to the tropics as a whole, if not for all organisms, then at least for corals? Connell (1978) has studied coral reef and tropical rainforest communities in Queensland, Australia over a nine year period and concluded that it can. Although his evidence is very convincing there are two difficulties that arise from trying to extend his relatively local observations to the whole of the tropics. First, he cites cyclones as the main agent of disturbance in his Queensland communities, and much of the tropics and zooxanthellate coral belt lies outside the influence of the two global cyclone belts. Second, and as Connell observes, some areas of high diversity on a coral reef lie in deeper water and would therefore be much less susceptible to cyclonic disturbance than the shallower, often lower diversity environments. This obliges us to look for other kinds of disturbance than cyclones to make the idea work. It is certainly not difficult to think of possibilities (reef slumping, sediment disturbance, and, in shallow water, the combined effects of several otherwise inconspicuous events coinciding, like heavy rain on a very low tide). On the global and longer term scale, Connell mentions climatic changes.

It is easy to imagine that these factors might all have ecological effects relating to equilibrium and diversity, but how does one show that they occur more often, or with greater effect on diversity in the tropics than in higher latitudes? On completely subjective considerations, the severity of environmental events and fluctuations related to the marked seasonality of temperate regions must be a more effective disturbance factor than anything in the tropics but cyclones. We are left with the possibility that disturbance must have a highly random (rather than seasonal) frequency to have an effect on diversity or that it must act in conjunction with other factors. Aspects of seasonality and non-seasonality are discussed further below.

Predation has also been suggested as a disturbance factor (Huston, 1979), and since there is apparently more predation in the tropics than in higher latitudes, this may be a more plausible way of applying the disturbance hypothesis to the tropics. It presupposes, however, that we can gauge how far predation is a routine occurrence within an ecosystem and how far it is a disturbance. Certainly the fluctuating effects of the Crown-of-Thorns starfish on corals look like disturbances, but it is difficult to say more than this.

Huston (1979) has recently extended the disturbance hypothesis to take into account the different population growth rates of different species. Whatever the competitive advantages one species might have over another, success or failure to occupy space requires time to attain a sufficient population size for resource competition amongst competing species to reach a critical level. Species that have a high potential population expansion rate can be thought of as fast competitors in this particular theory. To extend the idea to colonial organisms we also have to add fast growth rate of individual coral heads because both contribute to the occupation of space. Slow competitive rates amongst species allow potential competitors to coexist longer than fast competitive rates amongst species. Competitive differences moreover have to be more marked for earlier effect. A slowly competing group of species takes longer to reach equilibrium, so the chance of disturbance delaying the equilibrium stage is greater. The attraction of this suggestion for corals is that they are evidently slow competitors compared with many other groups of organisms. On the other hand the high diversity tropical distribution of corals consists largely of those corals that grow faster, that is, of zooxanthellates (Figs 9.2 to 9.4). It may be more appropriate to see zooxanthellate corals in the context of their fellow reef organisms, and to think of the tropics as being characterised generally by slow competitors. I believe that we also need to add considerations of seasonality to this possibility before we can make a more plausible speculation on the significance of disturbance.

More time

It will be clear by now that, as Connell (1978) acknowledges in his intermediate disturbance hypothesis, no single factor can be used as a basis for a self-contained explanation of tropical high diversity. It will have been obvious that in the course of this review so far I have progressively compounded elements of preceding explanations. Time, space, resources, competition and predation must all have some part to play in tropical high diversity. It is appropriate now to consider what factors might actually be favourable or adverse, and in what respect, to see if they act in concert to favour high diversity.

Warmth is the most striking feature of the tropics. It must be critical but favourable to zooxanthellates in the sense that for corals the algal symbiosis simply does not (cannot?) function below about 13 °C (Figs 9.3, 9.4). Above about 17 °C, Figure 9.4 shows that corals are however almost uniformly highly diverse (except where the extent of oceanic areas appears to inhibit full dispersal and where there are simply no suitable habitats (Fig. 9.1)). Within regions warmer than the threshold temperatures there are no obvious consistent patterns relating diversity to temperature (Figs 9.2, 9.7, 9.8). Thus the diversity–temperature gradients in the earlier coral literature are less apparent now that we have better distribution data, and have steepened to become step-like rises (Figs 9.2 to 9.4, 9.7, 9.8). Examples of within-tropical gradients like that in the northern Indian Ocean (Fig. 9.8) might be related to environmental gradients like any of those already discussed (e.g. seasonality) and not just to temperature. Warmth is, if anything, stressful to corals and is favourable only in the context of the symbiosis threshold. For organisms as a whole, warmth is probably stressful to other lower marine invertebrates (Beklemishev, 1969) but otherwise it should hasten growth and population expansion rates towards equilibrium conditions, so acting as a low diversity factor.

Sheer size of the tropical area has already been discussed as a factor favouring high diversity. It is not a specifically favourable factor affecting the well-being of individuals, but must exert its influence at the population level.

There is greater energy capture at the earth's surface in the tropics than in high latitudes (Stehli, 1968), but as it does not seem to be universally translated into higher primary productivity across the tropics, it is hard to see how this feature on its own can explain a universal ecological feature of the tropics. One would expect greater energy capture to be translated into greater biomass and hence more food, but availability of food is a paradoxical factor. Some parts of the tropics have the highest known primary productivity in the world, while the open tropical oceans have very low primary productivity. In terms of equilibrium arguments, the paradox is that higher primary productivity should favour lower diversity, unless other factors like disturbance are also significant.

Time has also been mentioned before, but largely in connection with equilibrium and disturbance. Southwood (1977) has emphasised time with respect to life-cycle patterns. The timing of reproduction, growth and of population fluctuations is closely related to patterns of resource availability. Valentine (1973) has discussed the lack of seasonality in the tropics and this would represent one way in which the supply of trophic resources might be stabilised. Newell (1971) has discussed seasonality for corals. In such conditions, species tend to have small niches, low abundances and low densities, and are restricted to a narrow habitat range or narrow prey specialisation. This is particularly apparent in tropical predatory gastropods (Taylor & Taylor, 1977 and references therein). In simplest terms the supporting theory is that predictability enables organisms to specialise on a particular resource, because although the resource may not be very abundant, it is not subject to annual or other periodic fluctuations and is in effect always available.

For corals, we have seen that trophic specialisation does not appear to apply. There is, as yet, no evidence for species being strongly seasonal in their breeding either, but there is good evidence of spatial specialisation. Perhaps lack of seasonality gives organisms more options, since their life-cycles do not all have to relate in some, presumably competitive way to the single dominating pattern of seasonal resource fluctuations familiar in temperate latitudes. In this way, one can think of the tropics as offering more time. Certainly many tropical organisms, including zooxanthellate corals, are long lived without obvious resource-storing strategies. It might be worth trying to relate this to the rates of population growth and hence to the kinds of disturbance arguments already discussed. It has been suggested for some tropical reef fish that the juveniles are so generalist in their needs that the occurrence of certain ecological combinations (guilds) of fish is a partly random 'who gets there first' effect (Sale, 1978). Thus a considerable number of species is maintained notwithstanding closely similar ecological characteristics. Lack of seasonality would give more time for a greater range of randomly sustained ecologically overlapping species to coexist, while it would at the same time lead to specialisations in enhancing occupation. Corals seem to act in the second way, so we might speculate that they are spatial–temporal strategists in the particular context of lack of seasonality in the tropics.

According to this interpretation, diversity in higher latitudes is low for corals because in shallow water they do not have the advantages conferred on them by the symbiosis with zooxanthellae for occupying different habitats, and possibly because they are unable to compete for the wide range of shallow water habitats against the rich algal colonisation that typifies temperate coasts.

Summary

In the absence of any clear single environmental factor causing tropical high diversity, it is only possible to construct a composite subjective model by linking together the more plausible elements from both environmental and biotic arguments. In general, biotic arguments derive from community ecology but environmental arguments apply to the tropics as a whole, so we cannot easily apply community concepts to the whole tropics. Yet environmental explanations invariably incorporate biotic explanations.

I am impressed by the implications of the size of the tropics. There are three interrelated elements of the more-area explanation: overall uniform area, geographical complexity and availability of inhabitable areas. They all have an effect on barriers to dispersal and hence on how many species ranges can be geographically accommodated. I am also impressed by the lack of seasonality in the tropics as a cause of resource stability, because it offers organisms a broader temporal framework (more time) for specialisation. There is certainly plenty of evidence for specialisation. In the case of corals, it is spatial specialisation that matters, supported by the strategies of slow growth and relatively little allocation of energy to production of offspring. It is important to understand corals' spatial specialisation as not just differentiated in habitat terms but especially as different ways of occupying space within a broad habitat range. Their spatial–temporal strategy would make them susceptible to intermediate disturbance and 'who-gets-there-first' factors.

The final question raised by these explanations, is that they throw no light on where the species come from in the first place. Is it an integral consequence of the more-area and more-time explanations? Or do we need to introduce further ideas to answer this?

Historical explanations

An historical argument seeks to explain where and how the species that constitute tropical high diversity originated. We have a pool of species within the tropics today that is presumably the net total of what number was there at some arbitrary point in the past, plus subsequent speciations and immigrations, less subsequent extinctions and emigrations (analogously to population dynamics). Migrations here would include extensions of the ranges of non-tropical species into the tropics and conversely the withdrawal of ranges from the tropics. It follows that an exhaustive discussion of historical explanations would have to consider all the possible combinations of these four factors in an effort to show that the tropics acquire a net gain of different species compared with higher latitudes. This is obviously not possible here.

One very important proposition, that of allopatric speciation, has to be accepted immediately in order to avoid digression on theories of speciation. There must have been several, or probably numerous events which divided previously single interbreeding populations of coral species so that each population became sufficiently separated from the other and for sufficiently long to develop new genetic characteristics. The principle is that two species thereby come into existence where there was one before. It is easiest to think of species populations being separated by the appearance of new biogeographic barriers (geographic allopatry). The possibility of ecological barriers (ecological allopatry) being a factor can also be considered. Other possibilities like the chance arrival of a few propagules at an otherwise isolated locality and their subsequent evolution into new species in isolation cannot be usefully discussed here for corals. It is reasonable to assume that in present-day coral distributions we are looking at the results of a long history of separations and recombinations of species' ranges overlapping each other through geological time and/or the results of ecological allopatric speciation.

Geological instability

It is very difficult to conceive of any ecological isolating factors that might apply to corals that would not also need to be linked with geographical barriers. A combination of climatic and tectonic events provides abundant possibilities for changes in geographical patterns of barriers and to changes in their effectiveness. Shifts of sea level and sea temperatures both associated with the extent of polar ice caps offer one group of possibilities. Sea level is also affected by mid-ocean ridge spreading rates. Continental movements together with uplift and subsidence of islands and smaller land areas offer another group of possibilities. These tectonic events are also thought to have climatic and oceanographic consequences (Valentine, 1973). There is good evidence for frequent changes of all these kinds throughout the last 70 million years, from deep sea drill cores and from the palaeomagnetic record of the ocean crust. There are also extra-terrestrial causes of climatic change like changes in the earth's orbit. So we no longer have to depend on Pleistocene glaciations alone to account for the necessary environmental changes. It is very doubtful, however, whether any of these events has a greater frequency or impact in its effect on the tropics than in higher latitudes, so they do not by themselves explain why there are more species in the tropics. On the other hand, if we accept any of the models already discussed for the way in which the tropics accommodate more species, then the historical aspect becomes a non-problem. We do not need to invoke any additional and distinct historical explanation specific to the tropics. All that needs to be postulated is a globally distributed continuum of speciation, with a lower proportion of extinctions in the tropics because more species can be accommodated there. Species may actually accumulate selectively in the tropics for the same reason.

Of the various accommodation explanations considered earlier the more-area argument has a greater appeal in a historical context than the more-time argument. It will be recalled that large areas are less likely to be fully occupied by a species unless that species is also an effective disperser, and that corals do not seem to be good dispersers. The resulting biogeographic mosaic of different species and their ranges would be especially susceptible to allopatric speciation caused by tectonic and climatic events. The application of the more-time argument in a historical context, however, carries with it special implications of natural selection, that is selection for narrower niche specialisation, and may be an unnecessarily more complicated argument.

For organisms whose diversity increases steadily towards the Equator, there would appear to be little more that can be said in a historical context. For groups like corals which show a step-like diversity rise at the fringe of the tropics, we still need to explain this particular pattern. For corals we have noted that it is entirely related to the distribution pattern of zooxanthellates. Our explanations should not only treat this step phenomenon, but also the actual fall in diversity of non-zooxanthellates within the tropics. We have seen that the zooxanthellates are especially restricted in their ecological limits, and so are even more susceptible to climatic–tectonic changes than many other groups of organisms. Their depth limit of around 100 m is well within the order of eustatic sea level changes for example. Since there is evidently a minimum threshold temperature (and hence, broadly, a maximum latitude) at which the coral–algal symbiosis exists, there must also be a concomitant sudden latitudinal rise in the susceptibility of corals to climatic–tectonic events.

Turning to non-zooxanthellates in higher latitudes there is a diversity increase towards the Equator which could be explained largely on the more-area argument, but on this explanation alone, we might still expect a continued gentle rise in non-zooxanthellate diversity within the tropics. Unless further distribution data on non-zooxanthellates correct the pattern shown in the present plots (Figs 9.2 to 9.4) to fulfil this prediction, we do have to suggest that the zooxanthellates occupy or compete more effectively for all the shallow water habitats in the tropics. Whereas in the non-tropics non-zooxanthellates occupy the complete range of habitats from shore to abyssal depths, in the tropics they are largely missing from the rich range of habitats found in shallow water, and their global diversity in low latitudes is actually diminished by this. Note that this does not depend on saying that the tropics are spatially more heterogeneous but that, more plausibly, the in-shore environment in all latitudes is universally more heterogeneous than the off-shore and deep sea floor environments. There may be nothing to explain, however, if we think of non-zooxanthellates being effectively diverse in the tropics through their evolutionary metamorphosis (from ancient stock or stocks) into zooxanthellates.

The geological record

Constancy of tropical high diversity Stehli et al. (1969) have presented data suggesting that tropical high diversity has been a biogeographic feature throughout much of the Phanerozoic. A closer look by Taylor et al., (1980) at one particular group – the predatory gastropods – suggests on the contrary that the tropics may not always have been characterised by high diversity, or that at least, different

groups have varied in geological time in their diversity patterns. In general, corals have been more diverse in the Tethyan region throughout the Mesozoic and Tertiary (Wells, 1956), and the Tethyan region lay in broadly tropical latitudes with an increasingly southward shift through this time. In fact, the relevant distribution data are difficult to compile because good stratigraphic control on sample faunas is wanting. We must also have lost many important sample regions through destruction of oceanic crust in crustal collision zones, because at the present time, many of the largest coral faunas occur in oceanic archipelagoes.

Evidence for constancy is thus equivocal. On theoretical grounds tropical high diversity should be constant as a feature on the strength of the uniform area part of the more-area argument. This argument also rests however on dispersal potentials in relation to geographical complexity and habitat availability. It is highly likely that the effectiveness and number of dispersal barriers have in the past been different for particular groups. This leads to the ideas put forward by Valentine (and others), considered next.

Diversity and provinciality Valentine (1973) has related changes in the global diversity of organisms through geological time to the various phases of continental collisions and separations. These ideas are now well known. In short, times of continental fragmentation appear to correlate with times of higher diversity, and times of unified land areas correlate with lower diversity. Although climatic and oceanographic changes, together with resource stability arguments are included in this model, a considerable part of it rests on the relatively straightforward idea that more fragmentation leads to more biogeographical provinces and hence to a higher total of species. It would follow that tropical high diversity should in part be related to the effect of these same events on both speciation and on species carrying capacity. The difficulty is that we have a very high modern coral diversity, much of it related to high diversity of zooxanthellate corals and yet provinciality is low in that they are distributed through only two faunal realms. We can resolve this in two ways. Either we must recall that a large apparently uniform biogeographical region probably consists of a distribution mosaic of species and species clines; diversity would in this way be related to the number and effectiveness of dispersal barriers rather than to generalised provinces defined by obvious land and sea boundaries. Or, less likely, we must conclude that the numerous species in the present-day tropics are the cumulative results of survival of previously isolated species brought together by climatic–tectonic events over geological time. They coexist by virtue of niche narrowing made possible by the nature of the tropical environment as discussed earlier. The stratigraphic record of corals throws some light on this.

The fossil coral record There is a great deal of good taxonomic revision necessary before we have an adequate knowledge of the stratigraphic record of living coral species. There are undoubtedly many synonyms between fossil and living corals. Even so, the present state of knowledge points to relatively few modern species being older than the mid-Miocene (15 million years ago) (Chevalier, 1977; Frost, 1977). Some of today's ecological assemblages can also be recognised in the late Tertiary. It appears as though many modern species (not the groups as a whole) of agariciids, poritids and the suborder Faviina originated in the Miocene (5·25 million years ago), but that many living acroporid and fungiid species may be more recent in origin, probably within the Pleistocene (1·8 million years). The Pleistocene was of course a time of major glaciations, and it was during the early Miocene that tectonic events separated the previously single Tethyan–Indopacific realm into the two modern realms. Isotope results from deep sea drill cores moreover indicate a general climatic cooling throughout the Tertiary (Newell, 1971), but there were sudden drops in temperature and more climatic instability in the mid-Eocene (45 million years ago), at the Eocene–Oligocene boundary (40 million years), in the mid-Miocene (15 million years) and at the Plio-Pleistocene boundary (1·8 million years) (T. C. Moore, personal communication). Zooxanthellate coral diversity rose during the Miocene by about 15 per cent and has risen since the Plio-Pleistocene by 80 per cent from a previous low diversity interval (Newell, 1971). Dispersal barriers would have changed a great deal during the Miocene climatic and tectonic events, and during the Pleistocene climatic events. These times represent the most likely periods when species' ranges were fragmented and redistributed, leading to the observed relatively sudden appearance of new zooxanthellate taxa during these times. A certain amount of 'background' speciation would have continued throughout these times, related to smaller less easily identified events.

Over this whole period, however, there was a progressive global climatic cooling. This would have narrowed the latitudinal belt favourable for zooxanthellate coral survival from the Miocene span when it was similar to the present day, to a belt narrower than the present one during the major ice advances of the Pleistocene (Newell, 1971; Stoddart, 1973). This climatic change would result in a decrease in area for zooxanthellates and might explain the observed drop in diversity between Miocene and Pleistocene times. The fossil record for corals can therefore be interpreted in terms of the more-area argument with effects of changes in geographical complexity superimposed on the longer term changes of total area.

Centres of origin The very different diversity of species in the two present-day zooxanthellate coral realms (Table 1) has been attributed by most authors to the relative difference in areal extent of the two realms. This is basically the more-area argument applied on a provincial scale. It is difficult to see how any of the other high diversity explanations discussed here could bear on these

particular provincial differences. This does at least represent some consistency of explanations.

Within the two realms there are three foci of high diversity (Fig. 9.1) of which the Indonesian and West Pacific focus, and the Caribbean focus have been interpreted as evolutionary centres of origin for zooxanthellate corals. Stehli & Wells (1971) have shown that amongst the living genera of the focal regions, a higher proportion are geologically younger than in a comparable breakdown of non-focal faunas. These focal regions therefore must have had histories and characters that in some way generate, accumulate and accommodate more species than do nonfocal regions. The fact that both foci correspond to the main island regions in their respective oceans immediately invites the geographical complexity explanation (a version of the more-area argument – Rosen, 1975). They must for the same reason be more sensitive to climatic–tectonic events leading to speciation.

The idea of centres of origin has been severely criticised in recent years (Croizat et al., 1974), but the age analysis by Stehli & Wells (1971) is very striking. Heck & McCoy's (1978), and McCoy & Heck's (1976) recent critique of the centre of origin argument as it applies to corals does not, however, adequately explain the possible objections, perhaps because they have concentrated on the particular issue of long distance dispersal in corals. It seems, however, that the most important criticism relevant here is that coral taxa could have originated in other regions than the foci, and that such regions have subsequently been destroyed or transformed by continental and sea floor tectonic events, such as the closure of Africa, Arabia and India against southern Asia. Analysis of the fossil coral faunas in the suture zones of such areas is obviously needed to complement Stehli & Wells' (1971) data, but the real evidence may have been destroyed altogether. The taxa which originated in such regions, however, could well have survived in regions elsewhere. They will in fact be most likely to accumulate in precisely those regions that have geographical complexity (more-area) and have long remained unaffected by complete tectonic destruction. In terms of present-day coral distributions, these refuge regions will be the two main archipelago foci above (Newell, 1971). If the true regional origins have been destroyed or the relevant fossil faunas not studied, the foci will thus appear to represent centres of origin by default of complete evidence. In actuality we can only say that they may or may not have been centres of origin. In addition, an archipelago region like that in the western Indian Ocean may also have been important but these islands have no investigated pre-Pleistocene stratigraphy to throw light on its possible evolutionary contribution during the Tertiary or earlier.

In conclusion, the diversity foci are undoubtedly regions of greater carrying capacity, and are probably regions of very recent and present-day speciation, but they also represent the only present-day refuges for taxa that may have originated elsewhere. Their historical role in causing high diversity is thus a combination of accommodation, refuge, accumulation and speciation.

Summary

Tropical high diversity in corals can be attributed most simply to the more-area explanation even in a historical context. The kinds of climatic–tectonic event that generate new species (by allopatry) presumably affect all latitudes with equal probability over long periods of geological time. The sudden rise in coral diversity in the tropics is due to the more-time explanation. In other words the explanations concluded earlier for the way in which the tropics accommodate more species would act through geological time to maintain the observed tropical high diversity. Note, however, that the more-area argument has three elements: area as a factor affecting dispersal capabilities and equilibrium, geographical complexity as a factor affecting dispersal, and availability of inhabitable areas. At the moment the marine tropics are relatively complex geographically, but we can postulate a marine tropical belt without any land-masses or islands, and in this instance zooxanthellate diversity in the tropics would be zero. We can therefore distinguish potential accommodation in the tropics from the fluctuating availability of inhabitable places through geological time. Tropical high diversity cannot therefore have been constant qualitatively or quantitatively through geological time. It is especially important to think of geographical complexity in broader terms than land–sea provinciality.

We can go on to postulate a positive feedback effect of geographical complexity on diversity because a complex region is more likely to be affected by smaller scale climatic and tectonic events. This is especially important for a group of organisms like zooxanthellate corals that occupy only the uppermost part of the water column. Geographical complexity should act both as a speciation factor and as an accommodation and refuge factor. At the present time tropical high diversity in zooxanthellates is due to the combined role of the three main oceanic archipelago regions or foci. Two main periods of zooxanthellate speciation appear to have been in the early to mid-Miocene and the Pleistocene. High diversity is not caused by centres of origin in its generally understood sense.

Conclusions

In the preceding reviews of the three main kinds of explanation of tropical high diversity I have progressively added the conclusions of one to those of the next to arrive at a synthesis. The result is a compound of propositions, some of which are testable and supported by data, and some of which appear to be testable but are as yet without evidence. Elsewhere, we are still at the stage of trying to translate broad ideas into something testable, and we even have data apparently supporting certain ideas but we are not really sure what the data demonstrate. This all represents the true state of this subject area of ecological biogeography. This may seem unscientific according to the strict criterion of falsifiable hypotheses, but we are still searching for a framework in which to formulate such hypotheses, and this search is also a valid part of science.

The accommodation of more species in the tropics can be related to three factors – more specialisation, more time and more area. Of these, more area appears to be most fundamental, but I have invoked more time (that is, non-seasonality) to help account for the observation of more specialisation. For a general historical explanation of tropical high diversity we should not need to look beyond these same accommodation factors. If they themselves have acted through geological time, they would always maintain higher diversity in the tropics. But two elements of the more-area explanation – geographical complexity and availability of inhabitable areas – cannot have remained constant within the tropics through geological time, and we can therefore predict from this that tropical high diversity must have fluctuated qualitatively and quantitatively (contrary to earlier views). There is already limited evidence for such fluctuations of tropical high diversity. The ways in which the more-area factor changes through geological time would also have an important effect on speciations and extinctions, and must also be considered when explaining past and present distributions both of particular taxa and of regional diversity patterns within the global pattern. The refuge role of geographically complex regions appears to be especially important. We can conclude by speculating that more ecological time in the tropics allows a greater proportion of species and/or their descendants to survive in such regions, by acting as a selective factor favouring greater specialisation in the tropics.

The status of some of these ideas can now be briefly commented upon. In resource partitioning we appear to have a framework for hypotheses to explain specialisation in the tropics (narrower niches), but do we need niche theory, especially the idea of limiting resources, to explain this? It has at least provided a framework for organising information on adaptive traits. Corals appear to specialise in space occupation but we should now try to translate this general observation into hypotheses derived for example from the ideas of Grime (1979) and Jackson (1979). It may be more useful to think of specialisation in the context of the more-time explanation, an intuitive idea that should be testable in terms of patterns of reproduction and growth, and of life histories. Using Grime's categories, we might predict that in the tropics there should be a higher proportion of competitors and stress tolerators compared with ruderals, than in temperate latitudes.

Within the more-area explanation we have three subsidiary elements: uniform area, geographical complexity, and availability of inhabitable areas. We do not yet have an adequate knowledge of coral species' taxonomy, nor of larval ecology to test these ideas, though the generalisation of the area effect is well established. In applying the more-area explanations in a historical context, we have what seems to be testable in the changes in geographical complexity (and hence by implication the availability of inhabitable areas) through geological time. Unfortunately, the incompleteness of the fossil record, the loss and tectonic transformation of large areas of the earth's crust through geological time, and the difficulties of accurately reconstructing ancient geographies and environments would together leave too much room for alternative interpretations. Moreover, we would not expect other factors affecting diversity (e.g. climatic change) to be in continuous phase with the geographical changes. It would be complicated if not impossible to weight and separate all these different causes. Notwithstanding these objections, the exercise of trying to match diversity fluctuations regionally and globally with palaeoclimatic and palaeographical changes remains heuristically attractive.

Finally, what implications does the foregoing survey of corals have for other organisms? In general, much of what has been concluded applies to other organisms, but not in a synonymous way. Although many other marine benthic organisms and many terrestrial plants specialise in space occupation, we also know that in other organisms there are more finely differentiated specialisations in diet, or more elaborate temporal strategies. The generalisation of more specialisation in the tropics has to be seen in terms of all these possibilities. Consideration of different modes of speciation will also be important in applying the historical explanations put forward here. For corals I have emphasised classical allopatric speciation in relation to changing geographical patterns, but although this probably applies to a greater or lesser extent to many other groups of organisms, it cannot be the universal nor the only mechanism. The time-scale of individual species' life histories will also be different from those of corals, and this will especially affect the more-time argument, as well as speciation considerations.

References

Beklemishev, W. N. 1969. *Principles of comparative anatomy of invertebrates. Volume 2. Organology.* 529 pp. Edinburgh: Oliver & Boyd. (translated by MacLennan, J. M. from 3rd Russian edition, Moscow: Nauka).

Cairns, S. D. 1979. Deep-water Scleractinia of the Caribbean Sea and adjacent waters. *Studies on the fauna of Curaçao* 57: 1–341.

Chevalier, J. P. 1975. Aperçu sur la faune corallienne récifale du Néogene. *Mémoires du Bureau de Recherches Géologiques et Minièves, Paris* 89: 359–366.

Coates, A. G. & Oliver, W. A. Jr. 1973. Coloniality in zoantharian corals, pp. 3–27 in Boardman, R. S., Cheetham, A. H. and Oliver, W. A. Jr. (Eds) *Animal colonies; development and function through time.* Stroudsburg, Pennsylvania: Dowden, Hutchinson & Ross.

Connell, J. H. 1978. Diversity in tropical rain forests and coral reefs. *Science* 199: 1302–1310.

Crame, J. A. 1977. *Succession and diversity in the Pleistocene coral reefs of the Kenya coast.* Unpublished Ph.D. Thesis, University of London.

Croizat, L., Nelson, G. J. & Rosen, D. E. 1974. Centers of origin and related concepts. *Systematic Zoology* 23: 265–287.

Faure, G. 1977. Distribution of coral communities on reef slopes in the Mascarene Archipelago, Indian Ocean. *Marine Research in Indonesia* 17: 73–97.

Frost, S. H. 1977. Miocene to Holocene evolution of Caribbean province reef-building corals. *Proceedings of the 3rd International Coral Reef Symposium* 2: 353–359.

Geister, J. 1977. The influence of wave exposure on the ecological zonation of Caribbean coral reefs. *Proceedings of the 3rd International Coral Reef Symposium* 1: 23–30.

Graus, R. R., Chamberlain, J. A. Jr. & Boker, A. M. 1977. Structural modification of corals in relation to waves and currents, pp. 135–153 in Frost, S. H., Weiss, M. P. and Saunders, J. B. (Eds). *Reefs and related carbonates – ecology and sedimentology.* Tulsa, Oklahoma: The American Association of Petroleum Geologists.

Grime, J. P. 1979. *Plant strategies and vegetation processes.* 222 pp. Chichester: John Wiley & Sons.

Heck, K. L. & McCoy, E. D. 1978. Long-distance dispersal and the reef-building corals of the eastern Pacific. *Marine biology. International Journal on Life in Oceans and Coastal Waters, Berlin* 48: 349–356.

Huston, M. 1979. A general hypothesis of species diversity. *American Naturalist* 113: 81–101.

Hutchinson, G. E. 1965. *The ecological theater and the evolutionary play.* 139 pp. New Haven: Yale University Press.

Jackson, J. B. C. 1979. Morphological strategies of sessile animals. *in* Larwood, G. P. & Rosen, B. R. (Eds) *Biology and systematics of colonial organisms. Systematics Association, Special Volume* 11, 499–555.

Laborel, J. 1970. Madréporaires et hydrocoralliaires récifaux des côtes brésiliennes; systématique, écologie, répartition verticale et géographique. *Annales de l'Institut Océanographique, Monaco* 47: 171–229.

Lang, J. C. 1973. Interspecific aggression by scleractinian corals. 2. Why the race is not only to the swift. *Bulletin of Marine Science. Coral Gables* 23: 260–279.

Lewis, D. H. 1973. The relevance of symbiosis to taxonomy and ecology, with particular reference to mutualistic symbioses and the exploitation of marginal habitats. *in* Heywood, V. H. (Ed.) *Taxonomy and ecology. Systematics Association, Special Volume* 5: 151–172.

Lewis, J. B. & Price, W. S. 1975. Feeding mechanisms and feeding strategies of Atlantic reef corals. *Journal of the Zoological Society of London* 176: 527–544.

Loya, Y. 1976. The Red Sea coral *Stylophora pistillata* is an r strategist. *Nature* 259: 478–480.

MacArthur, R. H. & Wilson, E. O. 1967. *The theory of island biogeography.* 203 pp. Princeton: Princeton University Press.

McCoy, E. D. & Heck, K. L. Jr. 1976. Biogeography of corals, seagrasses, and mangroves: an alternative to the center of origin concept. *Systematic Zoology* 25: 201–210.

Muscatine, L. 1973. Nutrition of corals. pp. 77–114 in Jones, O. A. & Endean, R. (Eds) *Biology and geology of coral reefs*, 2: New York: Academic Press.

Muscatine, L. 1974. Endosymbiosis of cnidarians and algae. pp. 359–395 *in* Muscatine, L. & Lenhoff, H. M. (Eds) *Coelenterate biology; reviews and perspectives*. New York: Academic Press.

Newell, N. D. 1971. An outline history of tropical organic reefs. *American Museum Novitates* 2465: 1–37.

Pielou, E. C. 1977. The latitudinal spans of seaweed species and their patterns of overlap. *Journal of Biogeography* 4: 299–311.

Porter, J. W. 1976. Autotrophy, heterotrophy and resource partitioning in Caribbean reef-building corals. *American Naturalist* 110: 731–742.

Rinkevich, B. & Loya, Y. 1979. The reproduction of the Red Sea coral *Stylophora pistillata*. II. Synchronisation in breeding and seasonality of planulae shedding. *Marine Ecology Progress* 1: 145–152.

Rosen, B. R., 1971a. Principal features of reef coral ecology in shallow water environments of Mahé, Seychelles. pp. 163–183 *in* Stoddart, D. R. & Yonge, C. M. (Eds) *Regional variation in Indian Ocean coral reefs. Symposium of the Zoological Society of London* 28.

Rosen, B. R. 1971b. The distribution of reef coral genera in the Indian Ocean. pp. 263–299 *in* Stoddart, D. R. & Yonge, C. M. (Eds) *Regional variation in Indian Ocean coral reefs. Symposium of the Zoological Society of London* 28.

Rosen, B. R. 1975. The distribution of reef corals. *Report of the Underwater Association* (New Series) 1: 1–16.

Rosen, B. R. 1977. The distribution of Recent hermatypic corals and its palaeontological significance. *Mémoires du Bureau de Recherches Géologiques et Minières, Paris* 89: 507–517.

Rosen, B. R. 1979. Modules members and communes: a postscript introduction to social organisms. pp. xiii–xxxv *in* Larwood, G. P. & Rosen, B. R. (Eds) *Biology and systematics of colonial organisms. Systematics Association Special Volume* 11.

Sale, P. F. 1978. Coexistence of coral reef fishes – a lottery for living space. *Environment Biology of Fishes* 3: 85–102.

Schuhmacher, H. 1976. *Korallenriffe; ihre Verbreitung, tierwelt und Ökologie*. 275 pp. München: BLV Verlagsgesellschaft.

Southwood, T. R. E. 1977. Habitat, the templet for ecological strategies? *Journal of Animal Ecology* 46: 337–365.

Stehli, F. G. 1968. Taxonomic diversity gradients in pole location: the Recent model, pp. 163–227 *in* Drake, E. T. (Ed.). *Evolution and environment*. New Haven: Yale University Press.

Stehli, F. G., Douglas, R. G. & Newell, N. D. 1969. Generation and maintenance of gradients in taxonomic diversity. *Science* 164: 947–949.

Stehli, F. G. & Wells, J. W. 1971. Diversity and age patterns in hermatypic corals. *Systematic Zoology* 20: 115–126.

Stoddart, D. R. 1973. Coral reefs: the last two million years. *Geography* 58: 313–323.

Sverdrup, H. U., Johnson, M. W. & Fleming, R. H. 1946. *The oceans; their physics, chemistry and general biology*. 1087 pp. New York: Prentice-Hall.

Taylor, J. D., Morris, N. J. & Taylor, C. N. 1980. Food specialization and the evolution of predatory prosobranch gastropods. *Palaeontology* 23: 375–409.

Taylor, J. D. & Taylor, C. N. 1977. Latitudinal distribution of predatory gastropods on the eastern Atlantic shelf. *Journal of Biogeography* 4: 73–81.

Valentine, J. W. 1973. *Evolutionary paleoecology of the marine biosphere*. 511 pp. Englewood Cliffs, New Jersey: Prentice-Hall.

Vandermeulen, J. H. & Muscatine, L. 1974. Influence of symbiotic algae on calcification in reef corals: critique and progress report. pp. 1–19 *in* Vernberg, W. B. (Ed.) *Symbiosis in the sea*. Columbia, South Carolina: University of South Carolina Press.

Vermeij, G. J. 1978. *Biogeography and adaptation; patterns of marine life*. 332 pp. Cambridge, Massachusetts: Harvard University Press.

Wells, J. W. 1954. Bikini and nearby atolls: (2) oceanography (biologic). Recent corals of the Marshall Islands. *Professional Paper of the United States Geological Survey* 260 (I): 385–486.

Wells, J. W. 1956. Scleractinia, pp. F328–444 *in* Moore, R. C. (Ed.) *Treatise on invertebrate paleontology Part F Coelenterata*. Lawrence, Kansas: Geological Society of America and University of Kansas Press.

Wells, J. W. 1957. Coral reefs. pp. 609–631 *in* Hedgpeth, J. W. (Ed.) *Treatise on marine ecology and paleoecology, Volume 1 Ecology. Memoirs of the Geological Society of America* 67.

Wells, J. W. 1967. Corals as bathometers. *Marine Geology* 5: 349–365.

Wells, J. W. 1969. Aspects of Pacific coral reefs. *Micronesica* 5: 317–322.

Yonge, C. M. 1973. The nature of reef-building (hermatypic) corals. *Bulletin of Marine Science. Coral Gables* 23: 1–15.

Zibrowius, H. 1976. *Les scléractiniaires de la méditerranée et de l'atlantique nord-oriental*. 302 pp. Unpublished Doctor of Science Thesis, University of Aix-Marseille.

CHAPTER 10

Coexistence and predation in aquatic microbial communities

C. R. Curds

Introduction

A detailed look at a natural microbial community in the aquatic environment soon reveals that the constituent populations of organisms are not static, they are dynamic, ever changing. Certain established species decline while others replace them. Many workers have attempted to relate such population changes to variations in the surrounding physical and chemical environment, few have succeeded and there are always serious doubts at the end of such investigations since there is usually a general lack of basic biological data concerning interactions between the organisms themselves. Even though the microbial populations are not static it is common to find that there are several species living together that are similar and apparently utilising the same food or nutrient sources. If this observation is correct then it is presumed to be an uncommon situation in nature, since species that have the same food and habitat preferences seldom survive together. Indeed it is a widely accepted biological principle that when two or more non-interbreeding populations, which utilise the same energy or food source, exist together in the same habitat, they cannot do so indefinitely – sooner or later one organism will displace the other. This is the principle of competitive exclusion (Harding, 1960).

The purpose of this essay is to describe the development of simple mathematical models that may be used to explore how microbial interactions might affect the coexistence and dynamic behaviour of simple microbial communities consisting of bacteria and protozoa. It further outlines how continuous cultures may be used by microbiologists to test the hypotheses produced by mathematical considerations and computer simulations.

The zoologist reading such an essay might ask the questions 'Why consider microorganisms at all? Why not look at some larger animals?' Well, in spite of, and in some cases because of, their size microorganisms do have several advantages over other invertebrate and vertebrate animal groups. For example many are simple to grow and there are several that may be grown on chemically-defined nutritional media in the absence of all other organisms. They are of course small, and therefore cheap to maintain in very large numbers, with fast rates of growth so that experimental data can be obtained quickly. Although sexual processes are known in some groups, microorganisms generally reproduce by simple asexual division which is considerably easier to model mathematically than the complex life-cycles and methods of reproduction found in 'higher' organisms. However, most of these advantages relate to laboratory experimental work and there are several real disadvantages when attempting to study microorganisms in the natural environment. Difficulties are encountered with counting, identifying, estimating biomass and measuring their rates of growth but these are also familiar problems to the ecologist studying larger animals in the natural environment.

Growth of a single-species population

Before considering the growth and interactions between mixed populations or organisms, we must first examine how a single-species population grows. As stated above, microorganisms commonly increase their numbers by some form of asexual binary fission. Generally, an individual grows to a certain body size and then divides into two, each progeny proceeds to grow and ultimately divides into two. Thus, at each division the population doubles its numbers. This is not to say that the division of all individuals in the population occurs simultaneously. After the first division some of the progeny will take a little longer or shorter time than the others to divide so that after a few divisions a sample will show that organisms in the population are at all division stages. This method of division enables one to use a simple method of evaluating the growth of a population which is known as the population doubling time.

If a single batch of sterile growth medium or substrate is taken, inoculated with a bacterium and incubated at constant temperature the growth of the resultant population

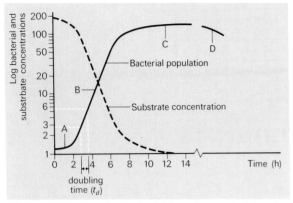

Figure 10.1 Growth of microbial species in a batch culture showing depletion of substrate (A – lag phase, B – log phase, C – stationary phase, D – death phase).

will follow the general curve as shown in Figure 10.1. Bacterial populations could be measured in terms of numbers of organisms per unit volume of culture but it is more meaningful to estimate both bacterial and substrate concentration in terms of dry weight per unit volume. Since microbial populations double their size at each division the increase is an exponential function and it is common practice to plot the population and substrate concentrations on a logarithmic scale against time on a linear scale. This procedure has the effect of producing a linear relationship between the two variables over much of the curve (Fig. 10.1) and the linear part is commonly referred to as the log phase of growth. It should be noted that the population reaches a maximum size (stationary phase). Although this could be due to changes in environmental parameters such as pH or the accumulation of metabolites, it is normally due to the depletion of the food supply and if left long enough, the population will then decline (death phase).

The population growth curve in Figure 10.1 may be used directly to measure the time taken for the population to double its size which is measured in units of time (commonly hours in the case of microorganisms). Doubling time can then be converted into a rate by the use of equation 1, where t_d is the population doubling time and μ is the specific growth rate of the organism,

$$\mu = \frac{\log_e 2}{t_d} \qquad 1$$

The specific growth rate, μ, represents the observed rate of growth per unit population (X) per unit time (t), that is $\frac{1}{X} \cdot \frac{dX}{dt}$, and has dimensions of reciprocal time. It can be said to be analogous to the compound interest on an investment so that the specific growth rate of 0.9 h^{-1} in Figure 10.1 is equivalent to a compound interest of 9 per cent per hour.

In Figure 10.1 it will be seen also that as the organism grows it utilises substrate or nutrient and, when plotted

on a log-linear scale, the curve for nutrient utilisation is a straight line during the logarithmic phase of growth. This illustrates that there is a simple relationship between the growth of an organism and its utilisation of substrate. As growth proceeds the concentration of substrate, s, decreases and the rate of increase of population is a constant fraction, Y, of the rate of substrate decrease. That is:

rate of change of organism concentration with respect to time = rate of change of substrate concentration with respect to time

that is:

$$\frac{dX}{dt} \equiv -Y \cdot \frac{ds}{dt} \qquad 2$$

where Y is known as the yield constant and is dimensionless. It follows from equation 2 that over any finite period of time,

$$\frac{\text{weight of organism formed}}{\text{weight of substrate consumed}} = Y$$

Among the many physical, chemical and biological factors that can influence the growth rate of a microbial population, the availability of nutrient or substrate is perhaps the over-riding basic factor and the relationship between the concentration of available substrate and the growth rate of the organism is of fundamental importance when considering microbial population dynamics. It is obvious that organisms cannot grow in the absence of substrate; at the other extreme, when food is in unlimited supply, certain early workers (see Curds & Bazin, 1977 for review and references) assumed that the rate of growth would also be unlimited. This is difficult for the modern biologist to accept, who would at least suspect that there would be an upper limit that an organism could not exceed and this has been borne out by experimental data.

Although many workers have produced detailed experimental data supporting slight variations, it is generally

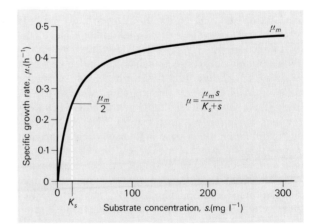

Figure 10.2 Relationship between the specific growth rate of an organism and the concentration of the substrate available. In the example, $\mu_m = 0.5 \text{ h}^{-1}$ and K_s (saturation constant) = 20 mg l^{-1}.

accepted (Monod, 1950; Herbert, Elsworth & Telling, 1956; Powell, 1965) that the relationship between the concentration of growth-limiting nutrient and the specific growth rate of an organism can be represented adequately by a Michaelis-Menten type of equation,

$$\mu = \frac{\mu_m s}{K_s + s} \qquad 3$$

where μ and μ_m are the specific and maximum specific growth rates, respectively, and s is the substrate concentration. The maximum specific growth rate is the rate of growth when the substrate concentration is in excess. The saturation constant (K_s) is numerically equal to the substrate concentration at which the specific growth rate is half its maximum (see Fig. 10.2). From equation 3, it will be noticed that the specific growth rate of an organism may be determined for any concentration of substrate provided the values of the growth constants, μ_m and K_s, are known.

Growth of a single-species population in continuous culture

The preceeding remarks about the growth of single-species populations of microbes have been limited so far to batch-culture systems. This is the classical method of cultivating and studying microorganisms in which, as we have seen, a single quantity or 'batch' of growth medium is inoculated with an organism that grows, isolated from the external environment, over a period of time. Initially the nutrients in such a culture are likely to be in excess and organisms few, but as time proceeds the population grows and the nutrients are depleted. At the same time, physico-chemical environmental factors such as pH and concentration of dissolved oxygen, for instance, change and there is an accumulation of potentially harmful metabolic products. Eventually the organisms cease to grow and finally die (Fig. 10.1).

If one considers an aquatic habitat such as a river, it bears little resemblance to a batch-culture system. The river is not isolated from the external environment, indeed the continual flow of water carries both nutrients and organisms along in its path. If one divides a river into an infinite number of small sections connected together in series (Fig. 10.3A), each section can be said to be, at least partly, analogous to the continuous-culture systems, known as 'chemostats', that are being used by may microbiologists to overcome certain problems inherent in batch cultures. In a chemostat, microbial growth may take place under steady-state conditions, that is to say growth proceeds at a constant rate in a constant environment. The advantages of continuous cultures are achieved through continually replenishing the nutrients by pumping fresh growth medium into the culture while removing cells and spent medium from the vessel at the same rate. In this way steady-state populations of organisms may be obtained and environmental factors may be maintained constant yet independently controllable by the operator. Continuous

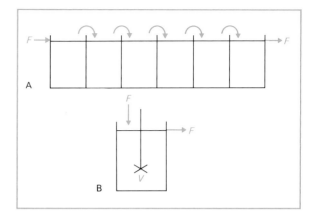

Figure 10.3 A diagrammatic representation of a stretch of river divided into many small compartments and B, the elements of a chemostat fitted with stirrer (F = flow rate, V = volume).

cultures are now widely used by microbiologists to solve a whole range of problems concerning the growth, metabolism, biochemistry and genetics of microorganisms and are becoming increasingly popular for the study of mixed microbial populations. Furthermore, the commercial production of microbial products such as those obtained by classical fermentations (for instance, beers, wines, dairy products), antibiotics, single-cell proteins, amino acids, insecticides and the aerobic treatment of sewage and industrial waste waters relies to a great extent upon continuous-culture systems.

Figure 10.3B represents diagrammatically the major features of a chemostat. It consists of a reactor vessel into which is pumped a constant flow, F, of substrate at concentration S_0. The reactor is vigorously stirred such that its contents are as completely mixed as possible and it is fitted with a constant volume device such that the culture overflows from the reactor at the same rate at which sterile nutrient media enters. If the volume, V, of the culture in the reactor is kept constant and the reactor is sufficiently well stirred, the residence times of the particles in the vessel will be determined by the ratio FV^{-1}, which is called the dilution rate, D, that is;

$$D = F/V \qquad 4$$

The dilution rate has units of reciprocal time, that is the same units as the specific growth rate of microorganisms, and is the number of complete volume changes per unit time.

If we now consider the fate of an organism in the reactor of a chemostat that receives a constant flow of growth-limiting substrate, it can be described by the mass balance equation:

rate of change of organism = growth in reactor − X with respect to time, t = output from reactor

that is:
$$\frac{dX}{dt} = \mu X - DX \qquad 5$$

where X is the concentration of organisms in the reactor. If the population reaches a steady state, that is, if there is no rate of increase of $dX/dt = 0$,

then $\mu X - DX = 0$ 6
or $\mu = D$ 7

Thus at steady state the specific growth rate of the organism equals the dilution rate of the vessel.

Similarly we can consider the fate of the growth-limiting nutrient in the reactor. If the concentration entering the reactor is S_o, and there is concentration s in the reactor, then the following mass balance equation may be written:

$$\text{rate of change of substrate, } s \text{ with respect to time, } t = \begin{array}{l} \text{Input to reactor} - \text{Output} \\ \text{from reactor} - \text{Consump-} \\ \text{tion by organism} \end{array}$$

that is: $\dfrac{ds}{dt} = DS_o - Ds - \dfrac{\mu X}{Y}$ 8

From equations 6 and 8, the steady-state values of microbial biomass and substrate concentrations may be calculated from,

$$\tilde{X}(\mu - D) = 0 \qquad 9$$

and

$$D(S_o - \tilde{s}) - \frac{\mu \tilde{X}}{Y} = 0 \qquad 10$$

where the tilde represents a steady-state value. In addition, we must use equations 3 and 7 in order to be able to calculate $\tilde{s}$, that is,

$$\mu = D = \frac{\mu_m \tilde{s}}{(K_s + \tilde{s})}$$

or

$$\tilde{s} = \frac{K_s D}{(\mu_m - D)} \qquad 11$$

Similarly we can calculate $\tilde{X}$ from equation 10 when $\mu = D$, that is,

$$\tilde{X} = Y(S_o - \tilde{s}) \qquad 12$$

Thus substituting equation 11 for $\tilde{s}$ in equation 12, then,

$$\tilde{X} = Y\left[S_o - \frac{K_s D}{(\mu_m - D)}\right] \qquad 13$$

From equations 11 and 13 it is now possible to calculate $\tilde{X}$ and $\tilde{s}$ at any dilution rate, provided that the values of μ_m, K_s and Y are known. This has been carried out for an hypothetical organism in Figure 10.4.

It will be seen from Figure 10.4 that the steady-state values of $\tilde{s}$ are unique at any dilution rate irrespective of the concentration of substrate S_o entering the reactor. However, the population of organisms is completely dependent upon the substrate concentration entering the system and this is why two population curves are given in Figure 10.4. The greater the concentration of substrate entering the greater the concentration of organisms that are supported. It may also be seen that as the value of the dilution rate of the vessel is increased such that it approaches the maximum specific growth rate of the organism, the population progressively becomes smaller

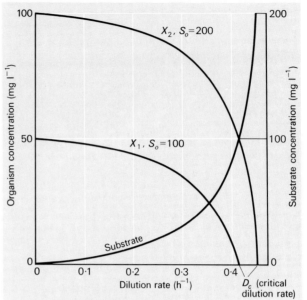

Figure 10.4 Steady-state solutions of organism and substrate concentrations in a chemostat. Results obtained by solving equations 11 and 13 ($\mu_m = 0.5$ h^{-1}, $K_s = 20$ mg l^{-1} and $Y = 0.5$). Two population curves are given, in X_1, $S_o = 100$ mg l^{-1}, and in X_2, $S_o = 200$ mg l^{-1}.

and eventually is washed out of the reactor since it is removed in the flow faster than it can grow – this point is known as the critical dilution rate. One can calculate the fastest rate of growth in the reactor by equating $\tilde{s}$ with S_o and substitution in equation 3. That is,

$$\mu = D_c = \frac{\mu_m S_o}{(K_s + S_o)} \qquad 14$$

where D_c is the critical dilution rate when $\tilde{X} = 0$. Inspection of equation 14 shows that D_c is dependent upon the value of S_o and this explains why in Figure 10.4 the biomass X_1 is washed out at a lower dilution rate than X_2.

There is now a wealth of practical data published in the literature that demonstrates that the above theory is usually adequate to describe the qualitative behaviour of a variety of single-species populations of microorganisms in continuous-culture systems. Since this is the case we may now extend the theory to consider the growth of mixed microbial populations in chemostats.

Growth of mixed microbial populations

When more than one species of microorganism are grown together they must interact according to one of the interactions which are listed and briefly defined in Table 1. Specific examples of all these types of interactions involving protozoa and bacteria may be found in Curds (1977), but let us here first consider competition.

Competition

Competition has been defined as the struggle between

Table 1 Microbial interactions

Interaction	Brief definition
Neutralism	No interaction
Commensalism	One member benefits, other unaffected
Mutualism	Each member benefits from other
Competition	A race for nutrients and space
Amensalism	One adversely changes the environment for other
Parasitism	One organism steals from another
Predation	One organism ingests the other

(After Bungay & Bungay, 1968)

organisms for an essential resource that is limited in the environment. The resource could be a nutrient, space or light, but most work has been carried out on the competition for nutrients. Two types of competition are commonly recognised – intraspecific competition which refers to rivalry between individuals of the same species and inter-specific competition where the competitors are of different species. I will confine my remarks to the latter but the same approach and conclusions are equally applicable to intraspecific competition where the variability of the individuals within a population of the same species results in a competitive relationship.

Theoretical work preceded experimental data on com-petition and several authors, particularly Gause (1934), developed similar mathematical arguments which sug-gested that when two organisms compete for a common nutrient, the one with the highest rate of growth under the prevailing environmental conditions would succeed and displace the other organism. The more recent analysis by Powell (1958) concerning the fates of contaminants or mutants within a chemostat system is equally applicable and rather more precise and detailed.

When two competing organisms are grown together in a chemostat there are two possible outcomes, either one survives and the other is washed out, or both survive. The actual outcome is determined by the relationship between the concentration of growth-limiting substrate and the specific growth rate for each species. If this relationship is identical for the two species then there would be no competition, both organisms would survive and the steady-state population of each species would depend solely upon the relative proportions of each in the original inoculum. However, this is an unlikely situation and most pairs of organisms would more commonly have a slightly different relationship according to one of two possible ways as is illustrated in Figure 10.5. In the first case (Fig. 10.5A), organism 1 has a higher maximum growth rate and lower saturation constant than organism 2. Thus organism 1 has the advantage at both high and low substrate concentrations

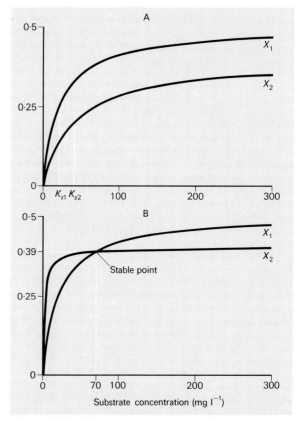

Figure 10.5 Relationships between specific growth rates of two organisms X_1 and X_2 with substrate concentration. A, Maximum growth rate of X_1 higher and saturation constant lower than X_2 ($\mu_{m1} = 0.5$, $\mu_{m2} = 0.4$; $K_{s1} = 20$, $K_{s2} = 40$). B, Both maximum growth rate and saturation constant of X_1 higher than X_2 ($\mu_{m1} = 0.5$, $\mu_{m2} = 0.4$; $K_{s1} = 20$, $K_{s2} = 2.0$).

which means that it will survive and displace organism 2 at any substrate concentration and hence at any dilution rate. In the second situation (Fig. 10.5B), however, the curves of the two species cross each other since although organism 1 still has a higher maximum growth rate, it now also has a higher saturation constant. Thus at low substrate concentrations (to the left of the cross-over point in Figure 10.5B) organism 2 has the advantage and will displace organism 1; but at higher substrate concentrations (to the right of cross-over point on Figure 10.5B) the reverse is true. The point at which both curves intersect indicates the substrate concentration at which the specific growth rates of both organisms are identical. At this point both organisms would, theoretically, survive indefinitely to-gether. The point may be calculated using the following equations where the subscripts refer to organisms 1 and 2, respectively. From equation 3,

$$\mu_1 = \frac{\mu_{m1} s}{K_{s1} + s}$$

$$\mu_2 = \frac{\mu_{m2} s}{K_{s2} + s}$$

thus when $\mu_1 = \mu_2$, that is at the cross-over point on Figure 10.5B,

$$\frac{\mu_{m1}s}{K_{s1}+s} = \frac{\mu_{m2}s}{K_{s2}+s}$$

or

$$s = \frac{\mu_{m2}K_{s1} - \mu_{m1}K_{s2}}{\mu_{m1} - \mu_{m2}} \qquad 15$$

If the two organisms in Figure 10.5B were grown together in a chemostat, it may be seen from equation 15 that both would survive if the substrate concentration were maintained at 70 mg l^{-1}, only organism 1 would survive at concentrations above this value and only organism 2 when below it. In practical terms a substrate concentration of 70 mg l^{-1} would be achieved at a dilution rate of 0.388 h^{-1} (this may be calculated from equation 3 where $D = \mu$ and $s = 70$), it follows that organism 1 will survive alone at dilution rates above it and organism 2 when below it.

While the above theory is now well accepted it is not always borne out in practice. For example, Tempest, Dicks & Meers (1967) found that the bacterium *Aerobacter aerogenes* outgrew several Gram-positive bacteria in a magnesium-limited mixed culture according to theory, but mixed cultures of the yeast *Candida utilis* and the bacterium *Bacillus subtilis* behaved paradoxically. In this case neither organism outgrew the other species unless the inoculum population exceeded a certain minimum concentration. Later, Meers & Tempest (1968) confirmed that the ability of *B. subtilis* to outgrow *C. utilis* depended upon the relative proportions of each in the inoculum. This result cannot be predicted by the theory outlined earlier since equation 3 does not contain a factor for population density. The most obvious explanation for these paradoxical results is that the bacterium produced substances that accumulated in the culture which inhibited the growth of the competing yeast, an example of amensalism, see Table 1. However batch-culture experiments showed that, in fact, the reverse was the case, the bacterium secreted a substance which *promoted* its own growth rate.

Meers (1971) also used *B. subtilis* and *C. utilis* in his investigation of mixed culture growth. In agreement with the theory, he found that he was able to select which species would survive in a chemostat. At lower dilution rates *C. utilis* became dominant and the bacterium was washed out, but changing to a higher dilution rate reversed the situation. Indeed he found that it was possible after several days operation to reduce one organism such that microscopic observation indicated its absence; but after suitable adjustment to the dilution rate, the apparently absent organism reappeared and became dominant over the other.

It is apparent from the above, that in some cases experimental data derived from pure-culture systems do not always agree with theoretical predictions. Thus, it is not surprising that one finds even less agreement when one looks at the growth of microorganisms in the natural environment. Theory predicts that, although possible, two

organisms are unlikely to survive together indefinitely when competing for a single resource. In nature, however, it is very rare to find microbial populations of a single species even though the organisms present appear to be competing for similar nutrients. Why is this? Some of the data above concerning bacteria and yeasts provide us with one possible explanation. In those cases it is not that theory is actually incorrect, it is more a case of theory being too simplistic; other factors, particularly other microbial interactions have not been taken into account. In my opinion there is little doubt that provided two organisms are studied whose only interaction is that of competition, then theory will be borne out in practice. Practically speaking however this is rarely the case and other factors over-ride the simple competitive interaction.

Let us now examine what factors could be over-riding competition in continuous cultures and in the natural environment. Table 2 summarises the conditions that would enable two species to survive together in a continuous pure-culture system. It will be seen that if the organisms are each competing for the same growth-limiting nutrient then there are three conditions that would enable both organisms to survive in a chemostat. The first condition has already been discussed above and the second and third cases are really variations on a single theme; that is, the faster growing organism either inhibits itself or stimulates the other organism by its product. In both cases it would be possible for both organisms to survive (see Pirt, 1975 for full mathematical treatment).

The second part of Table 2 considers the situation where the organisms are utilising different substrates and three factors are listed that would enable the two to coexist. The first simple case is where there is no competition at all since two different substrates are available. A situation such as this would usually be known in a pure-culture system but would not necessarily be apparent in the natural habitat. In the second case, one organism using the limiting substrate produces a compound that is

Table 2 Conditions that would enable two species to coexist in a chemostat culture

A. *Organisms utilising same growth-limiting substrate*
 i) When specific growth rates of two organisms are identical
 ii) When faster growing species is inhibited by its own product
 iii) When the faster growing species stimulates the slower by its product

B. *Organisms utilising different growth-limiting substrates*
 i) When two different growth-limiting substrates are fed into the culture
 ii) When the product of one species is the growth-limiting substrate for the other
 iii) When one organism preys upon the other

(After Pirt, 1975)

the growth-limiting substrate for the second organism and in this case both organisms would survive. This again is more unlikely to be known in the natural environment but perhaps one of the best known examples of this is the case of nitrification where the bacterium *Nitrosomonas* utilises ammonia to produce nitrites that are then used by the bacterium *Nitrobacter* to produce nitrates. Another example is that of the growth of propionic-bacteria on lactic acid derived from the growth of streptococci on lactose in milk and there are many other examples in the bacteriological literature. The final condition given in Table 2 is when one of the organisms feeds upon the other. In this situation both predator and prey can survive but predation can have other dramatic effects upon the populations. It generally induces violent oscillatory changes in the populations and can enable a slower growing organism to coexist with a faster growing competitor.

Predation

Until now the microbiological examples given have been particularly concerned with bacterial populations and we must now consider their predators. Protozoa are major predators of bacterial populations and it has been shown (Curds & Cockburn, 1969, 1971; Curds, 1971, 1974; Curds & Bazin, 1977) that the growth and feeding of protozoa can be modelled in a similar manner to bacteria. Like prokaryotes, protozoa reproduce asexually by binary fission so that the growth of a protozoan population is similar to that of bacteria; protozoa have a constant yield when feeding on bacteria and their specific rate of growth is related to substrate concentration in at least a similar manner to that of bacteria. The substrate for the protozoa under consideration here is of course bacteria and evidence is now gathering which suggests that the growth and feeding rates of protozoa (ciliates and amoebae) are related to the concentration of bacteria available per protozoan cell rather than to bacteria concentration alone. The former situation may be expressed mathematically as,

$$\mu_p = \frac{\mu_{mp}B/N}{K_p B/N} \qquad 16$$

where μ_p and μ_{mp} are the specific and maximum specific growth rates of the protozoa, B is the concentration of bacteria available, N the number of protozoa present and K_p is the saturation constant. Equation 16 describes a three dimensional surface made up of a family of curves of the type shown in Figure 10.2 and this is illustrated in Figure 10.6. Practical work carried out on ciliated protozoa (Curds & Cockburn, 1969) and amoebae (Owen, Dauppe & Bazin, 1979) indicate that the experimental data fit equation 16 and Figure 10.6 rather better than an equation of the Michaelis–Menten type as in equation 17 (see also equation 3) where,

$$\mu_p = \frac{\mu_{mp}B}{K_p+B} \qquad 17$$

the population of protozoa is not taken into account. In

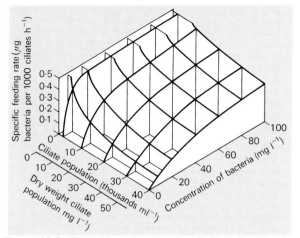

Figure 10.6 Three-dimensional model of effect of bacteria concentration and ciliate population density on the specific feeding rate of *Tetrahymena pyriformis*. See equation 16 where specific feeding rate is directly proportional to specific growth rate (after Curds & Cockburn, 1969).

spite of this it has been shown that for qualitative modelling purposes, equation 17 is adequate and will therefore, for simplicity, be used in the following treatment.

Let us first consider the simple predator–prey relationship in a chemostat, Table 2B (iii). In that system a chemostat receives a flow of soluble substrate that is utilised by the prey organism alone and the predator feeds only upon the prey (see Fig. 10.7A). The fates of the substrate, the bacteria and the protozoa of this simple food chain can be expressed mathematically by the use of the following mass balance equations,

$$\begin{array}{c}\text{rate of change}\\ \text{of substrate}\end{array} = \text{Input} - \text{Output} - \begin{array}{c}\text{Consumption by}\\ \text{bacteria}\end{array}$$

that is: $\qquad \dfrac{ds}{dt} = DS_o - Ds - \dfrac{\mu X}{Y} \qquad 18$

It will be seen that this equation is identical to equation 8,

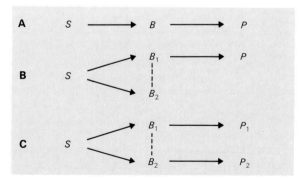

Figure 10.7 Three food chains mentioned in text, where S = substrate, B = bacteria and P = protozoan predator. Arrows represent nutrient pathways. Dashed lines represent competitive interaction between bacteria.

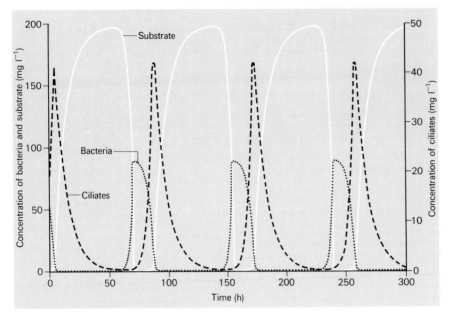

Figure 10.8 Limit-cycle oscillations of populations given by computer. Equations 3, 17, 18, 19 and 20 were used to simulate the populations where $\mu_m = 0.6$, $\mu_{mp} = 0.43$; $K_s = 4.0$, $K_p = 12.0$; $Y = 0.45$, $Y_p = 0.54$; $D = 0.1$ and $S_o = 200$ (after Curds, 1971).

$$\text{rate of change of bacteria} = \text{Growth} - \text{Output} - \text{Consumption by protozoa}$$

that is:

$$\frac{dX}{dt} = \mu X - DX - \frac{\mu_p P}{Y_p} \qquad 19$$

where μ_p, Y_p and P are the specific growth rate, yield constant and population size of the protozoan population respectively. The final equation is,

$$\text{rate of change of protozoa} = \text{Growth} - \text{Output}$$

that is:

$$\frac{dP}{dt} = \mu_p P - DP \qquad 20$$

Although it is possible to use equations 3, 17, 18, 19 and 20 to solve for the steady-state concentrations of substrate, bacteria and protozoa, practical experience with chemostats and computer simulation techniques have shown that a steady-state is not always practically and theoretically inevitable. The most usual solution is that all three populations pass through a series of repetitive oscillations (limit cycles) with constant amplitude and wavelength. An example of the type of limit-cycle behaviour that is predicted by our model is given in Figure 10.8. In addition to limit-cycle behaviour it is possible to achieve steady-state populations at higher dilution rates which may be approached via damped oscillations or asymptotically. The actual dynamic response is not only dependent upon the dilution rate of the reactor vessel but also upon other variables including the relative values of the kinetic constants of the predator and its prey. A full account of the effects of the various parameters upon the dynamics may be found in Curds (1971).

The above theory predicts that a predator and its prey would not achieve steady-state populations at the dilution rates at which a chemostat is likely to be operated. This has been borne out in practice by several workers who have studied a variety of microbial predators and their prey (for review see Curds & Bazin, 1977) although, to date, agreement is of a qualitative rather than quantitative nature. It follows therefore that the presence of a microbial predator in the natural environment might induce similar variations in the microbial flora and this may be one reason why populations of microorganisms vary with respect to time.

The influence of ciliate predators on other more complicated microbial food chains has been investigated by computer simulation techniques although no practical data are yet available. It should be pointed out here that these models assume that ciliated protozoa have some means of selecting their bacterial prey. While this is not yet proven evidence is gathering that certainly seems to indicate that this assumption is more than possible (see review by Curds & Bazin, 1977).

In the first of these simulations (Fig. 10.7B) the ciliate predator was assumed to feed on the slower growing of the two bacteria that were competing for the same nutrient supply. This resulted in both the slow growing bacterium and its predator being quickly washed out of the system; the remaining bacterium which was not preyed upon then achieved a steady-state population as expected. In two

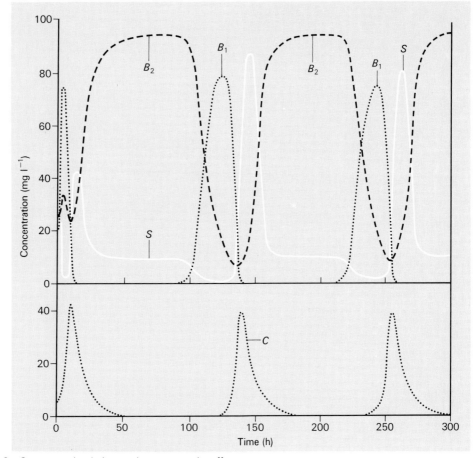

Figure 10.9 Computer simulation to demonstrate the effect of predation upon one of two competing bacteria. The food chain of the simulation is given in Figure 10.7B. (S = substrate, B_1, B_2 = competing bacteria, C = ciliate predator). After Curds (1974).

other simulations of the same food chain, the ciliate predator was assumed to feed exclusively upon the faster-growing of the two bacteria. In these cases when the maximum specific growth rate of the predator was assumed to be low, which would mean a low predation pressure, all populations passed through a series of damped oscillations and reached a steady-state within 170 hours simulated time. Here, both competing bacteria survived and the slower growing bacterium became dominant because the faster growing bacterium was preyed upon by the ciliate. When the predation pressure was raised by doubling the maximum specific growth rate of the protozoan, limit-cycle oscillatory behaviour was obtained. Apparently the increased predation pressure caused limit-cycles in its prey that subsequently induced fluctuating levels of competition for substrate between the two bacteria. Thus, the two bacterial species oscillated out of phase with each other and caused a double trough in the cyclic pattern of substrate concentration (Fig. 10.9).

In the final food chain simulated (Fig. 10.7C), two ciliates independently preyed upon two competing bacteria.

Two simulations of this food chain were carried out; in the first the more voracious predator was assumed to feed on the slower growing bacterium. Under these circumstances the slow bacterium was quickly reduced to a low concentration, due to a combination of predation and competition, which ultimately resulted in the complete washout of the voracious predator and the subsequent recovery of the slow bacterium. The system was therefore unstable and reverted to the food chain illustrated in Figure 10.7B in which a ciliate with low maximum rate of growth preyed upon the faster bacterium; steady-state populations were approached via damped oscillations. The results of the second simulation of the food chain in Figure 10.7C are illustrated in Figure 10.10. The voracious ciliate here was arranged to feed on the faster growing bacterium and the less voracious ciliate preyed upon the slower growing bacterium. In this case all populations survived even though the two side chains of the food web were competing for a common nutrient. In this simulation the populations did not oscillate in a regular manner even after 300 hours of simulated time.

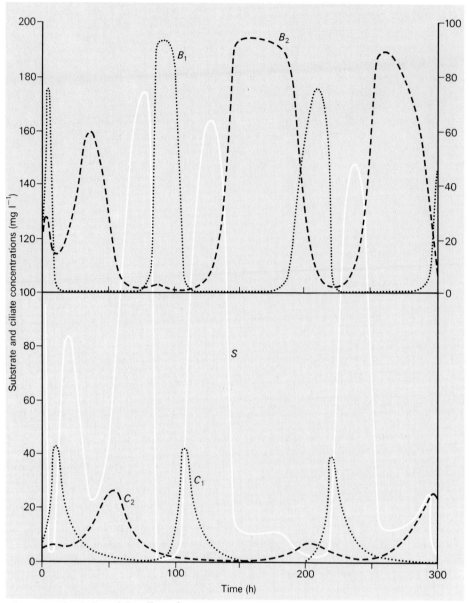

Figure 10.10 Computer simulation of the effect of two ciliates preying upon two competing bacteria. The food chain of this simulation is given in Figure 10.7 C (S = substrate, B_1, B_2 = competing bacteria, C_1, C_2 = ciliate predators). After Curds (1974).

The above examples therefore demonstrate that, theoretically, the introduction of one or two predators into a food chain containing competing prey can promote the coexistence of the competitors and secondly, it can lead to non-repetitive dynamic behaviour. Since these are relatively simple model situations, then perhaps it is not surprising to find complex dynamic behaviour in the microbial populations of natural aquatic systems which change radically for reasons that remain obscure.

Until now we have considered few factors other than interactions between the organisms themselves. Interac-

tions are those most frequently forgotten or overlooked by the ecologist but we must not also forget that there are many other possibilities too when attempting to explain differences between theoretical predictions and what is found in nature. For example, are the samples we are taking too big so that we are in fact really sampling several microhabitats? Has the flow rate, temperature and pH of the river changed significantly? Have the organisms themselves changed their mode of nutrition? Micro-organisms are well known for their nutritional flexibility – bacteria may utilise a variety of carbon and nitrogen sources and certain protozoa such a *Euglena* can even change their mode of nutrition from autotrophy to heterotrophy. These few examples are given to illustrate that one must take environmental, biological and sampling factors into account when studying ecology. When one thinks about the complexity of these interrelated factors in the natural environment perhaps we should be surprised to find a single-species microbial population living alone.

Finally, from the above considerations, it may be concluded that while theory supports the concept of competitive exclusion, practical observations on natural microbial communities, even experimental ones, rarely do

so. This does not necessarily mean that the concept is basically unsound; in my opinion it is precise and correct. But when considering microbial populations, the model of competitive exclusion is far too simplistic for theory to coincide with nature except perhaps on rare occasions. The physical, chemical and biological components of the natural environments are inextricably interwoven. These components interact and the complex parameters thus produced are those that always influence and often override the simplistic basic theory of competitive exclusion. Moreover, these components, like the populations, are also dynamic and ever changing.

References

Bungay, H. R. & Bungay, M. L. 1968. Microbial interactions in continuous culture. *Advances in applied Microbiology* 10: 269–290.

Curds, C. R. 1971. A computer-simulation study of predator-prey relationships in a single-stage continuous-culture system. *Water Research* 5: 793–812.

Curds, C. R. 1974. Computer simulations of some complex microbial food chains. *Water Research* 8: 769–780.

Curds, C. R. 1977. Microbial interactions involving protozoa, pp. 69–105 *in* Skinner, F. A. & Shewan, J. M. (Eds), *Aquatic Microbiology. Society of Applied Bacteriology Symposium, Series 6.*

Curds, C. R. & Bazin, M. J. 1977. Protozoan predation in batch and continuous culture. *Advances in aquatic Microbiology* 1: 115–176.

Curds, C. R. & Cockburn, A. 1969. Studies on the growth and feeding of *Tetrahymena pyriformis* in axenic and monoxenic culture. *Journal of General Microbiology* 54: 343–348.

Curds, C. R. & Cockburn, A. 1971. Continuous monoxenic culture of *Tetrahymena pyriformis*. *Journal of General Microbiology* 66: 95–108.

Gause, G. F. 1934. *The struggle for existence*. Baltimore: Williams & Wilkins (reprinted, 1969, New York: Hafner).

Harding, G. 1960. The competitive exclusion principle. *Science* 131: 1292–97.

Herbert, D., Elsworth, R. & Telling, R. C. 1956. The continuous culture of bacteria; a theoretical and experimental study. *Journal of General Microbiology* 14: 601–622.

Meers, J. L. 1971. Effect of dilution rate on the outcome of chemostat mixed culture experiments. *Journal of General Microbiology* 67: 359–361.

Meers, J. L. & Tempest, D. W. 1968. The influence of extracellular products on the behaviour of mixed microbial populations in magnesium-limited chemostat cultures. *Journal of General Microbiology* 52: 309–317.

Monod, J. 1950. La technique de culture continue; theorie et application. *Annales de l'Institut Pasteur, Paris* 79: 390–410.

Owen, B., Dauppe, A. & Bazin, M. 1979. Dependence of microbial predator specific growth rate on the ratio of prey to predator population densities. *Quarterly Journal of Society of General Microbiology* 6: 80.

Pirt, S. J. 1975. *Principles of microbe and cell cultivation.* 274 pp. Oxford: Blackwell Scientific Publications.

Powell, E. O. 1958. Criteria for the growth of contaminants and mutants in continuous culture. *Journal of General Microbiology* 18: 259–268.

Powell, E. O. 1965. Theory of the chemostat. *Laboratory Practice* 14: 1145–1161.

Tempest, D. W., Dicks, J. W. & Meers, J. L. 1967. Magnesium-limited growth of *Bacillus subtilis* in pute and mixed culture in a chemostat. *Journal of General Microbiology* 49: 139–147.

CHAPTER 11

Community structure and resource partitioning – the plankton

G. A. Boxshall

Oceans and seas cover about 70 per cent of the earth's surface to an average depth of 3700 m. This immense volume of water (1347 million km³) provides three major habitats for organisms – the intertidal zone on the shore, the sea bottom down to depths of 11 022 m and the pelagic zone. The pelagic zone, which consists of all the open water away from the bottom, is by far the largest of these habitats and can be subdivided into the shallow neritic waters found over the continental shelf and the deeper oceanic waters away from the continental land masses. The vast oceanic environment which extends uninterrupted from the Arctic to the Antarctic is, for the greater part, characterised by low rates of primary production of organic matter, similar to those of terrestrial deserts. Life, however, abounds in this watery 'desert' and complex communities of animals and plants have evolved that are well adapted for an oceanic existence. These communities contain plankton and nekton. The plankton includes all the plants and animals that are carried around by the movement of water masses whereas the nekton includes those organisms, particularly the larger ones, that are capable of moving independently of the water mass.

In contrast to terrestrial and other aquatic environments the open ocean contains a relatively small number of species in diverse groups of uni- and multicellular organisms. One of the factors involved in the explanation of this low diversity is the process of speciation itself. The origin of new species usually involves a period of reproductive isolation brought about by some geographical or topographical barrier between two populations. This is allopatric speciation (see Chapter 3). In the ever-moving, continuous oceanic environment there are few obvious permanent physical barriers to reproduction and gene flow between populations. The lack of obvious barriers and the relatively low species diversity of oceanic communities suggest that speciation may be a relatively uncommon event in the oceanic environment. It is possible that methods of speciation other than classical allopatric speciation are involved, such as sympatric speciation. It is

unlikely, however, that a method of speciation involving catastrophic changes in abundance, as invoked by the theory of quantum speciation (Chapter 3), would occur as the open ocean is buffered by size against extreme climatic influences. How, then, does speciation occur in oceanic environments? This question cannot yet be fully answered but it is apparent that the traditional concept of a physical barrier to gene flow between populations must be modified when applied to oceanic systems.

Oceanic organisms may perceive physico-chemical barriers in their environment which oceanographers cannot detect with their relatively crude instrumentation. There may be biological factors which influence spatial distribution that, in turn, can effectively provide barriers to gene flow. At least for the large numbers of species that are cosmopolitan or circumglobal distance alone does not appear to constitute a barrier to gene flow. It is important to stress, however, that the oceans are not simple, thoroughly mixed bodies of water. They are extremely heterogeneous at all scales from hundreds of kilometres down to a few millimetres. This complex physical structure will obviously influence both the ecology and evolution of oceanic organisms so it is appropriate, at this point, to consider some of the more important physical characters of the oceans.

The oceans consist of a number of water-masses, each with its own properties of temperature, salinity, depth and movement. In the eastern North Atlantic, for example, the outflow of high salinity Mediterranean water can be identified at a considerable distance from the Straits of Gibraltar. Similarly, the Antarctic Intermediate water-mass can be detected by its low salinity and temperature characteristics as far north as 20° N (Fig. 11.1). There is a pronounced latitudinal boundary for planktonic organisms between 38° N and 43° N in the North Pacific. This is associated with marked changes in the biomass of phytoplankton and zooplankton and corresponds to the boundary between Subarctic and Central Pacific water-masses. Some species may be associated with a particular

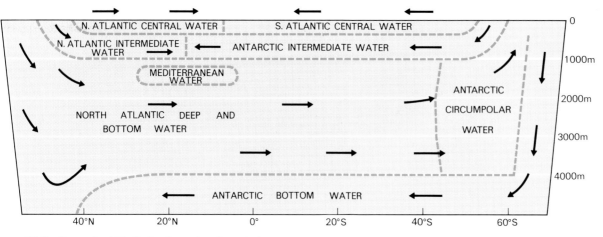

Figure 11.1 Generalised North–South section through the Atlantic Ocean showing approximate boundaries of water-masses. Direction of movement indicated by arrows.

water-mass but these indicator species, restricted to a single water-mass, appear to be the exception rather than the rule. Nevertheless, recent work suggests that there are indicator assemblages of species. The coexistence of these species can be linked to the presence of a particular water-mass demonstrating that a close correlation often exists, on the zoogeographic scale, between the physical structure of the ocean and biological patterns of distribution.

At a rather more local level (0·1–100 km) the oceans are similarly heterogeneous, comprising a mosaic of patches of water that can differ both physico-chemically and in their contained flora and fauna. This patchiness is one of the outstanding features of the oceanic environment and profoundly affects the oceanic community. It also causes immense sampling problems which make the careful design of sampling programmes imperative.

On an even smaller scale conditions of salinity, temperature and oxygen concentration have been shown to vary significantly over vertical distances of 1 m or less in surface waters. The nature and persistence of these microlayers can profoundly influence the distribution of planktonic organisms, particularly small ones. An examination of the effects of patchiness on oceanic communities forms one of the main themes of this chapter.

Temperature gradients within the ocean provide physical structure. There is a gradient of increasing temperature from the polar to the equatorial regions and there is typically a gradient of decreasing temperature with increasing depth at any particular locality. There is a marked contrast between the cold water surface fauna from the Arctic and the warm surface fauna from the tropics. Many planktonic organisms are almost entirely restricted to warm water, such as the copepod genera *Sapphirina* and *Corycaeus*, whereas other species, like *Calanus glacialis*, frequent only cold water. Some cold water species occur in lower latitudes but only in the colder waters at great

depths. This pattern of tropical submergence has been noted for several species, like the chaetognath *Eukrohnia hamata*, and is clearly related to temperature. The vertical temperature gradient also varies with latitude. In middle latitudes during the summer there is typically a warm, well-mixed layer of water from the surface down to 50–100 m depth, then a zone of relatively rapid temperature change (the thermocline) below which the temperature decreases very slowly. During the winter the surface layer cools and the seasonal thermocline typically disappears. Winter storms help the breakdown of the thermocline. In tropical areas the surface waters are warmer, the thermocline is often more clearly marked and, since surface cooling does not occur, it is permanent.

The thermocline is an important feature in the water column because little nutrient enrichment of the upper waters occurs from below a permanent thermocline. Also, the thermocline often marks the upper limit of the distribution of vertically migrating species, and so may act as a boundary in the spatial distribution of some planktonic organisms. The importance of the three-dimensional structure of the ocean in the speciation and evolution of pelagic species can best be assessed in ecological terms. The question 'how does speciation occur in oceanic environments?' is naturally complemented by the question 'how do species coexist in oceanic communities?'. These questions represent two sides of the same coin and by answering the second one can search for clues to the first.

Coexistence and niche theory

It has been recognised that each species has a special role or niche in the community to which it belongs. Niches are delimited by, and maintained against, competition from other species within the community. But, according to the principle of competitive exclusion, competitors for a given niche cannot coexist at equilibrium because one species

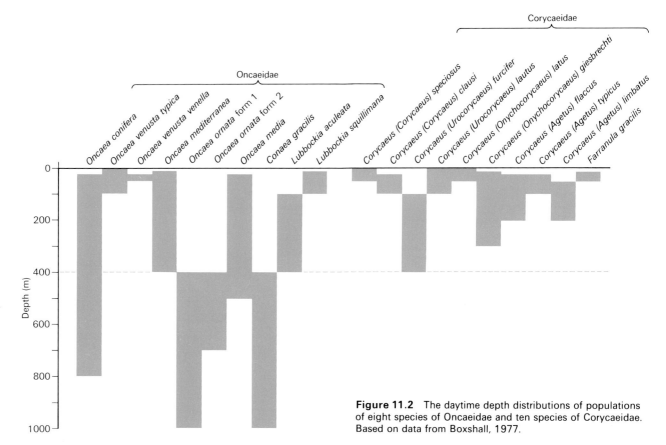

Figure 11.2 The daytime depth distributions of populations of eight species of Oncaeidae and ten species of Corycaeidae. Based on data from Boxshall, 1977.

will eventually displace the other (see Chapter 10). The corollary of this statement is that each species has a unique niche within the community. There may be, however, some overlap between niches of different species, especially closely related species, but in order to ensure their continued coexistence the degree of niche overlap must be limited. Niche overlap can be reduced by partitioning the resources available to the community, although there is obviously a limit to this process of subdivision.

Resource partitioning can affect any of the dimensions of the niches so that species can be segregated in space (spatial), in time (temporal), by their position in the food web of the community (trophic) and by their behaviour. I propose to examine community structure by looking primarily at the copepods of oceanic communities. Copepods constitute the greater biomass and are the most numerous animals of most planktonic communities and as such are one of the best indicators of community structure.

Spatial segregation

Species can be separated horizontally by occupying different geographical areas, or vertically by occupying different depth horizons within a water column. Many copepod species occur at particular depths. The depth ranges of 18 species of cyclopoid copepods belonging to the families Oncaeidae and Corycaeidae are shown in Figure 11.2. One obvious feature of these depth distributions is the difference

between the two families, with the Corycaeidae being restricted to the upper 400 m of the water column and the Oncaeidae ranging from the surface to 1200 m. More important are the differences between particular pairs of closely related species. Some species, such as *Oncaea venusta* and *O. ornata* occupy completely separate depth horizons whereas other pairs, such as *O. media* and *O. mediterranea*, have virtually identical depth distributions. Between these two extremes are the majority of species, each with a depth distribution that overlaps but does not completely coincide with that of any other given species. One interesting feature is that closely related species, such as *Corycaeus (Urocorycaeus) lautus* and *C.(U.) furcifer*, which belong to the same subgenus, often exhibit mutually exclusive depth distributions. Many species, however, coexist over large portions of their total range.

When considering the horizontal distribution of oceanic organisms one is faced with the problem of scale. On a large geographical scale a community exists within an area and all the members of that community might be assumed to interact. On a smaller scale oceanic communities have been found to be very heterogeneous with species or groups of species occurring in high density patches within the community with low density areas in between. This patchiness can greatly affect the interactions between populations within the community (see below).

The scale of sampling plankton populations should be

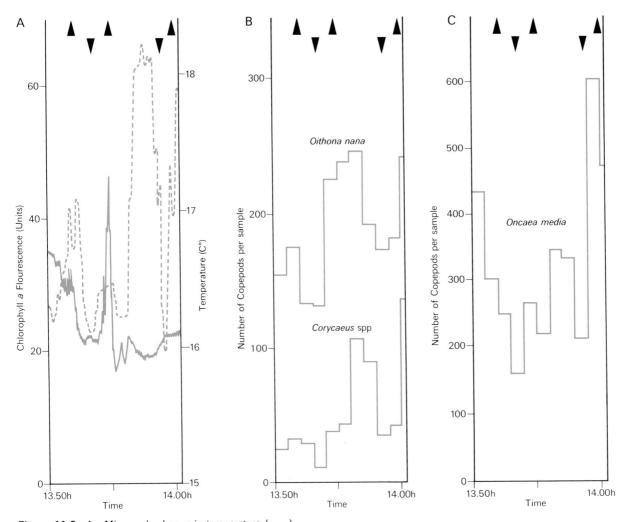

Figure 11.3 A – Microscale change in temperature (——) and chlorophyll *a* values (–––) off the West African coast, at a depth of 8 m over a period of 11 minutes; B – Changes in the abundance of *Oithona nana* and *Corycaeus* species over the same period; C – Changes in the abundance of *Oncaea media* over the same period. Based on unpublished data.

related to the size, mobility and microscale distributions (0·001–100 m) of the organisms under investigation. The importance of having disjunct vertical or horizontal distributions over distances of 10 m or less could then be interpreted in terms of competition and predation theory. Sampling the flora and fauna of microlayers with vertical dimensions of less than 1 m is not technically possible at present but the distributions of copepods and other zooplankton are responsive to small scale phenomena, as can be seen in the following experiment carried out off the north-west coast of Africa.

The intake nozzle of a pump was held at about 8 m depth, at the level of the thermocline. The continuous stream of plankton pumped on board was divided into minute-long samples. Horizontal movement of the water

is negligible over each sampling period so that changes in water quality reflect the rise and fall of water layers within the water column, particularly that layer representing the thermocline.

The temperature of the water is shown in Figure 11.3A, together with the chlorophyll α value, which provides an indication of phytoplankton concentration. The temperature and chlorophyll α value vary with microscale water movement within the water column and these values rise (▲) and fall (▼) more or less in unison. Figures 11.3B and 11.3C show changes in the abundance of the copepods *Oithona nana* and *Corycaeus* species and of *Oncaea media* respectively, over the same period of time. The increases (▲) and decreases (▼) in abundance of these copepods and of other groups (appendicularians and the siphonophore

Muggiaea atlantica) can be correlated with the changes in chlorophyll values and in temperature. The distributions of all these organisms are clearly linked to the microscale structure of the water column.

Temporal segregation

Species can be separated in time as well as in space. Copepods that coexist in the water column often exhibit different seasonal patterns of abundance and of resource utilisation. The temporal separation of the abundance peaks of the calanoid copepod genera *Clausocalanus*, *Calocalanus* and *Mecynocera* is shown in Figure 11.4. These three genera, all of which utilise similar food resources, coexist in the upper 500 m of the Sargasso Sea but their principle abundance peaks are found in June (*Mecynocera*), August (*Calocalanus*) and October (*Clausocalanus*). The partial temporal segregation effectively reduces competition between them at those periods during the year when resource requirements are at their greatest, that is, when reproduction is taking place.

Temporal segregation can also be achieved over short intervals of time. Different species may be found at the same depth but at different times during the day. The cyclopoid *Oncaea venusta venella* occurs between 0 and 300 m and the harpacticoid *Aegisthus mucronatus* occurs between 100 and 800 m. These distributions overlap as both species inhabit the 100–300 m layer, but during the day *O. venusta venella* is found principally at 100–200 m whereas *A. mucronatus* is concentrated at 200–500 m and at night *O. venusta venella* is concentrated at 0–50 m and *A. mucronatus* at 100–400 m. Neither species occurs at high densities at the same depth at the same time as the other.

Trophic segregation

Trophic segregation is dependent on the fact that different species can utilise different food sources. In oceanic communities there are only two principal modes of feeding found in zooplankton – particle feeding and predation. The oceanic environment is nutritionally dilute and particle feeders must employ a filtering mechanism in order to extract sufficient food particles from the medium. A range of particle sizes is present, from μ-flagellates of 0.002–0.2 μm diameter upwards, and primary trophic segregation can be achieved by the selection of different food particle size categories by different species. If body size is regarded as proportional to the size of the filter apparatus and therefore to the size of particle utilised, then differences in body size may be regarded as representing trophic segregation. Many authors have reported a body size ratio of 1.2 to 1.4 in pairs of coexisting species (see Chapter 13). This observation has led to the generalisation that a difference in body size of this magnitude (20–40 per cent) is sufficient to separate niches in the food dimension and allow coexistence.

Calanoid copepods are the dominant primary consumers in the oceanic community and they typically use their

second maxillae and maxillipeds for filter feeding. This filter feeding apparatus can be regarded as a sieve with a range of pore sizes. Nival & Nival (1976) found that in *Acartia clausi* mean pore size is proportional to body length (a function of absolute size) so that the spectrum of food particle sizes utilised changes during development. Their results suggest that the different developmental stages are each able to exploit a slightly different size class of particle most efficiently (see Fig. 11.5) thereby minimising intraspecific competition between young stages and the adults.

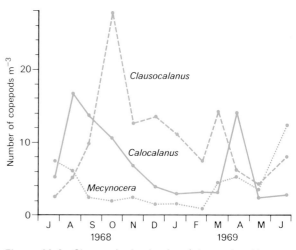

Figure 11.4 Changes in the density of three calanoid copepod genera in the upper 500 m of the Sargasso Sea, over a period of 12 months. Based on data from Deevey & Brooks (1971).

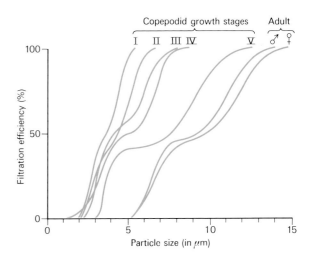

Figure 11.5 The filtration efficiencies of the filter feeding apparatus of copepodid stages I and V and adult ♂ and ♀ of *Acartia clausi*, over a range of particle sizes. Based on data from Nival & Nival (1976).

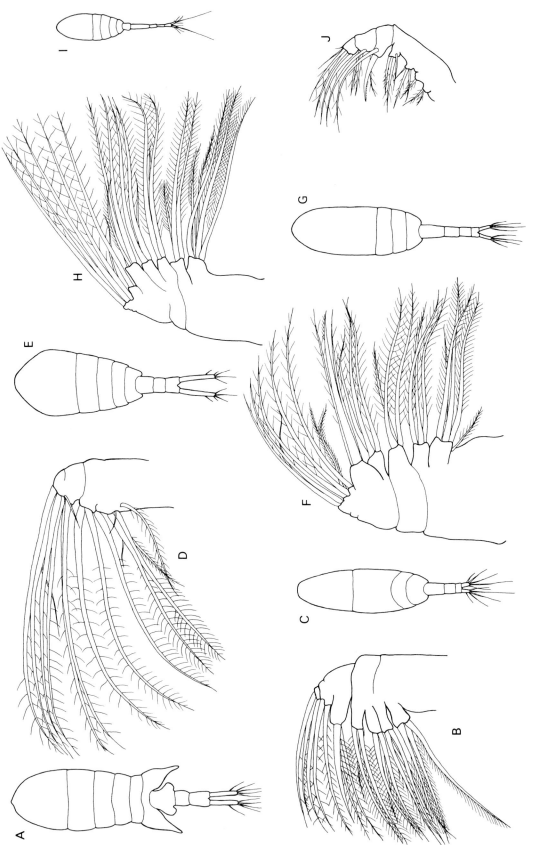

Figure 11.6 Variation in size of the whole animal and the size and structure of the feeding apparatus, the second maxilla, in five coexisting species of copepods. A, B, *Eurytemora herdmani*; C, D, *Acartia clausi*; E, F, *Temora longicornis*; G, H, *Pseudocalanus minutus*; I, J, *Oithona similis*. All species drawn to the same scale.

This illustrates, in theory, the potential of the filter feeding apparatus as a means of determining levels of competition between species. However, recent work by Poulet (1978) on coexisting species of copepods indicates that feeding behaviour may lead to severe interspecific competition. The calanoids *Pseudocalanus minutus*, *Acartia clausi*, *Temora longicornis* and *Eurytemora herdmani* and the cyclopoid *Oithona similis* all feed on suspended particulate matter. They differ in morphology and in the structure of their feeding appendages (Fig. 11.6) but all five species are non-selective with regard to particle size and all behave as opportunistic feeders by utilising the most abundant available particle sizes. Competition between these species occurs throughout the entire range of particle size (1·58 to 114 μm) and they do not segregate trophically with regard to particle size although there is some evidence that copepods are selective with regard to particle biomass and chemical composition and that particle feeding can be a chemosensory-determined behaviour process.

Copepods of the families Oncaeidae, Corycaeidae and Sapphirinidae are surface feeders and predators feeding on organisms larger than themselves. *Sapphirina* species exhibit specialised predator-prey relationships. *S. angusta*, for example, lives for much of its life within, and feeds on, the pelagic tunicate *Thalia democratica*. If a similar high degree of prey specificity is exhibited by other *Sapphirina* species then the coexistence of ten or more species of the genus in the same body of water can be readily explained in terms of trophic segregation. *Oncaea* species have mouthparts adapted for surface feeding, rather than filter feeding or raptorial predation.

The only natural surfaces available in the pelagic zone are other living organisms or aggregates of organic matter. Discrete organic aggregates, derived from pteropod feeding webs, abandoned appendicularian houses and other macroscopic mucoid or gelatinous particles, provide surfaces for feeding, for bacterial growth and for the absorption of dissolved organic matter. They also act as a source of food, as Alldredge (1972, 1976) has shown with direct underwater observations. Individuals of *O. mediterranea* were observed feeding on the abandoned houses of the appendicularian *Megalocercus abyssorum*. The copepods tended to prefer the inner filter apparatus where nannoplanktonic organisms are concentrated and form a readily obtainable food source. These macroscopic aggregates are of great importance to small copepods like *Oncaea* with its essentially benthic mode of feeding, as they provide both food sources and structural surfaces in an environment that is otherwise poor in structural surfaces. *O. mediterranea* is highly active and spends little time actually resting on the appendicularian houses and it has been observed on the mucous feeding web of the pteropod *Gleba cordata*. So, *O. mediterranea* is capable of utilising a wide range of food sources and it does not appear to be a very selective feeder. The rather uniform morphology of the feeding appendages within the genus *Oncaea* suggests a uniform feeding biology. It is probable that coexisting species such as *O.*

mediterranea, *O. venusta*, *O. conifera* and *O. media* overlap considerably with regard to food sources, showing that trophic segregation is of a low level.

Many copepods are predators but only a small proportion of the copepods in the water column are strictly carnivorous predators, most being omnivorous particle feeders. Field studies show that, even amongst those species that are true carnivores, there is a considerable degree of overlap in food preferences. Planktonic copepods in general appear to show little selectivity towards food items and hence little trophic segregation. Both particle feeders and predators feed opportunistically on available food sources.

Many aspects of behaviour are related to other dimensions of the niche. Vertical migration behaviour is related to spatial and temporal partitioning of the environment. Feeding behaviour and any selection of food items are related to trophic segregation. But the behavioural responses of planktonic organisms to dissolved substances in the environment may have a broader ecological and evolutionary significance for community structure.

Competition

Interspecific competition occurs when two or more species interfere with or inhibit each other, typically in the use of a common resources which is in limited supply. Considerable controversy exists regarding the importance of competition as a determinant of community structure, especially in aquatic communities. It can be assumed, however, that reproduction in a finite environment will ultimately lead to a certain level of competition and that this level will vary according to the degree of saturation of the environment (see Chapter 10). Competition within a community will tend to enhance resource partitioning, with component species occupying their respective niches and exploiting different parts of the resources spectrum along at least one of the many dimensions of the niche. The concepts of niche, competition and community have become very strongly linked in modern ecology but it is difficult to describe oceanic communities solely in these terms. Niche separation can be identified in terms of the spatial, temporal, trophic and behavioural segregation of species in oceanic communities. Competition is presumably involved in producing and maintaining this segregation. However, the continued coexistence of species with largely overlapping trophic requirements suggests that the effects of both direct and diffuse competition (competition with two or more other species) are moderated by other factors. It has become apparent that predation is a dominant factor in planktonic communities and that patchiness both in time and space may have a profound effect on the state of equilibrium attained by oceanic communities.

Predation

Predation can reduce competition between prey species or even reverse the outcome of a particular competitive situation. Predation may reduce the levels of resource partitioning by allowing prey species to coexist that have a high degree of niche overlap (by reducing the intensity of resource utilisation and hence of competition.). This is a quantitative effect. However, predation may affect resource utilisation qualitatively as well as quantitatively.

Many authors have suggested that resource utilisation should be discriminate only when resources are abundant relative to the requirements of the consumers. The greater the intensity of predation the more abundant the potentially limiting resources are likely to become. So, as predation intensifies, the level of resource partitioning may increase. The opposing effects of predation when considered quantitatively or qualitatively serve to illustrate how poorly the relationship between predation and competition is understood.

Predation may also affect resource partitioning by maintaining a polymorphism within a prey species. Predation, therefore, affects community structure both within and between trophic levels. It also exerts a strong selective pressure on prey, resulting in the evolution of predator escape strategies. In turn, this could lead to improved predation strategies and the initiation of a sequence of mutual counter adaptations analogous to the insecticide–insect resistance race. Before examining the ecological and evolutionary aspects of predation in oceanic communities it is necessary briefly to consider the predators themselves and their behaviour patterns.

Two groups of predators feed upon planktonic copepods – fish and invertebrates. Fish predators can visually recognise and discriminate between objects on the basis of size, colour, orientation, brightness and motion. They may also discriminate between objects on the basis of size, orientation and motion by means of their lateral line system. Recognition by means of colour is of little importance in oceanic communities because of the rapid extinction of most wavelengths of light with depth. At low light levels the visibility of an object depends upon the contrast between it and its background (luminance contrast). The main factors affecting luminance contrast perception are prey size, prey activity as well as prey and background contrast.

Fish predators are often highly selective of prey type and size, as shown by gut content analysis, typically selecting larger prey items. Invertebrate predators (excluding cephalopods) are primarily non-visual, relying on other methods of prey recognition, selection and capture. They may be less selective than fishes as regards prey type but often select according to prey size, typically selecting smaller and more readily handled prey. Arrow-worms (chaetognaths) are important in oceanic communities. They can detect and respond to mechanical vibrations produced by moving objects in the water and, although it has been shown that their maximum attack distance for a mechanical vibration stimulus is only 3 mm they appear to respond to the presence of prey at distances of up to 20 mm. It is possible that chaetognaths use chemosensory information for long-range prey location and that prey

vibrations provide the stimulus for accurate short-range strikes. Many medusae and seagooseberries (ctenophores) frequently capture and feed on copepods. There is evidence that some ctenophores, such as *Pleurobrachia pileus*, select for actively swimming prey types. Most predators employing mucous nets or passively drifting tentacles may be expected to have a broad range of dietary items and be non-selective. The ability of prey individuals to escape from these types of predators may increase with increasing prey size.

Many copepod species are themselves predators and can consume other copepods. Like many invertebrate predators, copepods are more successful at capturing and manipulating relatively small prey or particular developmental stages. In natural communities the range of prey of a small predator comprises the complete developmental series of small prey species and the younger stages of large species. Landry (1978) found that *Labidocera trispinosa* was much more successful at capturing nauplii than later developmental stages (copepodids and adults) of its prey and that *L. trispinosa* tended to feed selectively on larger nauplii. Similar results have been obtained for other omnivorous predatory copepods, such as *Tortanus* and *Sulcanus*. These results indicate that some invertebrate predators may feed more heavily on larger prey species although predation by the different developmental stages of the predator itself may moderate this effect. The mechanism of prey detection by predatory copepods is not known but they probably respond to vibrations and chemicals produced by the prey.

Midwater decapods such as *Sergestes sargassi* also feed on copepods. Remains of *Oncaea* and *Corycaeus* have been found in the foreguts of several mesopelagic decapod crustaceans (Foxton & Roe, 1974) although their presence was not thought to be due to selective predation. Foxton & Roe attributed their occurrence to accidental ingestion along with larger prey items upon which the decapod crustaceans were feeding. The presence of these and other small copepods may also be due in part to dietary contamination; that is, the *Oncaea* were already in the gut of the prey when it was eaten by the decapod crustacean. This is thought to contribute a significant predation pressure on copepod populations.

The predation cycle comprises the following sequence of events: locating, pursuit, capture, manipulation and consumption. Fish can locate their prey visually, or by employing chemosensory or lateral line systems, or by any combination of these means. Most invertebrates locate their prey by tactile (touch or remote sensing of vibrations) or chemosensory means. Fish predators are specialised feeders discriminating and selecting prey size and type, typically by luminance contrast perception. Invertebrate predators are often more generalised but may show some selection of prey type and size.

There are two basic strategies adopted by prey in order to escape predation by visually orientated predators – avoiding being seen by the predator, and being seen but being large enough, offensive enough or fast enough to avoid being eaten. Avoiding being seen by the predator is achieved by any of the following strategies: small size, transparency, cryptic coloration, cyclomorphosis, polymorphism and vertical migration. The last three phenomena are equivalent to being of suitable size to avoid predation at a particular time, place or level of illumination.

Predation and size. Many copepods escape predation by virtue of their small size. The genera *Oncaea* and *Oithona* are often the most abundant copepods in surface waters and few species of either genus exceed 1.2 mm in total body length. Their small size clearly reduces their visibility to visual predators. The effectiveness of small size is inversely proportional to ambient levels of illumination so that the threshold size for visual discrimination increases with depth as the level of illumination falls. Larger species may also gain protection as a result of their ontogenetically correlated descent, in which the small juveniles occur near the surface and the progressively larger stages occur progressively deeper in the water column at lower levels of illumination. Small size, however, may increase vulnerability to invertebrate predation and the overall advantage of small size relative to the level of illumination must be regarded as the outcome of the balance between selective advantages and disadvantages.

Predation and transparency. In the clear waters inhabited by oceanic communities many organisms are transparent and more or less invisible to visual predators. Representatives of many taxa have adopted this strategy including some coelenterates, ctenophores, polychaetes, chaetognaths, molluscs, crustaceans, tunicates and even some fishes. Transparency is a special case of cryptic coloration, reducing or eliminating contrast with the background. Size does not influence the effectiveness of this strategy and so there is no relationship between size and depth of occurrence within groups (such as the copepod family Sapphirinidae) that use transparency to escape predation (see Fig. 11.7). The diet and rate of digestion in transparent organisms here assume importance as conspicuous gut contents would reduce the effectiveness of transparency as a survival strategy.

Predation and coloration. An alternative method of camouflage is countershading, which reduces contrast between the organism and its background. This system of cryptic coloration is described below as it occurs in fishes which, although nektonic, provide the best examples of a system that gives an insight into the importance of light in the ocean depths. Fishes living near the surface are typically dark dorsally and pale and silvery ventrally, thereby reducing their visibility to predators approaching from above, below or the side. Midwater fishes typically have silvery reflective sides, dark dorsal surfaces and an arrangement of light-producing organs (photophores)

ventrally. The silvery sides reflect incident illumination and therefore mimic background levels of illumination. The dark dorsal surface conceals the animal from predators above it and the light produced by the ventral photophores helps to conceal the animal from predators below by breaking up the outline of its silhouette against the light background. This system is most effective in the gradient of sunlight penetrating from the surface and its effectiveness is greatly enhanced if the intensity of light produced by the photophores matches the ambient level of illumination at that depth. When the gradient of light from the surface is absent, for instance at night or at great depths, silvery sides may be disadvantageous as they would reflect any bioluminescent emissions and become very conspicuous. Most species must rely on their speed and alertness to escape, although *Valencienellus tripunctulatus* reduces reflectivity by expanding pigment cells over its silvered sides.

Predation and cyclomorphosis. Cyclomorphosis is the cyclic succession of different forms (morphs) within a population and is typical of species living in waters showing distinct seasonal changes. Recent work on freshwater zooplankton communities has suggested that predation is the major selection pressure involved in the evolution of this phenomenon. Cyclomorphosis typically involves the alternation between one or more generations of inconspicuous (small and/or less pigmented) and one or more generations of conspicuous (large and/or more pigmented) morphs. Visual predators are most active during the summer and select the conspicuous morph whereas during the rest of the year the reproductive superiority of the conspicuous morph enables it to dominate. Such differential selection could have shaped the observed pattern of seasonally alternating prey morphs.

Predation and polymorphism. Polymorphism is the simultaneous occurrence of two or more different, and genetically determined, morphs within a population at frequencies greater than can be accounted for by mutation alone. It has long been accepted, both in theory and in practice, that selection due to differential predation on the morphs of a polymorphic population can maintain a balanced polymorphism. Kerfoot (1975) has examined the interactions between the populations of two predators and one prey, the cladoceran *Bosmina longirostris*. The visual predators (fishes) selected against the large conspicuous morph of *B. longirostris* and the non-visual predators (copepods) selected against the small, easily manipulated morph. The fish also preyed upon the copepod. Kerfoot found that a line of demarcation was generated within a population of *B. longirostris* as a result of the selective advantage of the small morph inshore and that of the large morph offshore. The establishment and maintenance of this boundary in a freely mixing body of water is attributable to the interactions between the prey and its two different predators. In general terms a polymorphism

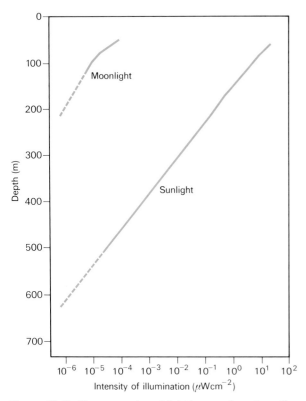

Figure 11.7 The penetration of light in oceanic waters off the Canary Islands in the eastern North Atlantic. Based on irradiance at 480 nm; data from Kampa, 1970.

as a result of the action of two opposing selection pressures, one of which is predation (the other is food availability).

The basic hypothesis is that small body size, as a predator escape strategy, is linked to the ambient levels of illumination so that progressively larger organisms fall below the discrimination threshold of visual predators as the level of illumination decreases. As the level of illumination in the ocean falls with depth (Fig. 11.7) this hypothesis would predict a gradient of increasing size with depth. Gradients of this type have been recorded in some zooplankton groups. The daytime depths of maximum occurrence of the 49 commonest calanoid copepods off the Canary Islands in the eastern North Atlantic have been plotted against mean body length of adult females in Figure 11.8. The general pattern is one of increasing body size with depth down to about 650 m. Below 650 m there is no significant correlation between size and depth.

The pattern of size and depth distribution related to light intensity, measured as a function of depth (shown in Fig. 11.8), is not a static one as light intensity varies on a daily basis. The level of illumination in the water column is much reduced at night, even in bright moonlight (Fig. 11.7). This reduced illumination would allow a prey organism of any given size to migrate vertically upwards following a particular isolume and still remain at or below the critical level of illumination (for its size) for avoidance of visual discrimination by a predator. Species occurring at around 600 m, for example, could migrate up to about 200 m at night and still remain at a level of illumination

may arise if there is a balance of selective pressures with adaptations for high predator escape ability being antagonistic to other aspects of overall fitness. Small size as a predator escape strategy may be balanced against concomitant reduction in birth rate (as fecundity is usually proportional to body size in copepods). Alternative strategies, either small size or high fecundity, may be selected for and a polymorphism generated.

Predation and vertical migration. The last of the predator escape strategies involves the phenomenon of daily vertical migration and is widespread and characteristic in pelagic communities. Populations of both planktonic and nektonic organism show the typical pattern of upward migration at dusk and downward movement at dawn. Numerous hypotheses have been advanced to explain the selective advantages of diurnal migration between daytime and night-time depths. Many of these hypotheses have been falsified or remain unaccepted and I do not propose to review them here. Vertical migration involves a substantial expenditure of energy and considerable selection advantage would be necessary for this behaviour to have evolved. Predation pressure in oceanic communities can profoundly affect the body size, morphology and even the spatial distribution of prey, and evidence is gradually accumulating which suggests that vertical migration has evolved

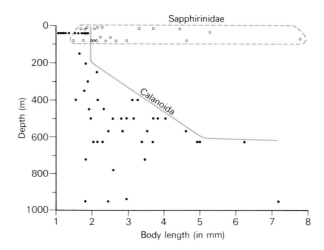

Figure 11.8 The relationship between body length of females and depth of daytime occurrence in two groups of copepods: the Sapphirinidae (species shown by ○) which have adopted transparency as a predator escape strategy and show no relationship between body length and depth, and the calanoids off the Canary Islands (species shown by ●) which have adopted body size relative to ambient illumination as a predator escape strategy and in which body length increases with depth down to about 650 m. Based on unpublished data and on data from Roe, 1972 respectively.

of about $10^{-6}\,\mu\mathrm{Wcm}^{-2}$. Some calanoid species, like *Pleuromamma xiphias* and *Chirundina streetsii*, undertake diurnal vertical migrations of about this amplitude.

If an organism migrates downwards during the day to avoid predation why does it migrate upwards again at night? The answer to this question concerns food availability. As all primary production in the ocean occurs in the strongly illuminated surface waters there is a nutrient gradient in the water column, with levels of food availability decreasing with increasing depth. Natural selection has resulted in the evolution of vertical migration in many species so that they can exploit the rich food source near the surface at night and return to the protection of greater depths during the day. Other species do not show vertical migration behaviour and are presumably adapted to remain permanently at a particular depth horizon. The larger the organism, the greater the daytime depth to which it must return in order to avoid predation by visual predators. Angel's (1979) studies on Atlantic halocyprid ostracods demonstrated that diurnal vertical migration occurred in progressively larger organisms with increasing depth, to a depth of about 650 m, below which diurnal migration ceased.

A depth of 650–700 m is a significant boundary in the vertical distribution patterns of many zooplanktonic and nektonic groups in the eastern North Atlantic. This is the depth at which the mesopelagic zone is subdivided into shallow mesopelagic (300–700 m) and deep mesopelagic (700–1000 m). This depth appears to represent an asymptote in the length/depth distribution of calanoid copepods (Fig. 11.8). It marks the level below which vertical migration largely ceases in calanoids and halocyprid ostracods. In addition, it forms a boundary in the vertical distribution of amphipods, decapods and midwater fishes marked by clear cut changes in the morphology of some members of these groups (see Table 1).

There are exceptions to these generalised patterns but the coincidence of morphological changes in three distinct groups suggests that this depth has a real biological significance. Changes in pigmentation and appearance are presumably related to vision and as pressure from visual predators appears to be intense in oceanic communities it is probable that the biological significance of the changes at 650–700 m is related to predation in some way.

The morphological changes represent changes in method of concealment. Above 650–700 m the organisms employ reflective surfaces or transparency often combined with ventral photophores. This method of cryptic coloration is effective in a gradient of downwelling light from the surface. Below 700 m body coloration tends to be uniformly dark red, brown or black. The absence of countershading suggests that concealment is no longer related to the gradient of sunlight from the surface. It is possible that the uniform dark coloration reduces surface reflectivity to biologically produced light. In the eastern North Atlantic at a depth of 650–700 m the intensity of illumination by sunlight from the surface is about 10^{-6}

Table 1 Changes in morphology of midwater animals occurring at 650–700 m

Depth	Amphipods[1]	Decapods[2]	Fishes[3]
Surface (0 m)	Transparent or reflective bodies	Semi-transparent/ semi-red bodies	Silvery reflective bodies
	Large eyes	Internal photophores	Large photophores
650–700 m			
	Dark pigmented bodies	All red pigmented bodies	Dark pigmented bodies
	Rudimentary or obsolete eyes	Dermal photophores	Relatively small photophores
	Thin delicate integument		
Bathypelagic zone (1000 m)			

[1] Data from Thurston (1976).
[2] Data from Foxton (1970).
[3] Data from Badcock (1970).

$\mu\mathrm{Wcm}^{-2}$ (see Fig. 11.7). This level of illumination may represent the minimum level of background illumination against which prey silhouettes can be detected by visual predators. There is evidence that fish detect and respond to much lower light intensities, down to about 10^{-9} $\mu\mathrm{Wcm}^{-2}$, as some fishes display diurnal migration from depths of 1000 m or more but it is possible that there is insufficient contrast for visual predators to recognise prey silhouettes against a background of 10^{-6} to $10^{-9}\,\mu\mathrm{Wcm}^{-2}$. They may, however, be able to detect localised spots of low level illumination against a darker background. The few data available on bioluminescence of deep mesopelagic organisms indicate that the intensity of bioluminescent emissions extends at least within the range 10^{-3} to 10^{-9} $\mu\mathrm{Wcm}^{-2}$. Below 700 m visual predators may use localised bioluminescent flashes, either produced or reflected by their prey, in order to detect their prey. Dark coloration is advantageous as it reduces surface reflectivity.

Light in the sea is either solar light or bioluminescence. Prey organisms have evolved morphological adaptions and predator escape strategies that are effective either in a gradient of sunlight (above 650–700 m) or in very low levels of illumination in which bioluminescence is the major source of light (below 700 m). The distinct changes in morphology observed at 650–700 m have evolved because any strategy intermediate between the two would be strongly selected against. The hypothesis that this particular faunal discontinuity is generated by visual predation pressure can be tested in the future by comparing the depth of the discontinuity in oceanic areas with different light penetration from the surface. It predicts that the morphological changes would occur at the depth at which illumination was about $10^{-6}\,\mu\mathrm{Wcm}^{-2}$, although some minor allowance may have to be made for differences in the visual acuity of predators from other areas.

The second basic predator escape strategy is to be readily visible but large enough, offensive enough or fast enough to avoid being eaten. Some planktonic organisms, like the colonial tunicate *Pyrosoma*, attain large size and these organisms have few predators. Many nektonic organisms are powerful swimmers and can avoid being caught either by their speed or by shoaling behaviour which affords some protection from predators. Other planktonic organisms are offensive by virtue of the toxic chemicals they produce. Many coelenterates, such as the Portuguese Man-of-War (*Physalia physalis*), belong to this category with numerous paralysing stinging cells (nematocysts) spread along their tentacles.

Patchiness and equilibrium

The continued coexistence of many competing species in both freshwater and marine planktonic communities has attracted much attention. Hutchinson (1961) termed this situation 'the paradox of the plankton' because it appeared that the principle of competitive exclusion was being violated. Hutchinson's hypothesis, designed to explain this paradox, was to propose that competitive equilibrium was never established because of temporal heterogeneity; that is, the rapid changing of environmental conditions continually affected the relative competitive abilities of the competing species and prevented the establishment of equilibrium in which one species displaces the other. Other non-equilibrium models have been suggested. A second hypothesis (the contemporaneous disequilibration hypothesis) was proposed by Richerson, Armstrong & Goldman (1970) who suggested that patchiness within the water mass of short- to medium-term persistence was sufficient to prevent the exclusive occupation of a niche by a single species, if the relative competitive abilities of the competing species varied from patch to patch. Riley (1963) proposed a third hypothesis, that competing phytoplankton species have evolved very similar adaptations and efficiencies and that their competitive abilities would be so similar that ecological displacement of one species by the other may proceed at an imperceptibly slow rate. These three hypotheses are not contradictory. The first stresses heterogeneity in time, the second stresses spatial heterogeneity and the third makes a more general point about evolutionary time scales. All three were initially proposed to account for phytoplankton diversity.

A composite model of patchiness, as a phenomenon in both space and time, enhancing the coexistence of competitors can be developed for both limnetic and oceanic communities. There are probably major differences between limnetic and oceanic ecosystems in terms of scale and persistence of patches. Both mesoscale (0·1 to 100 km) and microscale (less than 0·1 km) patchiness are likely to be ecologically significant in oceanic communities whereas microscale patchiness is probably more important in the generally smaller limnetic systems. Temperate limnetic communities are often heavily dominated by rotifers and cladocerans, with their highly opportunistic reproductive strategies. All temperate oceanic communities are usually dominated by copepods which are relatively less opportunistic than rotifers and cladocerans. This may indicate a basic difference between limnetic and oceanic systems in the persistence of patches. The relative importance of turbulent mixing, wind-generated langmuir circulations, horizontal diffusion and internal waves for the degeneration of patches may also differ between the two environments.

Oceanic communities have proved to be very patchy on the mesoscale. This patchiness is important ecologically as it has been shown that the density of prey organisms required as a stimulus for the onset of feeding (the feeding threshold) of many predators is higher than the average density of their prey so the continued survival of the predator population is dependent upon their locating and utilising the high density patches of prey. Few data are available on the critical factor, namely the longevity of the mesoscale patches, but their persistence is probably proportional to their size. The larger the patch, the longer it will persist. Conversely, the inter-patch areas could act as refuges especially for prey escaping predators with relatively high feeding thresholds.

The chance elements involved in the physical and biological generation of patchiness and in its breakdown by diffusion and migration undoubtedly have a significant effect upon community structure. The duration of the patch relative to zooplankton life-cycle duration and other ecological time-scales is also important because a moderately persistent patch could act as a refuge from a competitor or a predator. Patchiness may prevent the exclusive occupation of a niche by fragmenting the niche into a composite of temporary niches separated in space. This would introduce chance elements of colonisation into the system, preventing or retarding the establishment of competitive equilibrium. It may also provide environmental structure and heterogeneity which are important in predator–prey dynamics as prey distribution patterns significantly influence the optimal foraging strategy of the predator. Patchiness can be regarded as a spatial mosaic representing a series of local successional sequences that are out of phase and as such can increase diversity in the community as a whole.

Community structure

Planktonic communities, like any biotic communities, are structured by the biological processes of competition and predation. Competition, especially diffuse competition in which one species simultaneously competes with two or more different species, may be intense in the zooplankton because there are only two basic feeding modes – particle feeding and predation. At any given depth in the water column there are many species exploiting the available food resources and, as opportunistic feeding appears to be prevalent, there is little qualitative trophic segregation of the niches of the coexisting competitors (see above). Predation pressure can be implicated in the evolution of many phenomena exhibited by zooplanktonic organisms, such as cyclomorphosis and diurnal vertical migration. Predation may have another less readily observable effect by influencing or even reversing the outcome of competitive interactions, thereby allowing a greater degree of niche overlap between competing species. The relative importance of competition and predation in determining community structure varies between communities and even within a community according to taxon, trophic level or individual size, but both processes should be considered in any community analysis. Patchiness is a feature of most ecosystems, and its influence will be relative to the mobility, size and longevity of the organisms concerned. Patchiness may promote diversity by periodically changing competitive abilities in a temporally heterogeneous environment or it may promote diversity by fragmenting the niche spatially, allowing for the persistence of a competitively inferior species by virtue of its high dispersal and colonising abilities, despite pressures from a superior competitor.

Speciation in oceanic environments

Two lines of research can contribute to our understanding of speciation in oceanic environments. The first is concerned with determining the extent of genetic polymorphism within species, and the second is to identify the nature of barriers to gene flow in the oceans. Many planktonic species are polymorphic, with two or more morphs occurring in the same depth horizon. Most species of the copepod genus *Oncaea* are polymorphic and many have large and small forms that coexist and are ubiquitous. There are no other obvious morphological differences between the two forms. How are these polymorphisms maintained? It could merely represent a transient polymorphism in which the process of displacement of one form by the other is retarded because of patchiness, or it could represent a stable polymorphism maintained by a balance of selective pressures or by size selection by the predator. Information on the genetic basis of these polymorphisms can help to answer this question and such data can be indirectly obtained by studying the mate location and recognition systems of the species concerned.

Mate-seeking behaviour has been observed in some planktonic copepods and this behaviour appears to be chemically mediated. Katona (1973) observed that males of the calanoids *Eurytemora affinis* and *Pseudodiaptomus coronatus* could locate stationary females at distances of up to 20 mm. Males of *E. affinis* performed mate-seeking

behaviour towards female *E. affinis*, *E. herdmani* and *P. coronatus* but attempted to copulate most frequently with conspecific females (*E. affinis*), to a lesser extent with congeneric females (*E. herdmani*) and rarely with females of the other genus. The evidence available suggests that males detected other copepods of both sexes by means of diffusable chemicals and that relatively specific contact chemicals (pheromones) elicited copulatory behaviour.

The existence of relatively specific mate recognition systems based at least in part on sex pheromones provides one mechanism for preventing interspecific breeding. The sex pheromone system may also provide a clue to the possible mechanism of speciation within the oceanic environment if the chemosensory mate recognition system has enough flexibility to allow for some variation (polymorphism) in the chemical stimulus. Katona's findings that copulatory behaviour was induced differentially depending upon closeness of phylogenetic relationship, supports the assumption that different chemicals may elicit quantitatively different responses. The various morphs of a polymorphic species may also produce and differentially respond to sex pheromones that differ slightly in their chemical composition. The effect of such a system would be to reduce gene flow between different morphs, and create a tendency for like morphs to mate with like within a population (assortative mating), resulting ultimately in the genetic separation of the morphs.

This review of some aspects of the ecology of planktonic communities provides some indications of what can constitute a barrier to gene flow in oceanic communities. Although there is evidence that the boundaries to particular water-masses can act as faunal barriers to some species and that zoogeographical separation can reduce gene flow, it is apparent that neither distance alone nor the large-scale structure of the oceans constitute the only barriers to gene flow between populations in oceanic systems. The vertical structure of the water column is also a potential isolating mechanism and planktonic organisms respond to physico-chemical discontinuities, such as the thermocline, which can represent distributional barriers to some species but not to others. The barriers, ultimately caused by light levels, are however, neither as obvious nor as absolute as the barriers to gene flow in terrestrial environments. Biological barriers, such as that at a depth of 650–700 m in the eastern North Atlantic, may be as effective in preventing gene flow as physically determined ones. If allopatric speciation is found to be the dominant mode of speciation in the oceanic environment, as it appears to be in terrestrial, benthic and lacustrine environments, then these barriers may cause at least some of the necessary genetic isolation between diverging populations.

References

Alldredge, A. L. 1972. Abandoned larvacean houses: A unique food source in the pelagic environment. *Science* 177: 885–887.

Alldredge, A. L. 1976. Discarded appendicularian houses as sources of food, surface habitats, and particulate organic matter in planktonic environments. *Limnology and Oceanography* 21: 14–23.

Angel, M. V. 1979. Studies on Atlantic halocyprid ostracods: their vertical distributions and community structure in the central gyre region along latitude 30° N from off Africa to Bermuda. *Progress in Oceanography* 8: 3–124.

Badcock, J. 1970. The vertical distribution of mesopelagic fishes collected on the SOND cruise. *Journal of the Marine Biological Association of the United Kingdom* 50: 1001–1044.

Boxshall, G. A. 1977. The depth distributions and community organization of the planktonic cyclopoids (Crustacea: Copepoda) of the Cape Verde Islands region. *Journal of the Marine Biological Association of the United Kingdom* 57: 543–568.

Deevey, G. B. & Brooks, A. L. 1971. The annual cycle in quantity and composition of the zooplankton of the Sargasso Sea off Bermuda. II. The surface to 2000 m. *Limnology and Oceanography* 16: 927–943.

Foxton, P. 1970. The vertical distribution of pelagic decapods (Crustacea: Natantia) collected on the SOND cruise, 1965. II. The Penaeidea and general discussion. *Journal of the Marine Biological Association of the United Kingdon* 50: 961–1000.

Foxton, P. & Roe, H. S. J. 1974. Observations on the nocturnal feeding of some mesopelagic decapod Crustacea. *Marine Biology* 28: 37–49.

Hutchinson, G. E. 1961. The paradox of the plankton. *American Naturalist* 95: 137–145.

Kampa, E. M. 1970. Underwater daylight and moonlight measurements in the eastern North Atlantic. *Journal of the Marine Biological Association of the United Kingdom* 50: 397–420.

Katona, S. K. 1973. Evidence for sex pheromones in planktonic copepods. *Limnology and Oceanography* 18: 574–583.

Kerfoot, W. C. 1975. The divergence of adjacent populations. *Ecology* 56: 1298–1313.

Landry, M. R. 1978. Predatory feeding behaviour of a marine copepod, *Labidocera trispinosa*. *Limnology and Oceanography* 23: 1103–1113.

Nival, P. & Nival, S. 1976. Particle retention efficiencies of an herbivorous copepod, *Acartia clausi* (Adult and copepodite stages): Effects on grazing. *Limnology and Oceanography* 21: 24–38.

Poulet, S. A. 1978. Comparison between five coexisting species of marine copepods feeding on naturally occurring particulate matter. *Limnology and Oceanography*, 23: 1126–1143.

Richerson, P., Armstrong, R. & Goldman, C. R. 1970. Contemporaneous disequilibration, a new hypothesis to explain the 'paradox of the plankton'. *Proceedings of the National Academy of Sciences of the United States of America* 67: 1710–1714.

Riley, G. A. 1963. *Marine biology. I. Proceedings of the first interdisciplinary conference*, Riley, G. A. (Ed.), 286 pp. Washington D.C.: American Institute of Biological Sciences.

Roe, H. S. J. 1972. The vertical distributions and diurnal migrations of calanoid copepods collected on the SOND cruise, 1965. I. The total population and general discussion. *Journal of the Marine Biological Association of the United Kingdom* 52: 277–314.

Thurston, M. H. 1976. The vertical distribution and vertical migration of the Crustacea Amphipoda collected during the SOND cruise, 1965. II. The Hyperiidea and general discussion. *Journal of the Marine Biological Association of the United Kingdom* 56: 383–470.

CHAPTER 12

Mimicry and its unknown ecological consequences

R. I. Vane-Wright

It is 120 years since H. W. Bates wrote the manuscript in which he introduced the concept of mimicry. Since then thousands of papers have discussed the subject, and progress has been made on its physiological, genetical and behavioural aspects. But our understanding of the ecological consequences and possible effects of mimicry on community structure remains little better than guess-work. Can anything be done to rectify this?

Consider the familiar dronefly, *Eristalis tenax*. This harmless fly looks and sounds much like a honeybee (*Apis*), an insect avoided by many would-be predators because of its sting. It is suggested, with some experimental evidence to back the assertion, that bee-like but otherwise harmless flies gain protection through often being mistaken for stinging social Hymenoptera. The dronefly is thus a 'sheep in wolf's clothing', bluffing its way through life on the strength of a disguise. How could such mimicry originate?

Imagine a non-mimetic fly of about the same size as a dronefly, and having the habit of visiting flowers together with honeybees – where both would regularly be exposed to the same set of potential predators. If one of the flies was a melanic mutant, then its size and dark colouring might make it sufficiently similar to a bee to make some predators hesitate. Given the choice, they would tend to attack one of the familiar flies, not a melanic fly which might be a bee.

With this aid to survival, and assuming that the melanic fly survived the other exigencies of life long enough to reproduce (many would not, but there are millions of flies, and melanism often recurs), some proportion of its offspring would inherit the melanism-inducing mutation. These, assuming sexual reproduction, would vary among themselves, partly because of the modifying effects of their individual genetic constitutions. Some of these melanic offspring would probably be more bee-like than the original mutant, others would be less so. By the same token as before, there would be a tendency for the most bee-like of these to leave the greatest number of offspring, the resultant melanics again showing a range of bee-likeness,

and affording further opportunity for selection to favour the most bee-like amongst them. At each generation the initial approximate likeness to a bee would be improved.

In time, further mutations would arise capable of enhancing the bee-like appearance or behaviour (such as genetic changes giving rise to orange abdominal stripes or a more appropriate buzzing sound). These would be selected for in a similar way, leading eventually to a situation where the entire fly population, even the whole species, became very honeybee-like: the fly would have become a mimic of the bee.

This scheme contains, explicitly or implicitly, a number of factors: the production of more offspring than survive to reproduce; heritable variation; and differential survival of the variants resulting in, through successive generations, a steady improvement of a particular adaptation. These factors are precisely the elements of the Darwin/Wallace theory of evolution by natural selection, of which Bates' (1862) theory of mimicry is no more than a special case. As such it was mercilessly denounced by early critics of Darwinism and, to a gradually lessening extent, it has been ever since. The theory of mimicry survives because no alternative hypothesis has yet been advanced which explains more observations, or makes more potentially testable predictions about the many biological resemblances for which it has been invoked.

The structure of mimicry systems

Honeybees are adapted to gathering nectar and pollen to raise their brood. They use their stings for defensive purposes – collectively against parasites and predators that try to invade or rob their nests, individually against insectivores. Bees, as a result, are avoided by most vertebrates, including many insect-eating birds. Vertebrates recognise bees by a number of cues, including their dark, striped coloration and buzzing sound. These are very conspicuous features and have almost certainly been evolved as signals to advertise the repellent qualities of

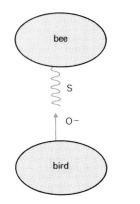

Figure 12.1 A bee produces a clearly defined warning signal (S). A bird, which finds stinging Hymenoptera unacceptable as food, perceives (learns through experience) the significance of this warning sign, and responds ($O-$) by avoiding the bee. Both organisms benefit from this coevolved, signal-based system.

bees, even though this now makes them easy targets for those few insectivores which specialise in eating them. We presume that, on balance, the degree of protection or lack of interference afforded by the warning signals more than offsets the disadvantages of being conspicuous.

Experiments with birds, toads and other vertebrates have demonstrated that, in general, the repellent nature of such insects is only appreciated through direct and painful experience, the significance of warning or aposematic coloration being learned rapidly by each individual through association. One or two experiences of being stung is sufficient to make most susceptible birds avoid bees on sight – often for months afterwards.

Such a signalling system (Fig. 12.1) is mutually beneficial to the birds and bees alike. But once such a signal-based system exists, it can be manipulated to the advantage of a third, mimetic organism, to a large extent regardless of whether or not this is to the benefit of the original signal-generating organism (the *model* – in the example, a bee) or the responding organism (the signal-receiving *operator* – in the example, a bird). How can a third organism, the *mimic*, manipulate such interactions?

In responding by avoidance to bee signals, the bird must in some way recognise the incoming information, and decide that the insect is a bee (one of those pain-inflicting things that it tried to eat five weeks ago), and not an acceptable fly or other presumed palatable insect. This is a problem in perception, and implicitly involves some rapid functional classification into two categories: acceptable *versus* unacceptable. Natural selection will, therefore, tend to promote the evolution of warning signals by well-protected species: those individuals most easily recognised and remembered by predators get most consistently avoided – and so tend to leave the most offspring. This selection process first promotes, and then stabilises the evolution of clearly defined warning signals. As

Silberglied (1977) has put it, in information theory terms, aposematic organisms are selected to increase their signal-to-noise ratio.

To take some advantage of a signal-based interaction, all another organism need 'do', as in the case of the dronefly, is simulate the signal properties by which an operator identifies a particular model. As a result, the operator will then also identify this mimic as that model, and respond to it in the same way. It is from the elicitation of this response that a mimic must gain some advantage.

In Figure 12.2 the bee is shown as before, producing its signal S_1, to which the bird responds negatively (O_1-), the appropriate operation. We have now added the dronefly mimic, producing a signal S_2 so similar to the true bee signal (S_1) that the bird responds to it in the same way (O_2-). This is, however, an inappropriate response, because the fly represents an acceptable food item. A positive term (R_2+) has been added, and a negative term (R_1-), to represent these true roles of the operator. The response actually evoked (O_2-) is necessarily an advantage to the dronefly, otherwise the mimicry would not be selected for or be stable (positive vector added to the mimic). However, as already noted, there is no general reason to expect that the presence of a mimic is necessarily advantageous or disadvantageous to the model or to the operator. In this case we can expect it to be a disadvantage to the model (negative vector on bee), because the presence of the otherwise harmless mimic devalues the bee's signal in its interaction with the bird – birds first encountering

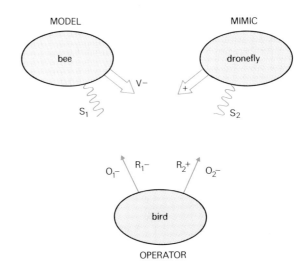

Figure 12.2 A bee and bird interact by signalling (S_1) and avoidance (O_1-), as in Figure 12.1. A third organism, a dronefly, produces a signal (S_2) so similar to S_1 that the bird responds to the dronefly (O_2-) as if it were the bee. However, the dronefly is a potentially acceptable food item, to which the true role of the bird is positive (R_2+), unlike its true role towards the bee (R_1-). In this mimicry system the bee is the model, the bird is the operator, and the dronefly is the mimic. The mimicry is a disadvantage to the bee ($V-$) because it degrades the signal value of the aposematic pattern.

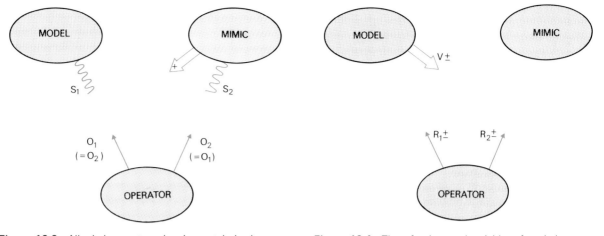

Figure 12.3 All mimicry systems involve certain basic components: three living organisms, model, mimic and operator. The model and mimic produce signals (S_1, S_2) so similar that the operator cannot distinguish them, and so it responds in a similar way to both (O_1, O_2; $O_1 = O_2$). For the mimicry to be selected for or to be stable, it must be of advantage to the mimic to evoke the normal response of the operator (O_1) in relation to itself (O_2).

Figure 12.4 Three fundamental variables of a mimicry system are the roles of the operator with respect to the model and the mimic (both independently positive or negative), and the effect of the presence of the mimic on the model (advantageous or disadvantageous). These interactive variables determine eight ($2 \times 2 \times 2$) different types of mimicry.

droneflies will be encouraged to attack bees, not discouraged.

If we abstract the invariables of a mimicry system, we obtain the elements shown in Figure 12.3, where the responses of the operator (O) to S_1 and S_2 are the same. The variables, shown in Figure 12.4, are the true biological roles (R) of the operator to the model and to the mimic, and the effect of the presence of the mimic on the model (V). Each of these three factors is independent, and could be positive or negative, thus defining eight (2^3) possible types of mimicry system. The dronefly example fits one of these, in which R_1 is negative, R_2 positive, and V negative – the system which describes the classical case of Batesian mimicry.

So far, we have developed this scheme as if the three components – model, mimic and operator – were each constituted by separate species. But this is not always the case, as can be seen by examining the aggressive mimicry of cuckoos and similar bird parasites. The host birds, failing to distinguish the eggs of the parasite from their own, nurture the inquilines at the expense of the lives of their own offspring. This is analysed, as before, in Figure 12.5. Now, as indicated by the dotted line, it can be seen that the model and the operator belong to the same species. Whether the components of a mimicry system are members of a single species or not could have profound genetical and ecological consequences.

It is possible to conceive five different ways in which the components of a mimicry system could be constituted: (*a*) in which all the components consist of different species; (*b*) where the operator and model are conspecific; (*c*) in which the model and the mimic belong to one species; (*d*) where the mimic and operator are conspecific;

and (*e*) where the three components are all comprised by members of a single species. This, combined with the eight categories already formulated, gives a total of 40 ($2^3 \times 5$) theoretically different mimicry systems – of which about half can be recognised in nature (Vane-Wright, 1976, 1980); hundreds, perhaps thousands, of examples are thought to exist in some of the categories.

Out of this variety, we must draw attention to two particular types: Batesian mimicry (already described; Fig. 12.2) and Müllerian mimicry (Fig. 12.6). The latter involves the convergence of signal pattern of an unsampled but nevertheless protected species on a model known to, and avoided by a predator. This is now thought to occur in a manner similar to the evolution of Batesian mimicry (Turner, 1977), the principal difference being manifest when an operator samples a mimic, which will reinforce the warning signal in a Müllerian situation, unlike the result of sampling a Batesian mimic. The selective advantage of Batesian mimicry is thus density-attenuated, whereas Müllerian mimicry is density-enhanced. The latter effect frequently leads to the establishment of Müllerian mimicry rings, in which a number of unrelated species in a local area come to share the same warning signal. There are many examples of this in the tropics where, for example, bitter-tasting beetles, poisonous moths and stinging wasps often all look superficially alike (Wickler, 1968, fig. 19).

Batesian and Müllerian mimicry were, historically, the first types of mimicry to be recognised. This has caused a seemingly permanent imbalance in the discussion of mimicry, so much so that most texts consider only these two forms. As a result only these two types have been given serious experimental or theoretical treatment in population biology terms. This has resulted in many general statements made about mimicry being misleading. Mimicry is a complex phenomenon, encompassing a very wide range of known examples (Wickler, 1968; Vane-Wright, 1976, 1980; Wiens, 1978). It is best characterised by the increase in reproductive fitness gained by the mimic through simulation and 'invasion' of the signal-based interaction between two other organisms. *Mimicry involves an organism – the mimic, which simulates signal properties of a second living organism –* the model, *perceived as signals of interest by a third living organism –* the operator, *such that the mimic gains in fitness as a result of the operator identifying it as an example of the model.* From this definition (Vane-Wright, 1980) it is implicit that mimicry systems are highly interactive.

In my opinion the subject of mimicry has yet to be placed firmly within a community biology framework. An objective of this paper is to suggest some ideas that might encourage the development of an appropriate theory – and the gathering of appropriate observational and experimental data – all of which appear to be lacking.

An example of mimetic coevolution

South American passion-flower vines (Passifloraceae), although chemically well-protected against most herbivores, suffer grazing by heliconids, a group of butterflies the larvae of which are entirely dependent on the vines for food. Many vines sustain serious damage, particularly where the larvae feed on the growing tips, sufficient to reduce plant growth and reproductive capacity. The vines offer patchy, limited resources, with the result that there is strong competition between individual butterfly larvae. This density-enhanced effect culminates in cannibalism, older caterpillars often devouring their younger brethren. Adult heliconids, well-protected against potential predators by poisons gathered from the passion-flower vines by the larvae, are brilliantly coloured, slow-flying aposemes, and often live for several months. Day after day, they seek out the scattered vines and, if conditions are right, each day they lay a few eggs (Benson, Brown & Gilbert, 1976). Any female tending to lay her eggs on 'oversubscribed' vines would be at a disadvantage, due to the larval competition already described. It is clear that females that are able to assess the density of eggs and larvae already present on a vine, and having a strong tendency to lay only where this falls below some threshold value, will have a higher fitness. This appears to be the case, because there is good observational and experimental data (Gilbert, 1975) to suggest that heliconid females do make such estimates. Well-protected by their aposematic life-style, they may spend minutes flying round a vine, 'counting' before laying even a single egg. To do this they must be able to recognise the early stages of their own species, and it seems this is mainly done by visual cues. In particular, the eggs, although small, are a conspicuous bright yellow, placed in ones and twos on young leaves or tendrils. The

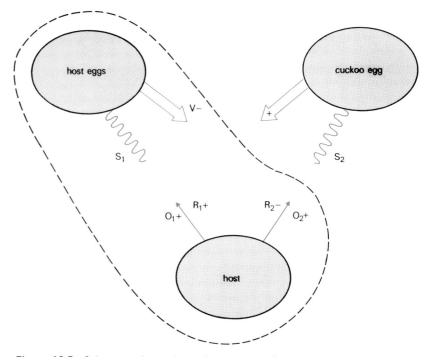

Figure 12.5 Scheme to show mimicry between a cuckoo and its host. The host's eggs are the model. The substituted cuckoo's egg is very similar, and the operator (brooding host) fails to distinguish between them, accepting the egg of the parasite (O_2+) as one of its own. In reality ($V-$), the host bird should reject the cuckoo's egg (R_2-). Note how in this case the model and operator are constituted by different individuals or stages of a single species (dotted line).

presence of eggs, then, is a deterrent to oviposition by female heliconids. Some passion-flowers have been able to take advantage of this system through the production of dummy eggs – small yellow swellings or outgrowths of the plants similar in size and distribution to the heliconid eggs. By this simple 'trick' the vines are able to inhibit the butterflies from laying (Fig. 12.7).

This development is of advantage to individual vines, decreasing their rate of being grazed, and so increasing their fitness. This in turn tends to increase the intraspecific competition amongst the butterflies, through directing their attack to those individual plants or vine species lacking the mimetic eggs. All these shifts in selection pressure are likely to evoke counter responses, including improved perception by the butterflies for distinguishing dummy eggs from the real thing – or totally new defences, such as the lethal trichome hairs of *Passiflora adenopoda* (Gilbert, 1971). Such dynamic, coevolutionary 'arms-races' are widespread in nature, and could be modelled or studied with many types of mimicry system.

Mimicry and community structure

The study of such two-species interactions, or even three species in a classic Batesian interaction, however instructive

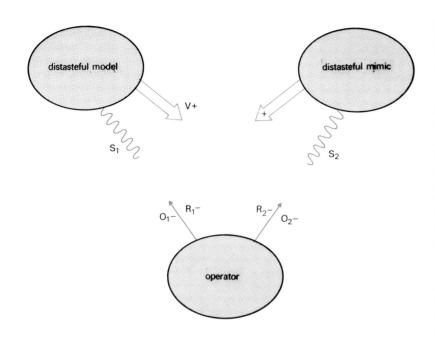

Figure 12.6 Where mimicry develops between two or more different species of protected organisms, to give a group of organisms all sharing the same warning signal, the system is beneficial both to the sampled aposematic species ($V+$) and the unsampled aposemes (functional mimics) alike. Such a Müllerian mimicry ring is also of benefit to the operator ($O_1 - = O_2 -$), to which all the aposemes are unacceptable ($R_1 - = R_2 -$).

in considering genetics or the evolution of adaptations (Turner, 1977), is nonetheless an abstraction. In reality such systems are embedded in a complex community or ecosystem, involving a vast number of interactions between its component organisms, many of which interactions contribute to the total array of selective forces acting on given individuals or species. These forces, acting on mutations and recombinations, evoke changes in the genetical structures of the populations, thus for ever altering, almost literally 'rewriting' the rules by which the system operates: the species and community evolve together in an essentially indeterminate fashion, moulded by many chances and many necessities. Subunits or elements can be studied at the expense of making many simplifying assumptions, but even simple models can give rise to formidable analytical problems. In this section I wish merely to sketch (Fig. 12.8) some potential interactions between a group of eight species, about the minimum one can consider in placing Batesian mimicry within a community framework.

Imagine an aposematic butterfly (A) which feeds on a poisonous plant (X), from which it also derives its defensive chemicals. A palatable butterfly (B) feeds on another plant (Y), and is a Batesian mimic of A. A predator (P) is specialised to feed on butterflies, but is unable to eat A, although B is potentially acceptable. As pointed out by Holling (1965), the idea of alternative prey is basic to Batesian mimicry theory, so it is necessary to introduce a third butterfly (C), a palatable, cryptic species, feeding on plant (Z).

In this system primary producers X, Y and Z are cropped by herbivores A, B and C respectively. At trophic level two, the numbers of the butterflies B and C are checked by P (if B becomes too numerous in relation to A, naïve P will associate the aposematic pattern with food, not non-food) – but the system does not yet include any check on A. This can be provided in our model by the addition of another biotic factor – hymenopterous parasitoid (H), specialised to feed on the larvae of A. This system, given the density-dependent processes built in, and given appropriate parameters to govern its dynamics, could be stable. The principal interactions are shown in Figure 12.8.

Even from this very simple model, the highly interactive nature of the community can be appreciated. For example, an increase in the abundance of H will tend to depress A and raise X. Any decrease in the abundance of A tends to result in P taking more B (due to the decrease in total protection afforded by the aposeme), potentially releasing Y from some grazing, but having an indeterminate effect on C and Z (beneficial to C if P does not increase in abundance, or potentially disadvantageous if P increases beyond a certain threshold; vice versa for Z).

In the absence of mimicry by B of A, changes in the abundance of H, A and X would all have little direct effect on P, B, C, Y and Z (the only direct interaction between P and A then being the initial testing of A by naïve P). Ideally, however, any such community model should be extended to allow for all possible interactions between each component and every other, and for intraspecific effects, such as competition. Levins (1975) has applied 'loop analysis' matrix algebra to this general type of problem, for the theoretical study of communities at or near equilibrium. However, the method is complex, and it remains to be seen if it could be applied to real mimicry systems – for which we are virtually without pertinent data.

Does the evolution of Batesian mimicry affect species abundance?

Nicholson (1927) was the first biologist to consider this question seriously. His original ideas were influential, both on the theory of mimicry in particular, and through extension by himself and others, to the regulation of animal numbers in general (see discussion in Clark et al., 1967).

Nicholson wanted to explain why Batesian mimicry is a comparatively rare phenomenon – if mimicry is good for some, why not all? Previously it had often been suggested that selection must, in some way, have acted more 'vigorously' on mimetic species, and that these species now owed their very existence to the 'protection' given by mimicry (Selous, 1909, p. 243). Nicholson pointed out that there was absolutely no evidence to support this, and went on to ask why mimetic insects are no more 'successful' than related non-mimetic forms. The main points in Nicholson's answer to his own question, developed in a verbal argument some 6000 words long, can be summarised as follows.

The population size or number of a species tends to remain constant over a period of time – if it were to increase or decrease progressively, 'the only alternatives' (Nicholson, 1927, p. 81), the population would either become extinct or over-run all else. Ultimately, numbers must be limited by the quantity of food available – but animals only rarely exhaust their food resources to starvation point. So the quantity of food must somehow determine numbers indirectly. To maintain stability this requires the action of some density-enhanced factor. For the control of butterfly numbers, used as an example, Nicholson identifies larval parasitoids and other early stage predators as the only suitable agents (he dismisses food, weather and diseases). The activity of the predators and parasitoids is tied to the quantity of host-plant in the following way: if the number of butterfly larvae increases on a fixed amount of host-plant (that is, the density of larvae increases), then the predators will encounter them more frequently, and so multiply. The increased number of predators will, in subsequent generations, reduce the butterfly larvae below their original density. The predators will now have more difficulty in locating larvae, having to search a greater area for each success, and so the survival of the caterpillars will improve again. Nicholson's model is of a density-dependent feedback type, mediated by the searching success of the predators, which becomes more and more efficient as prey density increases, and less and less efficient as prey density decreases.

Nicholson (1927, p. 85) then develops his argument further. 'It is evident from what has been said that if the numbers of an insect are caused to vary from the normal by any cause there is a definite mechanism which will tend to bring the numbers back to normal. Suppose, then, that the *major enemies, on which the numbers* of a particular insect *depend*, attack the larval stage, and that the *minor* enemy attacks the adult. Does it follow that this increased attack will cause a diminution in the numbers of the insect?' [my italics]. Nicholson concludes that it will not, because increased attack on the adults reduces the total number of eggs laid, but this then reduces the search efficiency of the larval-predators, so that more insects will now reach the adult stage, some compensating for the increased adult predation, and thus 'the numbers would remain practically unaltered'. Nicholson deduces from this that increased severity of the minor attack on the adult stage increases the proportion of insects that reach maturity, whereas decreased severity of this attack increases attack on the early stages, and then a smaller proportion of insects reaches maturity.

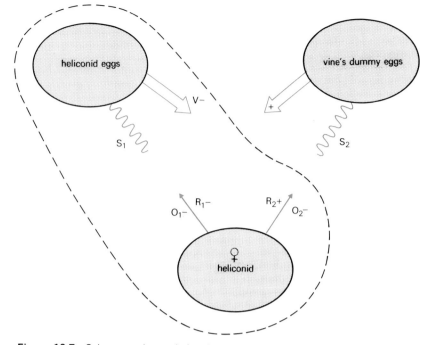

Figure 12.7 Scheme to show mimicry between the dummy eggs of a passion-vine and the real eggs of its heliconid butterfly parasite. Because of fierce larval competition, in which older caterpillars often eat their younger brethren, female heliconids avoid laying on vines which are already burdened with too may eggs or larvae. The presence of dummy, or false eggs on a vine can dissuade a female butterfly from laying, thus raising the level of competition on other vines ($V-$).

The next step in the argument is to apply these ideas to mimicry. 'From what has been said it follows that protection from attack in the adult stage can be of little, if any, importance to a species of butterfly: for reduction in the normal slight attack on the adults would only result in a slight decrease in the numbers of the earlier stages and the numbers of adults which emerged, the actual numbers of the species remaining practically unaltered. Therefore if a perfect mimetic pattern appeared suddenly in a non-mimetic species, giving complete immunity from attack, it would not increase the success of the species, which would be just as successful without the mimetic pattern' (Nicholson, 1927, p. 88). Nicholson (1927, p. 89) further concludes that although the adults would be free from attack, nevertheless 'less adults would be produced, on account of the increased severity of attack on earlier stages'.

By use of the type of argument developed above on how the evolution of bee-mimicry might have come about in droneflies, Nicholson next demonstrates how, despite this apparent paradox, mimicry will still evolve, and concludes (1927, p. 90) that 'mimetic resemblance, therefore, simply serves to fit the possessors more perfectly to their natural environment, without conferring upon them any material advantage.'

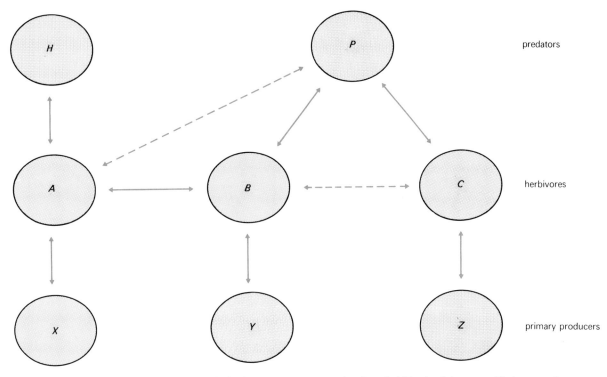

Figure 12.8 Simplified scheme to show principal interactions within a minimum community of eight species required to maintain a Batesian mimicry system. Three primary producers (plants X, Y and Z) are respectively the hosts for three butterflies (A, B and C). Species C is a cryptic butterfly preyed upon by an insectivorous bird (P), which will also eat B. Species B, however, is a Batesian mimic of aposematic butterfly A, a species unacceptable to P. The numbers of species A are held in check by a specific hymenopterous parasitoid (H). If A becomes rare, then mimicry is of little protective value to B, but when A is common, mimicry is a good defence for B against P. Through the actions of P, fluctuations in the number of B have an indeterminate effect on the numbers of A and C (shown as dashed lines, $P \leftrightarrow A$, $B \leftrightarrow C$).

This argument, in so far as it suggests that mimicry is of advantage to the individual but not to the species, has been widely accepted. Haldane (1953, p. 19), following Nicholson & Bailey (1935), was also of the opinion that 'if density is limited by food supply, a gene which makes the animals less conspicuous to a predator or more resistant to cold will be favoured by natural selection, and if it occurs by mutation will spread through the population. But it will not increase the food supply, or the density'. Sheppard (1961, p. 21) concluded from Haldane's comment that 'mimicry is certainly advantageous to the individual; in addition, it is often claimed that mimicry is advantageous to a species, allowing it to exist or to exist at higher densities than it would if it were non-mimetic. However, it is unlikely that this is often the case, especially in butterflies, since it would be necessary for the predators maintaining the mimicry also to be controlling population size…it is probable that the population is controlled by parasites, diseases and little, if at all, by…predation on the adults.' Turner (1977, p. 174) goes a little further: 'The…lack of influence of mimicry on population density is an integral part of the theory of mimicry, for as Batesian and cryptic mimicry become ineffective when the mimic is common, a sharp increase in population density resulting

from mimicry would ruin its adaptive value: in these circumstances Batesian and cryptic mimicry would not evolve at all.'

There are many problems, to my mind, in accepting Nicholson's model, which is unrealistic in a number of aspects. Many tropical butterflies have long-lived, over-lapping generations of adults, there sometimes being, at any one moment, greater numbers of adults than the short-lived early stages; no account is taken of the community interaction between model and mimic, which may have an over-riding influence on adult survival; no consideration is given to alternative prey, making Nicholson's original formulation of the problem doubtful; and there is little evidence that parasitoids do affect butterfly populations as required (Gilbert & Singer, 1975, p. 367). But the most important objection is simply the circularity of the argument. It is proposed that the numbers of a species must, in the long run, be stable, indefinite increase or decrease being 'the only alternatives'. Having postulated stability, the search is then on for a mechanism to provide it, which Nicholson 'finds' in density-dependent mortality caused by larval parasitoids. It is hardly surprising, therefore, that the addition of mimicry by the adults is found to have little effect on numbers, as

Nicholson devised a model system buffered necessarily to return to equilibrium, regardless of any such disturbance.

In fact, Nicholson's model is inappropriate for the task in question. It is not reasonable to presuppose that a species must remain constant in numbers because the 'only alternatives' are either to 'overrun the world' or become extinct. What we require to know about Batesian mimicry is whether or not its evolution is likely to increase, or decrease, or have no effect on the *equilibrium* levels of a species. Species sometimes fluctuate in number dramatically, but they are also known to remain more or less stable for moderate periods of time – could they not do so at successive levels, dependent on their current 'status' within the evolving community? Unfortunately, if we reject Nicholson's theory as based on a circular argument and inappropriate assumptions, we have nothing better to put in its place! For the present the Nicholson theory should only be retained as a basis for further discussion.

Does the evolution of mimicry affect species diversity?

The vast diversity of organisms (a recent estimate suggests there may be as many as six million insects alone) is an observable end-product, the evolutionary outcome of a myriad of biological processes – amongst which mimicry is just one factor. Could mimicry have an effect on this diversity? Ecologists normally measure species diversity by an index combining estimates of species richness together with relative abundance (Taylor, 1978). In the last part of this essay I will look at two spectacular and complex examples of mimicry in butterflies, to speculate about some possible effects of mimicry on abundance and on the formation of species. Although no firm conclusions can be drawn, I hope that the biological problems revealed will at least stimulate thought.

Convergence in the passion-flower butterflies

Heliconius is a genus of about 40 butterfly species confined to the Americas, being primarily a Neotropical group. As already described, the larvae feed on passion-flower vines, and the adults are slow-flying aposemes. Most of the species are restricted to rain forest. In different parts of South America various species of *Heliconius* tend to be very similar in appearance where they coexist, being members of Müllerian mimicry rings. That this is due to mimetic convergence, and not merely non-divergence within the genus, can be established from the many remarkable differences between allopatric races of *Heliconius* species. The races of wide-ranging species, like *H. melpomene* and *H. erato*, collectively exhibit a large number of the local aposematic patterns displayed by the genus as a whole (Turner, 1975).

To explain this remarkable situation, it is thought that differentiation occurred in forest refuge areas during recent dry glacial periods, when the Amazonian forest was broken up into small patches separated by drier woodland or savannah unsuitable for these butterflies (Brown, Sheppard & Turner, 1974; a similar explanation has been invoked for speciation in birds – see Chapter 3). Each patch or refuge would have maintained its own, unique 'species mix', and in each a different species would tend to be the most abundant amongst the various poisonous butterflies present. This dominant species would act as a focal model, 'capturing' the patterns of many of the other poisonous species by Müllerian mimicry. In relatively warm, wet periods, such as the present, the forest refuges extend to produce an almost continuous rainforest system. This leads to the situation we see now, in which large areas of South America can be characterised by patterns shared by many of the local *Heliconius* species. These areas are separated by narrow zones where the descendants of the original refugial populations now meet, and produce butterflies with an array of intermediate patterns through racial hybridisation.

If Nicholson's argument were correct, then the evolution of these common warning patterns, although beneficial to the individual butterflies, would not have any marked effect on the population size of the species, because this would be controlled by some density-dependent effect. And although some *Heliconius* butterflies are common, others are very rare – presumably restricted by the limitations of their scattered foodplant resources.

Predation on *Heliconius* early stages, particularly by ants or through intraspecific competition, is intense. Larval development in these butterflies is usually rapid (in the order of 10 days), whereas adult life is characteristically long (in the order of 100 days). This can result in up to four or more coexistent, overlapping generations of adults. Rapid larval development is apparently achieved at the expense of failing to gather sufficient food reserves for egg production. This material is obtained by the adults, which forage daily for nectar and pollen, amino acids obtained from pollen being incorporated into egg protoplasm (Gilbert, 1972). Newly emerged adults follow their parents, grandparents – perhaps even great great grandparents – to learn the whereabouts of these essential adult resources, eggs being laid by the females throughout their lives.

The aposematic life-style of adult *Heliconius* allows them to exploit scattered adult and larval food sources. The early stages are relatively vulnerable: the 'telescoping' of development has been made possible because the adults, through their ability to survive and feed over a long period, can make a significant contribution to the energy requirements of reproduction. The burden of survival has been partly shifted to the adults – the larvae 'escape by proxy'. As Müllerian mimicry increases the individual survival value of the adult aposematic signal (Benson, 1972), then it is very tempting to conclude that, in the case of *Heliconius* at least, the evolution of mimicry can increase population numbers and the ability of a specialised organism to survive.

Nicholson (1927, p. 90) foresaw this as a possibility where 'the selective and eliminative agents were actually the same natural enemy', but categorically denied that this could be the case in butterflies. Two further weaknesses in Nicholson's argument become apparent from this. His model makes no allowance for the significance of food gathering by adult butterflies, nor for the possibility that increased survival ability at any stage might lead to a change in life-history strategy and chronology.

Many of our ideas about the biology of tropical butterflies, of which mimicry is such a striking feature, have been constrained by false analogy with temperate species. These often have a single, short flight-period, and the adults are greatly outnumbered by their developmental stages throughout most of the year. As already noted, it now seems that the adults of many tropical butterflies are very long-lived (others may be exceptionally short-lived – and it is possible that in this group mimicry is very rare). The *Heliconius* situation may not, therefore, be atypical, and adult survival strategies may be of far greater importance than previously recognised by those population ecologists who insisted (with little pertinent evidence) that the numbers of such insects were largely controlled by parasitoids. Levins' (1975) work may also be important here. His theoretical analyses suggest that the evolution of predator avoidance systems may be of more importance for herbivores than increased feeding efficiency – a conclusion partly opposed to the Nicholson/Haldane/ Sheppard argument quoted above.

Leaving aside this question of abundance and 'success' of *Heliconius* species, we must also ask if their evolution into different local mimicry rings has brought about speciation, and so contributed to the species richness of the group. Despite the assumption that it has (Gilbert & Smiley, 1978, p. 99), there is no compelling evidence for this – indeed, the persistence of the allopatric patterns perhaps tells against it. One might look for infertility and barriers to gene flow evolving at the hybrid or 'suture' zones separating the different mimetic races, but Turner (1971) estimated that one particular hybrid zone had existed unchanged for 2000 generations. Such a finding is consistent with a view that although allopatric speciation is likely to occur and raise total species number, it does not contribute *directly* to the species richness of individual faunas (Vane-Wright, 1978). Allopatric speciation occurs by accidental processes, and does not involve any dynamic necessity for the new species so produced to be able to coexist. However, the impact of such geographic separation of *Heliconius* into different mimicry rings might have an altogether different impact on the evolution and diversity of *Batesian* mimics – but we have little idea about this at present either, and its possible significance can only be hinted at in the final section.

Divergence, polymorphism and the Mocker Swallowtail

Despite the amazing polytypism displayed by the heliconids, within a single area Müllerian mimicry must always tend to reduce the variety of aposematic patterns. However, all the protected species within a given area probably never

all share a single warning signal-pattern. Papageorgis (1975) reveals five major mimicry complexes of Lepidoptera in Peruvian rainforests, each complex being characteristic for a different forest stratum or 'layer', from ground level to above the canopy.

Batesian mimics, in contrast to Müllerian mimics, are often polymorphic. It is possible to imagine a palatable butterfly producing five mimetic morphs, one corresponding to each of the separate mimicry complexes in Papageorgis' scheme. Although such a case is not recorded for Peru, in Uganda the females of the Mocker Swallowtail (*Papilio dardanus*) are mimics of five strikingly different danaid and acraeid models (see Frontispiece) – and another African butterfly, *Pseudacraea eurytus*, has even more mimetic morphs.

What is the significance of this increase in intraspecific diversity, and what are its possible consequences? In the case of *Papilio dardanus*, genetical evidence suggests that its three most common morphs were added step by step, the most common first (Clarke & Sheppard, 1960), the least common last. If we were to accept a Nicholsonian view that the evolution of mimicry is unlikely to have any affect on population size, then we can only interpret the stepwise accumulation of further mimetic forms as a change that 'simply serves to fit the possessors more perfectly to their natural environment, without conferring on them any material advantage'. This follows because if the species were to have a fixed population size, then the development of polymorphic mimicry would require that some or all of the morphs were held below the number or density to which they would rise in the absence of the other forms. All individuals would benefit, and adult survival would increase.

Generally it has been assumed that the evolution of polymorphic Batesian mimicry is a response either to the need for (Rothschild, 1971, p. 205) or as a stimulus to (Wynne-Edwards, 1962, p. 443) increasing population size. But strict application of Nicholson's argument requires there to be a *diminution* in adult numbers the more polymorphic a mimetic species becomes. However, polymorphic mimic butterflies do seem, subjectively, to be relatively numerous as adults; many monomorphic Batesian mimics are very rare in collections. Comparisons of larval and adult abundances within those species polytypic for degree of mimetic polymorphism would be very valuable. *Papilio dardanus* itself would make an excellent candidate for study, having, for example, no Batesian mimic morphs on Madagascar, one in Nigeria, three in South Africa, and five in Uganda. Work on the ecology and behaviour of this species would be very desirable, so that we may gain more from the painstaking work already carried out on its formal genetics.

But to test Nicholson's (1927, p. 81) bald assertion that mimics are 'no more successful than related non-mimetic forms', we need to do more than study species in isolation. Patently, we need to identify the species, or groups of species, most closely related to our chosen example. To compare the effects of the evolution of mimicry within a particular swallowtail with the 'success' of the distantly related cabbage butterflies or an armyworm moth would be, to a systematist at least, a travesty of biology.

Some speculations about the evolution of P. dardanus Now it could well have turned out that the closest living relative of *Papilio dardanus* consisted of a whole group of swallow-tail butterflies, making comparisons onerous or even impossible. Equally, it might have been a species living in a totally different part of the world. Fortunately, there is evidence that the closest living relative of *P. dardanus* is a single and sympatric species, *P. phorcas*. These *sister-species* (species that appear to have arisen from a common ancestor that has given rise to no other living species) fly side-by-side in many forested parts of Africa. Both have polymorphic females, and are the only African swallowtails to show this peculiarity (see Frontispiece). The females of *phorcas* are either similar to the green male form, or like the related species *P. constantinus*; more rarely, interme-diate females occur. *P. dardanus* females, as we have already noted, are polymorphic mimics of a number of poisonous butterflies, throughout the central and southern parts of their range in Africa.

These two insects would thus seem to be ideal for comparison. *P. dardanus* occurs throughout almost the whole range occupied by *phorcas*, but flies additionally on the islands of Madagascar, the Comoros, Zanzibar and Pemba, and in north-eastern Africa. From the areas where the two overlap, there are 1033 wild-caught *phorcas* and 2224 wild-caught *dardanus* in the British Museum (Natural History) collection. *P. phorcas* has been recorded from four species of foodplant (three *Teclea* and a *Clausena* species – although in the laboratory it will also eat, with reluctance, *Citrus* and *Choisea*). *P. dardanus* has been recorded as developing on the same *Teclea* and *Clausena* species, and on six additional plant genera; it is a minor pest of *Citrus* groves in some parts of Africa.

From this information, despite its limitations, it might be tempting to suggest that *dardanus* was the more 'successful' butterfly – it is probably more common, and has a greater distributional and host-plant range. But the last factor could easily be the key to its 'success', rather than the mimicry. However, it is interesting to consider what happens where the two species do not coexist. In such areas, *P. dardanus* has a strong tendency to be non-mimetic, the females being male-like or largely so in Ethiopia, Somalia, Pemba, the Comoros and Madagascar – although the females in southern and north-western Africa are all mimetic. Where *P. phorcas* exists alone, in the uplands of southern Tanzania and northern Malawi, it is not poly-morphic, all the females found there being of the green, male-like form. Could it be that where the two species occur together, they do so by virtue of one species being mimetic, while the other is not? Even if the very existence of *P. dardanus* is not dependent on mimicry, perhaps its coexistence with *P. phorcas* is so dependent.

This would be a surprising conclusion in view of the much wider host-plant range open to *P. dardanus*. An alternative explanation can be made by extending the hypothesis that these two butterflies are each others closest relatives: perhaps the two species came into existence as a direct result of the mimicry now seen in *dardanus* evolving in the butterfly directly ancestral to both. The evolution of mimicry could 'precipitate' speciation (Vane-Wright, 1978). Superficially, this would appear to have happened in a number of butterfly groups, such as the genera *Pseudacraea*, *Hypolimnas* and *Elymnias*, which contain a high proportion of species mimicking a range of strikingly different models.

Male pattern, intraspecific mimicry and transvestites Both natural and artificial hybridisation between *Papilio dardanus* and *P. phorcas* are possible (Clarke, 1979). The different, very distinct male colour patterns (see Frontis-piece) of the two species could well act as species-specific signals for keeping the two separate (Vane-Wright, 1978). If this is so, we might expect that the evolution of these patterns accompanied the original speciation event, to give one species with all-yellow males and a mimetic female form (*dardanus*), and the second with green-banded males and a yellowish-striped female form (*phorcas*). But both species also have male-like females which, although not identical to the males, must be very similar in signal pattern – how might we explain the existence of these forms?

In areas where the females of *dardanus* are male-like, it is generally assumed that Batesian mimicry is locally ineffective, due to a lack of appropriate models or operators, or of both. Suppose the first *P. dardanus* to enter such areas had mimetic females, or that conditions previously favourable for mimicry were no longer so. If in such areas there were operators but no models, the mimetic females would be at a strong disadvantage, being conspicuous but with nothing to back-up their boast. Alternatively, if there were models only, or neither models nor operators, the defensive patterns would be of neutral benefit, and open to the action of other selection pressures. What other factors might affect the colour patterns of female butterflies?

Although many butterflies exhibit sexual dimorphism, the majority are monomorphic, or only weakly dimorphic, often having a species-specific colour pattern. Further, whereas polymorphism in females is quite common, it is almost unknown in males. Mating in butterflies, unlike moths, is generally initiated by the males, who fly out in pursuit of passing objects – falling leaves, birds, other insects, conspecific males and, always hopefully, virgins or other willing mates of their own kind. It has been noted to advantage by butterfly collectors that the *males* (but not the females) of brightly coloured species can often be lured by pieces of paper of the appropriate colour – blue paper for blue butterflies, red paper for red butterflies, and so on. Males are often aggressive, attacking conspecific males

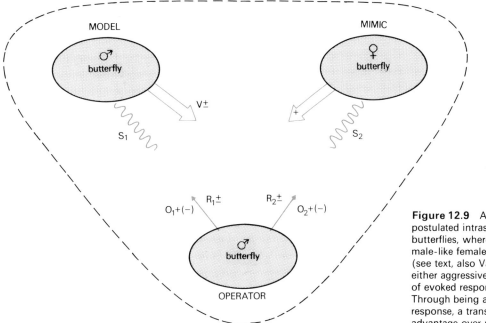

Figure 12.9 Attempt to develop a scheme for postulated intraspecific mimicry between the sexes of butterflies, whereby the colour pattern of non male-like females could become secondarily male-like (see text, also Vane-Wright, 1979). Males interact either aggressively or co-operatively (R_1). Both types of evoked response (O) involve rapid close-approach. Through being able to elicit this long-distance response, a transvestite female might gain a mating advantage over non male-like females.

on sight. At other times the males of a species will congregate peacefully, in large numbers, to suck salts or other nutrients from the ground.

All of the above suggests that male butterflies are very sensitive to the colour patterns of their own species – but how do they react if their females adopt a variety of different colour forms? In the few cases where mating behaviour has been investigated in species having both male-like and non male-like female forms, there is a suggestion that the males are more responsive to the male-like females – this is perhaps best exemplified by the work of Magnus (1958) on the European Silver-washed Fritillary (*Argynnis paphia*). The most plausible explanation appears to be that males readily recognise and are always in a 'mood' to interact with other males, and will fly up to them either for aggressive or co-operative purposes. If so, then male-like females will have a slight advantage over other types of female, in so far as males will tend to react towards them more vigorously, true sex being indicated on close approach by non-visual cues. Thus male-like females might tend to get mated more quickly, and so enjoy a longer post-mating life in which to lay their eggs. Where camouflage patterns or mimicry are effective, this slight advantage would be more than offset by the greater post-mating life-span the better protected forms could expect. But in an area where the non male-like patterns are of neutral value, or even a disadvantage, selection on females to become secondarily male-like could be effective (Vane-Wright, 1979). Such a process would appear to be taking place independently in several populations of *P. dardanus* (Pemba, and in the highlands of Ethiopia, Kenya and Rwanda), and may have

gone to completion in the Comoro Islands, Madagascar and Somalia. A similar process might also explain the existence of the male-like forms that occur in all races of *P. phorcas*.

Such processes could be interpreted as a form of intraspecific mimicry (Fig. 12.9), although the assignment of positive and negative values poses difficulties (mainly because the male-male role is variable). However, the signal value is easier to interpret: a male will always respond to the sight of another male by flying towards it, and this is exactly the response a receptive female needs to evoke – even at the expense of becoming a transvestite.

Conclusion

In such a short discussion about a complex subject, it is inevitable that many important points and current disagreements are not discussed – I am well aware that many of the ideas expressed here could give rise to heated debate (Clarke, 1979, for example, does not accept my heterodox views on *Papilio dardanus*). My main point in the last section has been to demonstrate that presumed sister-species (such as *P. dardanus* and *P. phorcas*) offer immense scope for investigating, in a truly comparative manner, how closely related species coexist, and what effects the evolution of mimicry might have on population and community structure. However, evolution is a complex, open-ended process, and mimicry is typical of that. In the hypothetical scheme outlined for the evolution of *dardanus*, it is suggested that the initial step of mimicry evolution could cause a speciation event. If this is coupled, for isolating purposes, with the formation of a new male pattern, this might subsequently lead, if Batesian mimicry by the females ceased to be advantageous, to convergence of the female pattern onto that of the male, giving mating advantage through a process of intraspecific mimicry.

The general purpose of this paper is to appeal for closer co-operation between phylogeneticists and ecologists. I hope that I have been able to demonstrate that evolutionary theory cannot yet cope with the problems posed by the effects that the development of mimicry could have on abundance and diversity. I also hope that I will have convinced at least some field ecologists that the study of sympatric sister-species could be a starting point for an attack on these problems, and the study of such pairs involving a contrast between a mimetic and a non-mimetic species offers an exciting field for investigation. Finally, although the 'old war-horses' of Batesian and Müllerian mimicry still offer much scope, it is time for other types of mimicry to be studied comparatively. There is no doubt that mimicry provides a wide array of both appealing and contrasting examples with which to study speciation, coexistence and coevolution.

References

Bates, H. W. 1862. Contributions to an insect fauna of the Amazon Valley. Lepidoptera: Heliconidae. *Transactions of the Linnean Society*, London **23**: 495–566.

Benson, W. W. 1972. Natural selection for Müllerian mimicry in *Heliconius erato* in Costa Rica. *Science* **176**: 936–939.

Benson, W. W., Brown, K. S. Jr. & Gilbert, L. E. 1976. Coevolution of plants and herbivores: passion flower butterflies. *Evolution* **29**: 659–680.

Brown, K. S. Jr., Sheppard, P. M. & Turner, J. R. G. 1974. Quaternary refugia in tropical America: evidence from race formation in *Heliconius* butterflies. *Proceedings of the Royal Society of London* (B) **187**: 369–378.

Clark, L. R., Geier, P. W., Hughes, R. D. & Morris, R. F. 1967. *The ecology of insect populations in theory and practice.* 232 pp. London: Methuen.

Clarke, C. 1979. *Papilio nandina*, a probable hybrid between *Papilio dardanus* and *Papilio phorcas*. *Systematic Entomology* **5**: 49–57.

Clarke, C. & Sheppard, P. M. 1960. Super-genes and mimicry. *Heredity* **14**: 175–185.

Gilbert, L. E. 1971. Butterfly – plant coevolution: has *Passiflora adenopoda* won the selectional race with heliconiine butterflies? *Science* **172**: 585–586.

Gilbert, L. E. 1972. Pollen feeding and reproductive biology of *Heliconius* butterflies. *Proceedings of the National Academy of Sciences of the United States of America* **69**: 1403–1407.

Gilbert, L. E. 1975. Ecological consequences of a coevolved mutualism between butterflies and plants, pp. 210–240 *in* Gilbert, L. E. and Raven, P. H. (Eds) *Coevolution of animals and plants.* Austin: University of Texas Press.

Gilbert, L. E. & Singer, M. 1975. Butterfly ecology. *Annual Review of Ecology and Systematics* **6**: 365–397.

Gilbert, L. E. & Smiley, J. T. 1978. Determinants of local diversity in phytophagous insects: host specialists in tropical environments. *Symposia of the Royal Entomological Society of London* **9**: 89–104.

Haldane, J. B. S. 1953. Animal populations and their regulation. *New Biology* **15**: 9–24.

Holling, C. S. 1965. The functional response of predators to prey density, and its role in mimicry and population regulation. *Memoirs of the Entomological Society of Canada* **45**: 1–60.

Levins, R. 1975. Evolution in communities near equilibrium, pp. 16–50 *in* Cody, M. L. and Diamond, J. M. (Eds) *Ecology and evolution of communities.* Cambridge, Massachusetts: Belknap Press.

Magnus, D. B. E. 1958. Experimentelle Untersuchungen zur Bionomie und Ethologie des Kaisermantels *Argynnis paphia* L. *Zeitschrift für Tierpsychologie* **15**: 397–426.

Nicholson, A. J. 1927. A new theory of mimicry in insects. *The Australian Zoologist* **5**: 10–104.

Nicholson, A. H. & Bailey, V. A. 1935. The balance of animal populations. – Part 1. *Proceedings of the Zoological Society of London* **1935**: 551–598.

Papageorgis, C. 1975. Mimicry in Neotropical butterflies. *American Scientist* **63**: 522–532.

Rothschild, M. 1971. Speculations about mimicry with Henry Ford, pp. 202–223 *in* Creed, R. (Ed.) *Ecological genetics and evolution.* Oxford: Blackwell Scientific Publications.

Selous, E. 1909. *The romance of insect life.* 352 pp. London: Seeley & Co.

Sheppard, P. M. 1961. Recent genetical work on polymorphic mimetic Papilios. *Symposia of the Royal Entomological Society of London* **1**: 20–29.

Silberglied, R. E. 1977. Communication in the Lepidoptera, pp. 362–402 *in* Sebeok, T. A. (Ed.) *How animals communicate.* Bloomington: Indiana University Press.

Taylor, L. R. 1978. Bates, Williams, Hutchinson – a variety of diversities. *Symposia of the Royal Entomological Society of London* **9**: 1–18.

Turner, J. R. G. 1971. Two thousand generations of hybridisation in a *Heliconius* butterfly. *Evolution* **25**: 471–482.

Turner, J. R. G. 1975. A tale of two butterflies. *Natural History* **84**: 28–37.

Turner, J. R. G. 1977. Butterfly mimicry: the genetical evolution of an adaptation. *Evolutionary Biology* **10**: 163–206.

Vane-Wright, R. I. 1976. A unified classification of mimetic resemblances. *Biological Journal of the Linnean Society*, London **8**: 25–56.

Vane-Wright, R. I. 1978. Ecological and behavioural origins of diversity in butterflies. *Symposia of the Royal Entomological Society of London* **9**: 56–70.

Vane-Wright, R. I. 1979. Towards a theory of the evolution of butterfly colour patterns under directional and disruptive selection. *Biological Journal of the Linnean Society*, London **11**: 141–152.

Vane-Wright, R. I. 1980. On the definition of mimicry. *Biological Journal of the Linnean Society*, London **13**: 1–6.

Wickler, W. 1968. *Mimicry in plants and animals* (English translation by Martin, R. D.). 255 pp. London: Weidenfeld & Nicholson.

Wiens, D. 1978. Mimicry in plants. *Evolutionary Biology* **11**: 365–403.

Wynne-Edwards, V. C. 1962. *Animal dispersion in relation to social behaviour.* 653 pp. Edinburgh and London: Oliver & Boyd.

CHAPTER 13

Coevolution of birds and plants

D. W. Snow

Birds interact with other organisms in many ways. Interactions that appear to be of long standing and importance to either or both of the partners, in affecting their survival or reproductive success, have been subjected to natural selection, so that structural, physiological and behavioural adaptations have evolved. In many cases the bird and the other animal or the plant with which it interacts have both become adapted to the interaction – they have evolved together, or 'coevolved'.

In practice, the term 'coevolution' is usually restricted to cases where the interaction is beneficial to both partners. Of course it need not be beneficial to both. Thus insectivorous birds are adapted in many different ways to catching and eating insects; and many insects have elaborate adaptations, such as camouflage, distasteful chemicals and acute powers of 'hearing' to avoid being taken by birds. The insect derives no advantage from this interaction, and we do not usually treat it as a case of coevolution though it could justifiably be considered as such. On the other hand, many fruit-eating birds have special adaptations for collecting, eating and digesting fruit; and the seeds of the fruits that they eat may be dispersed away from the parent plant, in a viable condition or even with their viability enhanced. In this way the plant benefits from having its fruit eaten. The characters of the fruit, such as its size, colour and nutritive content, may be adapted to attract the fruit-eater. It is to this kind of mutually beneficial, adaptive interaction that the term coevolution is usually applied.

In the evolutionary context, the idea of a 'benefit' or 'advantage' applies only to individuals or to their genes or genotypes, and not to the population or species to which they belong. Natural selection works on the individual; those which survive best and leave most descendants transmit more of their genes to future generations than those that survive and reproduce less well. No convincing way has been suggested by which natural selection could work on the population as a whole (though it may in certain circumstances work on groups of related individ-

uals – see Wilson, 1975); nor, in fact, is it obvious how such a benefit could be measured – whether by the length of time a population survives, its geographical range, the number of individuals, or some combination of these. One could visualise a situation in which an insect population might 'benefit' from predation by an insectivorous bird (for instance, by being held at a level low enough to escape serious attack by a parasite); but at the individual level natural selection would still tend to improve the adaptations to escape predation. Conversely, a coevolutionary relationship between a bird and a fruiting tree might enhance the survival of individual birds, but, by making them dependent on the tree, this might ultimately lead to the extinction of the population if other factors caused the extinction of the tree.

Birds have entered into two main kinds of coevolutionary relationship with flowering plants – as fruit-eaters that, at the same time disperse the seeds, and as nectar-eaters that at the same time pollinate the flowers. These two relationships are the main theme of this chapter. There is perhaps a third coevolutionary relationship, in which birds act as dispersal agents for plants by incorporating the fruiting parts into their nests; but the evidence for this relationship is still scanty, and in any one case it cannot involve more than a few kinds of plants in limited areas.

For the plant, the flower comes before the fruit; but in discussing coevolution between birds and plants it is convenient to begin with the bird-fruit relationship as it is more familiar to the general observer, especially in temperate countries.

Coevolution between birds and fruits

Some general and historical considerations

There is reason to think that fleshy fruits of the kind eaten by birds are primitive, that is, that they evolved early in the history of the flowering plants in equable, tropical conditions (Corner, 1964; Moore & Uhl, 1973); and that other kinds of fruits, such as those dispersed by wind, evolved later, perhaps in association with the climatic deterioration that took place during the Pliocene. This hypothesis is supported by the facts that the most important plant families that produce bird-dispersed fruits are of very wide distribution, and that several of them are generally agreed to be primitive (for instance, Lauraceae, Annonaceae). Several important genera are common to tropical America and Australasia. It is also significant that bird-dispersed fruits are found in greatest variety in the humid tropical forest, and that there are several plant families in which the tropical species have fleshy fruits, eaten by birds, whereas their temperate relatives have wind-dispersed fruits. A good example is the family Oleaceae, in which the ash (*Fraxinus*), with winged fruits, differs strikingly from tropical and subtropical relatives such as *Olea* with fleshy drupaceous fruits.

Birds are not the only animals that eat fleshy fruits, but they are generally the most important ones. Many kinds of tropical bats also take fruits from trees, as also do some other mammals (especially primates) and a few reptiles. Many mammals and reptiles eat fruit that has fallen to the ground, and in doing so they may disperse the seeds. There have undoubtedly been coevolutionary interactions between plants and these other vertebrate dispersal agents, but they are beyond the scope of this chapter. Two points only need be made. First, fruits adapted for dispersal by mammals are generally very different from bird fruits in that they are often dull-coloured but highly scented, instead of brightly coloured and relatively scentless. Second, reptile fruits probably reached their greatest development a very long time ago, and are now little more than a minor curiosity.

Although the plant families that have coevolved with frugivorous birds are ancient and mostly of wide distribution, they are not evenly distributed over the tropical regions. In particular, the two most important families, the palms (Palmae) and laurels (Lauraceae), are richly represented in the forests of tropical America and the New Guinea region, and well represented in south-east Asia, but very poorly represented in Africa. It is almost certainly a consequence of this fact that coevolution of birds with forest trees has reached a much higher degree of development in tropical America and New Guinea than elsewhere and that Africa is rather poor in specialised frugivorous birds (Snow, 1980). The tropical American bird families of the cotingas (Cotingidae) and manakins (Pipridae) and the birds of paradise (Paradisaeidae) of New Guinea and adjacent tropical Australia have no counterpart in Africa. As we shall see later, the extraordinary elaboration of ornamentation and of courtship behaviour in the cotingas, manakins and birds of paradise is connected with and is dependent on their fruit-eating habits.

A well-known characteristic of most bird fruits is that they are conspicuous; they advertise themselves and, to express it anthropomorphically, want to be eaten. This rather obvious fact has important implications, one of which has to do with the number of species of frugivorous birds. It is most easily appreciated by contrasting fruit with insects as a staple diet for birds, and is a consequence of the fact that there are many more ways of making oneself inconspicuous than of making oneself conspicuous. Insects have evolved innumerable strategies for avoiding predators, often involving elaborate camouflage and appropriate behaviour to make camouflage effective. Correspondingly, birds have evolved many different strategies for feeding on insects; hence the number of different feeding 'niches' for insectivorous birds in any area with rich vegetation is very high. Fruits, on the other hand, not only have fewer ways of advertising themselves than insects have of making themselves difficult to catch, but may even, it seems, gain a positive advantage by advertising themselves in much the same way as other fruits – that is, by evolving a common signal. Fruits therefore may behave like Müllerian mimics, only here the objective is to be eaten whereas the classic examples of Müllerian mimicry are quite the opposite. The result of all this is that there are far more different and specialised niches for insectivorous birds than there are for frugivores and this seems to be the main reason why there are more species of insectivorous birds in a tropical forest than of frugivorous birds. The following comparison, in round numbers (since the exact number of species in these families is still uncertain), makes the point for four important New World families of mainly tropical distribution:

entirely insectivorous	Formicariidae (antbirds)	220 species
mainly insectivorous	Tyrannidae (flycatchers)	370 species
mainly frugivorous	Cotingidae (cotingas)	80 species
mainly frugivorous	Pipridae (manakins)	60 species

Putting it in a slightly different way, we can say that fruit-feeding 'niches' are more difficult to subdivide than insect-feeding niches. It is therefore not surprising that closely related frugivorous birds tend to occupy separate geographical areas since they cannot easily coexist. For instance, among the 15 specialised frugivorous genera of the Cotingidae occurring in lowland forest in tropical America, there is only one (*Cotinga*) in which the ranges of different species overlap. In the flycatchers and antbirds, on the other hand, it is common for several congeneric species to occur in the same geographical area. This point is discussed within the context of speciation in Chapter 3.

Unspecialised and specialised frugivores

This is an important distinction. The difference between these types, although important, is not always absolute – there are borderline cases, as would be expected – but the degree to which a frugivorous bird species is dependent on fruit for its diet, or on particular kinds of fruit, has far-reaching consequences for the plant as well as the bird.

Corresponding to unspecialised and specialised frugivores, there are seemingly unspecialised and specialised fruits.

An unspecialised frugivore is one that regularly takes other foods besides fruit, and cannot subsist indefinitely on fruit alone. It is typically an opportunist, taking fruits as and when they become available. In order to attract unspecialised frugivores, it will be advantageous for a plant to have conspicuous fruits. They are usually borne in profusion, and linked with this, the nutritive content of individual fruits is not very high; they tend to be watery (succulent) and to contain mainly sugars and very little fat or protein (Table 1). They are thus a source of energy for the birds that eat them, but not a whole diet. They tend to be small, and to contain small seeds.

In the tropics, the small trees and shrubs that colonise clearings, edge habitats and other transient areas of secondary vegetation typically have fruits of this unspec-ialised sort. In Britain, elderberry is a good example. The 'strategy' of such plants is to produce large numbers of small seeds, investing little in each individual fruit, and to make the fruits as attractive to as many different kinds of unspecialised frugivorous birds as possible. As colonisers of temporary open ground, they tend to be in competition with one another, not only for ground space but also for dispersal agents.

The strategy of a plant that provides fruit for specialised frugivores is very different. In the first place, and this is probably fundamental, such plants are mainly trees of primary tropical or subtropical forest. Tropical forest trees typically have large seeds, with a reserve of food for the developing seedling, and the reason for this is probably that on the forest floor, with poor light and thin soil, seedlings need to be able to complete the first stage of their growth mainly on their own reserves, after which their further success depends on their being able to take advantage of light gaps caused by such relatively rare events as tree-falls and landslides. Now, if a plant with a large seed is to attract birds as dispersal agents it must coat the seed with something well worth eating, but at the same time not too bulky. The bird must swallow the fruit whole, not peck off bits of the flesh (Table 1). This seems to be the reason why many tropical forest trees have fruits consisting of a single rather large seed coated with an often surprisingly thin but dense and nutritious layer of flesh. In another common type of nutritious fruit (for instance, that of the nutmeg), the seed is partly or entirely covered by an aril, which is an edible outgrowth of the seed itself. A number of such fruits has been analysed and found to contain high percentages of fat and protein. Fruits of this kind are often not very conspicuous, as they need to attract birds of a rather different kind from those attracted to the more conspicuous and abundant but relatively unnutri-tious fruits that are discussed above. The specialised frugivore depends on the fruits on which it feeds. It is usually resident in a limited area and knows its food resources intimately, so that bright colours are not necessary.

Table 1. Protein and fat content of some bird-dispersed fruits

Family	Species	Protein %	Fat %
Tropical trees, dispersed by specialised frugivorous birds			
Lauraceae	*Ocotea wachenheimii*	3·5	8·5
	Ocotea sp.	5·1	16·7
	Ocotea sp.	2·4	8·1
Burseraceae	*Dacryodes excelsa*	4·4	9·6
Araliaceae	*Didymopanax morototoni*	3·8	10·7
Tropical and subtropical trees, dispersed by unspecialised frugivorous birds			
Ebenaceae	*Euclea natalensis*	0·7	0·2
Myrsinaceae	*Ardisia revoluta*	0·2	0·4
Euphorbiaceae	*Bridelia micrantha*	0·8	0·1
Temperate shrubs and climbers			
Rosaceae	*Rubus idaeus* (raspberry)	2·5	1·4
	Crataegus monogyna (hawthorn)	1·2	1·0
	Prunus spinosa (sloe)	0·7	0·3
	Rosa sp. (wild rose)	2·0	0·4
Araliaceae	*Hedera helix* (ivy)	5·3	3·4

Note. The percentages refer to the fresh weight of the fleshy part of the fruit (excluding the seed).

The specialised frugivore is typically a larger bird than the unspecialised opportunist, and it may have a very wide gape, enabling it to swallow whole the largest fruits possible. Since, from its point of view, the large seed that it ingests with every fruit is so much useless ballast, it must be able to process its food quickly, stripping the flesh off and regurgitating the seed without delay; and this in fact is exactly what specialised frugivores do. Finally, in order to survive, it must have a fruit supply that is available all year round, a point that is dealt with in a later section.

An important consequence of their exploitation of a nutritious and dependable food supply is that the more specialised frugivores can feed themselves in a very short period per day. Day-long watches on several species of specialised frugivores of the manakin and cotinga families in Trinidad and Guyana (D. W. Snow, 1962; B. K. Snow, 1970) showed that adults needed only about 10 per cent of the daylight hours for foraging. This is, it seems, one of the important predisposing factors enabling many species of specialised frugivores to evolve social systems in which the males spend almost all their time at display grounds, while the females perform all the nesting duties single-handed. The Greater Bird of Paradise (*Paradisaea apoda*) of New Guinea and the Cock-of-the-Rock (*Rupicola rupicola*) of the Guianas are two outstanding examples. The intense sexual selection that results from competition between the males for females has resulted in the most elaborate plumages and complex display behaviour. It is an evolutionary trend that is not confined to frugivorous

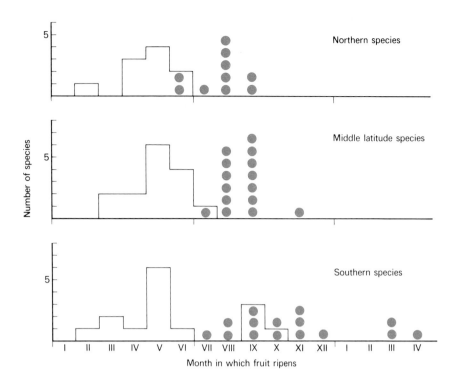

Figure 13.1 Flowering and fruiting seasons of 40 species of European trees, shrubs and climbers with bird-dispersed fruits. Histogram: months in which flowering begins. Circles: months in which fruit ripens. The data relate to middle latitudes in Europe (around 50° N). The plants are divided into three distributional categories: those whose range is mainly northern European, those of intermediate range, and those whose range is mainly southern European. It will be seen that the main peak of flowering is similar in all three groups, but northern species fruit on average earliest and southern species latest. (Data compiled from various European Floras.)

forest birds since it is also found in a number of species of grouse at high latitudes, a few species of waders, and some other birds of open country; but it is significant that in tropical forest it has taken place in some frugivorous bird groups but in none that is primarily insectivorous.

Seasonal aspects

At high latitudes where winters are severe, and at lower latitudes where there is a severe dry season, a strict seasonality of flowering, fruiting and leaf production is imposed on plants, and an ample year-round supply of fleshy fruits is not possible. In such environments only generalist, opportunist frugivores would be expected. In fact there are few specialised frugivorous species among temperate or boreal avifaunas and the most specialised among them tend to be migratory. In more equable environments, where the weather does not impose such severe constraints on plant seasonality, a year-round supply of fruits may be possible; but even in such environments it might be expected that some times of year would be better for flowering and fruiting than others, and so it is relevant to ask whether the bird–fruit interaction is, so to speak, at the mercy of the climate or whether it can itself exercise some effect on the seasonality of fruiting.

Let us consider a hypothetical case, of a group of plants of different species living together in a moderately seasonal environment, all of which produce fleshy fruits and share a number of frugivorous birds as their dispersal agents. Imagine that all of these plants, in response to rainfall, temperature or other climatic factors, produce ripe fruit at the same time of the year. It is clear that there would be intense competition among them for their dispersal agents; in fact there might well be a glut of fruits in the season and many would remain uneaten. Furthermore, no bird that was at all dependent on fruit could be resident in the area. Under these conditions any plant that altered its fruiting season, so as to be out of step with the others, would clearly be at a great advantage. Selection for dispersal of seeds should tend to lead to a staggering of fruiting seasons so far as might be compatible with the limitations set by climate.

This seems to have happened in the genus *Miconia*, the largest neotropical genus in the family Melastomataceae and one of the most important of plant genera that produce bird-dispersed fruits. In an area of forest in Trinidad, a year-round supply of fruit is produced by 20 species of *Miconia*, the fruiting seasons of which are individually quite short but staggered so as to cover the whole year (Snow, 1965). A similar explanation may account for the staggered fruiting seasons of a number of unrelated plants in coastal forest in South Africa (Frost, 1980), and further research may be expected to reveal similar cases elsewhere.

In north temperate latitudes, selection for effective seed dispersal by frugivorous birds has, apparently, led to latitudinal trends in fruiting season to conform with the timing of bird migration. Thus, northern European fruits tend to ripen early, in late summer, central European fruits in the autumn, and southern European (Mediterranean) fruits in the winter (Fig. 13.1). These differences correspond to the southward shift of the main populations of migratory frugivorous birds such as the thrushes. It is interesting that in the Mediterranean basin, where many northern frugivorous birds winter, bird fruits not only tend to ripen in the winter but they also show some of the attributes of specialised fruits in the tropics: they are drier and more nutritious than typical northern fruits, and so can provide a more complete diet for the wintering avifauna. Thus the distinction between specialised and unspecialised tropical fruits is paralleled at higher latitudes by the distinction between boreal and Mediterranean fruits. In this connection it is probably significant that the ivy (*Hedera helix*), the only member of a predominantly tropical family that has reached northern Europe, is unusual in three ways: it produces unusually nutritious fruits for a northern plant (Table 1), it ripens its fruit very late, not until early spring, and its fruits are regularly fed to nestling birds whose diets otherwise consist of animal matter. The family to which ivy belongs, the Araliaceae, is one of the most important families producing nutritious fruits for specialised frugivores in the tropics, and it seems that in its fruit ivy has retained some of the characteristics of its tropical ancestors.

Plants may adapt the timing of their fruit production in another way to the dispersal agents – by adjustment of the length of the period over which they bear ripe fruit. They may ripen all their fruit synchronously, or ripen it gradually over a more or less prolonged period. In general, the former method is adopted by plants whose fruits are eaten by unspecialised frugivores, and the reason is probably that such plants need to fruit copiously and conspicuously in order to compete successfully with other plants in attracting opportunist frugivores. Asynchronous fruiting, on the other hand, is the rule in some plants that produce fruits for specialised frugivores, probably because in this way they can provide a long-term food supply suitable for a smaller number of more certain dispersal agents.

Much more research is needed on these and other aspects of the seasonality of fruiting before the interpretations proposed above can be treated as anything but provisional and tentative. The general interest of coevolutionary interactions affecting the timing of fruiting is that the birds, by acting as the agents of competition between different plant species, thereby ensure the continuity of their food supply.

Coevolution between birds and epiphytic plants

A limited number of birds in all the main tropical and subtropical land areas are specialised feeders on the fruits of mistletoes (Loranthaceae). Some other epiphytic plants may also be involved, but they have been little investigated (bromeliads, aroids, and the epiphytic cactus *Rhipsalis*). The coevolutionary relationship between bird and plant is close, and there are some special features that set it apart from the typical bird–fruit relationship discussed above. The birds are mostly small, as are the fruits, with short and rather stout bills, very different from the typical large

wide-billed specialised frugivore. Effective dispersal depends on the seeds being lodged in a crack on a tree branch, and the fruits are adapted to this by having small seeds embedded in a sticky coating which resists digestion in a bird's gut. The bird either swallows the fruits whole and passes them very rapidly through the (in some cases highly modified) alimentary canal, voiding them with their sticky coating intact so that they adhere to the branch on which they land, or squeezes the fruit in the bill, swallows the skin and pulp, and wipes the seeds off onto the branch on which it is perched. The flesh of mistletoe fruits appears thin and unnutritious, but that of one species at least is oily (Frost, 1980) and it is certain that they must provide the birds that specialise on them with a complete diet..

The most important mistletoe-berry eaters are the mistletoe birds (Dicaeidae) of south-east Asia and Australasia, the small neotropical tanagers of the genera *Euphonia* and *Chlorophonia*, and the small barbets of the genus *Pogoniulus* in Africa. A few species from other families are also more or less specialised eaters of mistletoe berries, for instance the small neotropical flycatcher *Tyranniscus vilissimus* (Skutch, 1960).

Seed defences

Seeds are nutritious and are eaten by many kinds of birds, which may be called seed-predators to distinguish them from the legitimate fruit-eaters. Parrots and most pigeons, for example, are often loosely described as fruit-eaters but in fact are seed-predators. It is therefore pertinent to ask why the legitimate fruit-eaters do not digest the seeds as well as the flesh of the fruits that they eat. The question is not, however, altogether straightforward. If they did so, the coevolutionary relationship between bird and plant would break down; but even so it should be advantageous to an individual bird to get nourishment from the seed of a fruit as well as from the flesh. The more specialised frugivores are probably now so strictly adapted to stripping off and digesting the flesh, and getting rid of the seeds, that they could not break down and digest the seeds as well. Nevertheless, some less specialised frugivores at least might be expected to exploit the seeds if it were easy to do so. It is thus likely that the seeds of plants whose fruits are dispersed by legitimate fruit-eaters have in the past needed and may still need some kind of defence. That they have some defence is indicated by the fact that such fruits are generally not exploited by seed-predators. In Trinidad we found no evidence that parrots ever fed on the seeds of the many species of Lauraceae whose fruits were an important part of the staple diet of the specialised frugivores.

The defences of seeds against avian seed-predators have been little investigated (more attention has been given to defences against insect seed-predators), but it seems they are of two kinds, mechanical and chemical. Among the most important plant families for frugivorous birds, the palms have clearly gone in for mechanical defence of the seed. Palm seeds are protected by a hard, woody endocarp and no palm, as far as is known, contains toxic substances. The Lauraceae on the other hand probably rely on chemical defence. The seeds are often quite soft, so that it is sometimes surprising that the dense flesh can be stripped off in a bird's stomach without any damage to the seed. There is, however, a high incidence of alkaloids in the family (Li & Willaman, 1968), and the seed of the one cultivated member of the family, the Avocado Pear, is toxic.

Coevolution between plants and seed-predators

Among the seed-predators there is one group of birds that has, apparently, entered into a rather special coevolutionary relationship with the plants that they exploit. They are seed-hoarders or 'scatter-hoarders', birds of temperate or boreal regions that bury large numbers of seeds in late summer and autumn and dig them up and eat them, or feed them to their young, in the following winter and spring. All those that are known are nutcrackers and jays, members of the crow family (Corvidae).

Storing food for later consumption is a widespread habit in the Corvidae and in some other families, especially the tits (Paridae) and nuthatches (Sittidae). All these birds must occasionally and by chance act as dispersal agents for the seeds, for instance when a bird dies or forgets seeds that are buried in sites where they can germinate and grow; but in most cases there is no evidence that the plant derives any regular advantage from its seed-predators, or that its fruit and fruiting habit are adapted for dispersal by seed-predators. Recent research, however, on the food-hoarding behaviour of Clark's Nutcracker (*Nucifraga columbiana*) and other corvids in the mountains of western North America, and on the species of pines whose seeds they exploit, has given evidence that one species of pine, the Piñon Pine (*Pinus edulis*), has coevolved with the scatter-hoarders (Balda, 1980).

By comparison with other pine species growing in the same area, whose seeds are wind-dispersed, the seeds of the Piñon Pine have several features indicating that they positively 'offer themselves' to the nutcrackers. They are large and heavy (not small and winged) and so individually very nutritious; they have rather thin seed coats for their size and are therefore easy to de-husk; they have a characteristic colour change when they become edible (absent in the other species); they are more visible and exposed in the cone; and the cones are orientated upwards and not downwards, as in the wind-dispersed species. A further characteristic is that seed production in the Piñon Pine tends to be erratic but heavy when it occurs. Thus when seeds are available they are available in great quantities, and so especially easy to collect and store.

The efficiency of the system depends on the nutcrackers hoarding more seeds than they subsequently need to eat, at least on average. This they do, probably as an insurance, since their subsequent need for seeds is affected by the severity of the following winter and spring and cannot be predicted. They hoard the seeds at a time of year when food is abundant, enabling much energy to be devoted to activities other than satisfying their immediate needs. Furthermore, a unique modification of the oesophagus (the sublingual pouch) enables them to transport more seeds at a time than would otherwise be possible.

Similar coevolved systems undoubtedly exist, involving other species of nutcrackers and jays and other trees, but they have not been examined as thoroughly as the interaction between Clark's Nutcracker and the Piñon Pine. Nearer home, for instance, the European Jay (*Garrulus glandarius*) stores acorns, and is probably the main dispersal agent for oak trees. Bossema (1979) has shown that when jays are collecting acorns for storage they select the largest ones that they can swallow (slightly larger ones may be carried in the bill). If, as is likely, oak trees are propagated mainly from acorns buried by jays, it follows that jays may have been among the selective agents determining acorn size.

Coevolution between birds and flowers

Some general and historical considerations

The basis of the bird–flower relationship is simple, but its ramifications are complex and are only just beginning to be understood. The plant offers the bird food, in the form of nectar, and the bird in exchange pollinates the flower. There are some striking parallels with the bird–fruit relationship and some fundamental differences. Most striking among the parallels is the fact that there are unspecialised nectar-eaters, which opportunistically visit many different kinds of flowers, mainly small ones, and so are less reliable as pollinators, and specialised nectar-eaters which concentrate on a few kinds of flowers, mainly larger ones, and so are more reliable pollinating agents. A fundamental difference is that nectar does not provide a complete diet but is only a source of energy, so that all nectarivorous birds, even the most specialised, must also eat insects or other animal matter. Another difference is that nectar is a renewable resource; when a fruit has been eaten it is gone, but a flower continues to produces more nectar. One consequence of this is that a clump of flowers of sufficient size can supply a hummingbird with all the energy it needs as long as the flowers last. Such a clump is worth defending, and in fact many nectarivorous birds defend concentrated nectar sources, a behaviour that has no parallel in frugivorous birds.

In the absence of fossil evidence nothing certain can be said about the early origin of bird–flower coevolution. The subject is discussed within the general context of pollination by Proctor & Yeo (1973). It seems clear that the pollination of flowers by birds was a later evolutionary development than insect-pollination, and some of the bird-pollination systems that we see today may be quite recent. Thus Grant & Grant (1968) have shown that many of the hummingbird-pollinated flowers of western North

America belong to genera that are predominantly insect-pollinated. As already mentioned, the main plant families, and a few genera, whose fruits are dispersed by specialised frugivorous birds are widespread in the tropics and some at least are considered to be primitive among flowering plants; but most of the main plant families that are pollinated by birds, and all the main genera, are different on the different continents and the birds too are largely different, suggesting independent evolutionary developments subsequent to the establishment of the main faunal regions. Thus in the New World tropics the hummingbirds (Trochilidae) are the main nectarivores, and to a lesser extent some of the honeycreepers and their allies (Coerebidae); bird-pollinated plants are of many families, among the most important being the Musaceae (genus *Heliconia*), Bromeliaceae, Acanthaceae, Gesneriaceae, Passifloraceae, Rubiaceae and (in the Andes) Ericaceae. In North America a mainly different set of families provides nectar for hummingbirds, the chief being the Ranunculaceae, Scrophulariaceae, Labiatae, Polemoniaceae and Liliaceae. In Africa the sunbirds (Nectariniidae) are the main nectarivores, and the endemic sugarbirds (*Promerops*) of South Africa; the chief plant families include the Liliaceae, Iridaceae, Bignoniaceae, Leguminosae and in South Africa the Ericaceae and Proteaceae. In Australasia the honeyeaters (Meliphagidae) and brush-tongued parrots (Loriinae) are the main nectarivores, and the main plants that they exploit belong to the families Myrtaceae and Proteaceae and have open (non-tubular) staminate flowers quite different from the tubular bird-flowers characteristic of tropical America and tropical Africa.

The discussion that follows is confined to the coevolutionary interaction between hummingbirds and their flowers, as it has been the subject of a considerable mass of research in the last 15 years. Sunbirds have been less studied; it is clear that in some aspects of their ecology and behaviour they parallel hummingbirds closely, but they are a far more uniform group, showing nothing comparable to the diversity of the hummingbirds. Coevolution between flowers and the great group of Australasian nectar-eaters (honeyeaters and brush-tongued parrots) has been little studied, but the relationship must be very different, since in their structural adaptations both flowers and birds are quite distinct from those found in the other continents.

Characteristics of hummingbird flowers

The typical hummingbird flower has a tubular corolla with nectaries at or near the base of the tube and the anthers and stigma, when ripe, positioned near the tube entrance. The flower is formed by relatively tough tissue, and the base of the corolla tube may be additionally protected by a tough calyx or bract. The flower is typically scentless and is often red (or if it is not red itself, adjacent bracts or modified leaves are red). All these are clearly adaptations for attracting hummingbirds visually, ensuring pollination by hummingbirds without damage to the flower, and

excluding 'nectar thieves' which might break into the base of the flower and take nectar without effecting pollination.

The red colour of most hummingbird flowers has often been noted, and several early experiments showed that red tends to be preferred over other colours. But by no means all hummingbird flowers are red, and there is more recent experimental evidence that a hummingbird may be conditioned to the colour of the flowers on which it has been feeding, and that nectar quality over-rides colour in deciding which of two alternative nectar sources are preferred (Stiles, 1976). Furthermore, hummingbirds may discriminate hues of long wave lengths better than human beings, so that the question whether 'red' is preferred to other colours may to some extent be an unreal question based on our lack of visual discrimination. In addition to the birds' sensitivity to the longer wave lengths, contrast with the background probably plays a part. Red shows up conspicuously against green foliage. It may be significant that in a number of hummingbird plants with bluish foliage the flowers are yellow, the complimentary colour (Stiles, 1976).

The nectar of hummingbird flowers is typically a rather weak solution of sugars, in contrast to the more concentrated nectars characteristic of insect-pollinated flowers. Until recently it was generally assumed that nectar was purely a sugar solution, but small amounts of amino acids have now been found in many different nectars (Baker & Baker, 1973), and they may be important nutritionally to insects such as butterflies which have no other protein source. Hummingbirds, however, almost certainly do not make any significant use of such amino acids as may be present in the nectars that they eat; in fact recent experiments have indicated that if amino acids are present in detectable quantities the nectar is avoided by hummingbirds (Hainsworth & Wolf, 1976). Hummingbird nectar is thus essentially a source of energy which, since it flows freely and is produced in ample quantities, can be taken up rapidly by any hummingbird whose bill is long enough and the right shape to probe the flower's corolla tube but is not easily accessible to other organisms.

Unspecialised and specialised hummingbirds

Unspecialised hummingbirds have relatively short straight bills and exploit a variety of unspecialised smallish flowers with short corolla tubes (bill-lengths typically 12–20 mm; corolla-tube lengths up to a few millimetres longer). These types of hummingbirds feed on many different plant species, from large forest trees to herbs. This, at least, is true in the tropics, but in North America hummingbird plants are mostly perennial herbs or 'softwood subshrubs'; true shrubs with hard wood are less commonly utilised.

The unspecialised hummingbird-pollinated plant produces many flowers but invests relatively little of its resources in each flower. The striking parallel with unspecialised bird-fruits has already been mentioned, but it is clear that the plants' strategies are not the same in the two cases. The plant with unspecialised fruit achieves efficient dispersal of as many seeds as possible. For the plant with unspecialised flowers, it is presumably an advantage to attract as many different hummingbird pollinators as possible, since in this way it ensures that some pollinator is always likely to be available; but a consequence of producing flowers in a mass, as already mentioned, is that a hummingbird may find it worth while to defend a small group of plants, or a single plant, or even part of a plant. The effect of this must be greatly to reduce the dispersal of the pollen, and hence the chances of cross-pollination. Such an effect has been demonstrated, by means of marking the pollen with powdered dye, in four species of *Heliconia* (Musaceae) in tropical forest in Costa Rica (Linhart, 1973). For instance, the pollen of *H. imbricata*, a species that grows in clumps and supports territories of the hummingbird *Thalurania furcata*, is dispersed over very short distances. The pollen of *H. tortuosa*, a plant not growing in clumps and supporting no hummingbird territories, is dispersed widely and over long distances.

Specialised hummingbirds and their flowers are in many ways more interesting, as they exhibit some of the most striking coevolutionary adaptations. In the first place, the hummingbirds are mostly larger species with long bills that may be straight or curved, and the flowers are mostly large ones, with long and in many cases curved corolla tubes. The most obvious aspect of the specialisation is that the bill of the bird fits the corolla of the flower on which it feeds, often quite exactly and resembling a 'lock and key' (Fig. 13.2). The most striking cases of close fit between bill and flower are found in equable tropical areas, both lowland and montane. Typically the plant produces few flowers at a time but invests a lot in each flower. To the extent that a hummingbird species is specialised to visit one or a few kinds of the flowers in its area, the likelihood of pollination of those flowers is increased; the chance of cross-pollination is also increased, since successive visits are more likely to be to different plants of the same species than to plants of different species. In some specialised hummingbird-flowers the male and female parts of the flower mature at different times; this further increases the chance of cross-pollination, but may not assure it if flowers on the same plant are at different stages.

These adaptations, tending to increase the reliability of the pollinators, at the same time involve some risk to the plant, if for some reason the necessary pollinating species are not available locally.

The plants that have specialised hummingbird-pollinated flowers are mainly large herbaceous or semi-herbaceous forms, vines (many of which bear their flowers near the ground) and epiphytes, and thus are mainly of different life forms from the plants that have unspecialised hummingbird-flowers. They are in some cases less conspicuous than unspecialised flowers, probably because they do not need to advertise themselves to opportunist feeders but are exploited by birds that depend on them over relatively long periods and so learn their whereabouts.

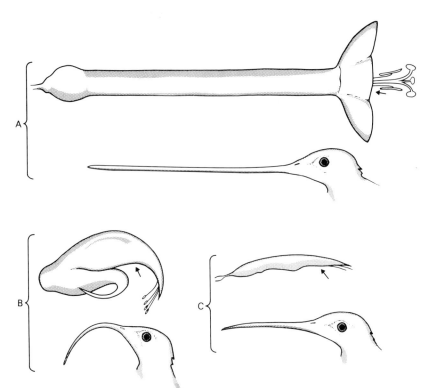

Figure 13.2 Three examples of coadaptation between the bills of specialised hummingbirds and their flowers. A – the Swordbill (*Ensifera ensifera*) and the passion vine *Passiflora mixta*. B – the Sicklebill (*Eutoxeres aquila*) and *Heliconia* sp. C – the Velvet-breast (*Lafresnaya lafresnayi*) and *Castilleja fissifolia*. Arrows indicate point of entrance of hummingbird's bill. Hummingbirds normally feed while hovering, but the Sicklebill when feeding on this *Heliconia* has to perch on a lower flower and tilt its head back in order to insert it into the corolla tube. The flowers are exactly spaced on the inflorescence to make this possible.

The parallel with the relationship between a specialised frugivore and its food plants is obvious.

Specialised hummingbirds do not normally maintain feeding territories. Their nectar sources are scattered and they must move from one to another; defence of any one of them is not worthwhile, in terms of energy expenditure. This feeding strategy, which in its fully developed form is quite distinct from the territorial strategy of unspecialised hummingbirds, has been called 'trap-lining'.

It is not obvious what factors have been involved in directing the coevolutionary paths of hummingbirds and plants along one or the other of these two main lines. One factor may have been the relative benefit to the plant of out-crossing or self-pollination. Another possibility is that herbaceous plants, with short generation times, can co-evolve more rapidly with birds than woody plants, especially large forest trees, so that specialised interactions can be more easily achieved. A third possibility is that plants that live in the reduced light beneath the forest canopy, or in small light gaps in the forest, cannot accumulate sufficient reserves to produce a mass of flowers at the same time, and instead devote their resources to producing a small number of flowers sequentially. Hence they cannot attract territorial hummingbirds, and pollination can be best assured by making sufficient nectar available in one or a few large flowers to attract a specialised trap-liner. Stiles (1975) has developed these ideas from studies of species of banana-like plants of the genus *Heliconia* which are fed on by specialised and unspecialised hummingbirds in an area of rain-forest in Costa Rica.

Seasonal factors

Except in equable tropical regions without severe dry seasons, flowering in any plant community is generally concentrated into a limited part of the year; although some flowers may be available at any time, there is usually a marked peak (or two peaks in areas with double dry and wet seasons) in the annual pattern of flower abundance. The more seasonal the climate, the more marked will be the peaks. Hence, in the more pronouncedly seasonal areas nectar-eaters tend to be migratory or nomadic, often strikingly so, their movements being related to the flowering of important food plants. Thus in western North America there is a well-marked migration of hummingbirds from the lowlands into the mountains in late summer, when flowers become scarce at low altitudes and reach their peak at higher altitudes. The seasonality of flowering may also influence the annual cycle of the birds as well as their movements. Several studies (e.g. Skutch, 1967; Snow & Snow, 1973) have shown that hummingbirds breed at the time when the flowers on which they depend for their energy are at their most abundant.

In most such cases it appears that the nectar-eaters adjust their movements and annual cycles to the flowers; but the possibility should also be considered that, at least in some situations, the interaction has been mutual, and that the plant may have adapted its periodicity to the behaviour of its pollinators. The most convincing case of this kind has been reported from a study of the Anna Hummingbird (*Calypte anna*) in chaparral country in the Santa Monica Mountains of California (Stiles, 1973). For reasons that appear to be connected with competition with other hummingbirds and the survival of the young, the Anna Hummingbird breeds several months earlier than most other chaparral birds, laying its first eggs in December or January. Among its food plants, two species of *Ribes* and one of *Arctostaphylos* depart from the seasonal pattern of most chaparral plants in exactly the same way by flowering in the winter. *Ribes speciosum* in particular, the more specialised for bird pollination of the two *Ribes* species, is absolutely dependent on the Anna Hummingbird for its pollination. These plants seem to have shifted their flowering season from the typical pattern in step with their main pollinator, probably deriving an advantage from the consequent avoidance of competition with other plant species.

In more equable climates, where sufficient flowers are available at all times for hummingbird populations to be sedentary, individual plant species nevertheless normally have well-defined, limited flowering seasons. In such conditions one might expect that interspecific competition for hummingbird pollinators should lead to a staggering of flowering seasons, just as competition for birds as dispersal agents has led to the staggering of fruiting seasons in *Miconia*, as discussed earlier. Stiles (1977) has reported a convincing case in an area of humid forest in Costa Rica, and it is probable that more cases will come to light when further long-term studies of hummingbird feeding ecology have been carried out.

Nectar thieves

Just as the coevolutionary relationship between fruit and legitimate frugivore is open to exploitation by fruit-predators that take the flesh without dispersing the seed, or by seed-predators that destroy the seed, so the relationship between flowers and their bird pollinators is open to exploitation by nectar thieves that take the nectar without pollinating the flower. As already mentioned, many of the better adapted hummingbird flowers are protected against nectar thieves by the tough, thick tissue of the corolla base or calyx, or of a protective bract, but many flowers are unprotected and are regularly exploited, to such an extent that the nectar thieves significantly affect the amount of nectar available to the pollinators.

In the New World tropics the main nectar thieves belong to two groups: the flower-piercers (*Diglossa*), a genus of small birds quite unrelated to hummingbirds, living in montane habitats in the Andes, Central America, and very locally elsewhere; and the hummingbirds themselves, some species of which have short and very pointed bills and pierce flowers at the base rather than enter the proper way. The main effect of nectar thieving on the coevolutionary hummingbird-flower relationship has probably been to promote the flower's nectar-protecting adaptations, but there is great variation in this. Some flowers – presumably those with a relatively short history of coevolution with hummingbirds – have no protection; others seem to have evolved complete protection for their nectar, for example the bromeliads, whose flowers are embedded in thick bracts.

Another possible solution on the part of the plant may be to divert part of its resources not to protecting the flowers but to producing extra flowers that will be exploited by nectar thieves, thus increasing the probability that the other flowers will be pollinated by the legitimate nectar-feeders. Colwell *et al.* (1974) have put forward this suggestion as a result of detailed study of the interactions between a hummingbird, a flower-piercer, and a bird-pollinated shrub (*Centropogon valerii*) in the highlands of Costa Rica. They found that, as a consequence of the hummingbird's aggressive behaviour to the flower-piercer, the latter fed mainly on the inner and lower *Centropogon* flowers, which set less fruit than the outer and higher flowers at which the hummingbird mainly fed.

Conclusion

The study of coevolutionary relationships between birds and plants is a comparatively new and rapidly expanding field. Only ten years ago hardly any of the ideas put forward in this chapter had been advanced, and very little of the information summarised here was available. We may be equally sure that in ten years' time our understanding of the subject will have advanced well beyond its present state.

The chief impression gained in trying to present a summary account of coevolutionary relationships between birds and flowering plants is that the main ideas put forward are valid and will stand the test of time, but that almost all general statements involve some oversimplification and the ignoring of particular, apparently atypical cases. There is a place for the occasional attempt at a general synthesis, and also for theoretical 'model-making', leading to predictions that can be tested by observation or experiment. But the important advances will come from detailed, painstaking studies of particular cases, in conditions that are as close as possible to those under which the coevolutionary relationship evolved.

References

Baker, H. G. & Baker, I. 1973. Some anthecological aspects of the evolution of nectar-producing flowers, particularly amino acid production in nectar. pp. 243–264 *in* Heywood, V. H. (Ed.) *Taxonomy and ecology*. London & New York: Academic Press.

Balda, R. P. 1980. Seed catching systems by temperate forest birds. *Proceedings of XVII International Ornithological Congress*.

Bossema, I. 1979. Jays and oaks: an eco-ethological study of a symbiosis. *Behaviour* **70**: 1–117.

Colwell, R. K., Betts, B. J., Bunnell, P., Carpenter, F. L. & Feinsinger, P. 1974. Competition for the nectar of *Centropogon valerii* by the hummingbird *Colibri thalassinus* and the flower-piercer *Diglossa plumbea*, and its evolutionary implications. *Condor* **76**: 447–452.

Corner, E. J. H. 1964. *The life of plants*. 315 pp. London: Weidenfeld & Nicolson.

Frost, P. G. H. 1980. Fruit-frugivore interactions in a South African coastal dune forest. *Proceedings of XVII International Ornithological Congress*.

Grant, K. A. & Grant, V. 1968. *Hummingbirds and their flowers*. 115 pp. New York: Columbia University Press.

Hainsworth, F. R. & Wolf, L. L. 1976. Nectar characteristics and food selection by hummingbirds. *Oecologia* **25**: 101–113.

Li, H. L. & Willaman, J. J. 1968. Distribution of alkaloids in angiosperm phylogeny. *Economic Botany* **22**: 239–252.

Linhart, Y. B. 1973. Ecological and behavioral determinants of pollen dispersal in hummingbird-pollinated *Heliconia*. *American Naturalist* **107**: 511–523.

Moore, H. J. & Uhl, N. W. 1973. Palms and the origin and evolution of the Monocotyledons. *Quarterly Review of Biology* **48**: 414–436.

Proctor, M. & Yeo, P. 1973. *The pollination of flowers*. 418 pp. London: Collins.

Skutch, A. F. 1960. Life histories of Central American birds II. *Pacific Coast Avifauna*, No. 34. 593 pp. Berkeley, California: Cooper Ornithological Society.

Skutch, A. F. 1967. Life histories of Central American highland birds. 213 pp. *Nuttall Ornithological Club, Publ. No. 7*.

Snow, B. K. 1970. A field study of the Bearded Bellbird in Trinidad. *Ibis* **112**: 299–329.

Snow, D. W. 1962. A field study of the Black and White Manakin, *Manacus manacus*, in Trinidad. *Zoologica, N.Y.* **47**: 65–104.

Snow, D. W. 1965. A possible selective factor in the evolution of fruiting seasons in tropical forest. *Oikos* **15**: 274–281.

Snow, D. W. 1968. The singing assemblies of Little Hermits. *Living Bird* **7**: 47–55.

Snow, D. W. 1980. Regional differences between tropical floras and the evolution of frugivory. *Proceedings of XVII International Ornithological Congress*.

Snow, D. W. & Snow, B. K. 1973. The breeding of the Hairy Hermit *Glaucis hirsuta* in Trinidad. *Ardea* **61**: 106–122.

Stiles, F. G. 1973. Food supply and the annual cycle of the Anna Hummingbird. *University of California, Publications in Zoology.* **97**: 1–109.

Stiles, F. G. 1975. Ecology, flowering phenology, and hummingbird pollination of some Costa Rican *Heliconia* species. *Ecology* **56**: 285–301.

Stiles, F. G. 1976. Coadapted taste preferences, color preferences, and flower choice in hummingbirds. *Condor* **78**: 10–26.

Stiles, F. G. 1977. Coadapted competitors: the flowering seasons of hummingbird-pollinated plants in a tropical forest. *Science, N.Y.* **198**: 1177–1178.

Wilson, E. O. 1975. *Sociobiology: the new synthesis*. 697 pp. Cambridge, Massachusetts: Belknap Press.

CHAPTER 14

Coevolution of plants and insects

V. F. Eastop

This chapter considers the outcome of plant and insect interactions during the 400 million years of their coevolution. The numerous relationships between insects and plants are not easily considered in discrete groups. Pollination and the provision of nectar intergrade with the development of extra-floral nectaries to provide food for insects. Some plants rely on ants for protection from grazing animals while others rely on them for nutrition. The provision of food for insects has similarities both with the provision of nectar for pollination and with parasitism. Plants may protect themselves with a thicker cuticle, shiny, hairy or sticky leaves, hooked hairs, pointed spines, poison or by various forms of camouflage or mimicry, sometimes making it appear that they are already infested. Some plants develop rapidly for a short time and remain dormant for much of the year. Other (perennials) produce vast quantities of seed sporadically, making life difficult for specialised seed parasites and predators. Many insects, living in or on plants, conceal themselves by resembling twigs, leaves or flowers.

The term coevolution has been used in different ways: some authors restrict its use to 'directed mutual adaptations' while others have applied it to almost any interactive adaptation developed between two organisms. It is used in the broader sense here.

The coevolution of insects and plants is the story of the way in which insects have lived for the last 300 million years, among and often directly on living or decaying plants, and the varied means used by plants both to escape the depredations of insects and to benefit by their presence.

About 250 000 species of vascular plants and 1 million species of insects have been described. In part this great number of species is the outcome of mutual adaptation between particular species of plants and insects. Feeding on leaves, stems and roots led to preferences both for particular parts of the plant and for particular species of plants. Feeding on protein-rich spores led to pollination and the evolution of flowering plants.

Plants as habitats

For many insects, plants constitute most of the vertical component in a habitat. Plants also provide a high uniform humidity, a sought after regime since desiccation is one of the more serious hazards faced by small organisms. Many insects lay their eggs on plant tissue. A diapausing egg must survive for months, it must not dry out, be eaten or be covered by fungus. A plant stem is a humid but well-drained site. Plants enormously increase both the amount and diversity of living space available to insects and provide microenvironments essential to aggregation, courtship and other territorial behaviour. Some insects absorb energy while basking on the plants in the morning, and will move into a sunny patch if they are shaded. Many insects spend much of their time on the trunks, branches or leaves of trees without feeding on the plants. Sometimes they feed on algae, pollen, honeydew or moulds found on the trees. Many predatory insects are regularly found on particular plants although, under experimental conditions, they can be reared on quite different prey.

Many insects live in the fruiting bodies of fungi and some ants and termites cultivate fungi in their nests, either for food or air-conditioning. Insects cannot digest cellulose and numerous insects that develop in wood have symbiotic relationships with wood-decomposing fungi, which may be carried in special structures on the insects. Insects are important in the decomposition of dead plants and animals, particularly by providing access for bacteria and fungi. Insects themselves can suffer from fungus diseases, most of which are fatal, but one group of fungi (*Septobasidium*) have domesticated scale insects and use them for obtaining food from trees, in much the same way that we get milk from grass, or ants obtain honeydew or manna from plant sap via aphids and coccids.

Insect larvae living in decaying vegetation benefit from the high constant humidity and availability of dead plant material, bacteria and fungi. Insects living in this way are likely to encounter roots, spores, pollen grains and fruits, and some will specialise in particular sorts of food. Some insects browse living plants, others bore into their fruits, leaves, stems or roots.

The distinction between predation and parasitism is obscure. Some insects can complete their development on many plants, while others are highly specific and can live on only one plant species. Nutrition is not simply a matter of completing development; a deficient diet may produce sterile adults. Some insects can be either winged or wingless as adults; commonly better nutrition produces wingless adults, but this may not always be the case.

Mutual benefits

Pollination and nectar

Pollination probably developed from early insects feeding on the food-rich spores that would have covered the soil of Carboniferous forests. It would seem a short step to climb the plants in search of spores and eventually become a distributive agent for pollen. The flowers of many of the 'lower' flowering plants (for instance Magnoliidae, Rosidae) are more open and visited by many different sorts of insects including beetles, flies and bees. Beetle pollination tends to occur in more arid climates and amongst the more primitive angiosperms. Beetles tend to damage the plants they visit and beetle pollinated plants often have inferior ovaries. Most (93 per cent) of the families of plants with inferior ovaries have bird or beetle pollination. Many of the more primitive flowering plants, including members of the Magnoliaceae, Degeneriaceae, Winteraceae, Eupomatiaceae, Calycanthaceae, Euryalaceae and Ranunculaceae are pollinated by beetles. There were numerous beetles during the Upper Jurassic/Lower Cretaceous when flowering plants appeared, but the higher Hymenoptera (bees, wasps and ants) and Lepidoptera (butterflies and moths) are not represented in the fossil record of that era.

The larvae of the higher insects (Endopterygota) are modified for feeding and the adults for reproduction and locomotion, and sometimes larvae and adults take quite different food. The long nectar-seeking tongues of bees, butterflies and moths, or the blood-sucking mouthparts of biting flies are quite different from those of their larvae. The evolution of the insect pupa resulting in an adult with a different life style, preceded and was probably essential to the evolution of flowering plants. The insect pupa evolved to allow a greater specialisation between more static immature stages largely concerned with feeding and more mobile adults concerned more with reproduction and dissemination. This greater mobility also allowed adults to utilise more inaccessible and ephemeral food sources such as spores produced sporadically at the end of shoots. The evolution of the insect pupa preceded and thus was probably essential to the evolution of flowering plants. Although some orthopteroid and hemipteroid insects act as pollinators now, it is difficult to envisage pollination

evolving with them. Active adult beetles and sawflies are much more effective pollinators.

Species are dynamic and may behave differently in different places, at different periods of their development and under different environmental conditions. It is not surprising therefore to find that plant groups with catkins that are mostly wind pollinated contain some insect pollinated species. For instance, among the Salicaceae, the poplars are wind-pollinated but many willows are largely insect pollinated.

Pollination, like all biological processes, becomes more effective as the energy expended on its execution is reduced. Some Rosidae such as Leguminosae and many of the higher plants, have their nectar and pollen concealed and accessible only to specialised pollinating insects, except for equally specialised robbers, which may steal the nectar by piercing the base of the flower. The early pollinators presumably ate the pollen and only pollinated plants by chance on their visit to a neighbouring plant. It may be more economical for the plant to provide nectar instead of surplus pollen and to provide petals as a nectar advertisement and landing platform. Nectar provision makes for easier cross-pollination mechanisms.

Other plants become still more economical, offering less food value in exchange for pollination, and in extreme cases mimicking the appearance and/or scent of the opposite sex of the pollinator. Some plants make traps to hold the pollinator while it is being dusted in appropriate places with pollen, after which it is released at a time suitable for cross-pollination. Plants may be adapted in a variety of ways to ensure cross-pollination. Insects may bear a variety of structures to carry pollen or to reach the nectar. It is not only the structure of plants and insects that is coadapted for pollination, but also the physiology of the plant. For example, much more nectar may be secreted at the time of day the pollinator flies. Many flowers have a self-pollination mechanism as a last resort, and the evolution of this is easy to imagine, the distortion of dying anthers or pistils can easily overcome a previous barrier to self-pollination.

In warm localities several plant species with similar flowers may flower in sequence throughout the year to provide a continuous source of nectar for a specialised pollinator. In other cases an individual plant, or even a single inflorescence (*Anguria*) may flower for a year or more.

The more elaborate pollinating mechanisms include various forms of traps, involving inclined hairs, liquid traps baited with sugar, and sliding traps.

The dusting of insects with pollen while they are temporarily restrained seems to have occurred fairly early in the evolution of plants. The family Aristolochiaceae trap small flies in this way. The flies are kept in the flower for a day or two after pollination until the pollen is shed, when the hairs barring the exit wither and flies can escape, to pollinate another *Aristolochia* flower. Some of the liquid traps (for instance, those seen in the orchid *Coryanthes*) drown the insect which, in order to effect pollination, must have visited a male flower first. Pollinators are drowned in similar traps to those used by pitcher plants to catch their prey.

Members of some families, such as Cucurbitaceae and Ericaceae, are pollinated by bees, and the nectar offered may be protected by the bell-shaped flower and its method of suspension. Flowers more specialised for pollination by bees occur in the Rosidae, the pea-flower being typical of bee-pollinated plants, in which the flower must be opened by the insect. Similar opening mechanisms are achieved in different ways by Labiatae and Scrophulariaceae.

Other plants have specialised in pollination by insects such as flies, moths and butterflies. Certain types of flower are characteristic of certain pollinators; for example, long nectar tubes for long-tongued pollinators, evil smelling plants for flies, strong, sweet-smelling night flowers for moths, landing stages for bees and butterflies. In a few cases the coadaptation is so great that there is probably only one species able to effect pollination. In many cases, however, there is a fail-safe arrangement whereby although a plant is particularly attractive to one pollinator, it will be visited by, and can be pollinated by other insects. Minor pollinators are probably important for the survival of species during periods when their regular pollinators are rare or absent. For instance, broad beans, usually pollinated by bees are visited regularly also by hawk moths, a much larger insect with a different method of obtaining nectar.

Plants that colonise temporary habitats have unspecialised pollination systems, as the new colonist is unlikely to find a suitable specialised pollinator already established there. Where specialised partners are required, they must travel together. The unspecialised nature of the pollination mechanism of weeds and other gypsies is an essential part of their make-up, maintained by selection just as much as the narrow corolla tubes of flowers.

Insects may be modified in a variety of ways that facilitate pollination. The female of the moth that pollinates *Yucca* has a prong on the maxillary palp to carry pollen to the stigma where she lays her eggs. The larvae eat some, but not all the ovules. There are points of similarity with one of the most extreme cases of interdependence of plants and insects, the pollination of figs, or caprification.

There is a vast and fascinating literature concerning pollination, of which *The principles of pollination ecology* by Faegri & van der Pijl (1979) is particularly readable.

Caprification

Figs are pollinated by 'fig insects', small wasps of the family Agaontidae structurally specialised for their role as pollinators. It seems likely that each species of fig is pollinated by its own species of wasp and that these insects feed only on the fig. Males spend their entire life inside one fruit, never venturing to the world outside. Fertilised females leave only in order to find another fig into which to lay their eggs. Similar-looking wasps belonging to related groups live in figs as parasites, without effecting

pollination. Janzen (1979) reviews the extensive literature concerning fig insects.

Extra-floral nectaries and ant-loving (myrmecophilous) plants

Many plants have extra-floral nectaries, that is, nectar secreting organs not associated with flowers or pollination. Unopened buds may also bear nectaries at the base of their sepals. These nectaries are usually for the benefit of ants which, in seeking nectar, may catch, disturb or deter browsing animals. Some plants produce conspicuous food bodies to attract ants. The ants may live in galls on the trees, as in the case of some species of *Acacia*, and the ants may feed also on the honeydew of coccids living in the galls. In extreme cases, as in *Macaranga triloba* (Euphorbiaceae) when the seedling is only 7·5–10·0 cm high, it starts to produce empty cavities in the stem and the stipules bear numerous fruit bodies. The isolated cavities may be each occupied by a queen ant after copulation. The first workers to be produced drive the females out of all the other cavities, so that eventually the tree is colonised by a population of ants descended from a single female.

Some trees growing in poor tropical soils have other roots growing in the middens of the aerial ants' nests constructed on their branches. While certain ants are largely predatory, others take honeydew from plant-feeding Homoptera. Buying protection by providing food bodies for ants probably also increases the likelihood of the plant being colonised by Homoptera. Where browsing animals are abundant, Homoptera may be the lesser evil.

Phoresy

The transport of pollen by pollinating insects is by far the most important phoretic aspect of plant/insect relationships. Ants specialised as seed gatherers have suitably modified mouthparts and associated bristles. It is not certain how much of the variation observed in plant seeds is concerned with insect transport. Seed gathering ants move large quantities of seed, but this may be of little advantage to many of the plants, except for some Ranunculaceae such as *Helleborus*, *Anemone* and perhaps *Ranunculus* and *Adonis*.

Insects regularly travel on floating trees, wind-blown seeds etc. Again the importance to the species concerned is problematical, but does speed local recolonisation. Colonisation is not just a matter of moving to new and distant areas. Populations are continuously being exterminated within the existing geographical range of a species, leaving uninhabited areas available for recolonisation. The first instar larvae of *Uroleucon taraxaci*, an aphid living in the rosette leaves of dandelions, are not uncommonly seen on wind-blown seed.

Pest pressure

Some of the earliest known fossil vascular plants (those found in the famous Rhynie Chert) show damage similar to that caused by biting insects today. Whether or not this 370 million year old damage is correctly interpreted, there is no reason to doubt that insects have been biting plants for hundreds of millions of years. Cockroaches and dragonflies appear in 300 million year old Upper Carboniferous rocks, together with giant club mosses, ferns and pteridosperms. Beetles occur in the Permian (about 250 million years ago) and flies and wasps in the late Jurassic and early Cretaceous (140 million years ago), together with flowering plants. Insects and plants have been coevolving for at least 300 million years and the relationship with flowering plants developed in the latter half of this time.

Antibiosis

Plants protect themselves from insects by sticky hairs, shiny leaves and stems, latex, repellent pigments, moats in connate leaves and a variety of other ways including the encouragement of aggressive insects (extra-floral nectaries and myrmecophilous plants, see above). Sticky hairs protect the young growth of many plants and a number of insects has exploited this glutinous habitat. It is a fairly safe place for a specialist to live, as the number of parasitic insects stuck to the young leaves often testifies, although there are predaceous bugs specialised for feeding on these insects. Related predators have taken to feeding on insects trapped by sundews. Insects living on smooth shiny leaves often have long tarsi, presumably to obtain a better grip. Aphids living on plants with sticky hairs usually have an elongate and/or hairy ultimate rostral segment. Insects on sticky plants may also walk very carefully, sometimes on short tarsi.

'Pest pressure' – the avoidance of natural enemies – seems to have been a significant factor in plant evolution, and one associated with the development of spines, stinging hairs and poisonous secondary plant substances. Some insects manage not only to cope with the poisons, but to use them for their own defence. Other plants seek to minimise insect attack by restricting growth to a short season, but some insects form galls, thus lengthening the period of active growth and hence the availability of metabolic products, both to themselves and other insects (inquilines) which are 'cuckoos' in their galls.

Other antibiotic devices of plants causing problems for insects include a thick cuticle and latex. Insects adapt to this hostile environment by changes in structure, physiology or behaviour. Once having adapted to deal with a particular defensive device, an insect is often able to colonise other plants protected in similar ways. Those with the ability to live among the sticky hairs of one group of plants may be able to colonise other groups of plants also defended by sticky hairs.

Phytophagous insects using plant poisons to protect themselves have led to specialisation among predators, that can then only complete their development on certain prey. The differing fecundities of predators eating pests feeding on different species of plants is one of the problems of biological control.

The physiological state of the plant is important in its relationship with browsing insects. Many insects are saprophytic, feeding only on dead plants. Trees under physiological stress are attacked by insects which would not otherwise feed on them. Examples of this are the recent outbreaks of the wood wasp *Sirex* and woolly aphids, *Pineus*, on *Pinus*. This environmentally induced variation in the effectiveness of plant defences against insect attack has probably been important in the evolution of the relationship between plants and insects. A plant growing at the edge of its range may be particularly susceptible to the attacks of new pests. This may both help insects to acquire new hosts, and limit the geographical distribution of plants.

Chemical defences against mammals involve both poisons and hallucigens which disrupt the normal behaviour of the eater. These chemicals, not physiologically essential to the plant, are known as secondary plant substances; they include alkaloids, quinines, essential oils, glycosides and flavenoids. Some of the better known alkaloids are nicotine, caffeine, quinine, marijuana, mescaline, cocaine and strychnine and magnoline. Insects using secondary plant substances may limit their evolutionary potential by becoming dependent on them. In one group of butterflies that has acquired new host-plants, the adults have to seek out members of the original host-plant family to obtain materials absent from the larval diet and which are needed for the production of pheromones.

Camouflage, aposematic colouring and mimicry

Many phytophagous insects are inconspicuous, presumably to escape predators. Small green aphids under leaves, brown scale insects on twigs, stick and leaf insects, mantids and the twig-like caterpillars of the Geometridae are well-known examples of this sort of adaptation to life on plants. There are many other correlations between the appearance of plants and insects which may be a form of camouflage against organisms relying on touch or with vision different from our own. For instance, the pupal cases of the aleyrodid *Bemisia tabaci* which develop on hairy leaves usually bear spines, while those which develop on smooth leaves do not. The aphid *Cervaphis quercus* bears branched projections resembling the branched hairs of its host. Such an arrangement could be a useful defence against predators relying on touch to distinguish prey.

Many caterpillars and greenfly are well camouflaged, but other species are blatantly conspicuous. Many examples of brightly coloured but poisonous insects (aposematic or warning coloration) are known. The cinnabar moth with its striped orange and black caterpillar feeding on ragwort, or the bright yellow aphid, *Aphis nerii*, feeding on Asclepiadaceae and Apocynaceae in warmer parts of the world are examples.

One of the problems in analysing information of this sort is to appreciate the sense likely to be used by a would-be predator. Some predators are nocturnal and some feed by day. Some insects are conspicuous in one part of their range and well camouflaged elsewhere. The orange pupae of the seven-spotted ladybird beetle, *Coccinella septempunctata*, are conspicuous and regarded as an example of warning coloration in western Europe, but in the grasslands of the Middle East the orange pupae hang from grasses during the flowering season of the orange-flowered wild Calendula, and are quite inconspicuous. Warning coloration and camouflage are not necessarily alternative evolutionary mechanisms: a colour pattern can be camouflage in one place and warning coloration in another, or to two different organisms.

Certain female butterflies appear to inspect a prospective food-plant before laying their eggs, and to reject plants that already bear any number of eggs or larvae (see Chapter 12). Some plants bear glands and stipules respectively resembling the eggs and caterpillars of butterflies. This seems to be a case of plants deterring the mothers of potential browsers by deceit. Some butterflies appear to recognise the shape of the leaves of their host plants, and this may have led to plants escaping predators by producing differently shaped leaves.

Plant galls

One reaction by plants to insects feeding on them is for cell division to continue, and to form a gall. This benefits the plant by localising the feeding damage, and the insect by providing a continuous supply of nourishment. Some galls are about the same size and colour as the berries of plants growing in the same area. Thus, plants may benefit from gall formation not only by limiting the damage, but also by encouraging predation of the causative insects by birds. The production of galls in response to feeding is a specialised activity and few insects produce galls on more than one or a few closely related species of plants.

In general gall formation is thought to indicate an old relationship, both because related gall-forming insects usually have related hosts, and because the galls may also contain specialised insects other than the primary gall former. These inquilines and parasites may be variously adapted to their life in the gall.

Many galls are formed in response to a feeding stimulus, and it seems that the level of response sometimes falls below that of the background, so that some galls may start to grow before the insect arrives (see myrmecophilous plants above). If a suitable insect does not arrive, the gall stops growing.

Oaks harbour many different species of insects, particularly gall-making cynipid wasps. The different species form galls on different parts of the plant as well as on different species of oaks. There is a similar situation with the less conspicuous gall midges on birch. It seems easier to partition a habitat than to acquire a new one. We must always bear in mind, however, that it is easier for the observer to detect the former situation. It is difficult to find satisfactory weighting mechanisms for these analyses, and insects may not confine themselves to just one species of host, but may form galls on several related hosts.

There would seem to be less nutritional advantage in forming a gall on a herb which is growing or senescing for most of its life, than on part of a tree which may be physiologically inactive longer. Physical protection may be relatively more important on herbs.

The coevolution of phytophagous insects and plants has generally not resulted in great structural adaptation, but has led to considerable host plant specificity, with related insects tending to feed on related plants. Sometimes genera and even families of insects live only or mostly on one family of plants. There are, however, many examples of taxonomically discontinuous host plant associations. For instance, mealy cabbage aphids and the caterpillars of cabbage white butterflies usually feed on cabbages and closely related plants, but also colonise the only distantly related *Pelargonium* (florists' 'geranium'). It may be significant that many Cruciferae, including cabbages and *Pelargonium*, produce a secondary plant substance, sinigrin.

Insectivorous plants

Carnivorous plants have developed in four major taxonomic groups. Their traps include snares, nooses, passive or active sticky surfaces, pitfall traps and spring traps. But despite these devices some phytophagous insects specialise in feeding on plants protected by sticky hairs; in turn these insects are preyed on by dicyphine bugs.

The digestion of trapped insects may be an important factor enabling insectivorous plants to colonise phosphorus-deficient soils. Carnivorous plants are most abundant in wet conditions, probably a reflection of both the low fertility of leached tropical soils and of bogs, and also the need for adequate water supply for external digestion.

Plant groups and their associated insects

Animals differ from plants in seeking food rather than making it for themselves, and this is reflected in the different ways the two groups have evolved. Plants of course are the ultimate food of all insects and the immediate food of many. There are also many more complex relationships. Predatory insects must lay their eggs, pupate, shelter and mate somewhere, and this is often on the food-plant of their prey. The most efficient way of finding prey may be to find its food-plant first. If the plant has protective structures against pests, such as sticky hairs, the predators which are beneficial to the plant must come to terms with the protective devices also. Data and critical reviews concerning the relationships between insects and plants can be found in van Emden (1973), Gilbert & Raven (1975), Hedberg (1979) and Mound & Waloff (1978).

Fungi and insects

Fungi and insects are associated in several different ways. Many insects, including beetles and flies are fungivorous. Some fungus gnats, for instance, are pests of mushrooms, some thrips and other small insects feed on fungus spores. The fruiting bodies of fungi often contain carnivorous insects preying on the fungivorous species.

Phytopathogenic fungi may act as physiological sinks and some insects feed at these sites. The aphid *Cinara cronartii* is commonly associated with pine rust lesions in North America. Several sooty moulds and other fungi grow on honeydew, the sugary excretion of many bugs. These fungi are utilising plant sugar made available by the insects.

Numerous insects are classed as 'detritus-feeders'. The decomposing plant material upon which they feed probably has a large bacterial and fungal content, but a distinction between the direct and indirect nutritional value of these organisms is seldom made. The majority of insects live in a world of living, moribund and dead plant material and wood has been important in the evolution of several groups. However, lignin and cellulose are resistant to degradation by insect enzymes and wood is deficient in sterols and vitamins which insects cannot synthesise. The original wood-living beetles probably fed on decaying wood, already part-digested by fungi. This led to the symbiotic associations between beetles and fungi, but which are always extra-cellular. Some groups of beetles live only on the fungus growing in their tunnels and not on the wood itself.

Most female wood-wasps have pouches containing fungus cells at the base of their ovipositor, some of which are deposited with the egg. These fungal cells give rise to a mycelium which permeates the wood and breaks down cellulose and lignin. The larvae thus feed on the wood modified by fungus. Young wood-wasp larvae are fungus-free but larger female larvae acquire fungus which is transferred to the base of the ovipositor of the adult female as she emerges from the pupa.

Termites use fungus gardens more for air conditioning than for food, although the fungi are eaten occasionally. The fungus may supply vitamins or other growth substances only required in small amounts. Ants of the tribe Attini culture saprophytic fungi which provide their sole diet. The fungi are cultivated on plant material brought into the nests by ants, and utilise the nitrogen from the ants' faecal material. How the ants manage to maintain pure monocultures of the fungi is not clear, unless it is just by assiduous weeding. Batra (1979) gives accounts of many relationships between insects and fungi.

Passage fungi

The spores of many coprophilous fungi must pass through the gut of a warm-blooded animal in order to germinate, but other species are associated with cold-blooded animals, particularly frogs. In some the conidia may be ingested by, or adhere to the body surface of insects, but in either case no further development occurs until the insect is eaten by a frog. The basidiospores of some Phallales may require passage through insects in order to complete their development.

The spores of certain fungi, including ergot, are distributed by insects attracted to a sugary secretion. Such insects also visit other sweet substances like honeydew or manna. Honeydew feeding is common in insects whose larvae eat the honeydew-producing Homoptera.

Laboulbeniales (Laboulbenomycetes) of the Ascomycotina are ectosymbiotic upon insects, mostly beetles, and also some other arthropods. Over 1500 species have been described. Infection is by ascospores that germinate to form a small thallus anchored by a small foot. In some species penetrant organs arise from the foot and reach the soft tissues, but the majority of species do not make any evident penetration. Many species seem to be highly specific, not only to a species of insect but also a particular sex or position on the insect.

Septobasidium

Fungi of the genus *Septobasidium* have domesticated scale insects. The lichen-like perennial fungus thallus occurs on trees and contains a labyrinth of chambers and tunnels in which the scale insects live. The chambers mostly contain single insects while the tunnels lead to the surface. Many of the scale insects live quite normally in these chambers, feeding on the bark of the tree. When the young scale insects are produced, some are infected by basidiospores produced on the surface of the thallus. A mycelium forms within the haemocoel of the infected scale insects. After a few weeks hyphae emerge from the insect and surround it to form a new thallus for an isolated insect or to anastomose with the hyphae of an existing thallus. Although tethered, the scale insect is free to move within its chamber and apparently feeds normally. Infected females produce no young but live longer than uninfected, fertile females. The fungus obtains nourishment from the infected scale insect and protects the uninfected females which will give rise to the next generation.

Various yeast and bacteria-like organisms are found regularly in insects and some have not been recognised anywhere else. Sometimes the organisms are free in the gut or the haemocoel and sometimes they are aggregated into 'symbiotic organs' or mycetomes. The organisms mostly occur in Homoptera and those Coleoptera feeding on low nitrogen or low vitamin diets (for instance plant sap). Work with antibiotics suggest that the organisms aid the nutrition of their host. Buchner (1965) gives an account of the endosymbiosis of animals with plant microorganisms, and Houk & Griffiths (1980) survey the intracellular symbiotes of Homoptera.

Algae and insects

Various groups of insects including Collembola (springtails), Psocoptera (book lice), some aquatic larvae of may-flies, the beetle family Haliplidae and some mosquito larvae feed on algae. It has been suggested (see review by Gerson, 1974) that the lichen association of algae–fungi developed to protect algae from predation by insects.

Lichens and insects

Lichens have both physical and chemical defences. The gelatinous covering of some species renders them almost immune to insect attack and lichens contain various chemicals known to inhibit insect development.

Lichens are common on tropical mountains where they may be associated with various insects. *Hospitalitermes* lives exclusively on crustaceous lichens and some Psocoptera and Lepidoptera also live on lichens. Lichen soredia may be distributed by Collembola and ants, and by lacewing larvae which camouflage themselves with small pieces of lichen. Some weevils are structurally modified to accommodate plants growing on them and may produce a sticky solution that promotes growth. Various insects are camouflaged to rest inconspicuously on lichen covered surfaces. Gerson & Seaward (1977) review lichen-invertebrate associations.

Bryophytes and insects

Collembola and various hemimetabolous insects including crickets, aphids and tingid bugs feed on mosses. Some species of aphid are always found associated with mosses even though most are associated with angiosperms. The Peloridae, a hemipterous family of uncertain affinities and biology, live among moss in the southern hemisphere. The species of *Gymnopholus* (*Symbiopholus*) weevils from New Guinea are specially modified to house and nurture plants, including liverworts and mosses. This they do by incorporating these bryophytes within grooves and pits on the elytra. Within these pits the plants develop, promoted by substances secreted by the insects. Many Splacnaceae grow on animal remains and droppings. Their spores are sticky and adhere in masses to the bodies of coprophilous flies apparently attracted to the capsules by secretions from special glands and/or by the coloured apophysis of the sporangium. The coprophilous flies are likely to move on to other habitats suitable for the moss.

Some dipterous larvae are camouflaged to resemble the aquatic mosses among which they live. Gerson (1969) reviews moss-arthropod associations.

Ferns and insects

Ferns contain many of the secondary plant substances protecting other plants but lack alkaloids and glucosinolates, and synthesise fewer substances within each class of chemicals than do angiosperms.

Of the insects recorded from ferns, 28 per cent are Homoptera, particularly aphids; 23 per cent are beetles, particularly weevils; 18 per cent are Lepidoptera, particularly noctuids (e.g. *Callopistria* and *Papipeira*) and 12 per cent are Heteroptera, particularly mirids (capsids). The flies and Hymenoptera are represented by only 7 and 6 per cent, respectively. These Diptera and Hymenoptera consist mostly of specialist fern-feeders and include the Cephidae, one of the most primitive families of Hymenop-tera. Most of the aphids found on ferns live only on ferns, and the mirid subfamily Bryocorinae is also associated with ferns. The 32 species of aphids belong to groups mostly associated with higher plants. Most of the whiteflies found on ferns live only on ferns. In contrast, most of the Coccoidea (scale insects and mealy bugs) found on ferns also feed on angiosperms.

Statistics on host-plant associations are always somewhat suspect, as host-plant data are recorded more carefully in some groups of insects than others. Several genera of stick-insects (for instance, *Carausius*, *Myronides* and *Neopromachus*) have only been collected from ferns in the tropics, but this sort of collection data rarely gets into host-plant catalogues. It seems likely that there are about 50 species of Phasmidae living only on ferns (*teste* Allan Harman). The adults in a number of genera of weevils (*Syagrius*, *Megacolobus*, *Rystheus*) feed on ferns but their larval biology is poorly known. The adults of a few other weevils with a very wide host range, also feed on ferns. The larvae of some sawflies bore into fern stems and the spores of certain tropical ferns are dispersed by symbiotic ants. Cooper-Driver (1978) and Gerson (1979) have reviewed fern-insect associations.

Cycads and insects

Pollen-eating beetles visit the male cones of the cycads *Encephalantos* and *Zamia*, and small bees collect the pollen of *Macrozamia*. Male and female cones emit strong smells, and beetles visit both, laying their eggs in the female cones. The beetles are probably important in transporting wind-blown pollen from the outside of the female cone to the micropile.

Gymnosperms and insects

The conifer-associated fauna contains primitive members of many insect groups, including coccids and aphids among the Hemiptera; however, these species may be derived from dictoyledon-feeding ancestors. The Xyelidae, whose present-day representatives live on conifers, are the earliest Hymenoptera in the fossil record.

The single sterile female flower in the male inflorescence of Gnetales exudes a sweet drop which attracts insects, suggesting that *Ephedra* and *Welwitschia* are descended from hermaphrodite ancestors which later became unisexual. The fruity odour of the male flowers of *Ephedra* suggests insect attraction. Insects associated with Gnetales are not particularly primitive, but are more characteristic of the fauna of xerophytes, suggesting that the association has been acquired recently.

Angiosperms and insects

The development of different pollination mechanisms was very important in the evolution of flowering plants. Many lower flowering plants, including Magnoliaceae, Degeneriaceae, Annonaceae, Eupomatiaceae, Euyralaceae and Ranunculaceae are pollinated by beetles, and it is supposed that beetles were feeding on spores before the appearance of flowering plants. Many of the higher plants are structurally adapted for pollination by insects with specialised structures or behaviour. Families such as Leguminosae have flowers specialised for bee pollination while related families like Rosaceae have kept a less specialised flower and consequently a greater variety of pollinating insects. The balance of advantage lies between ensuring some visitors, many of which will have the wrong pollen, and possibly not being visited at all in years when the specific pollinator is scarce. The change from insect pollination to wind pollination (anemophily) and probably *vice versa*, has happened many times. Plant families may contain both wind and insect pollinated genera; for instance, Polygonaceae with amenophilous *Rumex* and entomophilous *Rheum* and *Polygonum*; Saliceae with anemophilous *Populus* and entomophilous *Salix*. Of course, entomophily and anemophily are not mutually exclusive and in an apparently anemophilous genus like *Plantago* it may be a matter of how much pollen is blown by the wind and how much is transported by insects.

Any form of specialisation can either curtail further development or lead to other correlated changes. For instance, shedding the pollen before the stigma becomes receptive (protandry) facilitates cross-pollination. If, however, the pollination is to be enacted by pollen-collecting insects, the older flowers with receptive stigma but no pollen must deceive the pollinator into visiting them. Either the anthers can remain fresh and visually attractive, or the stigma can mimic the appearance or scent of the stamens. Any special means of attraction may have an unbalancing effect on other floral functions, requiring modification there. This is the way flowers evolved, and is the reason why, in general, flower structure is the key to the classification of the major groups of flowering plants.

Broad-leaved trees and insects Apart from the many phytophagous insect larvae feeding on various parts of trees, certain adult beetles (for instance, species of Synteliidae, Helotidae) are regularly found at tree sap. Wood-boring and bark beetles are well known and the association of wood-using insects with fungi has been discussed above. Many phytophagous insects living on the leaves of trees, move to other plants or other parts of the plant during the summer, or aestivate. The mature leaves often present problems to insects that thrive on the young growth during the spring and the senescing leaves during the autumn.

Palms and insects The palms, Palmae and Pandanaceae (screw pines), have a characteristic insect fauna, even in groups like Phasmidae which are not usually host specific. A characteristic aphid fauna is to be expected but the whitefly (Aleyrodidae) and coccid (Halimococcidae and Phoenicoccidae) fauna also consists of species found only on these plants. There is also an enormous fauna of beetles and moths associated with palms, but fewer flies and Hymenoptera. Palms seem to represent a distinct habitat to which members of many families of insects have become adapted. Palms are basically pollinated by pollen-feeding beetles. Lepesme (1947) documents their arthropod fauna.

Grasses and insects While many phytophagous insects are confined to Gramineae, fewer insects seem specific to particular genera of grasses than to genera of dicotyledons. This could be partly an artifact resulting from the inability of entomologists to identify grasses, but grasses may rely more on physical and less on chemical defences than many other plants. Bamboos generally have their own insect fauna, partly because of the restricted geographical range.

Herbs and insects The smaller size, continuous growth during their shorter life, and the phylogenetic youth of herbs have influenced their relationships with insects, as have their often specialised habitats, such as aquatic ones. Many herbs colonise temporary habitats and are thus dependent on generalised pollinators. Their annual or biennial nature precludes some defensive strategies, such as periodic fruiting. Herbs are commoner in temperate regions and deserts than in the forested warmer and wetter parts of the world. Insects sucking the sap of plants that are present only as seed for part of the year must develop a corresponding resting stage, or an appropriate change of behaviour. Just as some beetles are taken regularly at sap on trees, others regularly take sap from herbs. The common ladybird beetle, *Propylea 14-punctata*, feeds at the damaged flower heads of *Cirsium arvense* in August, as do red ants and some other insects.

Aquatic plants and insects The true water beetles (Dytiscidae) lay single eggs in the stems of water plants but pupate on the land. Whirligig beetles (Gyrinidae) lay their eggs and also pupate on water plants. Larvae of the mosquito *Taeniorhynchus* live in swamps and obtain oxygen by inserting their modified siphons into the roots of aquatic plants. Larvae of some ephydrid flies occur in the stems of water plants.

The aquatic larvae of some flies are strongly modified to resemble the water plants amongst which they live: concealment from predators is thus more important than structural adaptation to the plant.

Epiphytes, parasites and insects Epiphytic and parasitic plants have their own insect fauna, for instance the aphid genus *Tuberaphis* on Loranthaceae in eastern Asia. There are particular species of insects associated with mistletoe in Europe. Epiphytes are particularly common in the rain forests of South America but their insect fauna is only poorly known. Algae are common on the branches of many trees and are fed upon by Collembola (springtails) and Psocoptera (book- and bark-lice).

Xerophytes and insects Xerophytes have a special place in the evolution of insect/plant relationships. The first angiosperms were probably adapted to dry conditions and to disturbed areas in moister places, the flowers being attractive to insects already visiting the cycad-like Bennettitales. Insects feeding on xerophytes may have shorter appendages than relatives living on plants from moister areas, or different tarsi, but this is an indirect effect due to their subterranean way of life, or greater need to crawl over rocks, rather than an adaptation to life on xerophytes as such.

Insect groups and their associated plants

Life on plants has affected insects in a variety of ways. The ecological aspects of diversity of insects on different plants have been analysed by Southwood (1961) and Lawton (1978).

Apterygota and plants

Thysanura (bristle-tails) and Protura are found in soil, rotting wood, peat, turf, under stones or the leaf deposits of forest floors, and have little direct interaction with living plants. The Collembola are also basically soil dwelling, where they feed on detritus, fungi and algae. Some species commonly occur on the branches of trees, probably also feeding on algae. Pollen constitutes part, at least, of the diet of some Collembola, and they have also been observed feeding on wind-blown pollen on snow.

The most important contribution of Collembola to the evolution of plants is probably their persistent faecal pellets that play a part in soil formation. These faeces of algal-feeding Collembola not only improve soil structure, but also help retain nutrients in the upper layers of the soil.

Orthopteroid insects and plants

Among the orthopteroid orders the cockroaches and crickets are omnivorous, the stick- and leaf-insects, short-horned grasshoppers including locusts, and the largest subfamily of bushcrickets are phytophagous. Some of the other subfamilies of bushcrickets are omnivorous and others are carnivorous, as are the mantids. It appears that the phytophagous habit appeared early in the evolution of the group and is stable. Host-plant specificity is not well documented in orthopteroids, but there is evidence of seedling resistance in some Gramineae. Many stick-insects have a wide host-plant range on dicotyledons but some genera occur only on ferns, and others only on palms and screw-pines. Phasmids are well camouflaged and the group was present by the Permian, long before birds and mammals. Presumably their arboreal camouflage was protection from amphibians and reptiles. Mantids also resemble plants in order to catch other insects.

Hemiptera and plants

Many Heteroptera (capsids etc.) are predaceous on other insects but some families contain only phytophagous species. Other groups contain both phytophagous and predatory species, and the change from sap-sucking to blood-sucking has evidently happened many times. The individuals of some species may be predominantly phytophagous at one time of the year and predaceous on other insects when these are abundant at other times. Most phytophagous Heteroptera feed on leaves and shoots, but the Hyocephalidae and Lygaeidae are predominantly seed-feeding, and the Cydnidae may be root-feeders. The Aradidae live in cracks in the bark of trees and feed on fungal mycelia. Some of the Heteroptera that live on plants protected by sticky hairs, or on the insects trapped by them, may be important in pollinating the plants.

All Homoptera (leafhoppers and aphids) are phytophagous and feed on sap. A few species live on cryptogams but the great majority occur on phanerogams. The few species living on bryophytes and pteridophytes, and most of those on gymnosperms seem to have originated from dicotyledon-feeding ancestors. Aphids are mostly a northern group and appear to have developed from Lauraceae-feeding ancestors. The psyllids seem to be of southern origin and to have colonised Leguminosae from Moraceae-feeding ancestors.

Thrips and plants

Many species of thrips live on green plants but others feed on fungal spores and hyphae, the latter being the primitive behaviour. Thrips appear to be pollinators of Calluna and Erica (Ericaceae) and may be alternative pollinators for cocoa, lucerne and phlox. As the pollinating thrips also feed on the ovules, they may only be significant as pollinators in years when the regular pollinators are scarce.

Neuropteroid insects and plants

The predaceous Neuroptera have little evident direct effect on plants but some are nocturnal pollinators of sunflowers.

Beetles and plants

Beetles are primitively predaceous, both as adults and larvae.

Ground beetles and their allies (Adephaga) are usually regarded as the most primitive beetles and are mostly active predators on other animals, but the Haliplidae feed on algae, and some Carabidae are phytophagous (Harpalus, Zabrius, Pterostichus, Amara). Most of the Staphylinoidea are saprophagous or fungivorous, a smaller number are insectivorous, and a few are parasitic. Many of the remaining beetle groups are associated with living or decomposing wood, with fungi or with dried fruits or grain.

Some beetles are phytophagous and some of the carnivorous species come to sap and/or eat pollen. Hippodamia convergens pollinates the species of composites on which its aphid prey feed.

Gymnopholus (Symbiopholus) (weevils) from New Guinea are specially modified to accommodate plants, including fungi, algae, lichens, liverworts and a moss living on them, and appear to secrete a plant growth promoting substance. Most of the plants growing on the wing cases of the beetles also grow on nearby leaves and bark, and the value to the beetle seems to be camouflage.

Beetle-pollinated flowers tend to have few visual attractions, no special shape, depth effect or nectar guides. The flowers or inflorescences are generally large, flat, cylindrical or shallow bowls of easy access, with the sexual organs exposed but the ovaries protected. Their odour is described as fruity or aminoid. Magnoliaceae, Annonaceae and *Lithocarpus* with smelly, sticky pollen have beetle-pollinated flowers. The flat and smelly inflorescences of Umbelliferae and *Sorbus* are often also pollinated by beetles. Some carrion-beetles pollinate strong-smelling Araceae. A few beetles, such as *Nemognathus*, have long maxillae to reach nectar in the flowers they pollinate.

Butterflies, moths and plants

The caterpillars of butterflies and moths are well known as defoliators of plants but it is mostly only members of the more advanced groups that behave in this way. The most primitive Lepidoptera are detritus feeders and the larvae of many other groups mine the leaves or wood of living plants. Adults of the most primitive Lepidoptera have chewing mouthparts and some of these Micropterygidae eat pollen from *Caltha* and *Ranunculus* (Ranunculaceae).

The pollination of flowers by butterflies and moths is probably more important in the tropics than the temperate regions. Moth-pollinated blossom tends to be nocturnal, have long spurs and be strongly scented. *Cestrum nocturnum* is called Dama de noche because it is unobtrusive during the day, agreeable at night and disgusting in the morning. Night flowering plants often close for the day and open again the following evening. This serves to protect pollen in dry places and has developed into traps to restrain pollinating insects to ensure their thorough coating with pollen. The opening mechanism of the flower may be correlated with the flight period of a specific pollinator. *Gardenia* (Rubiaceae), *Brugmansia* (Solanaceae) and *Lonicera periclymenum* (Caprifoliaceae) are examples of strongly scented flowers pollinated by moths. Some Liliaceae and Amaryllidaceae are also pollinated by moths. Certain species of *Yucca* apparently have specific pollinating moths, the larvae of which eat the ovules of the flower pollinated by their mother. However, as some *Yucca* can set viable seed when grown outside the range of their apparently specific pollinator, there is probably also a 'fail-safe' pollinating mechanism involved.

Butterfly-pollinated blooms tend to lack odour and to have a narrow tube with flat rim (*Lantana*, *Buddleia*), the flowers arranged in a conspicuous dense mass. Several genera of orchids are pollinated by butterflies. Some tropical flowers are pollinated by actively flying, long-lived butterflies. This enables inconspicuous plants to survive at low densities, individual plants being up to 500 metres apart. The butterflies fly a regular daily route from plant to plant.

Sawflies, bees, ants, wasps and plants

The earliest known (Triassic) fossil Hymenoptera belong to the family Xyelidae. The larvae of present-day Xyelidae develop on pollen in the cones of conifers. Adult Xyelidae also eat pollen, and this was probably the stimulus that led to eggs being laid in conifer cones rather than on the ground.

The 'stem saw-flies' (Cephidae) are regarded as the most primitive living Hymenoptera. Their caterpillar-like larvae feed mainly on living plants. The host-plants of present-day Cephidae include many primitive plants such as ferns, horsetails, conifers, willows and birches. The larvae live in the stems and shoots of their hosts, and life in such enclosed habitats is thought to have led, through simple predation of other insects in the stem, to the ectoparasitic way of life typical of many groups of Hymenoptera. It is not uncommon in Hymenoptera to have pairs of closely related species, one of which is ectoparasitic on the other, and the second may complete its development on the gall caused by the feeding of the first, phytophagous species. The large sawfly family Tenthredinidae contains both gall-forming genera and free-living, leaf-feeding species. The orange coloured galls of *Pontania* on the leaves of willow are a conspicuous example from the north temperate region. This resemblance of galls to red berries on other plants at the same time of year may encourage birds to take the galls and protect the plants by reducing the insect population.

The Apocrita, in which the abdomen of the adult is constricted at the base, and the larvae are legless, contains the parasitic wasps, the social ants, bees and wasps, solitary bees and various groups of fossorial and predatory wasps.

Some of the major groups of 'Parasitica' such as the Ichneumoidea, Proctotrupoidea and Ceraphronoidea contain only species parasitic on other animals, mainly insects. The Chalcidoidea and Cynipoidea also contain groups that have returned to the phytophagous mode of life. These include some of the most remarkable plant/insect relationships.

Among the Chalcidoidea, the fig-insects, Agonidae, are all associated with, and many are apparently specific to, particular species of figs (see above). The males are wingless and bear little resemblance to their females. The fig-insects are essential to the development of the fig, which may also contain a unique parasitic fauna belonging to other families of Chalcidoidea.

Many Torymidae are parasites of gall-forming insects, the subfamily Idarninae being associated with fig-wasps as parasites or inquilines, while the members of other subfamilies are parasites of bees or inquilines. The family Eurytomidae also contains both phytophagous and parasitic species, some species which are usually parasites of stem-boring insects may complete their development on plant tissue alone if the egg is not laid near enough to an insect host. Certain *Tetramesa* are univoltine (single generation produced each year) and entirely parthenogenetic, while others alternate a spring wingless generation, in which a few males are produced, with a summer generation consisting entirely of winged females. Such alternation of generations seems much more common among phytophagous than predatory insects.

A number of plant-feeding Hymenoptera distort their host, and several groups are well-known gall formers. The galls of sawflies are caused by the action of enzymes secreted by females at the time of oviposition, while cynipoid galls result from the feeding of the larvae.

Bees (Apoidea) pollinate members of many families of plants. Several plant families have flowers specially modified for bee-pollination; for instance, Leguminosae (pea flowers), Labiatae (dead nettles), Scrophulariaceae (snap-dragons). Bees also 'work' less obviously specialised flowers and are often seen on the flat inflorescences of Umbelliferae, together with a variety of other insects, including fossorial Sphecoidea.

Flies and plants

Maggots, the legless larvae of flies are adapted to living in wet detritus, and the more primitive Diptera such as Tipulidae (craneflies) have larvae that feed on detritus, fungus and the roots of plants. Primitive Cecidomyidae (gall-midges) feed on fungi, others are free-living on plants while the more specialised cause plant galls. Some genera have carnivorous or even endoparasitic larvae. The soft bodied larvae of phytophagous flies often live concealed, being leaf miners (Agromyzidae, Cecidomyidae), gall-makers (Cecidomyidae, Trypetidae) or root-feeders (Tipulidae, Syrphidae, Micropezidae, Anthomyidae).

In temperate regions adult flies are active throughout a longer period of the year than bees and are also geographically widespread. They may be important in the pollination of plants that flower early or late (for example, ivy) and also in the arctic. Flies are important alternative pollinators, being active and almost ubiquitous in time and space. Except for phytophagous flies, large populations only develop when decaying organic material is available to the larvae.

Myophilous flowers are either simple, odourless, light but dull-coloured and patronised by nectar and pollen-feeding flies, or are smelly traps which retain the flies for a while before releasing them. Various orchids and other monocotyledons have 'sapromyophilous' blossom – they flower near the ground, smell and look like mushrooms and are pollinated by fungus gnats. These flies lay eggs in the flowers but the larvae do not develop. Sapromyophilous pollination depends on traps such as one-way bristles, slip-ways, see-saw petals, windows etc., similar to those of insectivorous plants. A number of 'highly evolved' groups of plants including Aristolochiaceae, Asclepiadaceae, Sterculiaceae, Rafflesiaceae, Hydnoraceae, Taccaceae, Araceae, Burmanniaceae and Orchidaceae are pollinated in this way.

Certain biting flies (for instance, the Ceratopogonidae) are also important pollinators for plants, such as *Theobroma*, Rubiaceae, the source of chocolate, and para-rubber, Euphorbiaceae. Some mosquitoes are specially adapted as pollinators, *Opistomyia elegans* and *Liponeura cinerascens* have mouthparts better adapted for feeding on nectar than blood.

Sometimes related species differ greatly in their effectiveness as pollinators. For instance, *Drosophila melanogaster* moved readily from plant to plant to pollinate 29 per cent of guayal flowers, while *D. hydei* pollinated only 7 per cent. Small behavioural differences are obviously very important in pollination and are thus a fertile evolutionary substrate.

Summary

The evolutionary history of flowering plants would have been different in the absence of insects for it is the power of flight that makes insects effective vectors. In turn, the great diversity of pollinating and phytophagous insects is an evolutionary response to the existence of flowering plants. Predation and parasitism have led to numerous evolutionary strategies. Plants have protected themselves with structural barricades and chemicals, have evaded would-be predators or have bought protection from other insects. Some fungi have parasitised insects, while others live on the honeydew produced by plant-feeding insects. Insects have used the defensive chemicals of plants to defend themselves, and have cultivated fungi for food and air conditioning.

Where the relationships between insects and plants are mutually beneficial, as in pollination, evident coadaptation can be seen. This is particularly obvious when the insects are using vision or scent within the human range. On the other hand, ants which protect and are fed by myrmecophilous plants are not noticeably different from other ants, but as we do not find the food bodies and extra-floral nectaries particularly alluring, we may not be using the right senses. The ants may recognise fruiting bodies by touch and taste, factors much influenced by size.

Mutual adaptation may develop between organisms, but commonly each member of a relationship adapts more or less completely to the most useful components of the other. The organism to which the relationship is most important is likely to alter most, and the adaptation will be in the appropriate organs or behaviour. The improved armour-plating of tanks and improved penetration of armour-piercing shells are an example of coevolution, but the relationship is not evident from examples preserved in military museums, particularly if all the tanks are in one place and all the shells in another. Most of the coevolution of plants and insects is of this sort, with the result that the functions of many structures and constituents of both plants and insects are not yet understood, and will not be until field and museum studies are carried out together.

References

Batra, L. R. (Ed.) 1979. *Insect-fungus symbiosis.* 276 pp. New York: Allanheld & John Wiley.

Buchner, P. 1965. *Endosymbiosis of animals with plant microorganisms.* 909 pp. New York: John Wiley.

Cooper-Driver, G. A. 1978. Insect-fern associations. *Entomologia Experimentalis et Applicata* **24**: 310–316.

Emden, H. F., van, (Ed.) 1973. Insect/plant relationships. *Symposium of the Royal Entomological Society of London* **6**: 1–213.

Faegri, K. & Pijl, L. van der, 1979. *The principles of pollination ecology.* 3rd Edition, 244 pp. Oxford: Pergamon Press.

Gerson, U. 1969. Moss-arthropod associations. *Bryologist* **72**: 495–500.

Gerson, U. 1974. The associations between algae and arthropods. *Revue Algologique* **11**: 18–41.

Gerson, U. 1979. The associations between pteridophytes and arthropods. *Fern Gazette* **12**: 28–45.

Gerson, U. & Seaward, M. R. D. 1977. Lichen-invertebrate associations. pp. 69–119, in Seaward, M. R. D. (Ed.) *Lichen ecology.* London: Academic Press.

Gilbert, L. E. & Raven, P. H. (Eds). 1975. *Co-evolution of animals and plants.* 246 pp. Austin: University of Texas Press.

Hedberg, I. (Ed.). 1979. Parasites as plant taxonomists. *Symbolae Botanicae Uppsalensis* **22**: 1–221.

Houk, E. J. & Griffiths, G. W. 1980. Intracellular symbiotes of the Homoptera. *Annual Revue of Entomology* **25**: 161–187.

Janzen, D. J. 1979. How to be a fig. *Annual Review of Ecology and Systematics* **10**: 13–15.

Lawton, J. H. 1978. Host-plant influences on insect diversity, the effects of time and space. *Symposium of the Royal Entomological Society of London* **9**: 105–125.

Lepesme, P. 1947. *Les insectes des palmier* pp. 1–903 Paris: Paul Lechevalier.

Mound, L. A. & Waloff, N. (Eds) 1978. Diversity of insect faunas. *Symposium of the Royal Entomological Society of London* **9**: 1–204.

Southwood, T. R. E. 1961. The number of insect species associated with various trees. *Journal of Animal Ecology* **30**: 1–8.

CHAPTER 15

Coevolution of digeneans and molluscs, with special reference to schistosomes and their intermediate hosts

C. A. Wright and V. R. Southgate

The Digenea or flukes are parasitic flatworms which have complex life-cycles involving at least two hosts. Most of the species for which the cycles are known have an adult sexually reproducing phase parasitic in a vertebrate and undergo a stage of asexual multiplication in a molluscan host. While the origins and affinities of the Digenea are still the subject of controversy and the precise nature of the 'asexual' reproduction in the intramolluscan phase requires further elucidation, there is now some general measure of agreement that the original hosts for the group were molluscs.

Arguments concerning the origins of the Digenea are hampered by the absence of fossil evidence and are clouded by such questionable concepts as 'parasitic degeneration'. This idea has frequently been invoked to account for the superficial morphological simplicity of some parasites but it ignores the great physiological specialisations that are necessary to succeed in a parasitic way of life. The 'simplicity' of some parasites is deceptive. Electron microscope and stereoscan microscope studies have revealed a huge array of complex sensory receptors and morphological specialisations not recognised at the light microscope level. Where parasitism started at a sufficiently early stage in evolution, selection would have favoured the necessary physiological specialisations rather than the morphological developments required for a free-living existence. The end product of such a process might be very difficult to distinguish from a form that had become secondarily modified by later adoption of the parasitic habit. It will become apparent in the course of this essay that in many respects parasites are not degenerate but are highly specialised.

The traditional view of digenean origins suggests that they evolved from free-living rhabdocoel turbellarians. How these turbellarians came to adopt a parasitic way of life is difficult to explain. Llewellyn (1965) put forward the suggestion that certain carnivorous species, which might have specialised in feeding upon the soft tissues of molluscs, eventually took to entering the wounds that they

had inflicted. This hypothesis sought to explain the absence of digeneans from the fast-moving cephalopod molluscs that might have been less susceptible to the attacks of turbellarians than would the more sedentary snails and bivalves. However, it does not account for their absence from the most sedentary of all molluscs, the chitons (Amphineura). An alternative hypothesis suggests that digeneans evolved from a common stock which gave rise also to the dicyemid Mesozoa (Wright, 1971). There are larval similarities between the two groups and the 'adult' dicyemids, which are endoparasites of cephalopods, might be considered as simple forms of the intramolluscan stages of digeneans. This theory suggests that there has been an association between the Digenea and the Mollusca going back in time to a point between the divergence of the Monoplacophora and Amphineura (neither of which groups has either dicyemid or digenean parasites) from the ancestral molluscan stock and the time when the Cephalopoda (with their dicyemids) separated, leaving the early digeneans to evolve with the Gastropoda (snails), Scaphopoda (tusk-shells) and Pelecypoda (bivalves) with which they are still associated. The intimacy of the mollusc–fluke association is greater than that between most adult flukes and their vertebrate hosts, for the digenea are tissue parasites of molluscs whereas the majority of the adults occur in the lumen of various vertebrate organs.

The digenean life-cycle

To facilitate discussion of particular aspects of the coevolution of digeneans and their molluscan hosts a brief summary of the stages in the parasite life-cycle is necessary (Fig. 15.1). The adult worms produce eggs that are voided from the body of the vertebrate host. Within the egg a larva develops which is known as a miracidium. This is the stage that is infective to the molluscan host. In some species the egg does not hatch until it has been ingested by an appropriate mollusc but in others the miracidium hatches in water and, propelled by its ciliated epidermis, seeks out

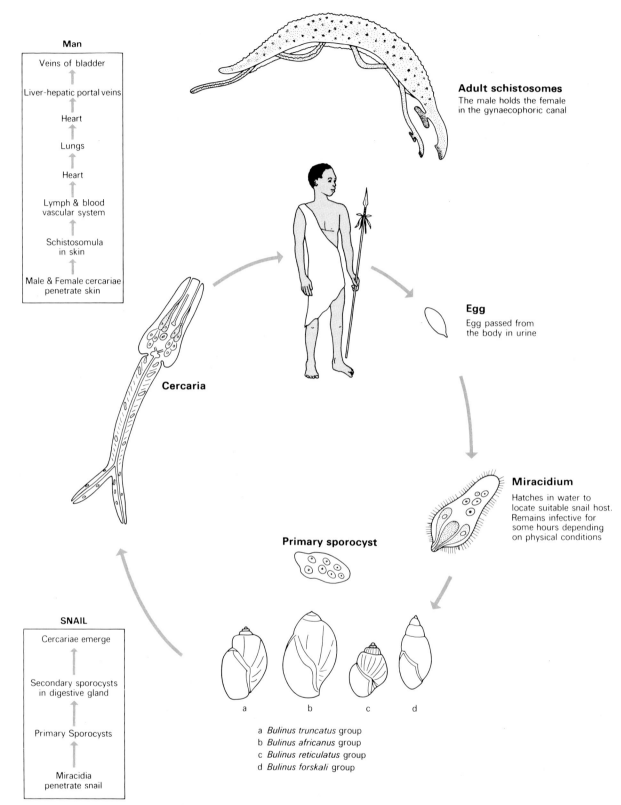

Man

Veins of bladder

↑

Liver-hepatic portal veins

↑

Heart

↑

Lungs

↑

Heart

↑

Lymph & blood
vascular system

↑

Schistosomula
in skin

↑

Male & Female cercariae
penetrate skin

Adult schistosomes
The male holds the female
in the gynaecophoric canal

Cercaria

Egg
Egg passed from
the body in urine

Miracidium

Hatches in water to
locate suitable snail host.
Remains infective for
some hours depending
on physical conditions

Primary sporocyst

SNAIL

Cercariae emerge

↑

Secondary sporocysts
in digestive gland

↑

Primary Sporocysts

↑

Miracidia
penetrate snail

a   *Bulinus truncatus* group
b   *Bulinus africanus* group
c   *Bulinus reticulatus* group
d   *Bulinus forskali* group

Figure 15.1   The life-cycle of *Schistosoma haematobium*.
This particular species uses any one of four species-complexes
of snail as the intermediate host.

and penetrates its host. It is important to note that, in general, the relationship between the parasite and the mollusc tends to be more specific than that between the parasite and definitive host and in some cases so specific that a particular species of parasite will only develop in a single species of host. Some parasites, however, appear to have evolved strategies that enable them to utilise a somewhat wider range of intermediate hosts.

Either at the time of entry into the mollusc or shortly afterwards the miracidium sloughs its ciliated epidermis and undergoes metamorphosis to become a mother sporocyst, a fairly simple sac-like structure within the lumen of which the next stage develops. The mother sporocyst usually remains near to the point of entry into the host but the subsequent stages, either daughter sporocysts which, like the mother lack a mouth, or rediae that have a muscular pharynx and a simple gut, migrate through the host tissues. The most common site in which these secondary intramolluscan stages settle is the digestive gland but some locate more specialised habitats such as the gonad or accessory reproductive glands and in heavy infections there is usually an overspill into tissues other than those which are the sites of choice. Within the daughter sporocysts and rediae develop the cercariae, the final larval stage that emerges from the mollusc. Cercariae are generally free-living stages with a body that sometimes has some features of the adult fluke and a muscular tail. Some cercariae are directly infective to the vertebrate host as in the various families of blood-flukes, others encyst on herbage or on the surface of other animals, and some actively enter another host before encystment and await ingestion by the final host. There are examples, particularly where the life-cycle has evolved away from its original dependence upon an aquatic environment, of the cercariae remaining within the mollusc until it is eaten by the final host.

Host-finding behaviour

The completion of any complex parasite life-cycle must depend upon successful encounters between infective stages and their hosts. In the life-cycles of those digeneans whose eggs do not hatch, the molluscan hosts are usually either coprophagous or detritus feeders. In the case of sedentary, filter-feeding hosts the parasite eggs tend to be of a size and density such that they will be readily taken up in the inhalant current of the mollusc. The snail hosts of some notocotylid flukes (mainly parasites of waterfowl) are algal browsers and the parasite eggs are equipped with long polar filaments which may become entangled in the epiphytic algae on which the snails feed. Success in entering the right mollusc in all of these cases is linked to the behavioural characteristics of the host and physical parameters of the egg.

In those species where the egg hatches in water and the miracidium is free-swimming, success in finding a host is correlated with the behavioural responses of the larva to environmental factors and to stimuli emanating from the host. These larvae are equipped with an impressive set of sensory receptors. Some miracidia have conspicuous, pigmented 'eye-spots' while others, which respond equally well to light, have none but possess complex structures in which an intracellular vacuole is occupied by lamellar stacks of ciliary membrane and it is believed that these function as photoreceptors (Brooker, 1972). Structures on the apical papillae of miracidia include simple ciliated nerve endings, believed to be touch receptors, and ciliated pit nerve endings which may be chemoreceptors. Many miracidia also have conspicuous lateral papillae whose structure suggests a function related to the larva's orientation with respect to gravity. However, there are many technical difficulties which prevent the direct recording of electrophysiological readings from receptor cells. Thus, it is not possible to assign definite function to most of the receptors, other than the obvious eye-spots.

Before the demonstration by electron microscopy of this assemblage of receptors, host-finding behaviour by miracidia had been the subject of experimental study. Due to the poor design of many experiments, some workers concluded that host-finding behaviour played little or no part in the completion of the parasite's life-cycle. Indeed, part of the original policy for the control of snail-borne diseases such as schistosomiasis was based upon the assumption that encounters between miracidia and snails were due purely to chance and as such were dependent upon the population density of the host snails and the larvae. This assumption led to the adoption of measures designed to reduce the snail population (for various reasons eradication had already been discarded as an unrealistic goal) and it is only recently, after more than two decades of these practices, that the results of the policy are being questioned.

A generalised pattern of miracidial host-finding behaviour has been outlined as follows. Initial responses to physical stimuli (for instance, light and gravity) bring the larvae into the broad area of the host environment. There follows a phase of apparently random movement which serves a distributive function and this in turn is followed by a klinokinetic response to stimuli emanating from the snails. Studies on a number of species from different parts of the world have tended to confirm this general pattern and have revealed some striking examples of refinements in particular details. The phototactic responses of the miracidia of the two common liver flukes of ungulates, *Fasciola hepatica* and *F. gigantica*, differ. *F. hepatica* develops in amphibious lymnaeid snails and the larvae respond positively to light stimuli, while *F. gigantica* uses fully aquatic species of the same snail family and the miracidia show a negative response to light. The miracidia of most blood-flukes of the genus *Schistosoma* show a negative geotactic response on hatching and rise to the water surface where the majority of their snail hosts occur, but in some areas where the snails live in deeper water the newly hatched larvae remain near to the bottom. The hosts

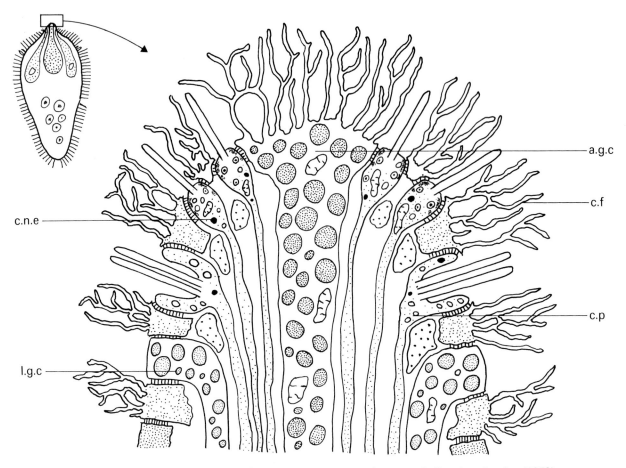

Figure 15.2 A sketch of a longitudinal electron microscope section through the miracidium of *Schistosoma* sp. showing apical and lateral glands, ciliated and ciliated pit sensory endings. The cytoplasmic folds of the surface layer form an anastomosing network. Based on Brooker (1972). Abbreviations: a.g.c – apical gland cell, c.f – cytoplasmic folds of surface layer, c.n.e – ciliated nerve ending, c.p – ciliated pit nerve ending, l.g.c. – lateral gland cell.

for the oriental blood-fluke *Schistosoma japonicum* are amphibious prosobranch snails of the genus *Oncomelania*. To avoid desiccation these snails move from the margins into the water during the heat of the day. The miracidia respond positively to light of any intensity at temperatures up to 15 °C but above this temperature they respond to reduced intensity of light and these behavioural changes correspond closely with those of their hosts, thus ensuring the maximum opportunity for successful contacts. One of the lung-flukes of the genus *Paragonimus* – *P. ohirai* – develops in estuarine prosobranchs of the genus *Assiminea*. In some localities three species of this host genus can be found and in laboratory experiments all three have been shown to be attractive to the parasite larvae although only one (*A. parasitologica*) is a really successful host. Where these three species occur together there is a vertical zonation of their respective habitats which appears to be determined by salinity differences resulting from effluent river water overlying the denser sea water. Laboratory experiments showed that the longevity and activity of

miracidia of *P. ohirai* are greatest at the salinity concentrations in which the good host species, *A. parasitologica*, occurs in nature.

The final phase of the host-finding sequence was the one originally most subject to debate. The development of techniques for measuring kinetic changes, particularly changes in angular velocity, and for estimating gradients of chemical stimuli have now established beyond doubt that miracidia exhibit a klinokinesis near their host. That is, they begin turning more rapidly and more frequently instead of swimming in straight paths. The turning behaviour tends to keep miracidia near the snails and this obviously increases the chance of direct contact. Once contact is made it seems likely that additional stimuli actually trigger the penetration process.

Much still remains to be done in this field, not only in the definition of the host substances involved but also in determining the role of the preceding phases of the behavioural sequence in releasing these later responses.

Penetration of the molluscan host

Entry into host tissues by larvae of those digenean species whose eggs only hatch after ingestion involves burrowing through the gut wall. Technical difficulties in observing this process have so far inhibited study of the subject. A few of the species with free-swimming miracidia probably also use this route after being drawn into the host's inhalant respiratory current but the majority achieve entry by penetration of the exposed surfaces of the mollusc. Miracidia are equipped with several anteriorly placed gland cells whose ducts lead forward to the apical papilla of the larva (Fig. 15.2). The miracidia of *Schistosoma* have an anterior centrally placed structure, originally referred to as a 'gut', which is flanked by two pyriform organs formerly called 'penetration glands'. Wajdi (1966) showed that during entry of miracidia of *S. mansoni* into snails of the genus *Biomphalaria* the contents of the 'gut' are completely discharged, thus revealing its true role in penetration. The contents of the 'penetration glands' could still be detected in larvae more than 24 hours after they had entered the snail, suggesting the possibility that they had some other function (see below).

The specificity of contact and subsequent penetration behaviour appears to vary in different host–parasite associations. Reports suggest that some miracidia make no attempt to attack and enter non-host species; others show some host restriction and will penetrate species related to their normal host even if these are not suitable for further larval development. Miracidia of *Schistosoma mansoni* appear to show little discrimination and have been seen (under experimental conditions) to attack almost any soft-bodied object including the tails of amphibian tadpoles and, in investigations on chemo-responses in this species using attractants incorporated into agar blocks, they have been found to leave marks on the agar where they attempted to penetrate.

Laboratory investigations can sometimes be misleading with respect to the field situation. Although it can be demonstrated in many host–parasite combinations that miracidia will penetrate snails in which no further development occurs, the chances of this happening in natural situations may be somewhat less. Indeed, biological control programmes have been suggested in which decoy snails might be 'seeded' into an environment with the idea that the 'decoy' snails will compete with 'natural hosts' and also attract miracidia. However, unless the decoy snails share exactly the same microhabitat as the natural hosts, they are unlikely to be as effective at attracting miracidia.

Kinoti (1971) showed that the tip of the apical papilla of the miracidium of *S. mattheei* was covered by a complex structure of anastomosing processes which he suggested might serve as the key in a lock and key mechanism with the host epidermis, thus determining the close specificity of this species for certain snails of the genus *Bulinus*. Later investigations on the fine structure of the apical papillae of other schistosome larvae showed that not only were these processes similar in a number of species using quite different snail hosts but that the processes appeared to be the edges of numerous small cups which might serve a suctorial function in attaching the larva to the host surface in order to facilitate penetration.

There is another possible explanation of these extensive membranous folds. The gland cells of the miracidia of *Schistosoma* spp. do not open directly to the exterior, but lead into the tegument covering the apical papilla. It is likely that the membrane-bound granules from the various gland cell types fuse with the surface plasma membrane of the apical papilla thereby releasing the contents onto the snail's epithelium. The membranous folds could be an adaptation for this type of secretory release in providing a considerable surface area of plasma membrane, ready for fusion with that of the secretory granules. Also, this type of release would ensure a high concentration of secretion being applied to a limited area, thus achieving the most effective action in breaking through the host epithelium.

Suggestions have been made that the composition of the molluscan body-surface mucus may play a part in determining host restriction. Differences in some chemical components of the mucus in closely related snail species have been demonstrated by chromatography but these have not been correlated with the normal host role for any of those species. Recently some correlation has been found between the types and composition of mucous gland cells in the skin of *Physa fontinalis* compared with those in lymnaeids and the entry of miracidia of *Fasciola hepatica* (McReath, personal communication). The fact that many miracidia do enter hosts that are not suitable for their further development suggests that, although surface interactions may exist, they are of less importance in determining the precise specificity of host restriction than are the subsequent intramolluscan interactions.

Host susceptibility versus parasite infectivity

Whether or not a parasite could develop in a particular snail host was at one time seen simply in terms of the suitability of the host's biochemical constitution as an environment in which development of the sporocyst and redial stages could take place. While this aspect is of great importance it is a simplification of a much more complex situation.

In 1952 Newton showed that miracidia of *Schistosoma mansoni* which penetrated into a strain of the snail *Biomphalaria glabrata* in which the parasite was known not to develop were rapidly encapsulated by host cells. This cellular reaction did not occur in strains of *B. glabrata* in which the parasite was successful and the process was regarded as a specific response to the parasite by the resistant hosts. Subsequent work showed that *B. glabrata* produced very similar, if not identical, responses to a variety of foreign bodies including pollen grains and polystyrene granules, thus revealing that the response to *S. mansoni* miracidia was not specific but merely a

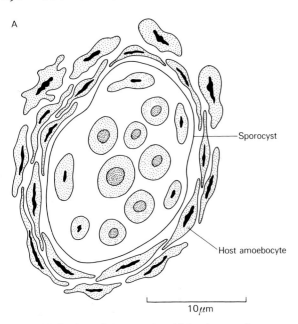

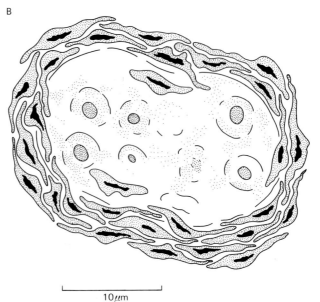

Sporocyst

Host amoebocyte

10 μm

10 μm

Figure 15.3 One of the ways in which a host snail reacts to invasion is to encapsulate the parasite by mobilising amoebocytes which surround and eventually break down the parasite. A. The early stages of encapsulation of a sporocyst. The plasma membrane of the sporocyst tegument is still intact and the germ cells are well organised. B. Later stage of encapsulation with thickening of host reaction and breakdown of the integrity of the sporocyst. Semi-diagrammatic.

manifestation of a generalised innate defence reaction. The obvious corollary of this conclusion was that the successful parasite must be one that was able to evade the innate host response.

Proof of the active, invasive role of the miracidium has been forthcoming from experiments using different species of *Schistosoma* and a single species of host snail and others using miracidia of interspecific hybrids of *Schistosoma* spp. (Wright & Southgate, 1976). In the first set of experiments one batch of host snails, *Bulinus crystallinus*, was exposed to miracidia of *Schistosoma haematobium* and another batch to larvae of *Schistosoma intercalatum*. *B. crystallinus* is a good host for *S. intercalatum* but not for *S. haematobium* and histological examination of snails 48 hours after exposure showed that the *S. intercalatum* parasites were developing normally whilst those of *S. haematobium* were encapsulated. The cellular responses to the incompatible parasites show that the snail's defence reaction is capable of recognising and reacting to foreign material while its lack of response to *S. intercalatum* suggests a failure to recognise this parasite as foreign. Whether this failure is due to blocking of the host recognition system by the parasite or by the parasite's adoption of some form of molecular camouflage or a combination of the two is discussed below.

In the second series of experiments, hybrid miracidia resulting from crosses between Cameroun strains of *S. haematobium* ♂ x *S. intercalatum* ♀ were used. These two species of parasite have mutually exclusive snail hosts. *S. haematobium* develops only in members of the *Bulinus truncatus* species complex and *S. intercalatum* in members of the *B. forskali* group. If the miracidia of one species enter the hosts of the other they are rapidly encapsulated but the hybrid larvae develop in both host groups. That the hybrid succeeds in the maternal hosts is perhaps not surprising but in this particular cross it shows greater infectivity to the paternal hosts than does *S. haematobium* itself. There is here, then, evidence that the hybrid has inherited from both parental species some factor or factors which enable it to evade the innate responses of both groups of snail hosts.

Molluscan cell responses

The responses to incompatible parasites in molluscs are not specific in the sense of vertebrate humoral antibodies which are produced by stimulation of the host's immune system by defined antigens (acquired immunity). The evidence strongly suggests that the cellular responses of molluscs are part of a generalised defence mechanism developed by specialisation of the amoebocytic system. Amoebocytes in molluscs perform a variety of functions. In the digestive gland they may assist in the uptake of food particles by pinocytosis and they are also active in certain phases of shell building and repair. The pinocytotic function is obviously easily adapted to 'phagocytosis' or the removal of foreign particles or damaged tissues within the body of the animal. The ingested particles may either be destroyed by intracellular degradation or, if the material is not amenable to this action, it may be sequestered within the cell and, if large numbers of particles are taken up, the carrier cell may migrate out of the host, thus removing the offending substance (diapedesis). Where the foreign material is too large for ingestion by individual amoebocytes they surround it in concentric layers and thus effectively encapsulate it, preventing contact with the rest of the host body. This is the response which is seen to incompatible digenean parasites (Fig. 15.3).

Much remains to be discovered concerning the nature of molluscan amoebocytes and the ways in which they function. In *Biomphalaria* spp. and *Bulinus* spp. they originate in a haemocytic organ located adjacent to the pericardium in the mantle of the animal, and are shed into a blood sinus. Whether there is more than one kind of amoebocyte in planorbid snails has been a subject for debate. Recent reports indicate that there are two, granulocytes which are involved in encapsulation reactions and hyalinocytes which are not. Other reports have suggested that there is only one kind of amoebocyte and that their appearance may change with age or the function which they are performing. In oysters three kinds of cells have been described, distinguishable by the staining reaction of their cytoplasmic granules – acidophilic, basophilic or agranular. Each of these cell types appears to have a distinctive role in wound repair but since, in the few pelecypods that have been studied, the mechanisms of wound repair differ quite markedly it would be dangerous to extrapolate from these molluscs to those that serve as hosts for schistosomes. Sminia, Pietersma & Scheerboom (1973) studied wound healing of the snail, *Lymnaea stagnalis*, and concluded that muscular contraction and the formation of thrombi of blood amoebocytes to form a large amoebocyte plug were the first stages in healing. About a day after wounding the amoebocytes differentiated into flattened cells, and approximately three days later these cells started producing collagen fibrils.

The means by which the host defence system is alerted to the presence of foreign material and how the amoebocytes are mobilised and find their target are still unknown. The rapidity with which histologically visible reactions occur in some incompatible host–parasite associations (within two hours) is a tribute to the efficiency of some host systems. That there must be recognition of foreign material before mobilisation of the response seems obvious, but whether this occurs through the direct medium of nervous perception at the point of entry or whether the intrusive object is encountered by an amoebocyte that can alert others by the release of a chemical signal is not known. There is evidence that once the system is activated the haemocytic organ undergoes enlargement and presumably increases its output of cells.

There has recently been a growing interest in the possibilities of recognition factors being secreted by haemocytes, circulating in the haemolymph, and binding to the surfaces of foreign materials. Once bound, it is thought that the foreign objects can thus be recognised and subsequently attacked by the host defence cells. Parish (1977) has proposed a generalised model for the defence mechanisms of invertebrates based on the production of recognition factors by haemocytes, and there is increasing evidence that in plants and animals glycosyl transferases have a role in cell–cell and cell–glycoprotein interactions. In his model Parish suggests that the recognition factors in invertebrate defence systems are composed of glycosyl transferases. Obviously, more work is required in this field to test Parish's model, and to identify specifically the recognition factors.

Parasite evasion of the snail host response

So far reference has been made to the encapsulation of incompatible parasites by host snail responses and the fact that successful parasites do not appear to provoke any host reaction. The impression may have been given that this is an 'all or nothing' situation but examination of different combinations of hosts and parasites reveals that there are variations in the timing of the host responses and these can throw some light on the mechanism of evasion by the parasite.

The basic immunological concept of recognition of 'self' and 'not self' in animals is now so well known as to require no further explanation. The question is how does a larval digenean manage to avoid recognition as 'not self' by a mollusc (or, alternatively, how does it pass itself off as 'self')? There are several possible strategies – it might be possible to secrete some substance that blocks the host's recognition system, or it might be possible to have an external covering (tegument) whose molecular structure closely resembles that of some host tissue so that it passes for 'self'. This second strategy could be achieved either by synthesising and secreting an appropriate layer of host material or by having a tegument capable of acting as a selective 'sponge', adsorbing materials of host origin.

Before examining the available evidence for and against these hypotheses it is important to note that two different types of ciliated epidermis have been described from digenean miracidia (Fig. 15.4). In both types the ciliated cells are separated from each other by unciliated ridges which are syncytial extensions of cells lying in the subdermal layer. In the first of the two types, exemplified by *Fasciola* and the echinostomes, the ciliated cells contain nuclei and are attached by septate desmosomes to the unciliated ridges. At the time of penetration into the snail host the ciliated plates are shed and the body becomes covered by a spreading of the unciliated ridges (Southgate, 1970). In the second type of epidermis the ciliated cells are, like the unciliated ridges, attached to cell bodies (containing the nuclei) in the subdermal layers. This structure occurs in schistosomes and the strigeids and the ciliated plates are not shed during penetration but are carried into the body of the host. Two hours after penetration the ciliated cells degenerate and become detached, and then the syncytial tegument of the young sporocyst develops from the unciliated ridges (Meuleman et al. 1978).

The differences between these two types of structure

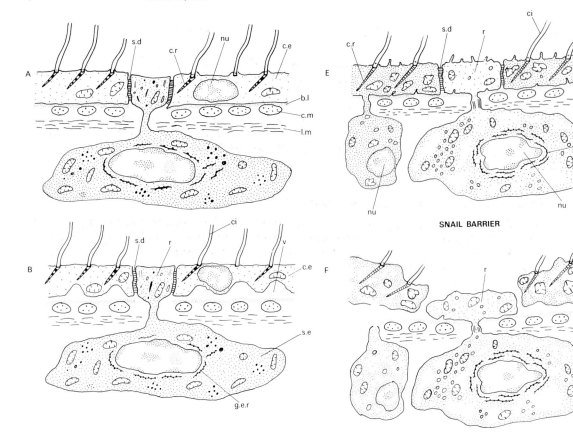

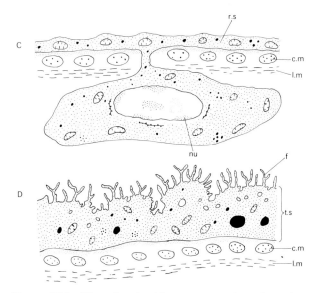

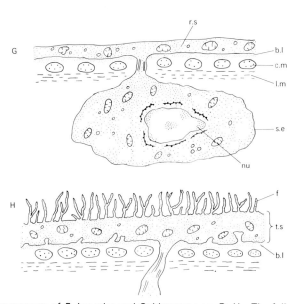

Figure 15.4 When the miracidium penetrates host-tissue the epidermal layer undergoes a transformation to adapt to an endoparasitic way of life and to form the sporocyst. A, E – Diagrams to show the relationship of the ridge layer with adjacent epidermal cells of the miracidia of *Fasciola hepatica* (A) and *Schistosoma* sp. (E). Note that the nuclei of the epidermal cells of *Schistosoma* sp. are sunken below the body wall muscles, acting as an 'anchor', whereas those of *F. hepatica* are above the body wall muscles. B, F – Early stages of the sloughing off of the epidermis. The epidermal cells of *F. hepatica* are cast before the miracidium penetrates the snail host, whereas those of *Schistosoma* sp. are cast inside the snail host. C, G – Diagrams to illustrate the origins of the sporocyst body wall from the subepidermal cells of the sporocysts of *F. hepatica* and *Schistosoma* sp. D, H – The fully developed tegument of *F. hepatica* and *Schistosoma* sp. Note the development of the folds of the surface plasma membrane. Based on electron-micrographs in Southgate (1970) and Meuleman *et al.* (1978). Abbreviations: b.l – basal lamina, c.e – ciliated epidermal cell, ci – cilium, c.m – circular muscle, c.r – rootlet of cilium, d.c.e – degenerating ciliated epidermal cell, f – folds of surface plasma membrane of sporocyst, g.e.r – granular endoplasmic reticulum, l.m – longitudinal muscle, nu – nucleus, r – ridge layer, r.s – ridge layer spreading over muscle layers, s.d – septate desmosome, s.e – subepidermal cell, t.s – tegument of sporocyst, v – vacuoles beneath epidermal cells.

and their consequences in initial presentation of the parasite to the host may call for different methods of evasion of the host response. It is possible to visualise the freshly secreted tegument of the fasciolid mother sporocyst either as a carrier of synthesised camouflage or as a sponge for immediate adsorbtion of host material. On the other hand the intact ciliated covering of the schistosome larva is unlikely to function in the same way. Indeed, the cast-off ciliated plates are immediately attacked by amoebocytes. Thus, it is not unreasonable to invoke the idea of some type of blocking of the immediate host recognition system, at least as an interim measure before camouflage can be adopted. A possible source of blocking agent might be in the secretions of the pyriform 'penetration glands' which histologically can be seen to be not fully discharged even 24 hours after entry of the miracidium.

There is, as yet, no direct evidence for a camouflage layer. Where the compatible relationship between host and parasite is very restricted there would be no great problem for the parasite to synthesise appropriately mimicked 'host' material. However, where the relationship is broader and the parasite develops in a wider range of hosts the synthetic repertoire of the parasite would have to be much greater and the adsorption of host material would appear to provide a more satisfactory solution. It is likely in this type of camouflage system that some larval digeneans would be incapable of adsorbing host antigens onto their surfaces simply because of the specificity of the receptor sites on the parasite surface. Of course, it is possible that the antigens in different species of snail vary in their receptor sites, and consequently in their probability of being adsorbed onto the parasite surface. The evidence from adult schistosomes in their mammalian hosts lends weight to the adsorption hypothesis (Smithers, Terry & Hockley, 1969). By a series of elegant experiments these authors showed that adult worms grown in mice or gerbils could be transferred to the hepatic system of rhesus monkeys. The parasites, after an initial set back, grew normally. However, worms transferred from mice to monkeys immunised against mouse erythrocytes were killed. These experiments suggested that the worms grown in mice had acquired mouse antigens on their surface.

Like most natural populations of animals, those of digenean parasites and their molluscan hosts tend to be heterogeneous, showing variation in various characters including, respectively, infectivity and susceptibility (Wright, 1971). One of the most marked pathological effects of parasitism in molluscs is a reduction in the fecundity of the hosts. Where a mollusc population is subjected to intense parasitism it is probable that there will be selection in favour of hosts which are more resistant to infection, provided that such resistance is not associated with other, deleterious effects which might reduce the biological fitness of the resistant individuals. Progressive evolution of resistance in the host population would eventually lead to elimination of the parasites unless they in turn evolved more effective mechanisms for evading the

host's defences. Indeed Clarke (1976) has suggested that cumulative host–parasite interactions of this kind may have played an important role in the origin and maintenance of many of the protein polymorphisms that occur in all animal groups.

Geographic distribution of hosts and parasites

It is with the results of this continuing process that the epidemiologist, concerned with problems of snail-borne diseases, is faced. The most extensively investigated intermediate host–parasite complexes are those involving some of the African schistosomes and their planorbid host genus *Bulinus*. The parasites include *Schistosoma haematobium* and *S. intercalatum*, both of which occur in Man; *S. haematobium* in the veins of the pelvic plexus, particularly those in the bladder wall, and *S. intercalatum* in the mesenteric venous drainage of the intestinal wall. *S. bovis* and *S. mattheei* are parasites of ungulates and both species are important causes of disease in domestic livestock. *S. margrebowiei* and *S. leiperi* are also parasites of ungulates but with a rather restricted host-range being found mainly in antelope which have a lot of water contact, especially the lechwe (*Kobus leche*). *S. margrebowiei* has been reported from Chad and Zaire, and both species occur in Botswana, East Caprivi and Zambia.

The genus *Bulinus* includes some thirty nominal species which are grouped in four species complexes – the *B. africanus* group which is confined to the Afrotropical region, the *B. forskali* group which is practically pan-African with representatives in Arabia and some of the Indian Ocean islands, the *B. truncatus/tropicus* complex, again pan-African with the tetraploid *B. truncatus* extending into the Middle East as far as Iran and northward onto some of the Mediterranean islands and the Iberian peninsula, and the *B. reticulatus* group with only two species, one having a patchy distribution in East, Central and Southern Africa and the other found only in somewhat isolated habitats in the Arabian peninsula. Schistosomes show varying levels of host-restriction in the range of bulinid species in which they are able to develop and the snails differ in their capacity to act as hosts for the various parasites. The degree of host-restriction is sometimes so marked that a given strain of parasite will only succeed in certain populations of a nominal species. Thus, while it is necessary to refer in general terms to the broad patterns of host-parasite relationships there are nearly always some local exceptions to complicate the picture.

The distributions of four species-complexes of host snails are shown in Figure 15.5. Distributions of four species of *Schistosoma* are shown in Figure 15.6.

Schistosoma haematobium is the most widespread of the parasites occurring throughout large areas of Africa and the Middle East. In the Afrotropical region the snail hosts belong to the *B. africanus* group but in the Mediterranean area and the Middle East the host is the tetraploid

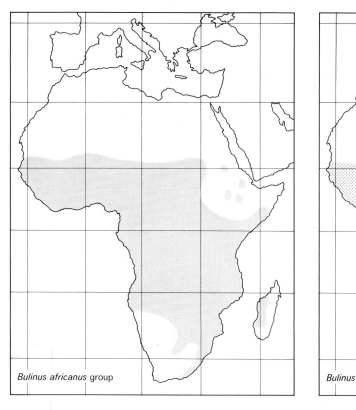

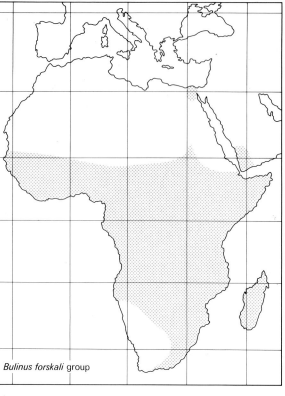

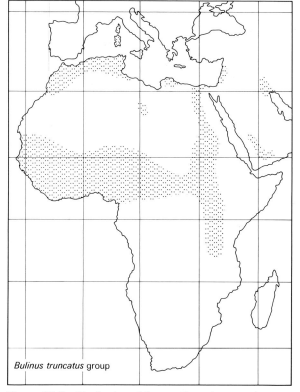

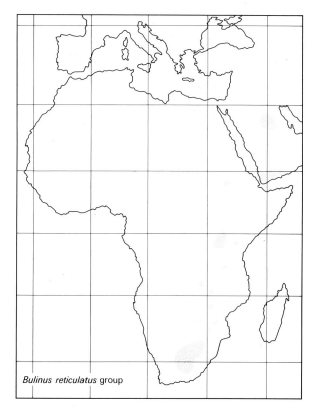

Figure 15.5 Distribution of four snail species-complexes used as intermediate hosts by *Schistosoma* throughout Africa.

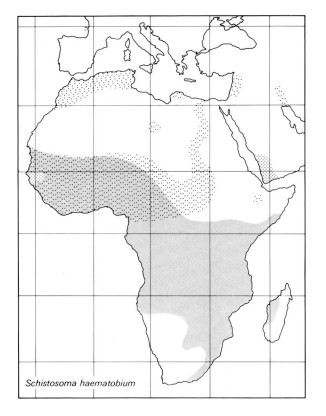

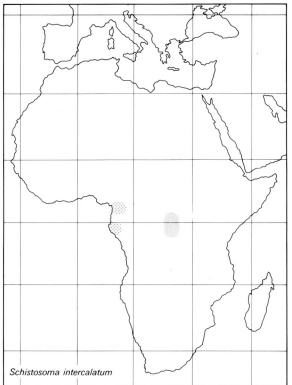

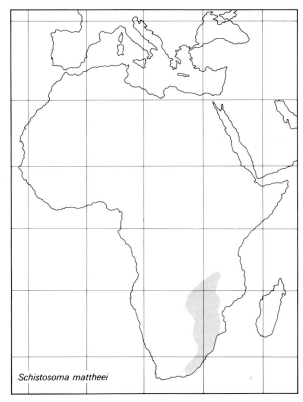

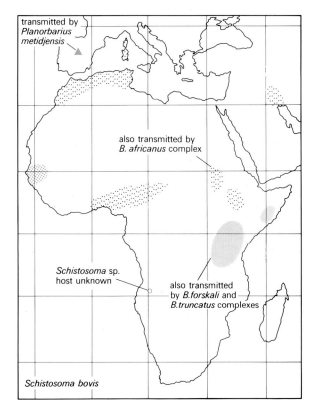

Figure 15.6 Distribution of four species of *Schistosoma* using different intermediate hosts throughout their range. The different shading identifies particular host-groups – see previous diagram.

B. truncatus. The parasites from the northern region cannot normally develop in *B. africanus* group snails and those from south of the Sahara cannot, as a rule, use *B. truncatus*. There are thus at least two kinds of *S. haematobium* with different snail hosts occupying different geographical areas. However, in West Africa both snail host groups and both parasites occur and in Ghana there appears to have been a break-down of the host restriction, with schistosomes from some areas able to develop in both groups of snails. Chu, Kpo & Klumpp (1978) have interpreted this as a simple mixing of the two parasites in the same human hosts but the evidence obtained by Wright & Knowles (1972) in a comparative study of *S. haematobium* strains, including two from Ghana, led Wright & Southgate (1976) to suggest that hybridisation between the two forms has occurred.

The origin of these two races of *S. haematobium* must be speculative. It is probable that the parasite which, in its adult stages, is almost entirely restricted to Man, evolved with the early hominoids in East Africa and this is also the most likely area from which radiation of the bulinid snails occurred. At some stage polyploidy developed in the *B. truncatus/tropicus* complex, possibly in the Ethiopian highland region where a complete series of diploids, tetraploids, hexaploids and octoploids is known to exist (Brown & Wright, 1972), and *S. haematobium* succeeded in establishing itself in the tetraploid *B. truncatus*. This snail colonised the Mediterranean region, where it appears better able to withstand the more sharply contrasting seasonal temperatures, and discrete northern and sub-Saharan races of the parasite became established. Tetraploid snails occur widely in East Africa to which area they have probably extended their range since the original separation of the two parasite stocks. Although many of these have been shown in laboratory experiments to be highly susceptible to *S. haematobium* of Mediterranean origin, that race of the parasite does not yet appear to have become established in the area. Should it be introduced in the future the possibility of hybridisation with the endemic *B. africanus*-borne race is a potential threat, for recent studies on some interspecific schistosome hybrids have shown that their infectivity for both intermediate and definitive hosts considerably exceeds that of either of their parental species (Wright & Southgate, 1976; Wright & Ross, 1980).

In addition to the broad division of *S. haematobium* into northern and southern races there are a number of other areas where the parasite is transmitted by neither *B. truncatus* nor *B. africanus* group snails but by members of the *B. forskali* group. In the Sahel region of West Africa the host is *B. senegalensis*, a species found mainly in temporary rain pools, in south-west Arabia it is *B. beccarii*, found in shallow, temporary, flowing streams, on Mauritius *B. cernicus* is the host, found in a variety of habitats, many of them temporary, and on Madagascar, *B. bavayi*, although not yet reported as naturally infected, has proved to be highly susceptible to infection in the laboratory. Also

in parts of Arabia *B. wrighti*, a member of the *B. reticulatus* group, transmits *S. haematobium*. All of these areas are peripheral to the main endemic regions of *S. haematobium* and all have some characteristics in common. The habitats tend to be rigorous and because of their temporary nature, the snails must be able to survive desiccation for prolonged periods; in some, such as the Sahel rain pools and some of the transmission foci on Mauritius, the water hardness is extremely low, down to 1 mg calcium per litre or less. By contrast *B. bavayi* on Aldabra lives in rain pools on coral limestone where the dissolved calcium at times reaches saturation and is well above the tolerance of most bulinid snails. With the exception of *B. senegalensis*, which is occasionally found in more permanent pools, all of these species occupy habitats in which no other bulinid snails occur. Under experimental conditions all of them have exceptionally wide ranges of host capacity for schistosomes and some carry very heavy parasite burdens in nature. The inference drawn from these facts is that these are species in which either the defensive responses are poorly developed or, alternatively, the host materials are particularly easily adsorbed by parasite larvae and, as a result, the hosts have been forced by parasite pressure into refuge habitats which have not, as yet, been adopted by more immunocompetent competing species (Wright, 1977). Preliminary observations on *B. wrighti* suggest that it may be the recognition mechanisms which are deficient for, although there is a reasonably well-developed haemocytic organ, this is the only known snail which can be infected with every 'terminal-spined-egg schistosome'; even 72 hours after exposure to miracidia of *Schistosoma mansoni* (a 'lateral-spined-egg' species which does not develop in *B. wrighti*) no sign of cellular reaction could be seen around the parasites.

Schistosoma intercalatum was described from an area near Kisangani (Stanleyville) in Zaire in 1934 and remained a somewhat obscure species until the mid-1960s when it became apparent that the parasite achieved a high prevalence in the Lower Guinea area (Cameroun and Gabon). In his original description Fisher (1934) reported that the parasite developed in *B. africanus* group snails but in the Lower Guinea area it was discovered that the host was *B. forskali*. Hypotheses were put forward suggesting that the parasite had been carried to the Lower Guinea area by migrant labourers and, by adopting the different snail host, had recently become established in the area. Since a parasite with similar characteristics had been reported from Gabon in 1923 it seemed more likely that the sudden discovery of the Lower Guinea foci had more to do with improvements in the diagnostic medical services than with any recent change in the intermediate hosts of *S. intercalatum*. A comparative study was undertaken in our laboratory and it was found that the Lower Guinea strain developed only in snails of the *B. forskali* group and the Zaire strain only in those of the *B. africanus* group (Wright, *et al.*, 1972). There was a single exception in that one *B. globosus* (*B. africanus* group) did become infected with the Lower Guinea strain but this snail (one of several hundred exposed) was from an unusual albino colony known to have an exceptionally wide susceptibility. The time required for development in this single individual was three times as long as that in the *B. forskali* group hosts and it was obviously an aberrant case without direct relevance to the general comparative study. Nevertheless, it does illustrate how even in severely restricted host-parasite associations an occasional breakthrough can occur and, although undoubtedly very rare, such an event in nature may have important consequences in the evolution of parasite strains.

Since the demonstration of the specificity of the inter-mediate host relationships of these geographically isolated forms of *S. intercalatum*, other studies have shown further differences between them. Frandsen (1978) found that F$_1$ hybrids between the two isolates showed a marked reduction in viability and Wright *et al.* (1979) have found that they differ in three out of eight isoenzyme systems. These facts point to a long period of isolation between the two main areas in which *S. intercalatum* is endemic, and do not support any of the earlier hypotheses of recent transfer. The question of whether the two forms should be regarded as geographical races or whether the level of differences between them justifies their recognition as distinct species remains open. When comparable investigations have been made on the remaining schistosomes in the terminal-spined-egg complex, it may be possible to view the *S. intercalatum* problem in a wider context.

Schistosoma bovis occurs in the Mediterranean region and Middle East, and is also found in East and West Africa. Like *S. haematobium* the northern (Mediterranean) *S. bovis* develops in the tetraploid *B. truncatus* except in Spain where it has been shown that *Planorbarius metidjensis* (a species apparently not closely related to the bulinids) serves as intermediate host. In East Africa *S. bovis* exhibits the widest snail host range known for any species in the terminal-spined-egg complex, developing in snails of the *B. truncatus*, *B. africanus* and *B. forskali* groups (Southgate & Knowles, 1975). Natural infections were found in all of these hosts and laboratory cross-infection experiments demonstrated that the parasite from each group of snails showed reciprocal infectivity for the other groups. However, all attempts to infect snails of the *B. africanus* group with isolates of *S. bovis* from Morocco, Spain and Sardinia failed. Thus this parasite also has geographically separate forms with different snail host characteristics. In this case a possible explanation exists.

The bovine parasite from Southern and Central Africa – *S. mattheei* – is restricted exclusively to *B. africanus* group snails. The northern range of *S. mattheei* and the southern limits of *S. bovis* overlap. Hybrids between an Iranian strain of *S. bovis* and *S. mattheei* (Taylor, 1970) and a Moroccan isolate of *S. bovis* and South African *S. mattheei* yield F$_1$ miracidia showing low infectivity for *B. truncatus* and greater infectivity for *B. africanus*. It is possible that the East African *S. bovis* originated from

natural hybridisation between the two species, the resultant parasite retaining the morphological characters of the northern form but acquiring, by introgression, the infectivity to *B. africanus* group snails of *S. mattheei*.

Two recent examples from Cameroun serve to illustrate ways in which human intervention may affect existing patterns of schistosome transmission through effecting changes in the endemic molluscan fauna.

In the rain forest area of Cameroun there are two isolated foci of transmission of *S. haematobium* in crater lakes about 16 kilometres apart (Barombi Mbo and Barombi Kotto). A control project was undertaken using a molluscicide (N-tritylmorpholine) and chemotherapy of the human population with the drug niridazole (Duke & Moore, 1976). In Lake Barombi Kotto there were two species of *Bulinus* present – *B. rohlfsi*, the principal host for the parasite, and *B. camerunensis* which was rarely found to carry a natural infection. *B. rohlfsi* proved to be reasonably sensitive to the molluscicide and its population density was drastically reduced, whereas *B. camerunensis* proved less amenable to control, even with increased dosage of the chemical at more frequent intervals of application. Control measures were continued for three years after which there was a one year follow-up during which it was found that the infection rate in *B. camerunensis* was three times as great as it had been in the pre-control period. At least two factors probably influenced this result. Removal of the principal intermediate host (*B. rohlfsi*) would have given a selective advantage to those miracidia capable of evading the host response and subsequently developing in *B. camerunensis*, while chemotherapy in the human population would have exerted a direct pressure upon the parasite population. There is evidence that at some time the strain of *S. haematobium* at Barombi Kotto has undergone some hybridisation with *S. intercalatum*. Earlier work on the effects of the drug niridazole on various schistosome species showed that *S. intercalatum* is less susceptible to the drug than is *S. haematobium*. Treatment of the human population would, therefore, have been likely to favour those parasites that were more resistant to the drug – those in which an appreciable part of their genome was derived from *S. intercalatum*. *B. camerunensis*, although only slightly susceptible to the Cameroun strain of *S. haematobium* is, under experimental conditions, an excellent host for *S. intercalatum*. Thus the dual pressures of mollusciciding and chemotherapy, the first giving an advantage to snails more suited as hosts to *S. intercalatum* and the second tending to favour parasites with *S. intercalatum* characteristics, appear to have resulted in a marked change in the local parasite strain, a change most obviously manifest in the altered relationship with the snail hosts.

The second example of the effect of human activity on schistosome transmission patterns in Cameroun comes from the town of Loum (Southgate *et al.*, 1976). In 1968 a survey of 500 schoolchildren revealed an infection rate of 54·2 per cent with *S. intercalatum*, the only species of schistosome found there. In 1972 some of the children were found to be passing schistosome eggs in their urine (*S. intercalatum* normally inhabits the mesenteric veins and eggs are passed in the faeces) and that these eggs ranged in shape and size from those of *S. intercalatum* to those of *S. haematobium*. Laboratory breeding experiments showed that hybridisation between the two species was taking place. Field surveys revealed that the snail hosts for both parasites (*B. forskali* for *S. intercalatum* and *B. rohlfsi* for *S. haematobium*) were present in the water-courses in the town and that their distribution corresponded closely to that of the two parasite species and their hybrid in the human population. *S. haematobium* has been known from foci within 20 kilometres of Loum for some years but, despite continual movement of people, did not become established in the town until 1972. Unfortunately there are no records of the snail fauna before this time but the pattern of distribution of *B. rohlfsi* in 1973 suggested that it was actively colonising the lower reaches of the streams and spreading upward. A further survey in 1978 confirmed this trend. *B. forskali* is more tolerant of shaded conditions than is *B. rohlfsi* and the invasion of this second species into the area has almost certainly been facilitated by forest clearance for agricultural development. Once *B. rohlfsi* became established it was possible for *S. haematobium* transmission to occur and for the process of hybridisation with the endemic *S. intercalatum* to begin. Laboratory studies and field observations in 1973 and 1978, suggest that the hybrid parasite has certain biological advantages over both parental species (infectivity to both parental snail hosts, increased infectivity to final host, possibly enhanced fecundity and more rapid development time) and that it will in due course replace *S. intercalatum*. Because *S. haematobium* appears to be dominant to *S. intercalatum* the hybrid resembles the first species in general characteristics but retains those features of *S. intercalatum* which are advantageous to it.

These are two reasonably documented examples of recent alterations in the relationships between African schistosome parasites and their snail hosts. Both have resulted in changes in the local strain of parasite and comparable events are undoubtedly constantly happening all over Africa. Because these events have been induced by human intervention we are more aware of their occurrence but they merely represent the most recent part of a continuing process.

Recently, Bousfield (1979) has touched upon the significance of human migration with relation to the development of new host–parasite relationships. In South America, schistosomiasis was more or less restricted to north-east Brazil in the 1920s, and most people in this

area were employed in the sugar mills. The mills were associated with river valleys providing alluvial fertilisation during winter, water for irrigation and consequently habitats for *Biomphalaria glabrata* (a host for *S. mansoni*). However, further inland the climate is drier giving rise to semi-arid regions, and here there are intermittant sources of water for which *Biomphalaria straminea* is better suited, another host for *S. mansoni*. A third species of snail, *Biomphalaria tenagophila*, occurs in the south and this species was originally thought to be incompatible with *S. mansoni*, explaining the absence of the disease in that part of the country. The plantations in the south are more progressive than those in the north, and have attracted people from the north-east in search of work, especially during the severe droughts of 1958 and 1970. The human migration was responsible for causing a rapid increase in size of the endemic area of schistosomiasis, and in the mid-fifties a number of foci were found in the Paraiba valley which were maintained solely by *B. tenagophila*. Subsequent laboratory work has shown that *S. mansoni* is gradually adapting to *B. tenagophila*, and this is now resulting in an even further expansion of the disease over a wide area of South America.

The African continent has been subject to geological movements and climatic shifts since long before the present assemblage of schistosomes and snails emerged. The most trivial of earth movements or changes in rainfall pattern can have major effects on the kind of small water bodies that serve as the main habitats for the snail hosts. Even those species that have colonised the larger lakes are not immune to these influences for, with few exceptions, their ecological requirements are such that they are restricted to the vulnerable marginal areas. To survive against this unstable environmental background it is scarcely surprising that the snail hosts exhibit a considerable degree of genetic heterogeneity at the population level and this in turn is reflected in their parasites. In this heterogeneity lies the resilience of the host–parasite relationship. The creation of new habitats by water conservation measures and irrigation systems, and the increased facility of movement of schistosomes within their definitive hosts, whether human or domestic livestock, are merely accelerating the micro-evolutionary process which has led to the existing complexes of parasites and their snail hosts. The application of modern techniques, both in the laboratory and in the field, continue to contribute to a further understanding of micro-evolutionary processes in the relationships between schistosomes and their snail hosts. Such understanding is essential for sound approaches to the control of the serious diseases of Man and animals caused by these parasites.

References

Brooker, B. E. 1972. The sense organs of trematode miracidia, pp. 177–180 *in* Canning, E. U. and Wright, C. A. (Eds) *Behavioural aspects of parasite transmission.* London: Academic Press.

Brown, D. S. & Wright, C. A. 1972. On a polyploid complex of freshwater snails (Planorbidae: *Bulinus*) in Ethiopia. *Journal of Zoology, London* **167**: 97–132.

Bousfield, D. 1979. Snail-paced parasite that is marching through South America. *Nature* **279**: 573–574.

Chu, K. Y., Kpo, H. K. & Klumpp, R. K. 1978. Mixing of *Schistosoma haematobium* strains in Ghana. *Bulletin of the World Health Organisation* **56**: 601–608.

Clarke, B. 1976. The ecological genetics of host-parasite relationships. *Symposia of the British Society for Parasitology* **14**: 87–103.

Duke, B. O. L. & Moore, P. J. 1976. The use of a molluscicide, in conjunction with chemotherapy, to control *Schistosoma haematobium* at the Barombi Lake foci in Cameroun. *Zeitschrift für Tropenmedizin und Parasitologie* **27**: 297–313; 489–504; 505–8.

Fisher, A. C. 1934. A study of the schistosomiasis of the Stanleyville District of the Belgian Congo. *Transactions of the Royal Society of tropical Medicine and Hygiene* **28**: 277–306.

Frandsen, F. 1978. Hybridization between different strains of *Schistosoma intercalatum* Fisher, 1934 from Cameroun and Zaire. *Journal of Helminthology* **52**: 11–22.

Kinoti, G. K. 1971. The attachment and penetration apparatus of the miracidium of *Schistosoma*. *Journal of Helminthology* **95**: 229–235.

Llewellyn, J. 1965. The evolution of parasitic platyhelminths. *Symposia of the British Society for Parasitology* **3**: 47–78.

Meuleman, E. A., Lyaru, D. M., Khan, M. A., Holzmann, P. J. & Sminia, T. 1978. Ultrastructural changes in the body wall of *Schistosoma mansoni* during transformation of the miracidium into the mother sporocyst in the snail host *Biomphalaria pfeifferi*. *Zeitschrift für Parasitenkunde* **56**: 227–242.

Newton, W. L. 1952. The comparative tissue reaction of two strains of *Australorbis glabratus* to infection with *Schistosoma mansoni*. *Journal of Parasitology* **40**: 352–5.

Parish, C. R. 1977. Simple model for self-non-self-discrimination in invertebrates. *Nature* **267**: 711–713.

Sminia, T., Pietersma, K. & Scheerboom, J. E. M. 1973. Histological and ultrastructural observations on wound healing in the freshwater pulmonate *Lymnaea stagnalis*. *Zeitschrift für Zellforschung und mikroskopische Anatomie* **141**: 561–573.

Smithers, S. R., Terry, R. J. & Hockley, D. J. 1969. Host antigens in schistosomiasis. *Proceedings of the Royal Society of London* (B) **171**: 483–494.

Southgate, V. R. 1970. Observations on the epidermis of the miracidium and on the formation of the tegument of the sporocyst of *Fasciola hepatica*. *Parasitology* **61**: 177–190.

Southgate, V. R. & Knowles, R. J. 1975. The intermediate host of *S. bovis* in Western Kenya. *Transactions of the Royal Society of tropical Medicine and Hygiene* **69**: 356–357.

Southgate, V. R., van Wijk, H. B. & Wright, C. A. 1976. Schistosomiasis at Loum, Cameroun; *Schistosoma haematobium*, *S. intercalatum* and their natural hybrid. *Zeitschrift für Parasitenkunde* **49**: 145–159.

Taylor, M. G. 1970. Hybridisation experiments on five species of African schistosomes. *Journal of Helminthology* **44**: 253–314.

Wajdi, N. 1966. Penetration by the miracidia of *Schistosoma mansoni* into the snail host. *Journal of Helminthology* **40**: 235–44.

Wright, C. A. 1971. *Flukes and snails.* 168 pp. London: Allen and Unwin.

Wright, C. A. 1977. Co-evolution of bulinid snails and African schistosomes, pp. 291–302 *in* Gear, J. H. S. (Ed.) *Medicine in a tropical environment.* Cape Town: Balkema Ltd.

Wright, C. A. & Knowles, R. J. 1972. Studies on *Schistosoma haematobium* in the laboratory: III. Strains from Iran, Mauritius and Ghana. *Transactions of the Royal Society of tropical Medicine and Hygiene* **66**: 108–118.

Wright, C. A. and Ross, G. C. 1980. Hybrids between *Schistosoma haematobium* and *S. mattheei* and their identification by isoelectric focusing of enzymes. *Transactions of the Royal Society of tropical Medicine and Hygiene* **74**: 326–332.

Wright, C. A., Southgate, V. R. and Knowles, R. J. 1972. What is *Schistosoma intercalatum* Fisher, 1934? *Transactions of the Royal Society of tropical Medicine and Hygiene* **66**: 28–64.

Wright, C. A. & Southgate, V. R. 1976. Hybridization of schistosomes and some of its implications. *Symposia of the British Society for Parasitology* **14**: 55–86.

Wright, C. A., Southgate, V. R. & Ross, G. C. 1979. Enzymes in *Schistosoma intercalatum* and the relative status of the Lower Guinea and Zaire strains of the parasite. *International Journal of Parasitology* **9**: 523–528.

CHAPTER 16

Meiofaunal dynamics and the origin of the metazoa

H. M. Platt

Most of the world's 360 million square kilometres of sea bed consist of sediments of one kind or another: from intertidal beach sands and the glutinous muds of quiet estuaries to the globigerina and radiolarian oozes of the deep sea, and from warm tropical lagoons to silty sub-zero polar seas. All of these sediments are inhabited by vast numbers of microscopic organisms, not only by monerans and protists but also by several groups of invertebrates. These small benthic metazoans, collectively known as the meiofauna, can occur in densities up to several million per square metre. But what part do these ubiquitous organisms play in the bionomics of the sea bed? And what influence has an awareness of this biome had on contemporary scientific thought?

Only a few years ago, these questions would have engendered little interest and 'meiobenthology' was relegated to the backwaters of academia. But recent research seems to indicate that the significance of these populations may have been under-estimated. Such concepts as the cycling of organic matter, resource partitioning, petroleum formation and even the origin of life itself all appear to be advancing as a result of an intensified investigation of this living system in the sea bed.

What is meiofauna?

Because meiofauna has so recently received detailed attention, it seems reasonable to begin with a brief profile. The term 'meiobenthos' (from Greek *meion*, smaller) was originally coined by Mare (1942) when describing that part of the mud fauna off Plymouth that was of intermediate size, 'such as small crustacea (copepods, cumaceans, etc.) small polychaetes and lamellibranchs, nematodes and foraminifera'. However, the term as now used includes the marine interstitial fauna (i.e. those animals living in the spaces between sand grains) which had earlier become known during the 1920s through the pioneering work of Adolf Remane, who is generally acknowledged as the father of meiobenthology. The reason why this assemblage

of minute forms took so long to be discovered was due mainly to the use of wide-meshed sieves by the early benthologists – a loose definition of the meiobenthos sometimes given nowadays is: those organisms able to pass through a 2 mm mesh but retained on a mesh of 40–100 μm.

Since almost all of these organisms are animals, the term 'meiofauna' is often used synonymously with 'meiobenthos'. Many additions have since been made to Mare's list, so that according to McIntyre (1969, p. 249) the meiofauna now includes 'nearly all Rotifera, Gastrotricha, Kinorhyncha, Nematoda, Archiannelida, Tardigrada, Harpacticoida, Ostracoda, Mystacocarida and Halacarida; many groups of Turbellaria and Oligochaeta; some Polychaeta and a few specialised members of the Hydrozoa, Nemer-

Table 1 Some comparative meiofaunal densities from various depths and geographical locations

Habitat	Depth (m)	Location	*Meiofauna (% nematodes)	Source of information
Sheltered mud	Intertidal	England	13·3 (93·5)	Warwick *et al.* (1979)
Sheltered sand	Intertidal	India	3·2 (99·9)	McIntyre (1968)
Sheltered sand	Intertidal	South Africa	1·9 (59·0)	McLachlan (1977)
Exposed sand	Intertidal	Scotland	0·3 (39·9)	McIntyre & Murison (1973)
Fine sand	35 m	North Sea	3·9 (96·5)	Juario (1975)
Fine sand	366 m	Atlantic	0·1 (92·0)	Wigley & McIntyre (1964)
Fine sand	567 m	Atlantic	0·1 (94·0)	Wigley & McIntyre (1964)
Fine silt	800 m	Atlantic	0·9 (59·7)	Coull *et al.* (1977)
Fine silt	4000 m	Atlantic	0·1 (30·2)	Coull *et al.* (1977)

* Number of individuals $\times 10^6$ m^{-2} Percentage of freeliving nematodes given in parenthesis.

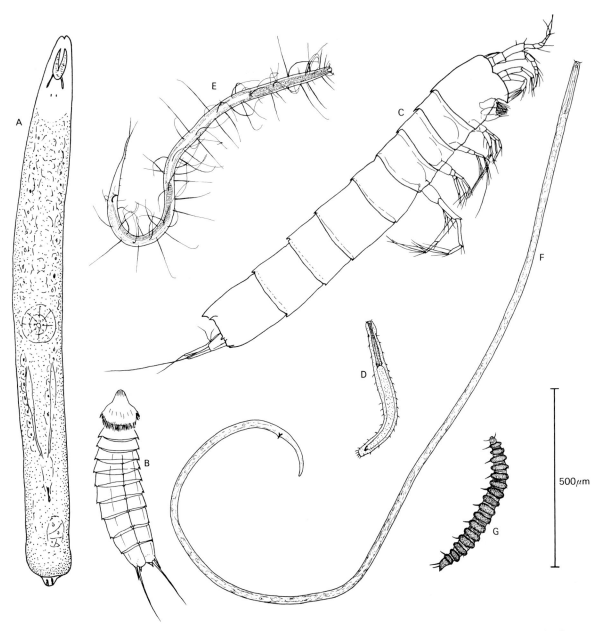

Figure 16.1 Some typical representatives of the meiofauna. A, *Schizochilus choriurus*, a kalyptorhynch turbellarian. B, *Echinoderes dujardinii*, a kinorhynch. C, *Cylindropsyllus laevis*, an harpacticoid copepod. D, *Paradasys cambriensis*, a gastrotrich. E, *Trichotheristus mirabilis*, a nematode. F, *Metalinhomoeus filiformis*, a nematode. G, *Desmoscolex* sp., a nematode. (A, after Boaden, B–G, original).

tini, Bryozoa, Gastropoda, Solenogastres, Holothuroidea and Tunicata'. In fact, almost all the known metazoan groups are represented, including unusual forms such as the Gnathostomulida known exclusively from this habitat. Figure 16.1 shows some of the typical meiofaunal organisms.

Meiofauna also encompasses an even more recently identified group of organisms – the thiobios. It was Fenchel & Riedl (1970) who first drew attention to the complex anaerobic environment under the oxidised cover of marine

sediments. Within these blackened layers, originally thought devoid of metazoa, they found a very specific faunal assemblage. Several of the usual meiobenthic taxa are represented, but notable exceptions are eukaryotic plants, most coelomates (including Crustacea) and Cnidaria. Quite unexpectedly, it seems these thiobiotic organisms may well assist our understanding of the primordial metazoa, as will be described below.

Finally, no description of the meiofauna can be considered complete without an appreciation of their numerical abundance and species diversity, particularly the nematodes which, in almost all sediments, are the most numerous metazoans (Table 1). When discussing food-chain relationships, Gerlach (1978) used, as a rough generalisation for a 30 metre deep habitat, a meiofaunal population of 4 million individuals per square metre, an approximation that may well turn out to be generally applicable to many shallow areas. The sparse data from the deep sea, however, appear to indicate that populations there are generally less dense.

A study of a Scottish beach carried out by McIntyre & Murison (1973) gave some indication of the number of species that may be expected. They reported 104 species of nematodes, 27 turbellarians, 14 copepods, 13 gastrotrichs and a few species of other groups such as coelenterates and tardigrades. New species are constantly being discovered and in some habitats, such as the deep sea, the percentage of undescribed species may be well above 50 per cent. Only about 4000 species of marine nematodes are known at the moment, but a conservative estimate would be that around 20000 taxa remain to be described.

The discovery not only of new species but of groups of higher taxonomic rank has clearly been of major significance to systematic zoology in the past few decades and further contributions in the future may confidently be expected.

The benthic food web

When meiofaunal ecology first received attention in any detail, the most widely held view was that these organisms were at the top of a food chain; that is, they were considered as a trophic dead-end, playing a part only in recycling nutrients or acting indirectly through competition with the macrobenthos for food resources. This belief can be attributed mainly to a lack of evidence demonstrating direct consumption of meiofaunal organisms by larger animals, which could thereby transmit meiofaunal production to higher trophic levels in the overall food web. Let us examine this proposition in more detail.

On average, meiofaunal biomass appears to be in the order of one tenth to one fifth that of the macrofauna, although it can be as high as 50 per cent in certain intertidal areas. But because of their small size, their metabolic activity and productivity per unit biomass is higher than that of the macrofauna (Table 2). The production to biomass (P/B) ratios of meiofauna and macrofauna are usually estimated to be about 10 and 2 respectively. If this is correct, then overall meiofaunal production should be at least half and sometimes even more than that of the macrobenthos. Clearly, this is an important resource. Would it not be strange for a billion years of evolution (perhaps more) not to have produced macrofaunal animals capable of utilising this resource? It is not surprising then, that evidence is now slowly accumulating to show that meiofaunal organisms are in fact fully integrated into the benthic food web (Platt & Warwick, 1980).

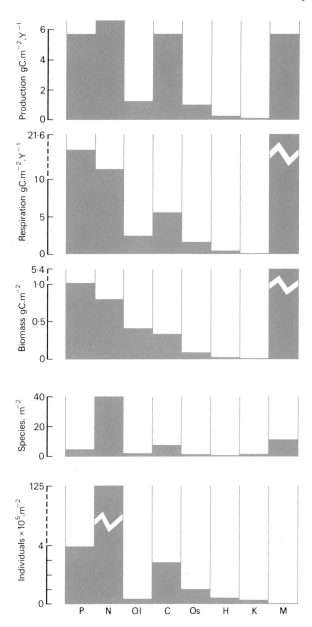

Figure 16.2 Relative importance of various meiofaunal groups compared with the entire macrofauna in the Lynher estuary, Cornwall. (After Platt & Warwick, 1980). Abbreviations: P, polychaetes; N, nematodes; Ol, oligochaetes; C, copepods; Os, ostracods; H, hydroids; K, kinorhynchs; M, macrofauna.

Warwick, Joint & Radford (1979) gave a very detailed analysis of the energy flow through a mud-flat community in the Lynher estuary, Cornwall, UK. Their results showed that the meiofauna could provide three times the output of the macrofauna, all of which could be available to 'mobile predators such as birds, fish and crabs' (Fig. 16.2). Their work also demonstrated both the complexity of the benthic system and the fact that much remains to be discovered. As Gerlach (1978, p. 63) remarked 'there is no reason to believe that meiofauna dies through old age...it may well be that meiofaunal populations are effectively controlled through macrofaunal predation, even if real evidence for this phenomenon is still lacking'.

A working model for discussion is presented in Figure 16.3. In this scheme, the difference between the inputs (consumption) and outputs (respiration and production) is lost as faeces.

Non-selective deposit feeders, such as certain polychaetes and sipunculids, cannot avoid ingesting meiofauna. The question is, if ingested, what proportion of the diet do they represent? A recent study of feeding in sipunculids and holothurians showed that food utilisation could be almost 100 per cent. Other observations on deposit-feeding oligochaetes and polychaetes suggest that assimilation efficiency (the proportion of the food consumed that is actually utilised for physiological purposes) in these organisms is much lower, in the order of 25–30 per cent. The nature of the organic matter will obviously have an important effect on the proportion assimilated, but considering the somewhat contradictory evidence available, it seems reasonable for the moment to assume that meiofauna form a dietary component proportional to their contribution to the total organic content of the sediment, say 10–30 per cent in many areas. However, the possible occurrence of selective digestion cannot be discounted.

If selective deposit feeders, such as terebellids and *Scrobicularia*, do not utilise meiofauna, then they must be able to sort them out before passing the preferred food to the mouth. Perhaps the vigorous motility of many meiofaunal animals contributes to their being rejected, although the eggs of organisms such as nematodes, which are produced in large quantities and are comparable in size to diatoms, could be consumed. Resuspension of surface sediments (i.e. the stirring up of the sea bed by organisms, waves or currents) may well play an important part in the nutrition of benthic suspension feeders, but since the distinction between suspension feeding and selective deposit feeding is essentially only one of degree, similar arguments should apply to both types of feeders. However,

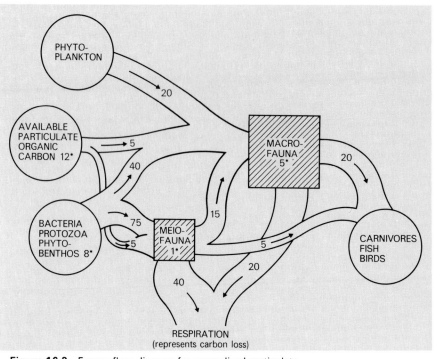

Figure 16.3 Energy flow diagram for generalised particulate sea bed. Numbers with an asterisk represent biomass in gCm^{-2}, those without reflect the annual rate of carbon production (gCm^{-2} y^{-1}): these numbers are relative. Based on the calculations of Gerlach (1978) and Warwick *et al.* (1979).

that meiobenthos are able to make some contribution cannot yet be entirely dismissed.

Mobile animals selectively taking individual meiofaunal organisms are predators in the strict sense. Very little direct evidence is available to show that this occurs to any marked degree, although adult shrimps have been shown under laboratory conditions to survive on a diet of nematodes. However, small mobile crustaceans, especially as juveniles, may well turn out to be the major meiofaunal predators: their highly active and delicate feeding mechanisms would seem to be ideal tools for manipulating small objects. In terrestrial soils, arthropods are known to feed on nematodes – perhaps crustaceans play a similar role in the marine environment.

Fish, especially in their young stages, are known to take large quantities of the meiobenthos. Most often implicated are flatfish, gobies and mullet. The stomachs of grey mullet from the Plymouth area have been found to be full of meiofauna, diatoms and sediment, but no macrofauna.

Finally, it is feasible that certain shore-feeding birds, such as shelduck and small waders, feed occasionally in a similar way although it is probably unlikely that meiofauna could provide anything other than an occasional 'make-weight'.

Thus, meiofauna may be expected to constitute a significant dietary component of many organisms operating at a higher trophic level. This scheme, as shown in Figure 16.4, includes a direct trophic connection from such

Table 2 Examples of food utilisation and production rates of selected macrofaunal organisms and a meiofaunal group

	1 *Nephthys* (*omnivorous polychaete*)	4 *Ampharete* (*deposit-feeding polychaete*)	28462 nematodes (*detritivores and herbivores*)
Total body weight (as mg carbon)	1·8	1·8	1·8
Food consumption (mgC y^{-1})	29·8	80·5	67·9
Production (mgC y^{-1})	3·4	9·8	15·1

(Based on data from Warwick *et al.* 1979).

groups as the nematodes and benthic copepods to deposit feeders, and from the meiofauna to filter feeders via resuspension. The actual amount of energy flowing between the various components will of course vary from place to place, depending upon the amount of 'dead' organic matter, water depth, the contribution from autotrophic organisms, and so on. Unequivocal substantiation of these relationships, however, may have to await the input of new data from such modern fields of research as immunology, biochemistry and physiology.

The possible role of the biome in the storage of organic material leading to the formation of petroleum lends added impetus to the study of the anaerobic thiobiotic layers, where much of the net deposition of more refractory material comes to reside. Thus, considerable interest is currently being shown in the initial chemical processes by which organic matter in the sea bed is converted into an insoluble complex polymer (kerogen) which, when subjected to thermal cracking at deeper levels in the earth's crust, results in the formation of oil and gas.

And finally, if this model can be corroborated, then a major consequence would be that detailed attempts to investigate short- and long-term effects of environmental disturbance, such as man-made pollution, ought to include an assessment of the meiobenthic component. Indeed, if the whole concept of using marine organisms to monitor or test for environmental conditions is viable, then meiofauna, and the nematodes in particular, should prove ideal candidates. Because of their conservative reproductive strategies (low gamete production, brood protection and absence of pelagic larvae), meiofaunal populations appear to be inherently stable. Short generation times and high diversity enable them to respond speedily to perturbations of the food supply; they span the whole range of polluted situations (nematodes often being the last metazoans to survive in grossly polluted sediments); and sampling for the numerically abundant groups is easier than for macrofauna. All these factors suggest that use of meiofauna as biological indicators is well worth serious consideration.

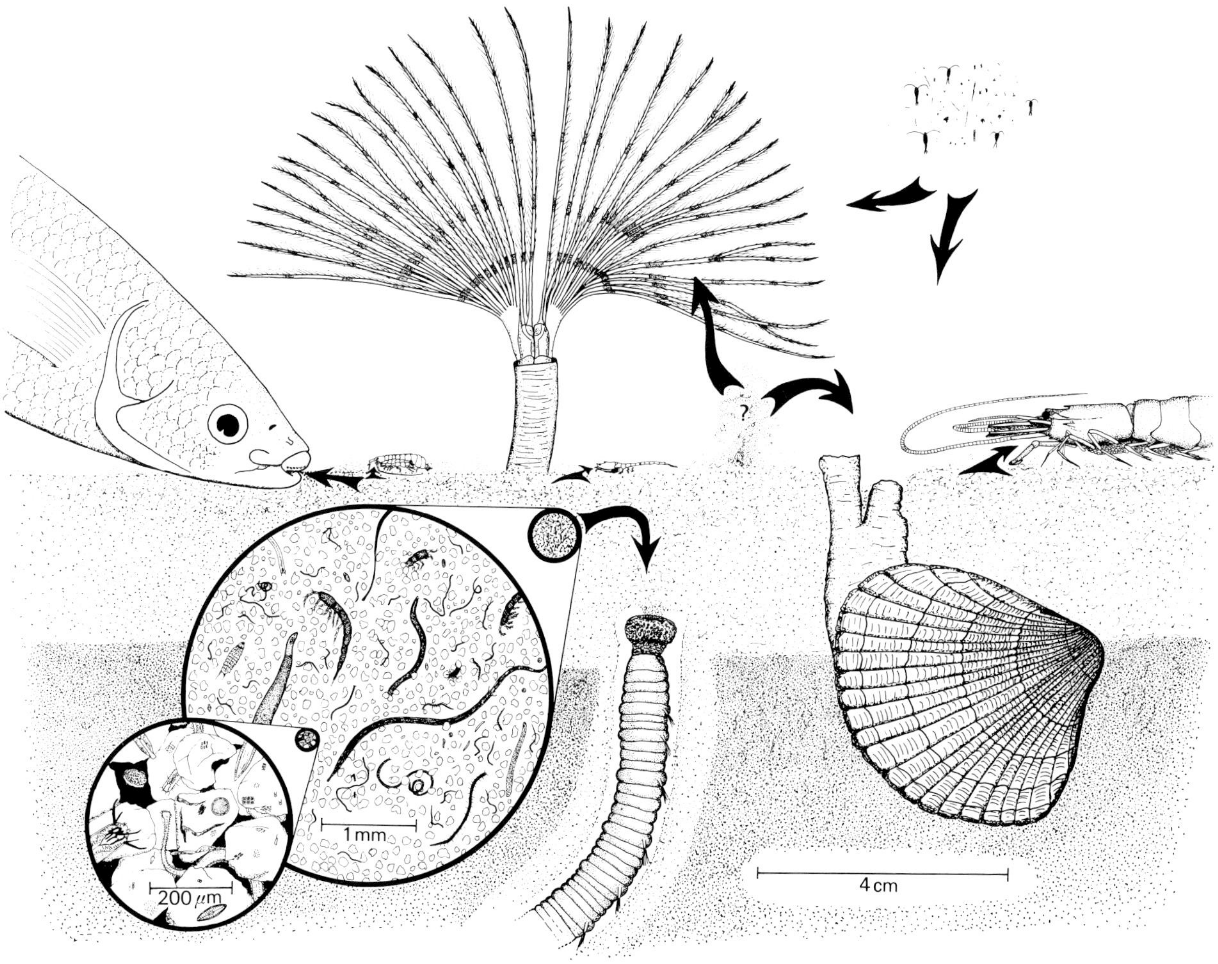

Figure 16.4 Hypothetical benthic food web, illustrating possible connections between the meiofauna and macrofauna.

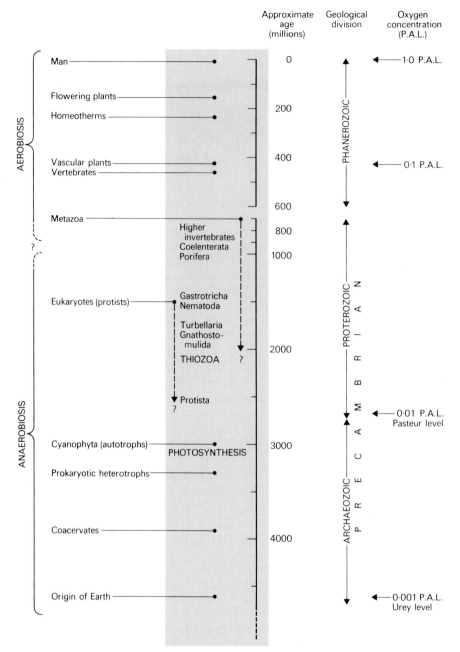

Figure 16.5 Geological column summarising currently accepted temporal sequence of evolutionary events and possible implications of the thiozoon hypothesis. P.A.L. = present atmospheric level. (After Bernal, Schopf, Rutten and others).

Meiofauna and the origin of life

Arguably, the most widely accepted views on the origin of life and the early development of multicelled organisms remain those based on the theories of Oparin, Haldane and Haeckel – with, of course, numerous subsequent refinements and modifications. The paradigm can be summarised as follows (Fig. 16.5).

After the solar system condensed from interstellar gas clouds (primarily hydrogen) some 4500 million years ago, a combination of high temperatures, electrical discharges and ultraviolet light on Earth promoted the formation of increasingly complex molecules. Chemical evolution

proceeded from water, ammonia and methane, through amino acids and peptide chains to the molecular associations known as coacervates. After a further 1000 million years, this chemical phase culminated in the primitive cell or 'eobiont'. Eobionts preceded the first prokaryotic organisms, which were anaerobic heterotrophs. The subsequent evolution of phototrophs, some 2–3000 million years ago, initiated one of the most vital phases – the development of photosynthesis – that led eventually to the production of relatively abundant free oxygen. Confidence in the timing of these events has gained strength from the discovery in the last decade of a far more impressive Precambrian fossil record than had originally been imagined (Schopf, 1975). These fossils are, however, of microscopic size.

Aerobic eukaryotic organisms, single-celled ancestral protists, are thought to have evolved between 1000 and 1500 million years ago. Some of these became secondarily heterotrophic and it is thought that, by forming colonial balls of cells, they thereby became organised into the ancestral metazoans. Most theories adhere to the view that the metazoa had a monophyletic origin, and their subsequent radiation began towards the end of the Precambrian, some 700 million years ago.

Therefore, eukaryotic cells are generally believed to be fundamentally aerobic. The abundant release of energy that this form of metabolism provided (compared with fermentation), coinciding with the development of meiosis, provided the spur to an outburst of evolutionary activity culminating in the 'sudden' appearance of complex macro-fossils, some with hard exoskeletons, at the beginning of the Cambrian.

Furthermore, the metazoan ancestor is generally conceived of as a free-swimming (pelagic or epibenthic), radially symmetrical organism feeding by intracellular digestion. Subsequently, some of these organisms adopted a benthic existence and bilateral symmetry but retained a pelagic larva.

There are, of course, numerous variations on this simplified theme, particularly in relation to the precise form of the earliest metazoan. Whether the original multi-celled animal was a syncytial flatworm-like creature, a hollow gastraea or a solid planuloid is a problem that has exercised biologists' minds for decades. Unfortunately, none of these theories seems to fit all the available evidence satisfactorily; each appears to be limited by the constraints imposed by assuming a temporal progression that places the development of metazoa *after* that of photosynthetic eukaryotes and the development of benthic metazoans *after* that of free-swimming ones. It is usually argued that the first invertebrates would have been pelagic since they required phytoplankton on which to feed.

But even the most cherished of hypotheses must remain open to criticism, such is the fundamental principle of scientific procedure. So let us now consider the arguments recently put forward by Fenchel & Riedl (1970) and Boaden (1975, 1977) to the effect that the earliest metazoans were in reality 'anaerobic, holobenthic, interstitial thiobionts feeding by epidermal absorption'. Much of the evidence so far assembled to substantiate this apparently controversial claim comes from the biochemistry, taxonomy and distribution of those organisms living today in the anaerobic layers of marine sediments; that is, the thiobios introduced earlier. Essentially, this hypothesis maintains first, that metazoans evolved as meiofaunal organisms in the sea bed at a stage of evolution far earlier than is currently accepted, perhaps as much as 2–2500 million years ago; second, that their anaerobic metabolic path-

ways were pre-adapted for the subsequent development of aerobic respiration when free oxygen eventually built up; third, that several of the taxa that evolved prior to oxic conditions left a relict fauna in the only remaining representative of that ancient habitat – the reduced sediments lying beneath the oxygenated surface layers.

Several years ago, Bernal (1967) suggested that the initial stages in the origin of life may well have occurred in or on the anaerobic sands and muds of Precambrian estuaries. It was here that adsorption of organic molecules and the process of polymerisation occurred and in this environment the primitive monerans and protists originated. These primeval life forms were probably bathed in a medium rich in dissolved or colloidal organic matter that was concentrated in the sea bed, and could therefore obtain their nutrient by absorption. It seems not unreasonable to propose that metazoans may also have evolved in this habitat. There is evidence to suggest that many primitive invertebrates, such as marine nematodes, still obtain much of their nutrition from dissolved organic matter.

According to Clark (1964) the first metazoans would have been too large (at 1 mm) to have lived interstitially and too weak to have burrowed. But these arguments may well be erroneous. Even if they were about 1 mm (and many meiofaunal organisms are smaller than this), many animals of a similar size today live quite successful interstitial lives. But the early meiofaunal environment may not have been sand, but soft flocculent muds and oozes. Consequently, although the primitive meiofauna may have been too weak to have heaved large sand grains around, they could easily have been able to burrow or perhaps, 'swim' through soft silt. Hence, it is not necessary to postulate that the first metazoans, if indeed they were benthic organisms, were bilateral if their habitat was essentially three-dimensional. This is because many of the relatively sharp vertical gradients in such parameters as redox potential, hydrogen sulphide concentration, and so on, which characterise most of today's marine sediments, would not have been developed. In the absence of these signals by which the modern meiofauna are able to recognise their preferred vertical horizon, the primitive metazoans were more free to roam in any direction. Perhaps only by recognising the increased compaction of the sediment with depth were they able to remain reasonably close to the sediment surface. Radially symmetrical, probably elongated organisms feeding by absorbing organic material could have exploited these ancient muds which, in addition, afforded them some degree of protection from the strong ultraviolet radiation. Only later, with increasing body size and/or colonisation of less flocculent substrates may it have been necessary to develop some degree of bilaterality.

What of the contention that early metazoans were aerobic? Aerobic animals today always have anaerobic metabolic pathways but the reverse is not true. It is feasible that primitive eukaryotic organisms may well have evolved *anaerobic* metabolic pathways that involved just

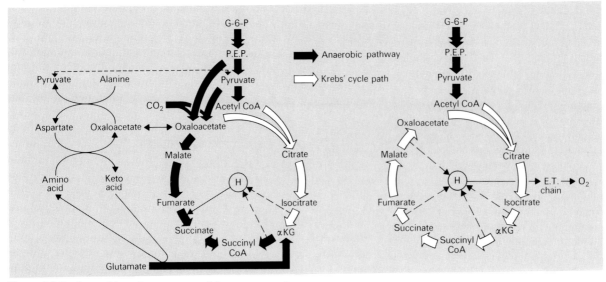

Figure 16.6 Anaerobic pathways as possible precursors of the Kreb's cycle, showing how the pathway could have evolved to serve an anaerobic function in primitive life forms and only later reversed to give the aerobic sequence (from Maguire & Boaden, 1975).

those enzymes and reactions for the aerobic tricarboxylic acid or Kreb's cycle. These pathways could, in effect, be reversed later to evolve the more efficient oxidative pathways familiar in aerobic organisms. Some invertebrates, respiring anaerobically, such as thiobiotic gastrotrichs, are thought to have retained this original reversed Kreb's cycle sequence (Maguire & Boaden, 1975) shown in Figure 16.6. Furthermore, none of the Platyhelminthes appears able to synthesise sterols or polyunsaturated fatty acids, and nematodes lack the ability to produce sterols; it is significant that these pathways require oxygen. Rather than having secondarily lost this ability, which is the usual explanation, it is simpler to propose that these groups evolved as anaerobic organisms and in fact never acquired this ability. One intriguing piece of physiological evidence concerning these primitive groups is that of Wieser, Ott, Schiemer and Gnaiger (1974) who found that a marine nematode they were studying was more temperature resistant under oxygen-deficient than in oxygen-rich conditions. This is the first marine metazoan in which it has been shown that a specific biological process is favourably affected by anoxic conditions.

If these primitive forms are indeed more facultatively anaerobic than previously assumed, and their ancestors did evolve under anoxic conditions, then they must have had their origin before the Pasteur level had been reached. The Pasteur level of about one hundredth of the present atmospheric level (0·01 present atmospheric level) is the point where facultative respirators are able to change their metabolism from fermentation to respiration. According to Rutten (1971) this occurred some 2750 million years ago (see Fig. 16.5). When oxidising conditions were established, the basic radiation of these early metazoans

began, sometime in the mid-Precambrian. Therefore, the oft-quoted 'explosion of life at the base of the Cambrian' would be no such thing, but simply an explosion of fossils!

Having thus introduced the concept of the hypothetical metazoan ancestor as a radially symmetrical, heterotrophic, benthic anaerobe (called the 'Thiozoon' by Boaden, 1977), we must now trace in outline its evolutionary development at least until it gave rise to the 'lower' invertebrate groups extant today. As these thiozoons began to deplete the reservoir of easily obtainable organic material, alternative trophic strategies would be required. To satisfy their requirement for nitrogenous food, the ability to supplement their diet by ingesting other organisms developed and this is a trend that may have led to the platyhelminth and 'aschelminth' grades of organisation. 'Solid' thiozoons provided the acoel ancestor whilst the development of a schizocoelous body cavity (to cope with increasing size?) led to the nematode/gastrotrich line. The concurrent changes in free oxygen concentrations had a stimulatory effect on the development of aerobic metabolism, at the same time affording some protection from ultraviolet radiation, so that epibenthic and pelagic existence became possible. The introduction of larvae and external fertilisation into the life cycle, together with marked bilateral symmetry to deal more effectively with mobile life in an increasingly two-dimensional habitat, provided the additional advances we can now recognise in the annelid line. Other thiozoons evolved into 'hollow' organisms but retained their primitive radial symmetry to be represented as the sponge and coelenterate lines. In other words, it is suggested that the various primitive metazoan lines arose at different times from a progressively evolving group of originally anaerobic thiozoons.

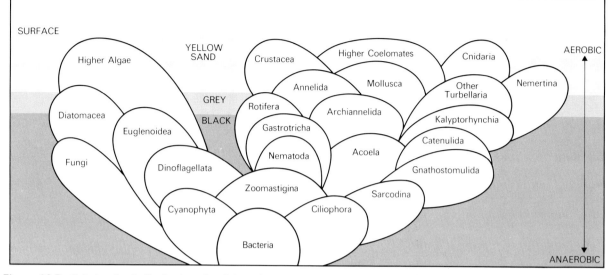

Figure 16.7 Relative depth distribution of meiobenthic taxa in relation to deoxygenation in marine sediments. The names are at the level of the known maximum depth at which each group occurs. The 'pebble-plant' outlines the relationship between depth and evolutionary origin (from Boaden, 1975).

Increasingly oxic conditions eventually separated the original thiozoons and/or their conservatively anaerobic descendants from the site of primary organic production in the photic zones. Most groups subsequently modified their metabolism, thus enabling their descendants to colonise the surface aerobic layers. But some remained associated with the sulphide system, perhaps unable to compete with the aerobic forms. Today the Gnathostomulida and the turbellarian families Retronectidae and Solenofilimorphidae are examples of taxa that remain restricted to this biome. Thus, with the full establishment of the aerobic environment, the 'golden age of the thiobios was over' (Boaden, 1977).

Perhaps this is as far as we dare go at this stage. Many of the links between the various invertebrate taxa are (and may always remain) a puzzle to us, being hidden behind some two thousand million years of evolution. But is this hidden history still reflected in the present-day distribution of meiobenthic organisms in marine sediments? Figure 16.7 shows the relative depth in marine sediments to which the various taxa are known to occur in relation to oxygen conditions. Notice that, as a generalisation, the most 'primitive' groups extend deepest into the reduced layers. Monerans and protists extend deeper than metazoans and of the latter, the nematodes, acoels and gnathostomulids

descend deepest. Segmented and coelomate groups are generally confined to the aerobic sands. This then, may be circumstantial evidence that the thiobios are relicts of 'the oldest biosystem on earth which preceded the aerobic biosphere' (Fenchel & Reidl, 1970). If it were otherwise, that is, if the reduced layers were secondarily colonised, would one expect *only* the most primitive groups to be successful?

Like all controversial ideas, the thiozoon hypothesis has been subjected to criticism, some of it quite severe. Reise & Ax (1979) even go so far as to deny the very existence of a meiofaunal thiobios! Despite this controversy, these recent insights into the important dynamic role played by the meiofauna of today's complex benthic system and those of the past, at the very origin of living systems, clearly will require considerable refinement if they are to take their place amongst the basic tenets of modern biology. And yet, whilst experienced biologists undoubtedly will be able to point to certain arguments that I have overlooked here, how much more rewarding is this continuing search for a fundamental understanding of our planet's living system than those investigations with more limited goals. Our search into the past must go hand in hand with our plans for the future if the latter are to be kept in perspective.

References

Bernal, J. D. 1967. *The origin of life.* 345 pp. London: Weidenfeld & Nicolson.

Boaden, P. J. S. 1975. Anaerobiosis, meiofauna and early metazoan evolution. *Zoologica Scripta* **4**: 21–4.

Boaden, P. J. S. 1977. Thiobiotic facts and fancies (Aspects of the distribution and evolution of anaerobic meiofauna) *in* Sterrer, W. & Ax, P. (Eds) *The meiofauna species in time and space.* Mikrofauna Meeresboden **61**: 45–63.

Clark, R. B. 1964. *Dynamics in metazoan evolution.* 213 pp. Oxford: Clarendon Press.

Coull, B. C., Ellison, R. L., Fleeger, J. W., Higgins, R. D., Hope, W. D., Hummon, W. D., Rieger, R. M., Sterrer, W. E., Thiel, H. & Tietjen, J. H. 1977. Quantitative estimates of the meiofauna from the deep sea off North Carolina, U.S.A. *Marine Biology* **39**: 233–40.

Fenchel, T. & Riedl, R. J. 1970. The sulfide system: a new biotic community underneath the oxidised layer of marine sand bottoms. *Marine Biology* **7**: 225–68.

Gerlach, S. A. 1978. Food-chain relationships in subtidal silty sand marine sediments and the role of meiofauna in stimulating bacterial productivity. *Oecologia*, Berlin **33**: 55–69.

Juario, J. V. 1975. Nematode species composition and seasonal fluctuation of a sublittoral meiofauna community in the German bight. *Veröffentlichungen des Instituts für Meeresforschung in Bremerhaven* **15**: 283–337.

Maguire, C. & Boaden, P. J. S. 1975. Energy and evolution in the thiobios: an extrapolation from the marine gastrotrich *Thiodasys sterreri. Cahiers de Biologie Marine* **16**: 635–46.

Mare, M. F. 1942. A study of a marine benthic community with special reference to the micro-organisms. *Journal of the Marine Biological Association of the United Kingdom* **25**: 517–54.

McIntyre, A. D. 1968. The meiofauna and macrofauna of some tropical beaches. *Journal of Zoology*, London **156**: 377–92.

McIntyre, A. D. 1969. Ecology of marine meiobenthos. *Biological Reviews* **44**: 245–90.

McIntyre, A. D. & Murison, D. J. 1973. The meiofauna of a flatfish nursery ground. *Journal of the Marine Biological Association of the United Kingdom* **53**: 93–118.

McLachlan, A. 1977. Studies on the psammolittoral meiofauna of Algoa Bay, South Africa. II. The distribution, composition and biomass of the meiofauna and macrofauna. *Zoologica Africana* **12**: 33–60.

Platt, H. M. & Warwick, R. M. 1980. The significance of freeliving nematodes to the littoral ecosystem, *in* Price, J. H., Irvine, D. E. G. & Farnham, W. F. (Eds) *The shore environment: methods and ecosystems.* London: Academic Press.

Reise, K. & Ax, P. 1979. A meiofaunal 'Thiobios' limited to the anaerobic sulfide system of marine sand does not exist. *Marine Biology* **54**: 225–237.

Rutten, M. G. 1971. *The origin of life by natural causes.* 420 pp. Amsterdam: Elsevier.

Schopf, J. W. 1975. The age of microscopic life. *Endeavour* **34**: 51–8.

Warwick, R. M., Joint, I. R. & Radford, P. J. 1979. Secondary production of the benthos in an estuarine environment, pp. 429–450 *in* Jefferies, R. L. & Davey, A. J. (Eds), *Ecological processes in coastal environments.* London: Blackwell.

Wieser, W., Ott, J., Schiemer, F. & Gnaiger, E. 1974. An ecophysiological study of some meiofauna species inhabiting a sandy beach at Bermuda. *Marine Biology* **26**: 235–48.

Wigley, R. L. & McIntyre, A. D. 1964. Some quantitative comparisons of offshore meiobenthos and macrobenthos south of Martha's Vineyard. *Limnology and Oceanography* **9**: 485–93.

CHAPTER 17

Competition, evolutionary change and montane distributions

E. N. Arnold

In discussions about interspecific competition, its more negative effects are often emphasised, particularly its role in reducing the resources available to species and in causing their extinction. Yet, a restriction of resources by competition from other forms could be one of the mainsprings of evolution, producing specialisation that may allow the invasion of new niches or the more efficient exploitation of niche space that was only partly occupied before. On occasions, a succession of such changes may lead to the development of markedly different types of organism that later become more widespread, perhaps eventually replacing some of the competitors that originally caused the precursors of the new forms to occupy restricted niches. Unfortunately such events take place on a far too extended time-scale for direct observation and the sort of continuous and unambiguous fossil sequence necessary to record them is rarely, if ever, available; so the role of interspecific competition in evolutionary change is difficult to confirm. However, many patterns of phylogenetic relationship and distribution can be interpreted in this context and, in some instances, it is not easy to see how else they could have arisen.

Character displacement

A common phenomenon that is often assumed to be the product of competitively induced change involves pairs of related, partly sympatric species. In such cases, it is often observed that the two forms concerned are morphologically more different from each other in their area of overlap than where they are allopatric (more rarely, the species are more similar in sympatry). Such situations were called *character displacement* by Brown & Wilson (1956), who suggest that they probably result most commonly from the first contact of two previously separated species which interact on meeting to produce the differences observed in sympatry. Interaction can be of several kinds. For instance, the species may diverge in their recognition features, so that mismatings are less likely, or they may converge in these

characters making aggressive displays of each form understood by the other, so that overlap of their territories will be avoided, thus reducing direct resource competition between them. Where resources are finite, species in sympatry usually restrict direct competition by exploiting different sections of them. This is true even when niches are very similar where the species occur alone. Such differentiation in ecology is likely to produce altered selective pressures which in turn result in divergent morphological changes in structures connected with obtaining and processing food and other necessities, and in features related to exploiting spatially or temporally different parts of the niche spectrum. It is such changes produced by competition that are liable to result· in evolutionary innovation.

The term character displacement was originally applied to any situation where species were more differentiated in sympatry irrespective of the origin of that difference. Grant (1972, 1975), however, has defined it more restrictively.' But such events occurred in the past and are now very difficult to substantiate adequately and Schindel & from the presence, in the same environment, of one or more species similar to it ecologically and/or reproductively.' But such events occurred in the past and are now very difficult to substantiate adequately and Schindel & Gould (1977) argue that it is more useful to retain the term as a description of observable situations and not restrict it to one probably frequent, but essentially unconfirmable, origin for them. Because of this, the process described by Grant will be called *competitive character displacement* below.

Possible examples of competitive character displacement occur in the lizard genus *Acanthodactylus* (Lacertidae). This is a group of about 25 species of ground dwelling lizards that occur in dry places from Spain and north Africa to north-west India. All the species are diurnal arthropod feeders and, when sympatric, appear to reduce interspecific competition by occupying different types of ground. In northern Arabia, however, two species usually associated with soft sand occur together. These are *A. scutellatus* with a large range in north Africa and *A. schmidti* which is found over nearly all Arabia. Outside their zone of overlap both are relatively small, rarely being much over 70 mm from snout to vent, and feed on similarly sized prey but, where they occur together, *A. schmidti* may reach 100 mm while *A. scutellatus* does not usually exceed 65 mm (see Fig. 17.1). Large *A. schmidti* take, on average, bigger prey than sympatric *A. scutellatus* and it seems possible that this species has been forced to feed in this range of prey size by *A. scutellatus* pre-empting a large proportion of small prey. If this were so, increased body size in *A. schmidti* would be expected since in lizards the upper size-limit of exploitable prey is determined largely by that of the predator.

However, as Grant (1972) points out, such supposed instances of competitive character displacement must be looked at critically. Indeed, this author dismisses many of the cases previously used to illustrate the concept. Two basic questions need to be asked: how did the sympatric situation arise; does sympatry itself cause some characters of the species to diverge? The first question could be rephrased: is the area of overlap due to two originally allopatric populations contacting each other? There is always the possibility that the species may have been initially sympatric and well differentiated but one or both later spread into allopatry and assumed a morphology intermediate between those of the overlapping populations. Such changes, resulting from losing a previous competitor, are known as character release. However, the occurrence of character release still implies that the differentiation of the two species in sympatry was maintained by competition and it does not exclude the possibility that it first arose from two allopatric species meeting. Indeed, the latter is very probable in the case of closely related species pairs that are likely to be the product of a single speciation event, for in most groups allopatric speciation seems to be the usual method.

In the case of *Acanthodactylus* described above, the area of sympatry is likely to be the result of simple primary contact between allopatric forms because the area of sympatry is very small compared with the allopatric ranges of the two species and each form is a member of two otherwise non-overlapping groups. *A. scutellatus* has its closest relatives in western north Africa whereas those of *A. schmidti* are in Iran and southern Arabia. Furthermore, there is circumstantial evidence that the desert reptile faunas of the Sahara and Arabia have been separated at times and that their present contact may be of comparatively recent origin. Information on climatic changes in the Quaternary supports this view and there are several cases where lizard species which are widespread in the Sahara meet related Arabian forms in the north of that peninsula, suggesting that original stocks were divided and became differentiated but have now re-established contact.

In some cases, such as the ground geckoes *Stenodactylus petrii* and *S. doriae*, the species approach without overlap

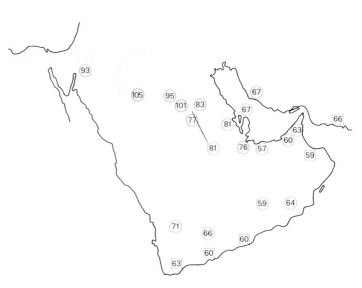

Figure 17.1 Possible character displacement in Arabian spiny-footed lizards. Like its close relatives, *Acanthodactylus schmidti* is small at many localities but is larger where it coexists with *A. scutellatus*.

shaded – approximate range of *A. scutellatus*
○ – localities of *A. schmidti*
□ – localities of close relatives of *A. schmidti*, viz. *A. blanfordii* and *A. arabicus*
Figures indicate maximum length from snout to vent in millimetres.

but other pairs of species show a broad area of sympatry and there may be marked morphological differences where this occurs. For instance, the north African *Acanthodatylus boskianus* is bigger where it overlaps with the small *A. opheodurus* of Arabia. They occupy the same habitats in some areas and may form a parallel to *A. scutellatus* and *A. schmidti*. Similarly, where the area occupied by the mainly African sand skink *Scincus scincus* contacts that of *S. mitranus*, the former species appears to occupy harder ground than usual. Perhaps in connection with this it loses the stream-lined head shape that facilitates 'sand-diving', the usual anti-predator technique of *Scincus* living on loose wind-blown sand.

It must also be asked whether the characteristics of sympatric populations might not have developed even in the absence of a competitor. Grant (1972) suggests that apparent character displacement may sometimes really result from independent clinal variation in the two species, the characters involved appearing to be most different in the area of overlap. But in such a case abrupt change near the borders of the region of sympatry would not be expected and this seems to be present in *A. schmidti*. Also the large size of *A. schmidti*, where it occurs with *A. scutellatus*, is very uncommon in the genus as a whole.

The Arabian lizard examples need further investigation before they can be fully accepted as examples of competitive character displacement, but Grant (1975) provides a very detailed case for the existence of this phenomenon

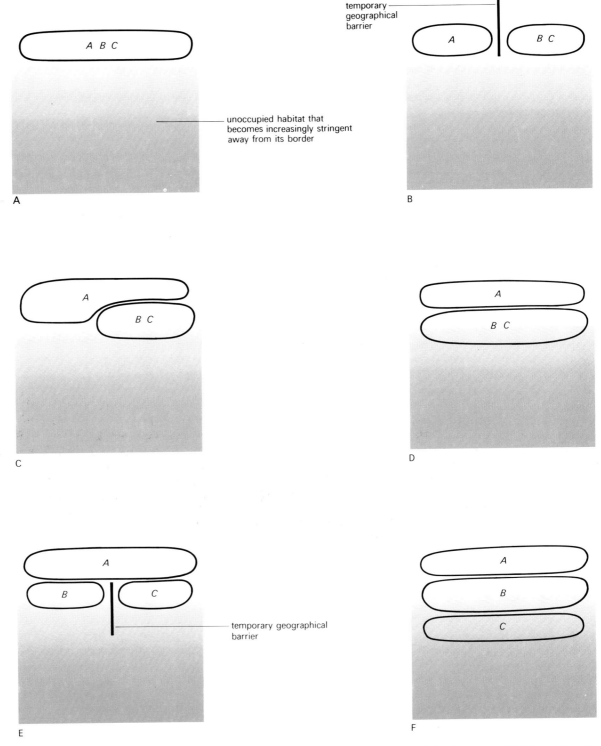

Figure 17.2 Competition-induced spread into unoccupied ecological space. Species *ABC* is distributed on the edge of an environment that becomes increasingly stringent away from its border (A). If allopatric speciation occurs (B) followed by contact of the two daughter species, one may be displaced into unoccupied areas (C, D). Adaptation followed by a second round of speciation in species *BC* (E) could result in daughter species *C* being pushed further into hostile ecological space (F). Allopatric speciation might be initiated by a variety of factors, for instance a long-term climatic cycle that caused the range of peripheral species to be split by the intermittent spread of unoccupied stringent territory.

in rock nuthatches (*Sitta neumayer* and *S. tephronata*) in Iran. Some other suggested instances involve not merely two largely allopatric species with a single area of overlap but situations where several allopatric populations of the two are interspersed with each other, and with a number of sympatric situations in which character displacement is shown. An example of this more complicated situation has been found in mud snails (Hydrobiidae) by Fenchel (1975) and Schindel & Gould (1977) report a case in Bermuda snails, *Poecilozonites*, where the samples are also dispersed through time. In such multiple cases, it seems more certain that the consistently greater differentiation in sympatry is due to the presence of two species together. It is also improbable in such cases that evolutionary change has always been in the direction of character release rather than competitive character displacement.

Although no completely convincing instance of competitive character displacement is available, examples like those discussed above suggest that it does take place and that, by forcing species into different selective regimes, competition may result in the development of innovations. While not very visible, because possible cases must involve certain patterns of distribution to be apparent, competitive character displacement seems likely to be widespread. Given the fact that sympatric species nearly always occupy different niches and are consequently subjected to differing patterns of selection, it would be surprising if this were not so.

Competition-induced expansion into new environments – advance through failure

Organisms are adapted to many kinds of extreme and, at first sight, hostile environments. In many cases these seem so stringent that it is difficult to envisage species that occur in less exacting habitats on their periphery expanding into them, for such conditions require considerable adaptation and it is unlikely that a species can adequately exploit both its own niche and the very different hostile niche space bordering it.

For this reason, certain niches may remain empty or grossly underexploited, unless a specialised form appears. For instance, in the deserts of Palaearctic Asia and Arabia, toad-headed lizards (*Phrynocephalus*; Agamidae) occupy extremely hostile niches. They hunt in very open, sandy, desert environments with low productivity and heavy predator pressure and often stay active in the hottest part of the day, long after other desert lizards have retreated. Yet in the Sahara, although a wide variety of lizard groups is present, none appears to exploit this set of niches very effectively. How do such specialists arise? If species on the periphery do not automatically expand into empty, hostile ecological space, how is it filled? As has been suggested, when two ecologically similar species become sympatric, one may eventually displace the other from a substantial part of its original niche, resulting in its occupying ecological space that it did not originally use and becoming

adapted to it. Such an event on the periphery of an unoccupied habitat that becomes increasingly hostile away from its border may enable its edge to be occupied. Thus, in Figure 17.2, species *A* has partly displaced species *BC* into previously unoccupied territory. This process probably usually involves relatively modest changes in niche, and small morphological modifications to cope with them, but further rounds of allopatric speciation and displacement may, by the same process, allow increasingly stringent ecological space to be colonised and for bigger structural changes to take place. Paradoxically, in such a situation, it is not the species best adapted to present conditions that breaks new ground and evolves the necessary adaptations to exploit it efficiently but the one which is, at least initially, the loser in competition and is pushed from its original niche.

As with simple competitive character displacement the invasion of new ecological space in this way is on far too long a time-scale to be directly observable but groups for which a credible phylogeny can be constructed suggest that it may have occurred. In the genus *Acanthodactylus* already discussed, the more primitive species are found in relatively mesic areas, such as open woodland and savannah, but the most derived are inhabitants of wind-blown sand and, although not so adequately adapted to such extreme conditions as *Phrynocephalus*, they extend far into the desert regions of Arabia and north Africa. In response to this environment, they have evolved many specialised features. For instance, the toes have extensive fringes of pointed scales that increase their surface area and enable the lizards to travel efficiently over loose sand. There are modifications of the nostrils, eyelids and mouth to exclude sand and the associated risk of water loss by capillary action and also reduce water loss by evaporation. The snout is narrow and pointed, allowing it to be used efficiently in probing the sand for arthropod prey, and the dorsal colouring enhances camouflage in open situations. The most likely phylogeny of this group of lizards suggests that these features may have developed gradually through a series of speciation events, in each of which one of the resultant species was pushed into a rather more extreme habitat and became adapted to it. Figure 17.3 shows a simplified version of the probable phylogeny of north African *Acanthodactylus* and the accompanying maps suggest that at each succeeding known speciation a more desert-adapted species has arisen. This hypothesised phylogeny of the group is based not only on characters functionally associated with the kind of habitat occupied, but also on structures such as genitalia where differences are unlikely to be directly related to ecological parameters. The speciation events involved may have been produced by the Quaternary climatic changes known to have taken place in the area. For instance, in a period of increasing aridity a species may be 'fragmented' into isolates based on the few areas with permanent water.

An example of the sort of event involved in the colonisation of increasingly desertic habitats is provided

Figure 17.3 A. Probable phylogeny of north-west African spiny-footed lizards (*Acanthodactylus*; Lacertidae) showing increasing adaptation to arid, sandy habitats. B, C. The most primitive species are confined to relatively mesic coastal and mountain regions in the north while the most derived extend deep into the Sahara desert.

by the lizard *Acanthodactylus erythrurus* and its close relative, *A. savignyi*, in northern Algeria and neighbouring regions (Fig. 17.3 A, B). In Morocco and northern Algeria, *A. erythrurus* itself is found in quite mesic areas on relatively hard ground types and is replaced in northern Tunisia by the extremely similar *A. savignyi blanci*. But on parts of the Algerian coast (around Oran) both species occur and here *A. savignyi* (as *A. s. savignyi*) is restricted to sandier habitats. Although by no means as obviously adapted to such environments as, for instance, *A. longipes*, it has clear modifications in this direction; the toes are slightly fringed, the snout is pointed and dorsal colouring tends to be more uniform than in related populations. If this apparent competitive character displacement had occurred on the desert edge instead of on the Mediterranean coast, it would be easy to envisage *A. s. savignyi* being able to expand further into sandy conditions than either *A. erythrurus*, which seems to have displaced it, or the common ancestor of both forms, and for subsequent speciation to have allowed further spread.

In the case of *Acanthodactylus*, a fairly complete sequence of forms links the primitive mesic species with the most derived xeric, sand-dwelling ones. In phylogenetic sequences of other lizard groups, however, there are abrupt steps in the morphological picture. This might sometimes be due to forms near the end of the sequence eliminating intermediates for, once all previously empty ecological space is filled, niche displacement must take resources from other species. More derived forms could be capable of replacing intermediates in this situation because adaptation to extreme conditions may produce features that are advantageous elsewhere. It may be that some characteristics useful in an intermediate niche are only likely to evolve in the more rigorous selective conditions found in extreme environments.

Change in evolutionary potential through competitive restriction – montane species

Although situations exist where interspecific competition seems likely to have powered the expansion of a group into empty or under-exploited ecological space, its results are probably less beneficial in many instances. Species may be displaced so that they occupy niches that are unstable and thus become liable to subsequent extinction or they may be totally replaced by a competitor. In other cases they may be restricted to what, at least in the short term, appear to be evolutionary culs-de-sac. The position of many mountain-distributed forms can be regarded in this way.

One of the commonplaces of biogeography is the tendency for mountain areas to have floras and faunas that differ markedly from those of the surrounding lowlands. In such cases, the *maintenance* of this difference is primarily due to topographic effects on climate – mountains, being cooler and usually moister than lowland areas, maintain populations which are adapted to such conditions while placing forms requiring a warmer, drier regime at a

disadvantage. The *origin* of such distributions has been discussed for birds in Micronesia by Mayr & Diamond (1976) who list the following possibilities:

1. *Jumping.* Mountains may be colonised by forms already adapted to high altitudes that reach them from other ranges by flight across the unsuitable intervening lowlands or are carried passively by wind or other vectors.

2. *Trickling through lowlands.* Mountain-adapted species may be able to reach new habitats by dispersing through intervening lowlands, even though they cannot breed there or survive for long.

3. *Spread during cool periods.* In many areas, the climate has been substantially cooler in the recent past so that conditions now restricted to mountain regions were more widespread. At such times, cold-adapted species would be able to spread to new ranges across the cool, intervening lowlands.

4. *Competition from a single lowland competitor.* Species may be restricted to mountain areas by a single competitor that displaces them from the lowlands.

5. *Diffuse competition.* Mountain restriction may result from competition by several lowland competitors.

6. *Pulling.* Mayr & Diamond (1976) suggest that species may sometimes shift into montane habitats because these contain unexploited resources. 'Because a species cannot be good at everything, these upward expansions and adaptations tend to come at the expense of competitive ability in the lowlands.'

Recognising competitive restriction

Distinguishing between the alternative origins listed above is not always easy and, undoubtedly, sometimes more than one plays a part. But jumping is confined to forms that can disperse aerially and, in the case of distant mountains, or ranges, trickling is likely to be limited to such forms with strong dispersal mechanisms. Organisms that cannot avail themselves of these methods will consequently be limited to spread during climatically favourable periods, by competitive restriction, diffuse competition and pulling.

Interpreting distribution in terms of climatically favourable periods is obviously more plausible in cases where there is good evidence of relatively recent climatic change, especially if a previous cool or moist period is identifiable. Plausibility is increased if related populations are found in mesic areas distant from the mountains under consideration – either in other ranges or in lowland areas at higher latitudes – and if this distribution pattern is repeated in a number of taxa. In itself, adaptation to cool conditions is no indication that a species reached a massif during a cool period for such features would also be expected to evolve after restriction by other means.

Competitive restriction is suggested if a species is replaced in the lowlands by another filling a similar niche. It is, however, also possible for such a restricting competitor to extend into the mountains and in such a case the niches of the two forms may not be very similar.

Figure 17.4 Position of the North Oman mountains. The broken line indicates the approximate course of the Tigris-Euphrates river when the Persian Gulf was dry.

Another indication of competitive restriction is the presence of populations related to the montane one in relatively warm, dry situations at low altitudes, outside the range of the presumed competitor. These suggest that the evolution of montane adaptations may have been secondary. Finally, in cases where the mountain species

and its supposed competitor are closely related, the fact that the latter has advanced features may suggest that it has competitive superiority. Restriction by diffuse competition is much harder to recognise and is only likely to be detected if a taxon occurs in a number of situations but is only confined to mountains in places where the lowlands contain a particular combination of forms, or where potential lowland competitors exceed some critical number. Recognition of 'pulling' also requires comparison of several different situations: it may have taken place when the species concerned is only confined to highlands where competitors are few, but is restricted to the lowlands where highland competitors are abundant.

The mountains of North Oman

The mountains of north Oman (Fig. 17.4) provide an example of a massif where both climatic and competitive factors are likely to have restricted its endemic populations of non-flying inhabitants. This highland area is steep, rising to 3000 m and is considerably cooler and moister, even at low altitudes, than the surrounding lowlands. It is now cut off from the other mountains by desert and sea, but within the last 100 000 years the Persian Gulf was dry and the massif connected by land to the Zagros mountains of Iran. At times, during that period of connection, the climate may have been cooler and moister than at present (Kassler, 1973). If this were so, mesic species may have reached the Oman mountains from the north to be cut off by the subsequent development of drier, warmer conditions, and the refilling of the Gulf. Even if the empty Gulf was generally too dry for such migrations, they could have taken place along tributaries of the Tigris-Euphrates which traversed its floor and received flow from both the Zagros and North Oman mountains.

Reptiles with isolated populations in the Oman mountains are listed in Table I. The ancestors of the two *Lacerta*

Table 1 Reptile species with isolated populations in the mountains of north Oman

	Known altitude range in Oman mountains (m)	Related northern populations	Possible direct competitors in areas surrounding mountains	Known altitude range of competitor in Oman (m)
Lizards				
Lacerta jayakari	0–1980	*L. cappadocica*	—	—
L. cyanura	60–1800	*L. cappadocica*	—	—
Ablepharus pannonicus	100–1700	*A. pannonicus*	—	—
Mabuya tessellata	160–760	*M. aurata*	—	—
Phyllodactylus gallagheri	60–?	*P. elisae, P. griseonotus*	—	—
P. elisae	mainly 460–750 (some lower isolates)	*P. elisae*	*Ptyodactylus hasselquistii*	0–460
*Hemidactylus persicus**	350–1860	*H. persicus*	*Ptyodactylus hasselquistii*	0–460
*Pristurus celerrimus**	0–1980	—	*Pristurus rupestris*	0–2130
Pristurus sp.*	760–1850	—	*Pristurus rupestris*	0–2130
Snakes				
*Pseudocerastes persicus**	1300–2150	*P. persicus*	*Echis coloratus*	0–850

* Related populations occur in warmer drier areas than Oman.

species, the species of *Ablepharus* and *Mabuya* and of *Phyllodactylus gallagheri* are likely to have arrived during a climatic amelioration and to have been isolated entirely by later weather changes for they have relatives in cooler areas to the north and no obvious lowland competitors.

Phyllodactylus elisae, *Hemidactylus persicus* and the snake *Pseudocerastes p. persicus* all have populations to the north and the Oman isolates almost certainly originated there and may have been partly limited by climatic change. They might also have been further restricted by competition, for they are largely or entirely replaced in the lowlands by species filling similar niches. Also, outside the ranges of these putative competitors, the *Hemidactylus* and the *Pseudocerastes* occur in hotter, drier places than in Oman, suggesting that competition may have driven them higher into the massif than climatic factors alone would have done. *Pristurus celerrimus* and *Pristurus* sp. have no northern relatives and may owe their montane distribution largely to competition from the generally similar *P. rupestris*. Species resembling *P. celerrimus* exist in warmer, drier areas on Socotra and this form is one of the most primitive in its genus. *Pristurus* sp. also has a relative in warmer drier areas, *P. flavipunctatus* of south-west Arabia and north-east Africa, although this is not entirely outside the range of *P. rupestris*. Summarising: as climatic changes are known to have occurred recently in Oman, these are certain to have had some effect on the distribution of all the mountain populations and in a number of cases they are likely to have been the main cause of colonisation and subsequent limitation. But competitive restriction also seems to have played a part in several instances, and in the two species of *Pristurus* this factor may have been paramount.

How highland areas maintain competitively restricted species

If a species is eliminated by competition from the lowlands why should this process sometimes stop in highland regions? In part, it may simply be due to the great differences in selective regimes between the two areas. For instance, suppose species *A* has an original range including an isolated massif and it is being replaced by species *B* which has initial competitive advantages unassociated with temperature or humidity tolerance. It is quite likely that when *B* reaches the borders of the massif these advantages will be counterbalanced by the superior adaptation of the remaining populations of *A* to montane conditions which has been developed because this species, unlike *B*, has been subjected to this environment for some time. That is: mountains, especially isolated steep ones, provide a situation of rapid selective change in which a species already established is liable to have considerable initial advantages over one invading from the surrounding lowlands.

The mechanism outlined above may be enhanced by the effects of gene flow. If species *A* is eventually totally restricted to an isolated massif which is relatively small and homogeneous, it is likely to become very closely adapted to montane conditions. In contrast, the close adaptation of species *B* to the abruptly different circumstances in the highlands may be inhibited by gene flow from lower altitudes. The absence of such flow to *A* could also allow it to extend higher than it originally did. Again these effects would be most marked in high, steep mountains that rise abruptly, for the higher *B* extends and provided the absolute distance from the lowlands is not thereby greatly increased, the larger the disadvantageous effects of gene flow. Conversely the more restricted species *A* becomes, the closer the adaptation of its populations bordering *B*, since its range will become still more homogeneous and the deleterious effects of gene flow from other areas will be further reduced. Species *B* would not necessarily have to be an essentially lowland species for *A* to gain advantages by this means. This could still happen even if *B* occurred in mountainous areas elsewhere, provided the lowland interval between these and the massif under consideration is large enough to prevent or severely restrict the flow of 'mountain' genes.

It might be objected that the role of gene flow in disturbing the adaptation of local populations is often disputed. In the absence of selection, only very low levels of gene flow are necessary to spread genes widely (Spieth, 1974) but many authors (for instance, Ehrlich & Raven, 1969) suggest that, in practice, selective forces will usually outweigh the effects of such spread, so that it will not cause a substantial loss of fitness. However, although this may be true where selective forces act continuously, it need not be so where they show great fluctuations through time. This seems to be a common occurrence: for instance, Berry (1978) indicates that house mice (*Mus musculus*) on small islands are subjected to conflicting selective forces at different seasons. An allele may be beneficial at one time of year but is disadvantageous at another. In such a situation, it could consequently spread rapidly at first but would later reduce the fitness of populations that it has reached. Such events also undoubtedly take place on longer time-scales, especially in the case of climatic fluctuations. It seems possible, for instance, that alleles from lower altitudes might flow into nearby montane populations because they are not usually selected against, and may in most conditions be advantageous, but would greatly reduce fitness in occasional years when weather conditions in the mountains were exceptionally severe. In such years these populations could be at a great disadvantage compared with those of a mountain species receiving no lowland gene flow.

Prevention of close local adaptation of the frontier populations of the restricting species by gene flow requires genetic continuity between these and the nearest lowland ones, so it is most likely to occur in forms not utilising microhabitats that are very discontinuous in the mountain area and which also have relatively high levels of dispersal. Ideally, a high proportion of individuals in a population should move in the course of their life-time and the distance covered by a significant proportion of these should be relatively large compared with the distance between the frontier populations and the nearest lowland ones. The topography of mountain regions may well increase the effects of gene flow. If an invading species does not extend over a whole massif, the higher parts of its distribution are often confined to valleys. When this is so, dispersal and therefore gene flow will be concentrated along valley axes instead of being more randomly directed, so a higher proportion of individuals are likely to move towards the frontier populations than in country where demes are more evenly spread. This effect would be enhanced in valleys that taper upwards appreciably since more lowland populations would contribute genes to fewer frontier ones. Conversely, in the case of a mountain-limited species that has the close adaptation of its frontier populations disrupted to some extent by gene flow from higher altitudes, the effect is less in such tapering valleys since fewer highland populations contribute genes to a larger number at lower altitudes.

The effects of lowland gene flow would probably be greatest during the early stages of any colonisation of the mountains by the restricting species, for at this time its distribution is most likely to be continuous; later some fragmentation of range may occur. But it is probably during the initial period that most advantage would accrue to the restricted species, since the reduced fitness of its competitor may give it time to improve its own montane adaptation sufficiently to survive.

Presumably the mechanisms outlined above need not be confined to mountain areas and could also occur in any abruptly different section of the range of a species in the process of being replaced by another: for instance, an isolate of forest in steppe, or a relatively moist area in an arid region. But mountain populations are one of the most likely situations in which such processes might be encountered, for they have greater permanence than many other natural isolates. Consequently mountains have more time to accumulate competitively restricted species. Not only do mountains exist for very long periods, they also 'control' their own weather to a considerable extent. This makes for greater habitat stability which is further increased because the continuity of a particular ecological association may be maintained by its moving temporarily upwards during a transient period of increased temperature and aridity and downwards into the lowlands when conditions are cooler and wetter. Only very extreme climatic swings are at all likely to eliminate it entirely in a high mountain range. The mechanisms are also liable to function in mountains because the degree of difference from the lowlands often increases steadily with distance from their perimeter. Thus, even if the competitor, which has ousted the restricted species from the lowlands, gains a peripheral foothold within the mountains and eliminates the restricted species there, this will not mean that it is adapted to conditions throughout the massif, so its further spread may still be limited. With many other types of discontinuity – for example, a forest isolate in steppe –

conditions near the periphery are often much like those in the centre, so colonisation of the former may well allow spread through the entire isolate.

Although mountains may 'store' relict forms for substantial periods, in many cases they probably only delay extinction, for the distinctive conditions they support are essentially islands and, as such, are subject to the relatively high extinction rates shown by MacArthur & Wilson (1967) to be typical of insular situations. As with oceanic islands, topography is likely to be important in determining likely duration of survival and this would be expected to be far less on small, relatively low massifs than on large, very elevated ones. Parallels with oceanic islands, however, are by no means total, especially of those colonised across water. For instance, in the case of non-flying lowland animals that become restricted to mountains, initial colonisation is by many individuals instead of usually a few, subsequent gene flow is non-existent instead of perhaps being possible at low levels, preadaptation to montane conditions may be slight whereas ability to reach warm oceanic islands at all usually involves some ability to survive in the hot dry conditions likely to be encountered there, at least in the coastal areas initially reached; finally, secondary invaders are likely to reach mountains in force and may consequently cause displacement in the first species but on oceanic islands such colonists are likely to be greatly out-numbered and may well suffer displacement by the original inhabitants.

In the previous, theoretical example, it was tacitly assumed that species A and B always occupied very similar niches and that the exclusion of B from all or part of the massif by A resulted in essentially non-overlapping distributions (Fig. 17.5A). But the same general mechanism, depending on differences in gene flow, that conserves A might still function even if B eventually extended throughout the mountains (Fig. 17.5B). Suppose A and B originally occupy more or less the same ecological space. In some critical niche parameter (for instance, food or temperature), the extremes of which are x and y, A functions best in competition nearest x and B nearest y, but in the lowlands the performance of B is superior to A throughout the parameter, resulting in the success of B and the elimination of A. Once the area of competition is limited to the mountain region, the advantages of more precise adaptation to local conditions gained by A may not allow it to retain the whole of the ecological space it originally occupied, but might enable it to survive in the portion nearest x on the parameter that is critical in lowland competition. If so, A and B could be sympatric in the mountains even though B has overwhelming superiority in the lowlands. Division of the montane equivalent of the niche originally occupied is not the only way A and B might survive together. The same effect might be achieved by one or other of them, or both, being displaced wholly or partly into previously unoccupied ecological space in the mountains.

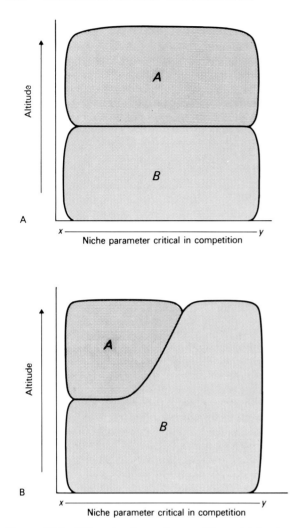

Figure 17.5 Different patterns of competitive restriction to high altitudes. A. Species *B* displaces *A* from lowlands, confining it to higher altitudes than *B* occupies itself. B. Species *B* confines *A* to high altitudes but also extends into them itself. For further explanation see text.

Change in evolutionary potential

Species that have failed in lowland competition may not only be enabled to survive, by restriction, to areas of high relief, but may undergo considerable subsequent change and their potential for future competition is liable to be profoundly altered. To begin with, restriction to such areas enables them to conform closely to montane conditions and this would usually include adaptation to relatively cool and moist regimes. In the fluctuating climatic conditions that have characterised the last million years, it is possible that such modification would eventually allow the restricted species to spread at the expense of the form that restricted it, even though their initial competition was quite unrelated to climate. Such a range extension might of course be transitory and could be reversed by a change back towards warmer and drier conditions. Secondly, in instances where

the restricting form extends its range into the mountains, the restricted species may be subjected to considerable selection pressures that are likely to elicit change. If the two species coexist in the massif, the montane equivalent of the original niche-spectrum that they each occupied exclusively, in sequence, in the lowlands might be simply divided or, alternatively (if circumstances allow) one or other of them, or both, may incorporate into its niche ecological space that was unoccupied or very poorly exploited before. This space could be a minor part of the altered niche or, for one species might even be the whole of it. As areas of high relief usually differ markedly from the surrounding lowlands, and often include microhabitats, food resources etc. that are absent or poorly represented in the latter, the fresh ecological space so occupied is likely to be characteristic of montane regimes. It might, for instance, be based on rock exposures and screes which are often far more abundant at high levels than in the lowlands. Such a shift in niche is more likely to occur in the restricted species for it initially became confined to high ground through its failure in the sort of niche occupied by the restrictor. In addition, gene flow from lower altitudes may tend to limit the ability of the latter to exploit new resources in the mountains. Such niche shifts are likely to result in character displacement in the restricted form.

In cases where the montane equivalent of the lowland niche is partitioned, the restricted form will be able to adapt very closely to the section that it occupies. Once such adaptation has taken place, it may be able to spread downwards into lowland areas previously exclusively occupied by the restricting species since here the adaptation of the latter to this section of the lowland niche will be limited because it also occupies the rest and so cannot specialise to the same extent. Such expansion by the restricted species is likely to be limited since during its montane restriction it will have become adapted to a relatively cool/moist regime which may well prevent extensive spread in conditions of climatic stability.

Rock lizards of Europe and north-west Africa – a possible example of competitive restriction to mountains

Some of the phenomena postulated in the last section may be discernible in west Palaearctic lizards assigned to the genera *Lacerta* and *Podarcis* (systematics discussed by Arnold, 1973). One of the groups concerned is *Lacerta* part II, an assemblage of about 30 relatively primitive species distributed widely but in a disjunct fashion in the Mediterranean region and neighbouring areas to the east. The species to be discussed here, which are mostly called rock lizards, form a fairly homogeneous unit within *Lacerta* part II and may all share a relatively recent common ancestor, although it is not possible to be sure about this. All are typical of fairly mesic areas and it seems probable that they have undergone some reduction in range during the warm period that typified the last stage of the Quaternary in the area. Notwithstanding this, they are quite widespread in

Figure 17.6 Distribution of wall lizards and rock lizards in the Mediterranean region.
blue area – combined distribution of the 14 recognised species of wall lizards (*Podarcis*).

grey areas – distribution of rock lizards (*Lacerta*, part II);
A – *L. dugesii*, B – *L. monticola*, C – *L. andreanszkyi*,
D – *L. perspicillata*, E – *L. bedriagae*, F – *L. horvathi*,
G – *L. mosorensis*, *L. oxycephala*, H – *L. praticola*, I – *L. graeca*.

the Caucasus region, in Asia Minor and parts of the eastern Mediterranean seaboard. But, in contrast, in Europe and north-west Africa they are confined to a few small, mainly mountain areas (Fig. 17.6). This restriction seems far too marked to be due to climatic change alone and, unlike many forms whose reduced distribution can be explained by increased temperature and aridity, they have no closely related populations to the north, where the climate is still relatively cool and wet.

Competitive restriction, on the other hand, seems more likely for a group with generally similar requirements that is widespread in lowlands. This group, the wall lizards (*Podarcis*), contains 14 species that occupy the same general sorts of habitats as many rock lizards and are comparable in size and diet. They may well be a more recent development than the rock lizards and their closest relative appears to be one of the latter, *L. andreanszkyi* of north Africa. Further possible indications of their recency of origin are their great similarity to each other and their compact and continuous combined range. Another pointer that rock lizards may have suffered competitive montane restriction is the presence of close relatives in warm areas outside the range of their presumed competitors, perhaps indicating that adaptation to cool conditions is secondary. Thus the highland north African *L. andreanszkyi* and *L. perspicillata* have a relative on Madeira and nearby dry islands, *L. dugesii* and, as stated, a number of south-west Asian and Caucasian forms have fairly wide distributions including quite dry areas.

With the exception of *L. praticola*, which is found in moist areas below 800 m, all European and north African

rock lizards are confined to massifs of considerable size and steepness, a prerequisite for the mechanisms suggested later. No studies of gene flow in lacertids have been made but it may occur to a significant extent in wall lizards extending into mountains, for they are often more or less continuously distributed up valleys and, although adults do not seem to wander much, there is some evidence that juveniles may disperse for some distance, even over unsuitable intervening territory. Certainly it appears likely that, where rock and wall lizards occur together, the latter are less well adapted to cool conditions. Usual body temperatures of active lizards at a number of localities suggest that rock lizards function at somewhat lower levels and presumably are able to maintain greater activity in cooler conditions. This is true in southern Greece, (*L. graeca*, *P. peloponnesiaca*, *P. taurica*), south-west Yugoslavia (*L. oxycephala*, *L. mosorensis*, *P. muralis*, *P. melisellensis*), north-west Yugoslavia (*L. horvathi*, *P. muralis*) and northern Spain (*L. monticola*, *P. muralis*). Whether this is partly sustained by gene flow effects or whether rock lizards have been merely displaced into cooler niches and become adapted to them is unclear.

Sometimes rock lizards appear to have been simply displaced upwards by *Podarcis* and the two taxa show little altitudinal overlap. This is true of *L. monticola* and *P. muralis* in Spain and in such cases, the two forms tend to occupy structurally similar niches and the lizards themselves are morphologically alike. In other areas *Podarcis* extends into the same altitudinal range as *Lacerta* (part II) and niche differentiation occurs. As their name suggests, rock lizards then typically occur on precipitous boulders, screes and cliffs while *Podarcis* remains in habitats more similar to those typical of the lowlands. When this happens the *Lacerta* part II species show a wide variety of anatomical modifications functionally associated with rocky habitats, particularly the use of narrow crevices as refuges (Arnold, 1973). As predicted earlier, it is the supposedly competitively restricted species that appear to have undergone displacement and modification. Three species strongly adapted to rocky habitats are found in the karst regions of western Yugoslavia. Two of them, *L. horvathi* and *L. mosorensis* are confined to high altitudes, not extending below 500 m, but the most extremely modified species, *L. oxycephala*, occurs down to sea level and reaches some quite dry off-shore islands. It is interesting to speculate, although impossible to confirm, whether this form has reinvaded the lowlands after developing superior adaptation to part of the usual niche spectrum of the restricting taxon in the way suggested previously.

Conclusion

Many systematic and distributional patterns can be interpreted in terms of past competition displacing species into new niches and fresh selective regimes to which they become modified, but such historical interpretations are difficult to confirm and alternative hypotheses are often possible. As in forensic science, the question 'were they pushed?', is often worth asking but, again as in that rather speculative discipline, answers must be accepted with caution.

References

Arnold, E. N. 1973. Relationships of the palaearctic lizards assigned to the genera *Lacerta*, *Algyroides* and *Psammodromus* (Reptilia: Lacertidae). *Bulletin of the British Museum (Natural History), Zoology* **25**: 289–366.

Berry, R. J. 1978. Genetic variation in wild house mice: where natural selection and history meet. *American Scientist* **66**: 52–60.

Brown, W. L. & Wilson, E. O. 1956. Character displacement. *Systematic Zoology* **5**: 49–64.

Erlich, P. R. & Raven, P. H. 1969. Differentiation of populations. *Science* **165**: 1228–31.

Fenchel, T. 1975. Character displacement and coexistence in Mud Snails (Hydrobiidae) *Oecologia* **20**: 19–32.

Grant, P. R. 1972. Convergent and divergent character displacement. *Biological Journal of the Linnean Society* **4**: 39–68.

Grant, P. R. 1975. The classical case of character displacement. *Evolutionary Biology* **8**: 237–337.

Kassler, P. 1973. The structural and geomorphic evolution of the Persian Gulf, pp. 11–32 *in* Purser, B. H. (Ed) *The Persian Gulf*. Berlin: Springer-Verlag.

MacArthur, R. H. & Wilson, E. O. 1967. *The theory of island biogeography*. 198 pp. Princeton: Princeton University Press.

Mayr, E. & Diamond, J. 1976. Birds on islands in the sky: origin of the montane avifauna of Northern Melanesia. *Proceedings of the National Academy of Sciences of the United States of America* **73**: 1765–69.

Schindel, D. E. & Gould, S. J. 1977. Biological interaction between fossil species: character displacement in Bermudan land snails. *Paleobiology* **3**: 259–69.

Spieth, P. 1974. Gene flow and genetic differentiation. *Genetics* **78**: 961–65.

CHAPTER 18

The evolution of predators in the late Cretaceous and their ecological significance

J. D. Taylor

The major extinctions that affected many animal groups, such as the dinosaurs, marine reptiles, ammonites, belemnites, rudistid bivalves and a number of other animal groups at the end of the Cretaceous Period, some 70 million years ago, have attracted much attention from palaeontologists and evolutionary biologists. This pre-occupation with extinction, however intriguing, has diverted attention from the biologically more interesting but frequently overlooked phenomenon, the massive diversification which was taking place in many marine animal groups at the end of the Mesozoic. This diversification formed part of a major reorganisation in the composition and structure of marine communities and has been called the 'Mesozoic revolution' (Vermeij, 1977). It stamped an essentially modern aspect upon faunas, so that animals of the early Cenozoic had general similarities with those living today. One remarkable feature of the late Mesozoic diversification was the appearance and rapid evolutionary radiation of groups of animals that today are major predators in marine environments. These animals include several important groups of teleost fishes, brachyuran crabs, stomatopod crustacea, coleoid cephalopods and predatory prosobranch gastropods.

These animals feed upon a wide variety of prey and today are known to exert considerable influence on the regulation and behaviour of prey animals and upon the composition of marine benthic communities. For instance, in coral reef ecosystems the heavy predation of invertebrates by fishes is thought to account for the generally cryptic behaviour and nocturnal activity patterns on the part of many prey species (Bakus, 1964). More often, the effects of predators are more subtle, but no less important. For example, predation by intertidal dog whelks (*Nucella* and *Thais*) allows the coexistence of species of barnacles which, in the absence of a predator, would compete for space and result in the exclusion of some of the barnacle species (Connell, 1961, 1970).

The rapid evolution and diversification of new predatory animals in the late Mesozoic had profound effects upon the structure of benthic communities and upon the adaptation and evolution of new species. For instance, it is probable that many features of gastropod shell ornament have been evolved in response to predation pressure (Vermeij, 1978), and it could be further reasoned that some of the extinctions at the end of the Cretaceous could have been caused by predation pressure. Predatory animals are, of course, known before the late Mesozoic. Groups such as asteroids, cephalopods and various types of fishes each have a long geological history, but they seem to have been present in nothing like the diversity and numbers as were the predators in the late Mesozoic and early Cenozoic.

This chapter is devoted to examining the diversification, and its implications, of one of the predatory groups – the prosobranch gastropods. This subclass of gastropods which includes most shelled, marine snails, is one of the most diverse groups of animals in the seas and includes a wide variety of feeding types including algal grazing, suspension and deposit feeding, parasitism and those which 'graze' upon other animals. Throughout most parts of the world about half of the gastropods are predators; that is, they actively seek and consume whole prey animals and are thus distinguished from other carnivores that graze on sedentary animals such as sponges and ascidians. There are approximately 6000 species of predatory gastropods living today and these are classified into some 26 families. The history of these gastropods has not been studied in detail but there is no doubt that the appearance and adaptive radiation of these families in the late Cretaceous is one of the most spectacular of Mesozoic diversifications. In order to give some idea of the range of habits of these gastropods and their contributions to benthic communities, I shall briefly review some of the facts about the biology and ecology of living species.

Modern predatory gastropods

Predatory prosobranchs occur in most marine benthic habitats, from the high intertidal zone down to hadal

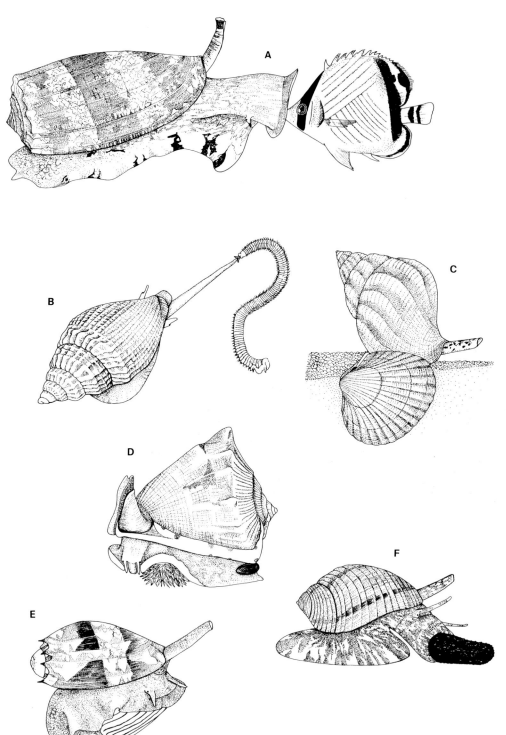

Figure 18.1 Various predatory gastropods attacking their prey
A *Conus geographus* Linnaeus enveloping a small chaetodontid fish
B *Lora trevelliana* (Turton) impaling a polychaete
C *Buccinum undatum* Linnaeus wedging open a cockle
D *Cassis tuberosa* (Linnaeus) eating an echinoid

E *Melo amphora* Solander enveloping another volutid gastropod *Zebramoria*.
F *Tonna perdix* Linnaeus eating a holothurian.
(All figures redrawn from various published and unpublished sources.)

depths, and from the equator to the highest latitudes, with a few species even found in freshwater habitats. They live on or in most kinds of substrate from rocky shores to fine muds. They are of course not equally abundant in all habitats but in some, such as the intertidal rocky shore, they are often the major predator present.

When compared with other feeding methods such as algal grazing or deposit feeding, the act of predation involves a complex sequence of behaviour on the part of the predator. Usually this involves the processes of search, evaluation, pursuit, capture, immobilisation and consumption. Predators should thus differ anatomically and behaviourally from gastropods that feed in other ways. The structural adaptations often found in a prosobranch predator are a well-developed siphon and chemosensory osphradium for locating the prey, an often highly extensible proboscis for probing and enveloping the prey, radula teeth modified for tearing, scraping, drilling or impaling the prey, oesophageal glands to provide toxic secretions for immobilising the prey, and glands in the foot or proboscis to provide secretions to aid in the dissolution of shelled prey (Ponder, 1973; Taylor et al., 1980).

As might be expected from such a diverse group, predatory gastropods show considerable variation in feeding behaviour and eat a wide variety of prey including members of most invertebrate groups and even a few vertebrates. Some families contain species which are relatively specialised and eat only a single or a few species of prey animals, whilst other families contain species with more catholic diets. For instance, all species of the largely tropical family Mitridae so far examined have been found to feed upon sipunculid worms, whilst the common whelk (Buccinidae) feeds upon species from seven or eight different invertebrate phyla. Polychaete worms and molluscs are probably the most abundant macro-invertebrate animals and these are the items most commonly eaten by predatory gastropods. But to give some idea of the diversity of trophic adaptations found in gastropods a few examples are described below; further details and references may be found in Taylor et al. (1980).

Holothurians are avoided by many predators because of the toxic secretions produced in the skin, but they form the main food of the large gastropods in the family Tonnidae which envelop them in the highly extensible proboscis (Fig. 18.1 F). A related family, the Cassidae, feed almost exclusively on echinoids and will even eat such spiny forms as *Diadema*. The echinoid is held down by the large foot whilst a hole is drilled through the test by the proboscis, which is then inserted to scrape the flesh inside the test (Fig. 18.1 D). Members of another related family, the Cymatidae, and, in particular, species of *Charonia*, regularly eat asteroids, including the notorious Crown-of-Thorns starfish (*Acanthaster*); some workers have attributed the occurrence of large populations of *Acanthaster* to the removal of the large and decorative shells of *Charonia* by shell collectors.

Large Crustacea are generally too active to be caught by gastropods but representatives of the family Harpidae seem able to catch crabs and prawns which they immobilise in the large active foot, at the same time releasing copious amounts of mucus. A similar technique is used by species of Volutidae which feed mainly upon bivalves and other gastropods which are enveloped and asphixiated in the large foot (Fig. 18.1 E).

Representatives of two families, the Naticidae and Muricidae, are known for their ability to drill through the shells of their molluscan prey. The drilling process combines mechanical scraping by the radula and the secretion of acids and chelating agents by glands in the proboscis of naticids and the foot of muricids. The holes produced are neat and circular – those of the Naticidae usually having a countersunk rim. Naticids generally forage, and drill their prey, beneath the surface of the sediment, whilst the Muricidae feed upon epifaunal or shallow burrowing prey. The Muricidae are particularly diverse upon rocky substrates whilst the Naticidae are restricted to soft substrates.

Some gastropods in the families Buccinidae and Fasciolariidae wedge open the shells of their bivalve prey with the outer lip of the aperture of their own shells (Fig. 18.1 C) and in some species, particularly of *Busycon* (Melongenidae), this behaviour has been elaborated, so that the predator uses leverage of the outer lip of its own shell to chip pieces off the edges of the shells of its bivalve prey, thereby opening up a gap sufficiently large for the insertion of the proboscis.

Other groups of predatory gastropods use toxic secretions from modified oesophageal glands to immobilise their prey; some squirt this toxin at the prey from the proboscis, whilst others touch the prey with the tip of the proboscis and release the toxin. Probably the most advanced feeding behaviour is found in many species of the Turridae, Terebridae and all species of Conidae where the radula teeth, which in most gastropods are normally attached in rows upon a ribbon, are secreted and used individually. Each tooth is highly modified into a harpoon and frequently has complex barbs at the distal end. The teeth are stored in a quiver at the base of the proboscis until needed; one of the teeth is then moved to the distal end of the proboscis and held by a sphincter muscle. When feeding, the radula tooth is stabbed into the prey and at the same time a neurotoxic venom is injected down the hollow tooth, immobilising the prey within seconds. Polychaetes are the most common prey of these gastropods (Fig. 18.1 B), although some species of Conidae eat other molluscs, and a few species catch and eat small fish. The proboscis expands considerably to accommodate the prey, but the fish are usually too large to pass far down the alimentary canal and are usually digested whilst in the extended oesophagus (Fig. 18.1 A). Many gastropods feed upon moribund or dead animals, e.g. the Nassariidae, many species of which are opportunistic feeders, which forage actively and are attracted from great distances by chemical stimuli from moribund prey.

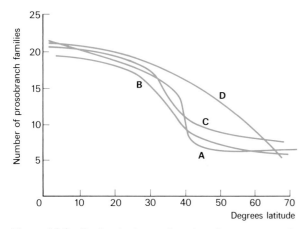

Figure 18.2 Decline in the number of predatory gastropod families along a number of continental margins. A north-east Atlantic, B north-west Atlantic, C south-west Atlantic, D north-east Pacific. Full data in Taylor *et al.* (1980).

As in many animal groups there is a steep latitudinal gradient in the diversity of predatory gastropods, with the largest number of families and species found at low latitudes (Fig. 18.2). As might be expected gastropods show adaptations to the different environmental conditions found along the gradient. Out of those families that occur at high latitudes, two, the Buccinidae and Turridae, become relatively much more important than at lower latitudes, so that around latitudes 40–70° North and South, species of Buccinidae comprise about 40–50 per cent of the predatory gastropod species, compared with less than 5 per cent in tropical regions (Fig. 18.3). The Buccinidae are a family which includes the European edible whelk *Buccinum undatum*; the high latitude species have generalist feeding habits when compared with their tropical counterparts. The whelk may feed on species from a variety of invertebrate phyla as well as eating carrion, and it obviously has considerable behavioural flexibility in dealing with different food types. These behavioural features are thought to be adaptations to the fluctuating, less predictable, food supplies found at higher latitudes (Taylor & Taylor, 1977). The other family which is very important at high latitudes is the Turridae. Members of this family seem to feed largely upon deposit-feeding polychaetes. The food available for deposit feeders at high latitudes is likely to be present in fairly constant amounts throughout the year and the polychaetes therefore form a stable resource for the gastropods in otherwise strongly seasonal environments.

If we examine the detailed ecology of groups or guilds of coexisting, functionally similar gastropod species, particularly in tropical shallow-water environments, then we find that the species frequently show small but clear differences in the microhabitats they occupy, in their diets and in their size. These observations are consistent with the idea that the differences seen between the coexisting species are the results of competitive interactions between the closely similar taxa. The evidence for this is generally circumstantial and difficult to test, but some support for the idea comes from 'natural experiments' in which predatory gastropod species, in the absence of their usual putative competitors, show character shifts and may take a wider range of prey, live in a greater range of habitats and become more abundant. For instance, at Easter Island on the edges of the tropical Indo-Pacific province *Conus miliaris*, in the absence of other *Conus* species has a diet of which 77 per cent consists of species of polychaetes which it never or infrequently eats in other parts of its geographical range, where many other coexisting species of *Conus* may be present. Moreover many of these polychaetes eaten by *Conus miliaris* at Easter Island are important components in the diet of *Conus* species elsewhere (Kohn, 1978). Similarly at Hong Kong, also on the edge of the tropical Indo-Pacific province, a muricid gastropod *Cronia margariticola* has a much greater depth range and feeds upon a much greater range of prey than in other parts of its range where closely related species are also abundant (Taylor, 1980). It must be stressed, however, that there are no controls to the 'experiments', and that factors other than competition could have caused the observed differences.

In many areas of the world, but particularly in the tropics, species from a number of predatory gastropod families may coexist in the same habitats, but generally the food webs involving members of the different families are largely independent with little overlap. It is probable that many of the predatory families are descended from a common ancestor and we thus might speculate that their dietary adaptations and specialisations may have arisen by competitive interactions. Selection will tend to favour those phenotypes that feed most efficiently upon particular prey types thereby producing anatomical and behavioural modifications enabling them to cope more effectively with those particular foods. It is clear that this will have the effect of 'canalising' the gastropods into feeding upon particular prey types so that, after the initial radiation, evolution has consisted largely of diversification in taxa specialising upon one or few prey types. For instance, the adaptations that enable *Cassis* to catch and eat echinoids would not be very effective for the capture of free-living polychaetes, as practised for example by representatives of the Conidae.

History of predatory gastropods

If we accept that the morphology of the animal and its shell is correlated with its life habits, then it appears that, from an examination of gastropod faunas throughout the Cenozoic, the specialisations and adaptations of the predators were evolved and established a considerable time ago. If we trace the lineages of predatory gastropods back through geological time, then we find that all but two taxa first appeared during the Cretaceous Period. The two exceptions are members of the superfamily Naticacea, which have a history as far back as the early Jurassic, and

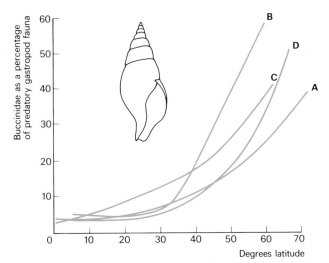

Figure 18.3 Increase in importance of whelks (Buccinidae) at higher latitudes along a number of continental margins. A north-east Atlantic, B north-west Atlantic, C south-west Atlantic, D north-east Atlantic.

representatives of the Harpidae whose earliest records have been found in Eocene rocks. It is therefore clear that the main adaptive radiation and the probable acquisition of the broad feeding habits adopted by the various families took place during a rapid phase of an essentially late Cretaceous evolution.

Apart from the Naticacea, predatory prosobranchs are first recognisable in the Aptian stage of the Cretaceous (approximately 110 million years ago), but by the later Albian and Cenomanian stages representatives of five families were present. Even though the Naticacea have a longer geological history in the Mesozoic, the first unambiguous traces of their distinctive countersunk drill holes in bivalve shells are not seen until the Albian (some 105 million years ago). This is not to say that the pre-Albian naticids were not predators, but they must have fed in a different way. The other gastropod family that drills molluscs is the Muricidae. Drill holes of muricids also appear in bivalves of Albian age contemporaneously with the appearance of the first muricid gastropods. Holes bored into echinoids, probably by members of the Cassidae, are not seen until the Eocene.

We can thus say that predatory gastropods as we now know them first appeared in the upper part of the Lower Cretaceous (some 110 million years ago). The phylogenetic relationships to other prosobranch gastropods have yet to be worked out, but from a knowledge of living gastropods we can outline some of the changes in feeding behaviour that may have led to the evolution of predatory habits. As previously mentioned, prosobranch gastropods show a range of feeding behaviours, from omnivorous browsers feeding upon sessile colonial invertebrates as well as plant material, to browsing carnivores feeding upon sedentary colonial animals such as sponges, coelenterates and ascidians, to predatory feeding upon sluggish or sessile

invertebrates and thence to hunting predators, feeding upon much more mobile prey such as free-living polychaetes and fish. The acquisition of certain anatomical and behavioural characters allowed the prosobranchs to feed upon more mobile prey, a mode of feeding previously unexploited by gastropods. This, in turn, allowed a rapid evolutionary radiation with the development of new specialisations for coping with particular prey. The constraints upon the extent of this adaptive radiation, other than the anatomical and functional limitations, probably were and are determined by a number of factors that include competition with other predatory groups, the success of prey species in avoiding predation, and predation upon the gastropods themselves by fish and crabs.

The main adaptations of each predatory gastropod family were probably established by the late Cretaceous-early Cenozoic and subsequent evolution has largely been a process of refinement and of diversification within particular major types. Examples are the five hundred or so polychaete-eating species of *Conus* living today, which are all fairly similar morphologically, and the multitude of species of intertidal muricids feeding upon barnacles and sedentary molluscs.

An analysis of gastropod faunas from the Cretaceous and Cenozoic (Taylor *et al.*, 1980), shows the number of predatory families to have an almost exponential rise during the late Cretaceous and early Cenozoic, with a flattening of the curve towards the middle of the Cenozoic (Fig. 18.4). This sort of rise in diversity is characteristic of animal groups invading new adaptive zones. Similar curves are seen, for example, in the adaptive radiation of the mammals in the early Cenozoic and the initial radiation of invertebrate animals in the Cambrian (Van Valen, 1971; Sepkoski, 1978, 1979). The curves are similar to the logistic growth model used by population biologists to describe the growth of single species populations (see Chapter 10). That is, where the population increases sigmoidally with time, the early parts of the curve approximate an exponential rise, but the curve later reaches an asymptote towards an equilibrium population size, which is the maximum that the environment can support in a given area.

Sepkoski (1978, 1979) explored the use of the same logistic model in describing the diversification of species and higher taxa. After an initial radiation, diversity increases exponentially until an equilibrium condition is reached, which is diversity dependent. That is, the origin of new taxa is balanced by the extinction of others. Thus, with increasing diversity, the extinction rate increases faster than the rate of origins of new taxa. There are probably many reasons for this. One is that as diversity increases there will be competition causing exclusion and extinction of some species. Additionally, the colonisation of new habitats will be more difficult because of the species already in occupation; also, where 'species-packing' is greater, population sizes may be reduced, resulting in a greater probability of extinction for the rarer species. It

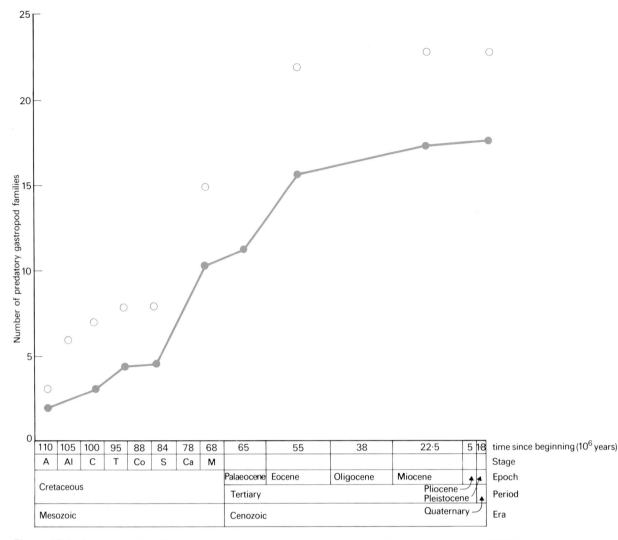

Figure 18.4 Increase in diversity of predatory gastropod families from the late Cretaceous to the present day. Solid line represents the mean number of predatory gastropods present in various Cretaceous and Cenozoic faunas. Open circles represent maximum number of families reported in faunas for each age. Full data in Taylor *et al.* (1980).
Key to stage names for Cretaceous Period
A – Aptian, Al – Albian, C – Cenomanian, T – Turonian, Co – Coniacian, S – Santonian, Ca – Campanian, M – Maastrichtian.

needs to be stressed that such sigmoid growth curves occur only where populations are growing into unconstrained, open systems where ecological resources such as food and space are sufficiently abundant for organisms to multiply until their own numbers limit the increase. This reinforces the idea that predatory gastropods in the late Cretaceous were establishing trophic niches which, initially, overlapped very little with those of other organisms. That is, there were apparently no ecological equivalents of the predatory gastropods present in the Cretaceous.

The first families to become numerically dominant during the earlier part of the gastropod radiation were, by analogy with their present-day relatives, probably predators upon other molluscs or upon sedentary polychaetes. Later, towards the end of the Cretaceous, various gastropods appear to have eaten a wider variety of prey and utilised more specialised capture techniques. By the Eocene some of the specialised families had become very diverse and abundant in benthic faunas. For instance, the Turridae, perhaps the most diverse of the predatory families (330 genera), had reached approximately their present-day levels of importance by the lower Cenozoic. Certain families such as the Volutidae and Fasciolariidae, which were relatively important in Cretaceous faunas were eclipsed during the early Cenozoic. In most parts of the world today members of these families form only relatively minor parts of predatory gastropod faunas. The present-day domination of high latitudes by trophically generalist whelks did not seem to begin until the late Miocene and their radiation at those latitudes was probably a response to polar cooling.

Other predatory groups

Two other major predatory groups of animals, the teleost fishes and crabs also underwent major adaptive radiations in the late Cretaceous and early Cenozoic.

Fishes are today clearly the most important predators of marine benthic communities and studies of fish diets (for instance, those of Randall, 1967; Hiatt & Strasburg, 1960; Hobson, 1974) reveal the great range of invertebrate and vertebrate animals eaten. Many fishes are food generalists: two species of puffer fish (*Sphaeroides*) from Florida (Targett, 1978) feed mainly upon crabs, bivalves, gastropods, and amphipods, but may also take smaller quantities of isopods, barnacles, tunicates, ophiuroids, polychaetes, stomatopods, echiurans, sipunculids, copepods, sponges, pycnogonids, bryozoans and holothurians. However, there are many examples of more specialised fishes; for instance Hobson (1974) reports that *Forcipiger flavissimus*, a chaetodontid fish, which has an extended snout and jaws that act like snipe-ended pliers, tears off pieces of sessile invertebrates. A related species *F. longirostris* uses the highly elongated snout to probe deep into coral crevices for crustaceans. The labrid fish *Coris gaimardi* overturns small stones with its snout and feeds upon the animals exposed beneath.

The acanthopterygian order Perciformes, which includes a very large number of present-day marine fishes, showed an apparent major radiation at the very end of the Mesozoic, so that by the end of the Eocene nearly all of the present-day families had appeared and probably, therefore, so had most of the major types of feeding adaptation.

Schaeffer & Rosen (1961) attribute the radiation of higher teleosts to improvements in jaw mechanics, which allowed the diversification into a seemingly infinite number of feeding specialisations. Recent work is showing that many of the higher teleost taxa which apparently appeared so suddenly in the early Cenozoic have their origins in earlier Cretaceous forms usually assigned to more primitive teleost groups. Many of these may have been predators upon invertebrate animals. However, whatever the details of phylogenetic relationships it is apparent that large changes in the composition of marine fish faunas occurred in the late Cretaceous.

Crabs, another major predatory group in contemporary benthic communities, are an extremely diverse group comprising some 4500 out of about 8300 decapod crustacean species. They live in a wide range of marine habitats, but most species occur in the shallow sublittoral zone. All show a high degree of mobility and have a wide range of feeding adaptations including deposit feeding, suspension feeding and herbivory. However, a large proportion of species are carnivores and many have chelae specialised for breaking and crushing mollusc shells. The effects of crab predation on molluscs have been extensively documented by Vermeij, (1978; see also Zipser & Vermeij, 1978; Hughes & Elner, 1979) who shows that a variety of crabs such as *Cancer*, *Daldorfia*, *Eriphia*, *Carpilius*, *Calappa* and *Carcinus* can effect extensive damage upon even large thick-shelled gastropod prey. The crabs will usually try to break the apex of gastropod shells or crack the body whorl; but if this fails then the crab will progressively break the outer lip of the shell aperture. The tropical sand-living family of crabs, Calappidae, have elaborated this last technique, and cut a jagged spiral into the shell of their prey. This 'tin opener' like peeling can extend for as much as two whorls around the shell.

Portunid crabs are renowned for their generalised feeding habits. The Blue Crab (*Calinectes sapidus*) for instance, feeds upon live or dead fish, bivalves, gastropods, crustacea, insects, polychaetes, bryozoa, hydroids and plants (Tagatz, 1968).

Crabs are generally rare in the early Cretaceous but by the early Cenozoic they had become very diverse, abundant and geographically widespread; most of the families, containing predatory species, such as the Xanthidae, Portunidae, Cancridae and Calappidae, appeared during a late Cretaceous-early Cenozoic radiation.

Another group of predators, the cephalopods, have a long geological history extending back to the late Cambrian (some 510 million years ago), but fundamental changes in cephalopod faunas took place at the Cretaceous–Cenozoic boundary. As is well known, both the ammonoids and belemnites became extinct and the groups that survived into the Cenozoic were the coleoids (squid), sepioids (cuttlefish), octopods and a few nautiloids. Packard (1972) has emphasised that there are many convergent features between living cephalopods and fish and suggests that those groups that survived and diversified across the Cretaceous–Cenozoic boundary (about 65 million years ago) were the most fish-like. Many cephalopods eat fish and some fish eat cephalopods, and Packard documents some probable instances of competitive interactions between them, suggesting that there is some broad-scale partitioning of the oceans by the two groups. It is probable that many of the shallow water habitats where cephalopods were abundant during the Mesozoic were increasingly occupied by fish in the late Cretaceous and early Cenozoic.

Effects of predation

Although predation has undoubtedly been an important factor in the evolution of benthic communities throughout the last 600 million years, the sheer diversity and numbers of new predators which appeared towards the end of the Mesozoic suggests that the level of predation, and the diversity of prey eaten, increased markedly at that time. Therefore it seems reasonable to conclude that the rapid diversification of predators would have had considerable and far reaching effects on invertebrate populations and communities perhaps not previously exposed to such an intensity of predation. The strong selection pressures imposed by predation would have produced widespread evolutionary responses in the prey populations, which might be expected to fall into three broad categories – morphological, behavioural and those concerning the reproductive biology.

Responses of organisms to predation are coevolutionary in the sense that both predator and prey are involved in an escalating 'arms race'. For example, any development of more elaborate armour and avoidance devices by prey animals could lead to the development of more effective prey location and capture devices by the predator. Predators will, in normal circumstances, tend to choose the most profitable prey, that is those which supply its energetic and nutritional demands and which can be caught with the least expenditure of effort and the least risk. Selection will favour those adaptations of the predator, whether morphological or behavioural, which contribute towards these ends. At the same time, prey organisms are evolving various sorts of defence mechanisms to reduce the effects of predation.

These interactions involving predator and prey are never simple, because prey animals usually have to cope with a number of different predators having different feeding methods. The predators themselves usually feed upon a variety of different types of organism, each with diverse defence mechanisms. Moreover, the predators themselves are nearly always the prey of some other predator, and many of their own adaptations are concerned with predator avoidance. In present-day marine ecosystems the effects of predation upon prey populations can be tested by experimental analysis and a good deal of such work has been done with intertidal rocky shore communities. However, these studies, although very instructive, tell us only about relatively short-term effects and any ideas we have about the effects of predation over evolutionary time are necessarily inferential.

Predation and morphology

It is only relatively recently that biologists and palaeontologists have been investigating the evolutionary effects of predation upon benthic communities. The most notable of these workers is Vermeij (1978) who has provided an immense amount of evidence that many features of the gastropod shell such as spines, nodes, ridges, apertural dentition, elongate apertures, stout columella and thick shells are adaptations to resist shell crushing predators, in particular, crabs and fish. Recently, some of these ideas have been confirmed experimentally by Palmer (1979) who showed that if the spines of several species of muricid are removed, vulnerability to predation increases markedly. In a study of shore crabs (*Carcinus maenas*) eating dog whelks (*Nucella lapillus*), Hughes & Elner (1979) showed that the crabs attack all sizes of dog whelk encountered, but reject

those that do not break within 2·75 minutes; there is thus a persistent selection against individuals with weaker shells.

The proportion of gastropods possessing strengthening features is greatest in the tropics, which suggests that the intensity of predation is higher there. Vermeij (1977) has correlated an increase in the frequency of gastropods with such strengthening features in fossil communities from the late Mesozoic onwards. This is accompanied by a general decrease in the numbers of gastropods having such features as open and planispiral coiling (coiling in a single plane) and the presence of an umbilicus, which are considered to be mechanically weak features. Many of the gastropods that possess strengthening devices are themselves predators on other invertebrates. Morphological compromises are obviously necessary and those in species of the family Conidae are particularly interesting (Kohn *et al.*, 1979). This family, which first appeared in the Upper Cretaceous, has cone-shaped shells with long, narrow apertures and almost planispiral coiling, so that the thinner shell of the inner whorls (those laid down early in growth) is not exposed externally. The shell material of the last whorl is of nearly equal thickness all round. Additionally, the inner whorls are resorbed thus reducing shell weight and increasing the living space, factors important in gastropods which generally swallow their prey of polychaetes, fish or molluscs whole. All these features contribute to an extremely strong shell which is nevertheless still vulnerable to attack by fish and crabs, and the broken remains of cone shells are common in coral reef habitats.

Bivalve molluscs are some of the most abundant of the larger invertebrates in most marine benthic habitats, and a large proportion of the mortality in many populations is due to the activities of predators, such as fish (Schwartz & Porter, 1977), crabs (Hughes & Elner, 1979), gastropods (Ansell, 1960; Edwards & Huebner, 1977) asteroids (Schwartz & Porter, 1977) or shore birds (Goss-Custard *et al.*, 1977). These predators feed largely upon epifaunal (those which live on the substrate) or shallow-burrowing bivalves, deeper burrowing forms being much less heavily exploited. One of the features of Mesozoic molluscan faunas was the great rise in the numbers of siphonate (and thus burrowing) suspension feeding bivalves and a concomitant decrease in the proportions of epifaunal species and those which live attached but half-buried. At the present day between 80–90 per cent of suspension feeding bivalve species are siphonate burrowers (Stanley, 1977). This increase in the numbers and diversity of deeper burrowing forms with siphons may be associated with the increase in predation pressure during the late Mesozoic. To express this anthropomorphically, bivalves were effectively driven underground by predation pressure.

Predation and behaviour/habitat

Large aggregations of epifaunal sessile bivalves are today generally found in intertidal (for instance mytilids), or estuarine habitats (oysters and mytilids), where the diver-

it. The history of biotas is viewed as one of fragmentation of ancestral biotas into daughter (vicariant) biotas. Speciation is viewed as being allopatric with allopatry caused by the fragmentation of the original area. In other words speciation is the direct result of and is contemporaneous with the formation of a barrier (Fig. 3.2, compares the differing views of allopatric speciation according to dispersalist biogeographers and cladistic vicariance biogeographers). Since vicariance biogeographers place little emphasis on dispersal across barriers then the identification of a centre of origin becomes irrelevant.

Our attempts to explain the distribution of organisms are not governed by an agreed set of methodological rules. Ecological and dispersal biogeography are similar to one another in principle, if not in scale, and explain distribution of organisms in terms of cause. They differ from vicariance biogeography in the kinds of questions asked. Vicariance biogeography is, in many respects, diametrically opposed to the traditional methods since it concentrates on the identification of pattern rather than cause. This is definitely an untraditional approach but our inclusion of two such chapters in this book reflects a growing interest in vicariance biogeography over the last decade and, we believe represents a comment on biogeography in the seventies.

References

Croizat, L., Nelson, G. & Rosen, D. E. 1974. Centers of origin and related concepts. *Systematic Zoology* 23: 265–287.
Darlington, P. J. Jr. 1957. *Zoogeography: the geographical distribution of animals.* 675 pp. New York: Wiley & Sons.
Darlington, P. J. Jr. 1965. *Biogeography of the southern end of the world.* 236 pp. Cambridge, Massachusetts: Harvard University Press.
Ekman, S. 1953. *Zoogeography of the sea.* 417 pp. London: Sidgwick & Jackson. (English translation of 1935 German edition).
George, W. 1962. *Animal geography.* 142 pp. London: Heinemann.
Hesse, R., Allee, W. C. & Schmidt, K. P. 1951. *Ecological animal geography*, 2nd edition. 715 pp. London: Chapman & Hall.
MacArthur, R. H. 1972. *Geographical ecology.* 296 pp. New York: Harper & Row.
MacArthur, R. H. & Wilson, E. O. 1967. *The theory of island biogeography.* 198 pp. Princeton: Princeton University Press.
Matthew, W. D. 1915. Climate and evolution. *Annals of the New York Academy of Sciences* 24: 171–318.
Mayr, E. 1944. Wallace's line in the light of recent zoogeographic studies. *The Quarterly Review of Biology* 19: 1–14.
Nelson, G. 1978. From Candolle to Croizat: comments on the history of biogeography. *Journal of the History of Biology* 11: 269–305.
Nelson, G. & Platnick, N. I. 1980. A vicariance approach to historical biogeography. *Bioscience.* 30: 339–343.
Patterson, C. 1980. Methods of paleobiogeography. pp. 446–500. *in* Nelson, G. & Rosen, D. E. (Eds) *Vicariance biogeography: a critique.* New York: Columbia University Press.
Rosen, D. E. 1975. A vicariance model of Caribbean biogeography. *Systematic Zoology* 24: 431–464.
Rosen, D. E. 1978. Vicariant patterns and historical explanation in biogeography. *Systematic Zoology* 27: 159–188.
Simberloff, D. S. & Wilson, E. O. 1970. Experimental zoogeography of islands. A two-year record of colonization. *Ecology* 51: 934–937.
Simpson, G. G. 1962. *Evolution and geography.* 64 pp. Eugene: Oregon State System of Higher Education.
Vermeij, G. J. 1978. *Biogeography and adaptation.* 332 pp. Cambridge, Massachusetts: Harvard University Press.
Vuilleumier, F. & Simberloff, D. S. 1980. Ecology versus history as determinants of patchy and insular distributions in High Andean birds. *Evolutionary Biology* 12: 235–379.

P. L. Forey
British Museum (Natural History)

CHAPTER 19

The land snails of islands – a dispersalist's viewpoint

J. F. Peake

'*Somewhere among the note-books of Gideon I once found a list of diseases as yet unclassified by medical science, and among these there occurred the word Islomania, which was described as a rare but no means unknown affliction of the spirit. There are people, Gideon often used to say, by way of explanation, who find islands somehow irresistible.*' Lawrence Durrell (*Reflections on a marine Venus*).

Small isolated islands hold a fascination for many people and they have figured in our literature to the extent that stories such as *Robinson Crusoe* and *Treasure Island* now occupy prominent places in our cultural heritage. The reasons for these responses are complex, but islands obviously generate considerable emotion and evoke curiosity amongst a wide range of people. Biologists have reacted in a similar manner, for there is a history of individuals who have been stimulated to visit and describe the inhabitants of many of the more remote microcosms. A desire to organise this information led from the initial inventory approach to a more synthetic or narrative phase. Here the objectives were both to recognise patterns in the distribution of organisms and to provide plausible explanations for their origin and persistence. The explanations (or arguments) became increasingly rigorous and open to experimental investigation as more information became available. A landmark along this pathway was provided by the theoretical studies of MacArthur and Wilson (1967). Nevertheless it still remains true that, for many groups of organisms and for many areas of the world, progress in the analysis of distribution patterns has been limited by the lack of relevant or adequate data.

Both Darwin and Wallace recognised that the distribution of land snails on islands challenged their concepts of dispersal. Here was a group which it might be supposed from direct observation had very poor powers of dispersal, but were nonetheless ubiquitous inhabitants of islands, even oceanic islands which had always been separated from larger land-masses. Surely the snails must have reached such islands by transoceanic dispersal. Moreover, Wallace was aware that the molluscan faunas of these islands were frequently so rich in both numbers of species and endemic taxa, that he was moved to suggest 'if we take the whole globe, more species of land shells are found on the islands than on the continents' (Wallace, 1892, p. 80). Nevertheless, these faunas can also be regarded as depauperate or disharmonic, that is unusual in their particular combinations of species, for many of the taxonomic groups found in continental areas are either absent or poorly represented on islands.

There are, therefore, two groups of problems associated with the distribution of land snails on islands. First, are those concerned with the geographical range of taxa and the role of dispersal in determining these distributions. Second, are those associated with problems of diversity in the numbers of species, and hence in the evolution and ecology of the animals. In both groups of problems the search for geographical patterns and explanations is largely empirical. It is impossible, however, to consider adequately such problems without a brief review of some broader biogeographical issues.

In contrast to some of their contemporaries, both Darwin and Wallace accepted the overall permanence of the oceans and the continents, although they insisted that the terrestrial environments of many regions had been radically changed during geological times. Hence, the emphasis was placed on dispersal as an important mechanism in the establishment of distribution patterns and comparisons were drawn between the occurrence of sedentary and vagile (mobile) organisms on different land masses. While we can be critical of such 'assumptions', because there is more information available now, their views were derived from personal experiences and studies. Both men were concerned with problems and paradoxes which still exercise the minds of biologists; for example, problems associated with the nature and permanence of barriers that separate land masses. Indeed, Wallace expounded his approach forcibly in the introduction to *Island Life* (1892, p. 10), '...the permanence of oceans and the general stability of the continents throughout all geological time, is as yet imperfectly understood, and seems, in fact, to many persons in the nature of a paradox. ...it is certainly the most fundamental question in regard to the subject we have to deal with: since, if we once admit that continents and oceans may have changed places over and over again (as many writers maintain) we lose all power of reasoning on the migration of ancestral forms of life, and are at the mercy of every wild theorist who chooses to imagine the former existence of a now-submerged continent to explain the existing distribution of a group of frogs or a genus of beetles'. Certainly, many workers on land snails have erected land bridges or constructed hypothetical continents to account for the presence of a particular species or genus on an isolated island.

Wallace was questioning the underlying principles of biogeography and the methods that should be employed in analysing the distribution patterns of terrestrial organisms. Indeed a vigorous debate is still in progress regarding the methods and 'rules' or premises which are applicable, and whether to accept a largely narrative approach, or an entirely deductive one whereby one hypothesis can be tested, falsified and rejected in favour of an alternative. While the ability to falsify is eminently desirable, the power to do so frequently depends on maintaining only a limited number of factors as variables; the remainder being held constant. In biogeography the constants may be established by the 'rules', but if we agree to the inclusion of both dispersal and continental drift as variables then difficulties arise in establishing falsifiable statements to separate the influences of these two factors. A certain degree of piquancy is added to this situation if we consider Wallace's statements (see above).

It is hardly surprising that sharp divisions have been drawn by the protagonists of the two models that find the widest acceptance today, namely the dispersalist and the vicariance models. The dispersalist model, it is suggested, is concerned with the movement of taxa across or around existing barriers, while the vicariance model emphasises the splitting or division of biotas through the development of barriers, that is vicariance events (see Fig. 3.2). Vicariance biogeography does not deny the existence of dispersal, rather it questions its importance in explaining the present distribution of biotas. It equates dispersal with stochastic or random processes and consequently anticipates that any form of generalised pattern cannot be derived from such a process. There is a distinction, however, between a process and the product; a pattern can be superimposed, for example, by the nature of the taxa involved, the arrangement of the land masses and the ecological conditions prevailing in both the source and the recipient areas. Furthermore, our ability to recognise patterns is influenced to a large extent by the geographical size or scale of the distributions being examined.

Vicariance biogeography is concerned solely with the relationships of biotas, that is their genealogy, and not with overall similarity even though this may be extremely important in an analysis of the ecological factors influencing distribution patterns. Therefore, widespread taxa that could have attained their present distribution either by dispersal or through vicariance events are not considered, as these will provide no information concerning genealogical relationships. Instead, attention is paid to endemic taxa and consequently areas of endemism, for only these groups enable a hierarchical scheme of biological relationships to be developed for geographical areas. The aim is to establish a shared distribution pattern for a number of different taxonomic groups, although this alone does not enable a distinction to be made between dispersalist and vicariance explanations. The recognition of congruence between the phylogenetic reconstructions and the geological history of the region encompassed by the pattern suggests that the two patterns are closely linked and a causal relationship can be inferred. Historical events provide an explanation of the biological patterns. In establishing congruence, the geological history must also be arranged in a hierarchical manner. Only then can comparisons be made between phylogenetic reconstructions for both biological and geological events. Indeed some of the more successful analyses have been carried out on the distribution patterns of higher taxonomic categories and especially those of vertebrate groups which have a low propensity for dispersal without drifting continents.

By contrast, dispersalist biogeography does not have an agreed set of rules but there has been a long history of practitioners who have made their individual contributions. It is concerned with the distribution of the whole biota of a region, although in practice particular groups are often selected for detailed study. To varying degrees dispersal is invoked as an important factor in determining these distributions, but there have always been critics who remain unconvinced that certain groups of organisms could cross particular barriers. Consequently attention is frequently paid to the nature and permanence of barriers, in relation to other possible explanations, such as continental drift. Herein lies the weakness of the approach, for if one explanation fails there is always another; thus the programme can be self-fulfilling. Yet, paradoxically, the strength of the dispersalist model also lies in a detailed analysis of the nature and permanence of barriers. This examination will consider carefully both the spatial and temporal scales in assessing the probability for successful dispersal.

Ironically, proponents of vicariance biogeography have suggested that evidence of dispersal can be derived from two sources, but both must be considered to some degree equivocal. The first concerns the presence of sympatry amongst closely-related taxa; if only an allopatric model for speciation is accepted then subsequent coexistence of two sister-groups can only result from dispersal subsequent to divergence. While the assumption that speciation tends to follow an allopatric model may be true for many vertebrate groups, doubts have been expressed about the ability to extrapolate to invertebrates. Indeed models for

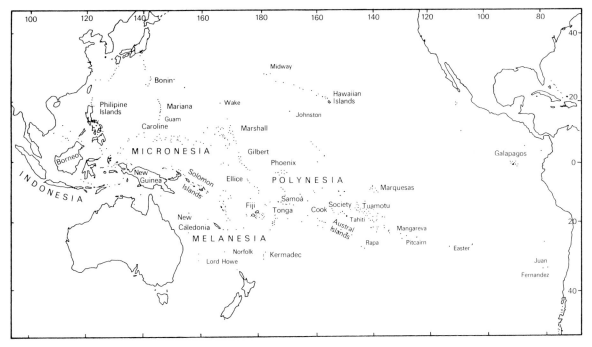

Figure 19.1 Base map for interpretation of figures 19.2 and 19.4.

sympatric speciation (Chapter 4) are becoming more widely accepted, although their application in nature may be very limited. The second source concerns the mixing of terrestrial biotas to provide a 'uniform' distribution pattern on the single Mesozoic continent of Pangaea. It is necessary to postulate such mixing if the derivatives of a Mesozoic biota are to be recognized on the drifted continents following the break-up of Pangaea. However, the distribution patterns of recent taxa on large continents should make us slightly wary of accepting a concept of widespread biotas covering a single land-mass.

The importance of oceanic islands in any dispersalist paradigm should now be obvious, as they provide unique opportunities to investigate complete biotas derived unequivocally by dispersal.

At this point a digression is necessary to clarify two distinct phenomena that have been recognised under the term – dispersal. One is concerned with the regular movement of individuals, often over short distances, in response to such factors as population pressure. This has been termed dispersion (Platnick, 1976). It is basically an ecological concept and frequently no extension of the taxon's range is incurred. In contrast, chance dispersal has been treated as a rare event or an historical concept that often involved the extension of the species' range. Proponents of vicariance models have attempted to nullify the importance of dispersal in biogeography by suggesting that many of the examples quoted are related, in fact, to dispersion tactics. But we must ask ourselves the question: can the two phenomena be distinguished in practice? It

is doubtful if there is a clear distinction based on distance moved, for the relationships between distance and numbers of individuals are random. Indeed dispersion will merge into dispersal as factors extrinsic to the organism become increasingly important.

A more illuminating classification of migration phenomena has been proposed by Baker (1978) and this places the problems of dispersal in a clearer context. Here the important distinction is between accidental and non-accidental migration. If the animal initiates the movement then it must be considered non-accidental migration even though some environmental factors may prolong the initial movement. Yet the same environmental factors, for example wind, may be involved in accidental migration, but in this case the environmental factor initiates and maintains the movement and it is outside the control of the animal. In this Chapter the dispersal of land snails is equated with accidental migration and the faunas of islands offer an opportunity to explore the phenomenon.

Dispersal of land snails

Many inquiries about the nature of distribution patterns for any group of organisms begin with the question – how did they get there? Unfortunately it is impossible to provide a simple answer for land snails on islands even though there is a wealth of anecdotal information and speculation. Certainly no single method of dispersal is likely to be the prerogative of any particular taxon and the only approach to a discussion of the subject must be

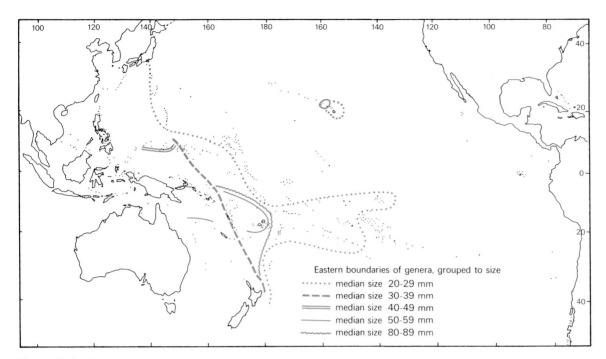

Figure 19.2   Present distribution of the Pacific land snail genera classed according to median size. The smallest classes occur throughout the region, the larger ones only on those islands lying relatively near to the mainland (after Vagvolgyi, 1975).

inferential. Dispersal, or accidental migration, of land snails involves mechanisms like wind drift, which are exploited in the non-accidental migration (dispersion) of other organisms. In snails the non-accidental phase involves the animal crawling, but there is no reason to suggest that such behaviour is involved in initiating dispersal.

Furthermore, the features which can be correlated with successful dispersal and subsequent colonisation are those that would enhance the survival of the individual and its progeny in stress situations. These features may be as simple as the presence of a shell, which reduces water loss, or ovoviviparity, where protection of the developing young is provided at a stage when the highest mortality occurs in oviparous forms. The preponderance of these attributes amongst the land snails on oceanic islands could be attributed, therefore, to the strong selection pressures that must be imposed during dispersal and colonisation.

Amongst the wealth of anecdotal information which is often quoted to emphasise the success of dispersal is the example of Krakatau in Indonesia. A series of volcanic eruptions devastated the island in 1883, destroying a large proportion if not all of the biota. Land snails rapidly recolonised the island together with a wide range of other organisms, but the species composition of, at least, the snail fauna was different from that found prior to the eruption. The number of species of land snails recorded 44 years after the catastrophe closely corresponds with that predicted from the area-species curve for the islands in

that region (see Fig. 19.5). Krakatau, however, does not provide a very instructive example of transoceanic dispersal as it is only separated from the islands of Sumatra and Java by about 24 km.

By contrast the island of Aldabra in the western Indian Ocean is isolated from the nearest land-masses (Africa, Madagascar, Comoro Islands) by distances greater than 400 km, and Pleistocene changes in sea-levels would have produced no material reduction in these distances. A recent study of this raised atoll has provided data concerning the development of the terrestrial fauna from late Pleistocene to Recent times. On at least two occasions during this period the island was completely submerged and the pattern of subsequent recolonisation is reflected in the fossil record. Giant land tortoises colonised the atoll on a minimum of three occasions together with various species of lizards and land snails. However, the same species of snails have not been involved during each wave of colonisation and some of the species represented only as fossils can just be differentiated, at the subspecies level, from forms at present extant in other areas. The present fauna has become established on Aldabra only during the last 80 000–100 000 years.

In spite of the many anecdotes regarding the colonisation of oceanic islands it was only the eloquent advocacy of Zimmerman (1948) and his collaborators in a detailed analysis of the Hawaiian biota that really added respectability to the concept of transoceanic dispersal. Prior to

1948, and even later, there was a tendency to discount transoceanic dispersal or accidental migration and place reliance instead on land-bridges and hypothetical continents. But such accounts failed to consider the information which was available. The uniqueness of Zimmerman's study was the breadth of the evidence reviewed, not only for the biota but also the geological history of the islands. The problem was thus transposed from a local to a much wider framework, where the evolution of the Pacific, including both the islands and the biota, was considered. The importance of island stepping stones as links along a chain uniting now widely separated archipelagoes was clearly recognised. Here Zimmerman was complementing earlier studies of Darwin and Wallace, for in 1859 Darwin had written 'I freely admit the former existence of many islands, now buried beneath the sea, which may have served as halting places for plants and for many animals during their migration. In the coral-producing oceans such sunken islands are now marked, as I believe, by rings of corals or atolls standing over them'. This viewpoint was vindicated nearly a century later by the discovery of land snails in deposits of Miocene age (about 15 million years old) at a depth of 550 m on the Pacific atol of Bikini in the Marshall Islands and by the recognition that the submerged flat-topped seamounts (guyots) had once been above sea-level. The corollary, the absence of stepping stones, would consequently limit opportunities for successful dispersal.

While these narrative accounts provide evidence for the success of transoceanic dispersal, they also beg the question – what are the distances over which it is or could be successful? Of course the majority of propagules of any organism which move over the sea will be lost, but dispersal is stochastic and so no simple answer can be given to the question posed for there is no absolute solution. The chance of success is related to the nature of propagules available, the frequency of the dispersal events and the size of the recipient island. The most important factor is the form of the probability distributions for dispersal. Such distributions can be divided into three groups (MacArthur & Wilson, 1967): (a) normal distribution where the propagules move in a random manner constantly changing direction and the numbers surviving will decrease with distance travelled in a 'normal' manner; (b) exponential distribution where the propagules move in a constant direction and the numbers which survive decrease with increasing distance in an exponential manner (this being, as it were, negative 'compound interest'); (c) uniform distribution where the ability to disperse long distances is very great and the chance of reaching an island, whether it is isolated or close to a land mass is almost equal. While these theoretical patterns are probably never wholly realised in nature, they do indicate that the form of the probability distributions for dispersal as well as the dispersal power of the species are important.

The exponential form of the probability distribution is usually considered the most appropriate for situations where the propagules are passively dispersed by wind drift, and the normal one when they are carried on floating sea-going rafts that have a limited life. Since the former has a considerably greater probability of success, it might not be unreasonable to expect a bias towards aerial dispersal for snails, whether directly by wind or assisted by other animals like birds. The problems of becoming airborne may be expected to restrict the number of propagules. Although once airborne the probability of remaining airborne could be high and decrease only slowly over long distances for very small animals. The limit would be influenced more by the tolerance of the animal to the physical conditions during dispersal. Here land snails exhibit a major advantage over many invertebrate groups in their ability to withstand desiccation.

The size and geographical distribution of the islands will impose a pattern on the products of dispersal, so that the resultant distributions are not random. Differences in the dispersal powers and probability distributions will have profound effects, but taxa similar in these properties will tend to form congruent geographical patterns. Here an island's size will influence not only its effectiveness as a target for the products of dispersal, but also the ability to maintain viable populations and hence the production of propagules being dispersed from it. This, it is suggested, is the situation for land snails of Pacific islands.

The feasibility of aerial dispersal has been supported by some interesting evidence derived from surveys of aerial plankton using nets attached to aeroplanes and to the masts of ships. Although the catch is composed largely of insects belonging to the families that dominate the fauna of islands, large mineral particles have also been collected. Particles up to $5 \cdot 4 \times 2 \cdot 5$ mm in size have been discovered at altitudes of about 3000 metres, being of greater size and weight than many adult snails found on Pacific islands.

The form of the probabilistic distributions for dispersal and the predominance of small forms of snails on isolated islands (see Fig. 19.2) have recently been combined in a more detailed analysis of the distributions of land snails on Pacific islands (Vagvolgyi, 1976). It is argued that the smaller snails would be favoured in aerial dispersal over long distances because of their small size and lighter weight. In addition, their higher population densities would lead to increased numbers of potential propagules. If water transport by drift was important then size would probably be of no particular advantage or disadvantage. This hypothesis was constructed upon a model of a stable Pacific. The distribution of 'medium-sized' forms on the isolated archipelago of the Hawaiian Islands, which one might expect to be inhabited by 'small' species, was explained on the basis of their origin elsewhere from smaller forms or aerial dispersal of juvenile individuals. The latter might not present the difficulties encountered by many other groups of animals where more than a single individual is required for successful reproduction. Pulmonates are hermaphrodites and, at least, some species can undergo self-fertilisation.

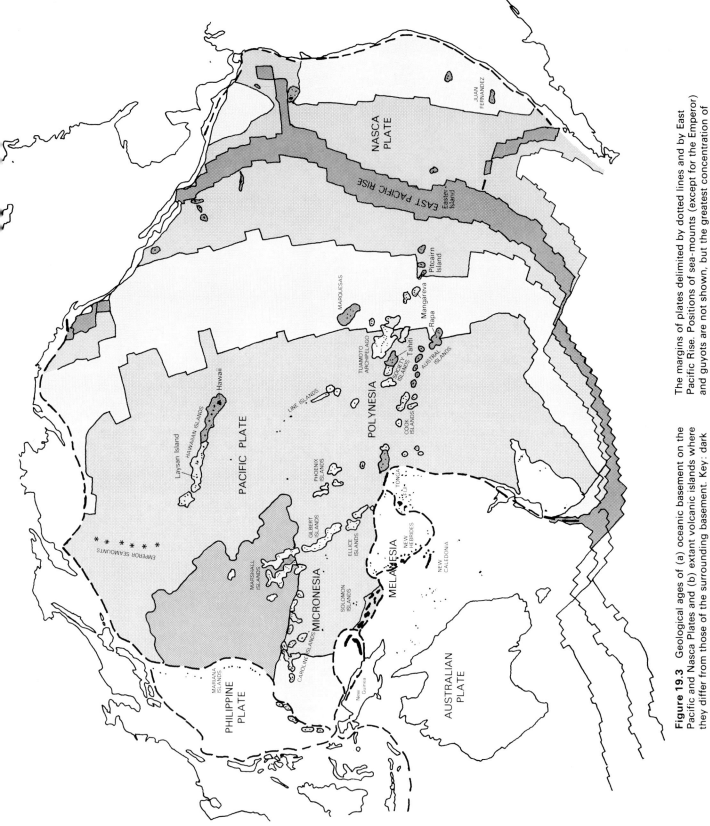

Figure 19.3 Geological ages of (a) oceanic basement on the Pacific and Nasca Plates and (b) extant volcanic islands where they differ from those of the surrounding basement. Key: dark grey – Pleistocene and Pliocene; light grey – Miocene; light blue – Oligocene, Eocene and Palaeocene; mid-blue – Cretaceous, and dark blue – Jurassic.

The margins of plates delimited by dotted lines and by East Pacific Rise. Positions of sea-mounts (except for the Emperor) and guyots are not shown, but the greatest concentration of these is to the south and west of the Hawaiian Islands.

An examination of the maps (see Figs. 19.2 and 19.4) demonstrates, however, that medium- and large-sized forms are divided into two distinct groups. One centred on the northern Hawaiian Islands and the other among the southern archipelagoes extending from Micronesia and Melanesia to the Marquesas in Polynesia; the former being composed of species belonging to the Amastridae and Achatinellidae, the latter including the Partulidae (see Fig. 19.4B). Here the geological history of the Pacific Basin could provide an important clue to understanding distribution patterns.

Pacific Islands

Geology

The Pacific region has been the testing ground for many theories concerning island biogeography and this is obviously related to the number and wide diversity of islands found there. These extend from the large islands of Indonesia and Melanesia situated on continental crust close to the land masses of south-east Asia and Australia, to the smaller and more scattered islands of Polynesia and Micronesia located on oceanic crust in the central basin. The result of the Deep-Sea Drilling Programme is providing data allowing us to reconstruct the geological history of the Pacific Ocean and is beginning to help us reconcile various divergent views for the origin and evolution of the terrestrial biota.

In the Pacific new crust is being produced continuously along the summit of the East Pacific Ridge (see Fig. 19.3). As the crust migrates away from this region it gradually, but continuously sinks until it is ultimately consumed as it plunges beneath the island arcs of the western Pacific or near the margins of the American continents.

The oldest areas of crust still extant are found in the northwestern sector of the Pacific. These are of Jurassic age, but a complete age sequence is preserved up to and including the present. Although the movement of the Pacific Plate (see Fig. 19.3) is primarily westward, variations have given rise to a northerly component as is clearly seen in the deflection of the Hawaiian–Emperor chain of islands and sea-mounts. Scattered across the plate are series of large volcanoes. Although a few are still active, the majority are extinct and many now exist either as guyots or as the bases for atolls and reefs.

The importance of this dynamic model for the evolution of the Pacific is that it combines both horizontal and vertical movements of the plate. Further variations in the vertical component have been produced by changes in sea level, movement of the plate across irregularities in the underlying mantle and by eustatic changes in the crust produced by the load of volcanic material and by its subsequent erosion. The effect of the irregularities in the mantle could be considerable and amplitudes of ± 300 m persisting for 10–15 my have been quoted (Menard, 1973). Many of the islands and guyots form linear patterns across the Pacific Plate, the most conspicuous of these being the Hawaiian–Emperor chain. It has been suggested that these chains were formed as the plate moved, in conveyor-belt fashion, across a hot spot fixed to the mantle beneath the crust. As the volcanoes are formed at points some distance from the Pacific Ridge they must always be younger than the crust on which they are moving. Although there is considerable speculation and argument concerning the nature of the hot spots and whether they are fixed, such details are possibly unimportant biologically. Furthermore, it must be possible for new chains and consequently hot spots to arise, for island groups like the Marquesas do not fit any established linear pattern.

The Hawaiian–Emperor chain demonstrates the continuity of volcano or island production from 70 million years ago at the northern end of the chain to 20 million years ago at Midway near the western end of the Hawaiian sector; volcanoes are still active at the eastern end of the island of Hawaii. It was thought that analagous situations could be recognised in other areas of the Pacific but unfortunately these hopes have not been confirmed. There is not a simple linear age relationship in the rock samples obtained from other island lines.

Much still needs to be explained, but some points of interest to island biogeographers can be made. First, within an archipelago or island group the islands are of different ages and in different stages of evolution. Second, the islands in the west of any group are generally older and smaller than those in the east. Last, the histories of the various archipelagoes or island groups can be very different, both in terms of the relative ages of the constituent islands and their geographical context, that is, their position in relation to other island groups.

Pacific Islands: distribution patterns

The sequence from island formation to extinction is analogous to a flotilla of 'Noah's Arks' sailing across the sea, but with the more ancient gradually becoming waterlogged, settling and finally sinking. In this situation the survival of the cargo would depend on the successful transfer of items from one ark to another. Transfer would need to be effective across water gaps of varying distances and this would influence to a considerable degree the nature of the cargo being transported. The biological pattern appears to follow this crude analogy. It is the smallest snails which have been 'transferred' across the widest gaps and consequently have the broadest distributions.

The problems associated with constantly having to 'transfer' to new areas probably account for some of the curious features of the Pacific snail faunas. Many of the taxa characteristic of continental regions are absent. Conversely, there are three families and two subfamilies either restricted to this area or poorly represented on any of the surrounding land masses. Other families have extra-territory associations; for example, among the archipelagoes extending from Samoa to the Society Islands

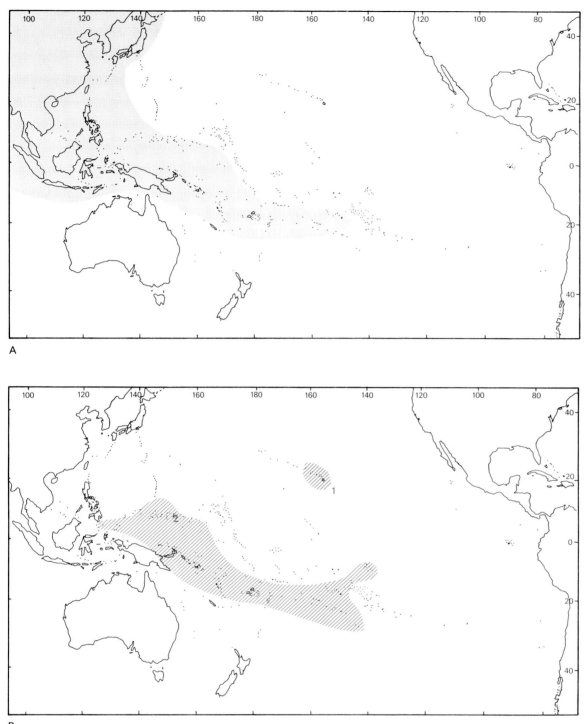

A

B

Figure 19.4 Maps illustrating different distribution patterns of land snails in the Pacific Region.

A. Sub-family Trochomorphinae illustrating the distribution along the southern chain of Pacific Islands of a group of medium- and large-sized snails with affinities to taxa occurring widely in New Guinea, Indonesia and continental Asia.

B. Three groups of medium- and large-sized snails with ranges which are either restricted to the Pacific Basin or extending to only a limited extent onto adjacent islands. 1: Family Amastridae and family Achatinellidae; restricted to Hawaiian Islands. 2: Family Partulidae with a more southerly distribution and reaching highest diversity in the Society Islands where there are 69 species.

C. Genus *Tornatellides*, a widespread group of small-sized snails belonging to the family Achatinellidae, but absent from all islands in the central Pacific.

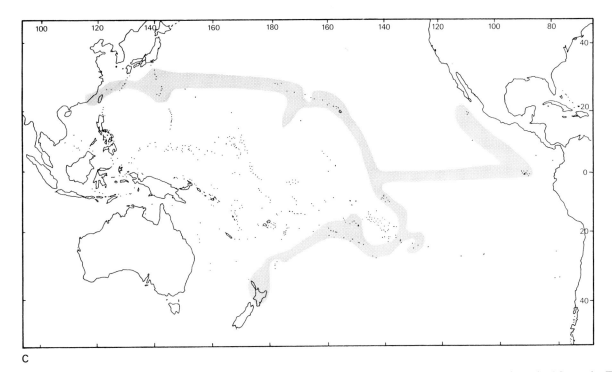

C

there are many species whose affinities are largely with taxa found in New Guinea and Indonesia (see Fig. 19.4A). Indeed it is possible to trace a common distribution pattern for these groups back to continental Asia, with no major biogeographical break in the pattern, even in the region of Wallacea. This is in close accord with the pattern for many families of plants, but quite unlike that recorded for many vertebrate groups. Here then are two general distribution patterns which are not congruent, and possible explanations would need to consider the dispersal and demographic properties of the taxa in the two groups, in addition to the geological history of the region (MacArthur & Wilson, 1967).

Of the three families and two subfamilies of snails which are characteristic of Pacific Islands three belong to a single order, the Orthurethra. This group is considered to be the most primitive of the three or four orders of pulmonate land snails. The majority of the included families have wide distributions extending over more than a single continent. It is, therefore, significant that in a recent examination of fossil molluscan faunas from North America, Solem & Yochelson (in Solem, 1978) have demonstrated that one of the now characteristic Pacific families, the Achatinellidae, was present on that continent during the Palaeozoic. Furthermore, the fossil species can only be separated at the generic level from extant forms. So at least some elements of the Pacific fauna must be regarded as older than the islands or the region in which they exist. Thus the intriguing question is raised – is there enshrined within the Pacific region a relict of a much wider distributed and older continental fauna?

The geological record indicates that the history of the

Pacific Basin extends back at least into the Mesozoic. This does not imply that all of the organisms found in the region have had such a long history. Indeed such a presumption would be unwise, for the snail fauna is composed of taxa with very different phylogenetic relationships. There would also appear to be no evidence to support a suggestion that it has been impossible for new immigration into the central Pacific to be successful during any geological period. At present, therefore, it is only possible to comment on the primitive nature and relict distribution of a limited number of taxa.

Although some groups of small snails have extensive distributions it would be erroneous to extrapolate and suggest that no regional patterns are discernible. A revision of the family Achatinellidae (Cooke & Kondo, 1960) provides excellent examples of local radiation at the specific and generic levels, and these are congruent to a considerable degree with patterns found in other organisms. Three centres of radiation are recognised; one in the Hawaiian Islands where there are eight genera and 152 species, another in the Austral Islands with 11 genera and 35 species and a third in Juan Fernandez (close to South America) where there are two genera and 19 species. The radiation in the Hawaiian Islands is typical for many groups of snails, but the extension to Juan Fernadez is exceptional, although parallels can be found amongst the plants. The extraordinary diversity in the Austral Islands is confined largely to the small island of Rapa, where there are nine unique genera confined to a single subfamily.

In contrast there are genera with wider distributions, but these represent only four out of the 26 genera included in the Achatinellidae. Two of these genera are found on

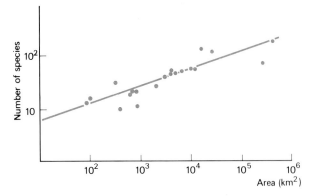

Figure 19.5 Relationship of area to numbers of molluscan species; the islands, shown here as dots, have been selected to illustrate a wide variety of sizes and geographical distributions in the region bounded by Sumatra in the west and Samoa in the east (see Peake, 1978 for details).

virtually all of the islands in the Pacific basin with occasional records from other islands or continental regions. This pattern is repeated in another genus except that here there is wider representation in areas outside the Pacific Basin. The remaining genus has a most extraordinary distribution. It is almost entirely absent from the central islands in the basin and has a peripheral range extending in the north to Taiwan, in the south to New Zealand and in the east to the Galapagos Islands and two small islands off Mexico (see Fig. 19.4C). Although man has been involved in disseminating some of the widespread taxa, sub-fossil deposits and the fact that populations are differentiated locally demonstrate that this is not the complete story.

The overall picture for the distribution of most groups of snails in the Pacific Basin is one that confirms the success of dispersal. The taxonomic relationships appear to follow a geographical pattern, with close relatives found either on the same island or within the same archipelago. There are, however, links which transgress these close geographical boundaries and genera with species in the Hawaiian Islands and the Marquesas or Society Islands can be found. Perhaps our understanding of these rather disjunct distributions will be enhanced by the discovery of further fossil material akin to that already described from North America and the Marshall Islands.

Pacific Islands: diversity and numbers of species

The second group of problems recognized by both Darwin and Wallace were concerned with the large numbers of land snails found on islands. Many factors must influence the numbers of plant and animal species found on islands. The problem is to decide which factors are the most important. Is it the size and the degree of isolation of the islands, the biological properties of the biota, environmental conditions, or a combination of these

factors? Initially, the ability to make any meaningful comments will depend on the availability of inventories of the biota. Since often these reflect the intensity of collecting and the state of taxonomic knowledge it is hardly surprising that there is some confusion and disagreement concerning the nature of the underlying patterns. Under these circumstances it is frequently difficult to test or assess the quality of the available data and assumptions are made regarding its internal consistency. Armed, however, with species lists for different islands some progress can be achieved by using the comparatively simple relationship that exists between species number and island area, this is the area-species curve. Here numbers of species provide the simplest measure of diversity. Furthermore, by restricting the comparisons to islands existing within the same geographical region, the influence of some of the more imponderable variables is eliminated. In this context it has been found that variations in the numbers of species are largely accounted for by two parameters, island area and isolation, even though it is accepted these parameters must reflect to varying degrees a suite of properties exhibited both by the islands and the biotas. Island age is usually unimportant. The relationship shown by the area-species curve is, therefore, one of correlation not necessarily causality.

Comparisons of the patterns of diversity exhibited by different taxonomic groups reveal some rather surprising differences. The richness of the snail fauna on many of the high islands (that is, mountainous as opposed to low atolls) in the Pacific region, for example, contrasts markedly with the impoverishment recorded for some of the other groups such as birds. For these other taxa, the numbers of species are often considerably less than would be anticipated on the basis of comparisons with the fauna of the continents or less isolated islands. This is the converse of the pattern noted for molluscs, where the numbers of species, at least on the high islands, is considerably greater than on less isolated islands. The distinction is not absolute; for example, amongst the biota of the Hawaiian Islands some taxa, like the genus *Drosophila*, exhibit exceptional diversity which rivals or even exceeds that of land snails.

The relationship between the numbers of species and island area has been described by the equation $S = CA^z$, where S is species number, A is island area, C a constant giving the number of species when area is 1, and z is a constant related to the slope of the regression and is usually between 0·20 and 0·35 (MacArthur & Wilson, 1967). The factor z is independent of the taxonomic group or region of the world being studied, but is related to the degree of isolation of the islands and to the mathematical relationship of this type of logarithmic curve; it rises with increasing isolation. The equation has a linear relationship if logarithmic transformations of the data are used, but the latter tends to obscure any minor variations in the information. Figure 19.5 illustrates an area-species curve for land snails; a value of approximately 0·35 for z serves to emphasise the influence of isolation on faunal size.

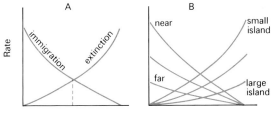

Number of species present

Figure 19.6 A Diagrammatic representation of equilibrium model for faunal size on a single island. The point of intersection between the two lines, representing rate of immigration of new species and extinction of existing species, indicates the equilibrium between two processes on the particular island.

B Equilibrium model expanded to cover a number of parameters; immigration rate being influenced by varying distances from source area, that is near to far, and different sizes of islands, small to large. Variations in these factors will alter the predicted equilibrium point, for example, both reduction in the size of the island and increase in isolation will reduce the predicted faunal size (after MacArthur & Wilson, 1967).

The simple relationship between species numbers and area does not appear to operate universally, and there has been some illuminating speculation on the underlying reasons for this. Nevertheless, the extent to which it does apply is impressive and this has provided the impetus for the development of models for analysing the structure and evolution of island biotas.

One model, the equilibrium theory, relates species numbers to a balance between immigration or colonisation and extinction (Fig. 19.6), faunal size being determined by the balance between these two opposing processes (MacArthur & Wilson, 1967). The rate at which the equilibrium point is approached will depend on properties of the taxa and the environment, hence the difficulties in extrapolating from one taxonomic group or one geographical region to another. Indeed it would be wrong to presume that the rate would be similar for even closely related groups, albeit widely divergent ones, and some taxa may exist in a non-equilibrium state for considerable periods of time. The balance must also be dynamic, for immigration and extinction are continuous processess and even though the number of species may remain approximately constant the composition of the fauna will change.

If new taxa are formed through the splitting of the original colonising species (cladogenesis) or new combinations of taxa arise which are better co-adapted, then an increase in faunal size could be anticipated. The importance of competition in producing these new combinations is problematical. Even amongst those vertebrate groups where there is a wealth of information on distribution patterns it is, however, difficult to produce irrefutable evidence for interspecific competition leading to competitive exclusion. The decision, therefore, rests between the application of a deterministic or a stochastic model to the process of colonisation. The former is favoured largely by vertebrate zoologists who suggest that sets of rules can be applied to the process of colonisation and that these determine the pattern of species acquisition. The other model, which tends to be accepted by invertebrate zoologists and botanists, denies the existence of rules, colonisation being probabilistic (see discussion in Simberloff, 1978).

Wilson (1969) applied a simple classification to the stages in the evolution of island biotas and the establishment of an equilibrium. The initial balance solely between immigration and extinction being described as a 'quasi-equilibrium' which exists for a comparatively short period during 'ecological time'. The appearance of new combinations of species will give rise to an 'assortive equilibrium'. Then, over a much longer period (that is, in 'evolutionary time') new and additional species will appear thereby producing a still higher equilibrium. Presumably these new taxa would be closely adapted to the environment and thus reduce the extinction rate. Although the importance of establishing a balance is stressed in this scheme, it may never occur in 'ecological time' and if it occurs within 'evolutionary time' it will then include the products of evolution besides colonisation and extinction.

The close correlation that exists between the island area and the number of species of land snails on islands in the southwestern Pacific is clearly illustrated in Figure 19.5. Many of the islands in this geographical region are comparatively large and lie close together on continental crust. Similar results have been obtained for the molluscan faunas of islands in the western Indian Ocean and the Atlantic Ocean, but those of the central Pacific stand in contrast. Here, the islands are small and exist on oceanic crust, often in archipelagoes isolated by long distances from other islands. For these islands there is no clear or simple correlation between island areas and the numbers of land snail species (Cooke, 1928; Solem, 1973), even though these exist for other taxonomic groups in the same region.

The picture is one where the numbers of species are higher than would be anticipated, and there is also considerable variation between individual islands, even those within the same archipelago. Some of this variation can be associated with differences in elevation, but a considerable proportion is not accounted for by this or any other obvious physical feature. The highest diversity is found on the smaller islands with an elevation greater than 400 m and with areas between 13 and 38 km^2, and on some of the more isolated islands such as Rapa and Mangareva (Solem, 1973; see Table 1). The number of species restricted to individual islands is high, with over 90 per cent of species unique to single islands, thus emphasising that the present faunal size is probably the product of evolution subsequent to colonisation.

A study of species diversity on the islands of the central Pacific raises two questions. The first is concerned with the evolution of the fauna; how have high levels of

Table 1 Mean number of land snail species per island according to island area and altitude (after Solem, 1973)

Island area (km²)	Under 300 m elevation	Over 400 m elevation
13–21	9·5	34·3
26–39	9·5	31·5
47–73	12·5	20·0
88–155	7·0	16·5
259–583	8·0	21·8

diversity arisen and consequently why are there major differences in the levels between adjacent islands? The second deals with coexistence and enquires: how can so many closely related species live together on a single island?

Solutions to the first question attempt to relate the evolution of species diversity to concepts of speciation and the geological history of the islands. The sequence of geomorphological changes that occur during the life of a single island or archipelago indicates that there must have been a continuous process of dispersal, colonisation and speciation (see Fig. 19.6). Speciation may have occurred through two distinct processes: anagenesis (phyletic speciation) where gradual changes occur within populations throughout geological time, and cladogenesis which involves the splitting and subsequent divergence of populations. The former results in basically no change in the number of species on islands while the latter produces an increase. Both processes will occur during the development of island faunas, but for diversity to increase cladogenesis is necessary, with the splitting of the populations being the result of both dispersal and vicariance events. Dispersal would give rise to populations which were derived from one or very few individuals; this, the 'founder principle', has been strongly advocated as an important element in speciation, particularly in the evolution of the genus *Drosophila* in the Hawaiian Islands (White, 1978). Vicariance events could also lead to the fragmentation of the original populations into smaller subsets consisting of a limited number of individuals. However, this is not invariable and the populations may not suffer major reductions in size. Observations on the island of Hawaii of the division of habitats into smaller units by successive lava flows from active volcanoes provide a clear illustration of vicariant events. Furthermore, the survival of many populations of the smaller species in restricted and discrete areas would tend to support this largely allopatric model. Unfortunately the only detailed studies that have been undertaken on the population genetics of the Pacific land snails suggest a different model!

These investigations were made by Clarke and Murray on species and subspecies of *Partula* found on Moorea in the Society Islands (see Clarke *et al.*, 1978 and White, 1978, for summaries of the results). *Partula* differs from many other elements of the molluscan faunas in belonging to the group of medium-sized snails; species are also extremely numerous (at least in some areas), arboreal and often highly polymorphic. In their analysis of the situation on Moorea, Clarke and Murray drew attention to two phenomena, namely 'area effects' and the presence of steps within clines of variation. The term 'area effects' is used to describe regions within widely distributed species where local populations are characterised by the predominance of particular genetic characters. Typically there is a sharp change in the frequency of such characters across the boundary between these populations and those of the surrounding regions. Various terms have been applied to such situations, for example, 'semi-geographic' and 'parapatric', the latter emphasising the contiguous nature of the populations involved and the absence of any major geographical barriers.

An interesting feature of the situation on Moorea is that although some sympatric species are reproductively isolated in parts of their range, yet under similar conditions in other areas they interbreed sufficiently, in certain cases, to integrade completely. This has been treated as an example of 'incomplete speciation' by Murray and Clarke; that is the different species and subspecies originate during periods of geographical isolation within a single island, but merge to varying degrees upon subsequent breakdown of the geographical barriers. The model for the geological history of these islands would suggest there has been ample opportunity for the isolation and subsequent reunification of populations to have occurred, not during a single point on the time-scale, but as a continuing process.

The degree to which diversity is influenced by local evolution or by dispersal patterns is difficult to assess. As White (1978) so aptly stressed in the first chapter of his book, there is no single case of speciation where the information exists for an assessment to be made of all the factors involved in the process. Indeed there has not been an assessment of the number, direction and chronology of the dispersal and colonisation events involved in the evolution of any group of molluscs which is comparable to that for the species of *Drosophila* in the Hawaiian Islands. The reason is simple: the absence of a reliable reconstruction of the phylogeny of any group equivalent to that produced by biochemical and chromosomal investigations of *Drosophila*. Clarke has suggested that on those oceanic islands which have a depauperate fauna speciation could proceed at an accelerated rate compared to continental or less-isolated areas. Diversification through sympatric speciation would result from strong disruptive selection, that is selection for more than one morph, and this would lead to species occupying different niches. However, if the group had low probability for dispersal or non-accidental migration over comparatively short distances, as do many land snails, then the opportunities for allopatric speciation would be considerably greater than for the sympatric form of disruptive selection. Such radiation would not necessarily lead to the occupation of radically different niches.

The geological model draws attention to another correlation, one between island age and variation in species

Table 2 Geological and geographical data for two archipelagoes, together with numbers of land snail species recorded for each island

	Island age (my)	Island area (km²)	Island altitude (m)	Number of species
Society Islands				
Tahiti	1·4	1000	2231	49
Moorea	2·6	132	1212	29
Huahine		34	710	20
Raiatea ⎫	2·8	241	1033	48
Tahaa ⎭		41	590	16
Bora Bora	4·0	38	772	16
Hawaiian Islands				
Hawaii	0·84	10438	4214	68 (129)
Maui	1·30	1886	3058	125 (156)
Lanai	—	365	1027	53 (70)
Molokai	1·84	673	1515	89 (118)
Oahu	3·6	1564	1228	244 (282)*
Kauai	5·6	1437	1576	79 (103)

The numbers of species are derived from recent taxonomic revisions. The quality of the information available for the Hawaiian Islands is variable, therefore two figures are given; the first is for only that part of the fauna for which recent revisions are available, the second figure, in brackets, is for the whole fauna. The important comparisons are within archipelagoes and then between the largest and youngest island in the east and the rest.
* Solem (1973) has suggested that total number of species for Oahu should be 395.

diversity in land snails. In many examples of the area-species curve which have been investigated, island age has not been considered an important variable, except during the initial stages of colonisation. A few studies of other groups of organisms have suggested that an equilibrium would only be established during 'evolutionary time' on some of the more isolated oceanic islands, so the balance would be between colonisation plus speciation and extinction. Cooke (1928) realised, however, that a knowledge of island age was crucial for understanding the diversity of snail faunas on Pacific islands. His conclusions were drawn from a series of simple observations on the Society and Hawaiian Islands, namely that the largest and youngest islands at the eastern limits of the archipelagoes did not have the number of species concomitant with their size. Indeed the number was comparable with that found on adjacent and smaller islands; recent data are summarised in Table 2. Variations in diversity between archipelagoes cannot be explained simply by age and other factors will need to be considered. Geographical context and numbers of different taxa colonising an archipelago are important.

Rapa and Mangareva, two isolated islands with exceptionally high diversity, cannot be fitted into a similar comparative scheme. Nevertheless they do provide some corroborative evidence for the importance of age, since both are considerably older than any of the elevated islands in the Society and Hawaiian archipelagoes. Moreover,

both have complex topographies with altitudes greater than 400 m, and with Mangareva at a stage in island evolution comparable with the older islands in the Society group; that is, with a fringing reef enclosing a lagoon containing a number of separate islets.

Although there is a correlation between island age and variation in species diversity within archipelagoes the causal agents in this relationship are difficult to establish. Various explanations can be suggested to account for this variation, but unfortunately there is very little evidence for selecting one factor or any combination of these in preference to another. A list of the factors which could be important must include the following.

1. A slow rate of colonisation, so that an equilibrium fails to be established when island size is at a maximum, then diversity could continue to rise even though island size was decreasing. But there is no way an equilibrium point could be recognised using the available information, nor is there any reason to presume that a balance will be established.

2. Speciation, as an adjunct to colonisation, could continue to give rise to new taxa through 'evolutionary time'. The evidence from *Partula* on Moorea adds some credence to this suggestion, for the 'incomplete speciation' which was discovered would then be completed.

3. An extremely slow rate of extinction, so that there was a lag or delay in the reduction of species diversity to correspond with decreasing island size and habitat diversity. The products of this type of lag may be seen on the Laysan atoll in the Hawaiian chain, where the presence of a snail belonging to the genus *Tornatellides* has been regarded as a relict from the period when the atoll was a high island (Schlanger & Gillett, 1976). Confirmatory evidence for a slow rate of extinction may be derived from a study of Lord Howe island to the east of Australia. The high diversity here may be a relict of a period when the island was larger, for although this island lies on continental crust it is in a region which has been subjected to considerable geological changes throughout the Cenozoic.

4. The geographical position and context of an island would vary with age; for example, it could change from being at the end of a linear chain to being incorporated within the sequence until ultimately it disappears at the other end. In this model diversity would be related to a continuing process of invasion, differentiation and reinvasion. It would be anticipated that the rates of colonisation for islands like Hawaii and Tahiti would be lower than those of Maui and Moorea, while islands like Mangareva, which are not in a simple linear chain, could have been the target for dispersal from a range of high islands lying to the west.

Indeed Zimmerman (1948, p. 128–129) recognised that a process of invasion and reinvasion could even be important for the isolated island of Rapa, for 'almost 50 miles to the east of it is situated a group of about ten almost unknown rocks called Marotiri all that is left of an ancient island, probably similar to Rapa...found there a dozen species of plants, some of them distinct species, and on some of those plants were found new species of insects and spiders. The insects belong to genera which are characteristic of the well watered forest of high islands. I consider these species to be the last survivors of a fauna which had its beginning on the slopes of a high, densely forested island...Could not part of the unique biota that is found on Rapa today have had its origin on ancient Marotiri?' These quotations from Zimmerman provide an indication of the flavour of the problem and suggest that more information is required on the geological and geographical scenarios before invoking sympatric speciation to account for the extraordinarily high diversity on isolated islands like Rapa.

The second question regarding problems of species diversity on the islands of the Pacific deals with coexistence and investigations of how can so many closely related species live together on the same island. This line of enquiry usually implies that some level of competitive interaction leads to the exclusion of one species by another. Evidence for such a process occurring amongst land molluscs is rare or largely inferred from distribution patterns or examples of character displacement. Character displacement introduces many conceptual difficulties (see Chapter 17) and although some patterns of variation may seem to support the principle of competitive exclusion, they are not without counter-example, and both may be explained by any one of several possible mechanisms. Information from detailed analysis of distribution patterns is often equally illusive, and just as inconclusive.

Recently, however, evidence for intra-specific interactions has been produced for a common British snail, *Cepaea nemoralis*, under both field and laboratory conditions (see Peake, 1978, for references), but it is doubtful if we can extrapolate from an intra- to an inter-specific situation. Yet if higher diversity does involve closer packing or compression of the ecological range of individual species, and this implies that extensive niche differentiation must occur, then competitive interactions are perhaps involved, although these may not lead to replacement amongst the competing species. In such situations small size could be correlated with increased compression, thus following a pattern common in many regions where many more small, rather than fewer large, species are found coexisting.

It may be unnecessary, however, to invoke the principle of competitive exclusion in these areas of high diversity, for the evidence derived from field collecting suggests that many of the small snails exist in very local and discrete populations, which are often isolated from those of closely related taxa. Indeed a recent hypothesis (Huston, 1979) concerning diversity began with the observation that complete exclusion of one species by another appears to be a rare phenomenon in nature, even though it has been demonstrated under laboratory conditions. It was suggested that many ecological communities exist in an unstable state; fluctuations in various environmental factors will often produce significant reductions in the populations of

the component species and thereby reduce or eliminate competition for a limited resource. In these situations, although some form of competition may exist, complete exclusion of species is less likely than the establishment of a balance between the frequency of environmental perturbations and the degree of competitive replacements.

Whether such a model is applicable to the molluscan fauna of the Central Pacific Islands is debatable, as there is virtually no information on the stability or otherwise of any populations of the indigeneous fauna. Nevertheless some of the model's predictions are interesting, for they suggest that the 'accumulation of new varieties and species should be favoured in situations where the rates of increase of all competitors are low' (Huston, 1979) and consequently where competitive replacement is reduced or prevented. Furthermore 'extensive niche differentiation may not be necessary for survival'.

Many of the small species found on the islands of the central Pacific are ovoviviparous and probably have a comparatively low reproduction rate, with only small numbers of eggs maturing simultaneously.

There is also some evidence to suggest that, in at least a few species, individual growth rates of the snails are slow, with a concomitant long life span extending to 10–15 years. In these taxa the rates of population increase could be very slow. Correlations, however, with other environmental factors exist; for example, the absence of indigenous species of predatory ants from many of these islands (Solem, 1973) must have had a profound effect on life history phenomena and the maintenance of viable populations. It is difficult to comment in detail on the levels of niche differentiation, but certainly the results of field collecting suggest that it is difficult to correlate any differences in fine-level distribution patterns of closely related taxa with any major features of the environment. So there are indications that the molluscan faunas of these islands could fulfil at least some of the predictions of Huston's hypothesis.

Discussion

The importance of dispersal in determining the distribution patterns of biotas has been championed by biologists who have studied islands or areas which have been subjected to catastrophic changes. The tendency to extrapolate from this context has been encouraged by the realisation that our environment is a mosaic of isolated habitats. Indeed the analogy with islands can be applied extensively.

The attention paid to island biotas by both Darwin and Wallace, and many others, reflected their views on problems associated with barriers to the distribution of organisms. Although dispersal was one mechanism whereby such obstacles could be overcome, alternative explanations were not dismissed. The interest in a group of organisms, that is land snails, which did not appear to conform to their concepts is, therefore, understandable. In this chapter the distribution of the same group has also provided the vehicle or opportunity to consider and discuss some of these biogeographical problems.

Although the mechanisms involved in the initial colonisation of Pacific Islands by land snails are probably associated extensively with aerial dispersal, the observed patterns of distribution and high diversity can only be understood when viewed against a suite of interacting factors. The latter include the geographical arrangement of the islands, the geological history of the region and the biological properties of the organisms. Indeed the patterns are the result of a game of chance, change and challenge played on a chequerboard of islands. Where chance is provided by dispersal, a stochastic process, the labile history of the islands are the changing scene and the biological properties of the snails provide a unique response to the challenge. The problem is to assess the influence of each factor when major limitations are imposed by the lack of detailed biological information on, for example, the ecology of the snails. This is a common difficulty in studies of natural situations, where there is frequently a dearth of data with which to test a plethora of models. The conclusion must be, therefore, that although a dispersalist model operates for land snails on Pacific islands and that various hypotheses can be erected to account for the high species diversity, no clear distinction can be made between the effects of colonisation, speciation and extinction.

The geographical pattern of islands in the Pacific region emphasises the importance of scale in a discussion of dispersalist biogeography, and by inference, vicariance biogeography. Scale is relative to the vagility of the organisms and it is also correlated to some degree with the taxonomic levels of the groups used to define the patterns. There is a tendency for the analysis of small-scale patterns to be concerned largely with species or genera, although the converse is not necessarily true, for there are many examples where major discontinuities exist even within the same species or between closely-related taxa. The degree to which smaller scale patterns produce only modifications to broader or more general patterns will vary according to the situations being examined.

The decision whether to accept a dispersalist or vicariance explanation for a particular pattern can only be infallible for those taxa that occur on oceanic islands or for those that have no propensity for dispersal, respectively. For all other taxa, indeed the majority, any analysis will involve a choice, as both dispersal and vicariance events will be involved to varying degrees in establishing common or general distribution patterns. The question is, therefore, which factors have had the greatest influence?

If a dispersalist explanation is selected, then some workers would suggest that the resultant patterns are not worthy subjects for biogeographical analysis. Certainly dispersalist models have frequently been poorly formulated, but the models which have been developed by MacArthur & Wilson (1967) and subsequent investigators have provided the basis for some progress. It is doubtful

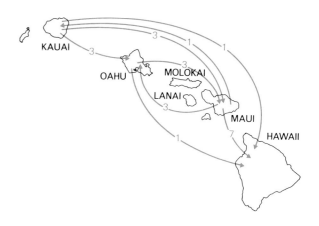

Figure 19.7 Minimum number of inter-island dispersal events among Hawaiian *Drosophila* species (see White, 1978).

whether the distribution patterns of mobile organisms can be analysed by relying on deductive methods of reasoning, although plausible arguments can be advanced to explain widely disjunct distributions of such organisms on the basis of vicariance events and continental drift. Here again, scale is the crucial factor in deciding whether dispersal mechanisms explain a particular distribution.

A vicariance explanation can only be considered to account for the distribution of all the groups included within a particular pattern if it is unlikely that a general pattern could have occurred in all of the constituent taxa by chance alone (Rosen, 1978). There are major difficulties in providing an assessment of chance in such situations, other than one derived from a biologist's imagination.

A convincing reconstruction for the phylogeny of any group of Pacific land snails is not available and therefore it is impossible to follow the detailed pattern of invasion and reinvasion within and between archipelagoes. The quest for a reconstruction is constantly exacerbated by the great amount of minor morphological divergence in proportion to the total land mass of the islands in the central Pacific; their combined area is probably little larger than Switzerland. Yet there is only a slight degree of profound morphological differentiation between many of the groups as compared to the extent of their range over the area of the central Pacific (Baker, 1941). Analogous situations have been described in other phyla, noteworthy examples being found amongst the *Drosophila* species in the Hawaiian Islands. Here, confusion produced by parallelisms amongst minor morphological characters in closely related species have only been recognised by using evidence derived from chromosomal studies. This form of parallel evolution probably arises through homologous changes in that portion of the genome which must be shared amongst closely related taxa (Rensch, 1959).

The studies of Carson and his co-workers on the *Drosophila* fauna of the Hawaiian Islands provide some of the clearest indications of the manner in which future

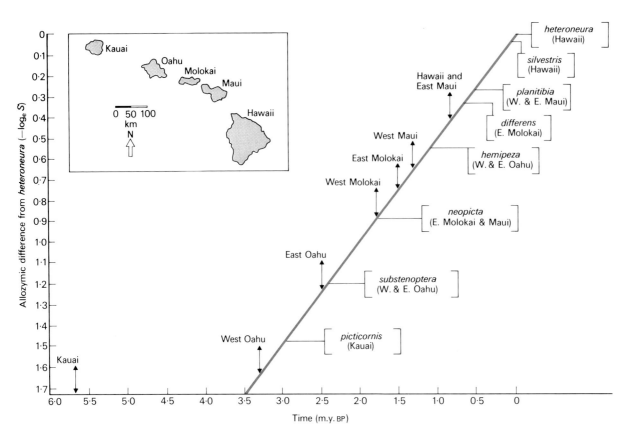

Figure 19.8 Allozymic differences, measured as divergence from *heteroneura*, and inferred time of species origin in eight species of Hawaiian *Drosophila*. The volcanoes on the islands from a sequence from north-west to south-east. The ages of the oldest rocks from each volcanoes are plotted. The eight species of *Drosophila* belong to a single sub-group with *heteroneura* probably the newest species in the series. *D. hemipeza* is considered a relatively recent reinvader of Oahu from Molokai (after Carson, 1976).

progress in the study of dispersal may be achieved (see White, 1978 for summary). Utilising data obtained from cytological and biochemical investigations they have been able to produce phylogenetic reconstructions in some groups of closely-related taxa not only at the species level, but also extending down to the population level as well. Moreover, they have integrated this information with the distribution patterns, and suggested for the group known as 'picture-winged flies' that a minimum of 22 inter-island colonisations were necessary (see Fig. 19.7). Here dispersal has been predominantly in a south-westerly direction from the oldest to the youngest islands as might be anticipated, even though this was against the north-east trade winds.

Carson has stressed that although there have been at least nine 'founder events' on the youngest island of Hawaii, speciation must have been rapid. No intermediate 'subspecies colonisers' are known, all of the invaders being 'full' species. The importance placed upon 'founder events' for speciation of Hawaiian drosophilids is interesting, but is it unique to island faunas or can we extrapolate to continental areas?

Carson (1976) has extended the analysis of distribution

patterns by adding another dimension, time, to the process of speciation and subsequent divergence. This is feasible as a number of detailed studies of protein evolution have suggested that amino acid substitutions might occur at a constant rate per site. These changes are under genetic control and it is possible to follow the substitutions through variations in the allozymes. Thus if there is an inbuilt 'clock' in the rate of evolution, a measure of genetic similarity will serve as a direct measure of time since the divergence of two species. The only problem is calibrating the clock; this is achieved by correlating the degree of genetic similarity to the distribution of the taxa and the ages of the islands. It is assumed that at the time of divergence the two sister-species would be identical for allozymes. Anomalies which could arise through reinvasion of older islands by species derived from younger islands are recognised from the chromosome studies and are easily accommodated. The results are shown in Figure 19.8; the regression line assumes that genetic differences accumulate at a rate of 1 per cent in 20000 years. This fits the observed pattern closely, but it is 15 times faster than has been suggested in any other examples. One wonders what patterns would be visible if such information was available for other taxonomic groups and whether in future it will be possible to provide a much clearer picture of cladogenesis by using this biochemical clock.

While islands offer unique opportunities to study dispersal phenomena, they obviously reflect only part of a much wider spectrum of biogeographical situations. Both dispersalist and vicariance models have roles to play in the analysis of these situations, but the importance placed on any hypothesis will depend on the questions being asked and the taxa involved. Similarly, the approach adopted, whether it is descriptive, narrative or analytical, will reflect to a considerable degree the available information. While there is always a need for hypotheses which are consistent, predictive and testable, it would be wrong to presume that the information is available to formulate or test them or indeed that problems which have an historical component can necessarily be formulated in such a manner. Yet, whichever approach is employed it is important that the analysis or discussion is broad-based and is consistent with the available information. In many instances our only recourse must be to clarity in presenting the data and the reasons for our conclusions.

A revolution may be in progress within the field of biogeography, but only a retrospective analysis of the present situation will be able to decide. Whatever the final judgement the present debate is important for we are all being forced to consider critically many of our accepted or preconceived ideas. The origins of this revolution can be traced to the developments in the field of geophysics and the consequential impact of continental drift. Thus biogeographers were able to deduce causal agents for certain major distribution patterns of comparatively sedentary animals, but the extent to which such arguments can be extended to more mobile groups is questionable. Revolutions are probably always traumatic events, whether they are political or scientific; certainly in biogeography many strong and irrational statements have been made. In this climate the polarisation which has occurred between some of the protagonists of both the dispersal and vicariance models is not unexpected, but it is hardly conducive to the recognition of the achievements of both groups. The idea that 'distortion is not only routine in science, it is a hallowed method of debate' (Hull, 1978) appears to have been realised, but to many biologists this is both surprising and unacceptable. There would appear to be a need for pluralism in defining the underlying philosophies of biogeography, for there is no reason to presume that the 'truth' is enshrined in any one model or any single philosophical discipline. We still need to remember Wallace's concern for the 'power of reasoning'.

References

Baker, H. B. 1941. Zonitid snails from Pacific Islands, Parts 3 and 4. *Bulletin of the Bernice Pauahi Bishop Museum* 116: 202–370.

Baker, R. R. 1978. *The evolutionary ecology of animal migration.* 1012 pp. London: Kent, Hodder and Stoughton.

Carson, H. L. 1976. Inference of the time of origin of some *Drosophila* species. *Nature* 259: 395–396.

Clarke, B., Arthur, W., Horsley, D. T. & Parkin, D. T. 1978. Genetic variation and natural selection in pulmonate molluscs. pp. 219–270 *in* Fretter, V. & Peake, J. F. (Eds) *Pulmonates*, Volume 2A. London: Academic Press.

Cooke, C. M. Jnr. 1928. Notes on Pacific land snails. *Proceedings 3rd Pan-Pacific Science Congress:* 2276–2284.

Cooke, C. M. Jnr. & Kondo, Y. 1960. Revision of Tornatellinidae and Achatinellidae (Gastropoda, Pulmonata). *Bulletin of the Bernice Pauahi Bishop Museum* 221: 1–303.

Darlington, P. J. Jnr. 1957. *Zoogeography.* 675 pp. New York: John Wiley.

Hull, D. 1978. The sociology of sociobiology. *New Scientist* 79: 862–865.

Huston, M. 1979. A general hypothesis of species diversity. *American Naturalist* 113: 81–101.

MacArthur, R. H. & Wilson, E. O. 1967. *The theory of island biogeography.* 203 pp. Princeton: Princeton University Press.

Menard, H. W. 1973. Depth anomalies and the bobbing motion of drifting islands. *Journal of Geophysical Research* 78: 5128–5137.

Peake, J. F. 1978. Distribution and ecology of the Stylommatophora. pp. 429–526 *in* Fretter, V. & Peake, J. F. (Eds) *Pulmonates*, Volume 2A. London: Academic Press.

Platnick, N. 1976. Concepts of dispersal in historical biogeography. *Systematic Zoology* 25: 294–295.

Rensch, B. 1959. *Evolution above the species level.* 419 pp. London: Methuen.

Rosen, D. E. 1978. Vicariant patterns and historical explanations in biogeography. *Systematic Zoology* 27: 159–190.

Schlanger, S. O. & Gillett, G. W. 1976. A geological prospective of the upland biota of Laysan atoll (Hawaiian Islands). *Biological Journal of the Linnean Society* 8: 205–216.

Simberloff, D. 1978. Using island biogeography distributions to determine if colonisation is stochastic. *American Naturalist* 112: 713–726.

Solem, A. 1973. Island size and species diversity in Pacific Island land snails. *Malacologia* 14: 397–400.

Solem, A. 1976. *Endodontid land snails from Pacific Islands (Mollusca: Pulmonata: Sigmurethra). Part 1, Endodontidae.* 508 pp. Chicago: Field Museum of Natural History.

Vagvolgyi, J. 1976. Body size, aerial dispersal and origin of the Pacific land snail fauna. *Systematic Zoology* 25: 465–488.

Wallace, A. R. 1892. *Island life* (Second edition). 563 pp. London: Macmillan.

Wilson, E. O. 1969. The species equilibrium. *Brookhaven Symposia in Biology*, No. 22: 38–47.

White, M. J. D. 1978. *Modes of speciation.* 455 pp. San Francisco: W. H. Freeman.

Zimmerman, E. C. 1948. Introduction. *Insects of Hawaii* 1, 206 pp. Honolulu: University of Hawaii Press.

CHAPTER 20

The development of the North American fish fauna – a problem of historical biogeography

C. Patterson

Questions about the development of regional faunas or floras pose the problems of historical biogeography. Biogeography, which deals with the distribution of organisms, can be split into ecological and historical biogeography. Ecological biogeography tackles the distribution of species in terms of their interaction with the environment and with each other. These ecological properties come down to estimates of the relative dispersal abilities of individual species, and of the results of chance events of dispersal. Such events can be investigated experimentally, as in recent work on colonisation and species-turnover on artificial islands. Historical biogeography tackles general patterns of occurrence of species, and explains them in terms of history. It can operate with large areas and the geological time-scale, but as a historical science it must rely on inference rather than experiment. The aim of this article is to use the fishes of North America to illustrate the kinds of inference that constitute historical biogeography.

The fish fauna

The basic data of historical biogeography are ranges of species. In asking about the North American fish fauna, we are really interested in species endemic to that continent; that is, fishes found in North American fresh waters and nowhere else. George Myers (see Darlington, 1957, p. 41) provided a way of classifying fishes for biogeographic analyses. He recognised four main categories:

1. Primary freshwater fishes – found in fresh water and nowhere else (an ecological observation), and assumed always to have been so confined by problems of osmoregulation in sea water (a historical interpretation).

2. Secondary freshwater fishes – normally found in fresh water, but thought to be able to survive in sea water, because individuals have occasionally been taken at sea, or because of experimental evidence of salt tolerance.

3. Peripheral freshwater fishes – able to move from salt to fresh water and vice versa. This category includes

several subgroups. One is diadromous fishes, which migrate between rivers and the sea, usually for spawning. Some, such as salmon (*Salmo*), spawn and hatch in fresh water, and spend most of their life at sea. Others, such as freshwater eels (*Anguilla*), spend most of their life in fresh water but spawn and hatch at sea. Another subgroup is sporadic freshwater fishes, which seem to wander between the sea and rivers at random.

4. Marine fishes – found only in the sea, and assumed always to have been so confined, by problems of osmoregulation in fresh water.

The fish fauna of North America includes primary, secondary, diadromous, sporadic and other varieties of peripheral freshwater fishes. These categories are part of ecological biogeography rather than historical biogeography, but to prevent this article spilling over into the biogeography of the north-western Atlantic, Caribbean and north-eastern Pacific, I will accept them and deal only with continental fishes, which include Myers's primary and secondary categories. The distinction between these two is to some extent arbitrary, and seems originally to have been made as an *ad hoc* solution to certain biogeographic problems. For example, the gars (Lepisosteidae) are confined to continental American fresh and brackish water, except for an endemic species in Cuba (see below). The problem of the Cuban gar was 'solved' by records of gars in brackish and salt waters along the coast of the Gulf of Mexico, showing that gars are secondary fishes and supporting the belief that Cuba was colonised by gars swimming across the Gulf.

Today, North American fresh and brackish waters contain almost 600 species of primary and secondary fishes. The main groups, and the number of species in each, are shown in Table 1, and Figure 20.1 shows a theory of relationships between the groups. Some species are widespread, others local, but there is no space here to discuss details of distribution *within* North America, and the history of different drainage basins there (Hubbs, Miller & Hubbs, 1974, and Hocutt, 1979, are recent

Table 1 Classification and number of species of Recent North American primary and secondary freshwater fishes

Subclass Actinopterygii (rayfinned bony fishes)	
Infraclass Chondrostei	
Family Polyodontidae (paddlefish)	1
Infraclass Neopterygii	
Division Ginglymodi	
Family Lepisosteidae (gars)	5
Division Halecostomi	
Subdivision Halecomorphi	
Family Amiidae (bowfin)	1
Subdivision Teleostei	
Supercohort Osteoglossomorpha	
Family Hiodontidae (mooneyes)	2
Supercohort Elopocephala	
Superorder Ostariophysi	
Order Cypriniformes	
Family Cyprinidae (carp, minnows)	200
Family Catostomidae (suckers)	57
Order Characiformes	
Family Characidae (characins)	1
Order Siluriformes	
Family Ictaluridae (catfishes)	37
Superorder Protacanthopterygii	
Order Salmoniformes	
Suborder Esocoidei	
Family Umbridae (mudminnows)	4
Family Esocidae (pikes)	4
Suborder Salmonoidei	
Family Salmonidae (salmon, trout, whitefish)	38
Superorder Paracanthopterygii	
Order Percopsiformes	
Family Percopsidae (troutperch)	2
Family Aphredoderidae (pirateperch)	1
Family Amblyopsidae (cavefishes)	5
Superorder Acanthopterygii	
Order Atheriniformes	
Suborder Cyprinodontoidei	
Family Cyprinodontidae (killifishes)	34
Family Poeciliidae (viviparous killifishes)	9
Order Perciformes	
Suborder Percoidei	
Family Centrarchidae (sunfishes)	30
Family Percidae (perch, darters)	104
Family Cichlidae (cichlids)	1
Total	536

examples of such discussions, the first concentrating on the fishes, the second on geology). The plan of this article is to discuss the groups of fishes in the order given in Table 1 using each as an example of the approaches that may be used in historical biogeography. Before beginning that, it is worth commenting on the geography and palaeogeography of North America.

The area

On today's globe, North American fresh waters are clearly circumscribed to the north, by the arctic ice; to the west, by the Pacific, the Bering Sea and the narrow (less than 100 km) Bering Strait between Alaska and Siberia; and to the east, by the Atlantic and the Davis Strait between arctic Canada and Greenland. To the south, North America extends into the tropics and is linked with South America by the Central American isthmus, and more tenuously by the island chains of the Antilles. In terms of biogeographic realms, North America is the main component of the Nearctic Region, so named by Philip Sclater in 1858, on the basis of bird distribution, later extended to mammals and other groups by Alfred Russel Wallace and others. The Nearctic Region includes Greenland and the islands of arctic Canada, and continental North America south to the central Mexican plateau (not to the narrowest part of the isthmus). The southern boundary of the region is necessarily arbitrary. Here the boundary will be drawn below the Rio Grande basin, in northern Mexico. The fish fauna of Middle America, south of this boundary, is reviewed by Miller (1966) and Myers (1966), and from a different point of view by Rosen (1978).

Figure 20.2 shows sketch maps of North America illustrating some of the geological events which might be expected to influence fish distribution in late Mesozoic and Cenozoic times. Remember that these are sketches, and that palaeogeographic reconstruction is by no means beyond argument. Passing back in time, the most recent momentous events are the Pleistocene glaciations, when ice covered much of the continent well south of the Great Lakes (Fig. 20.2A). The ice sheet is generally thought to have sterilised most of the northern half of the continent, so that the fishes now found there have arrived by dispersal since the ice retreated, during the last ten thousand years or so. Yet the idea that present distributions are the result of postglacial dispersals has been overinfluential in biogeography, and it might be better to treat the sterilising effect of ice as a hypothesis to be tested, rather than an axiom. As Figure 20.2A shows, the water locked up in the ice-cap caused a world-wide drop in sea level, so that 40 000 years ago Alaska and Siberia were continuous and unglaciated, and were separated from the rest of North America by an ice barrier.

Since the Miocene, the Pacific drainage of North America has been an area of intense geological activity, with the Coast Ranges being thrown up, and north–south shearing motion along the San Andreas Fault and similar systems extending into Alaska. These recent, and continuing, disturbances followed a series of mountain-building episodes in the West (the Cordilleran orogeny) which

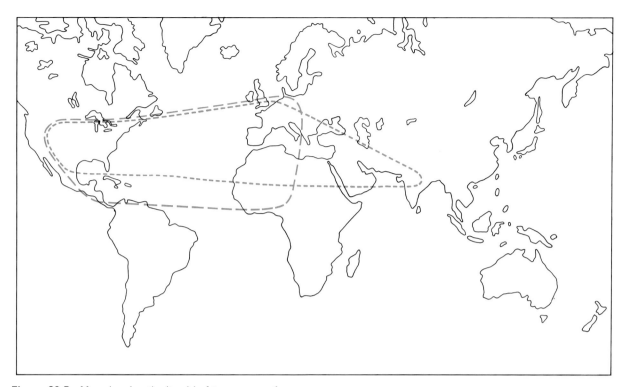

Figure 20.5 Map showing the 'track' of two genera of gars, *Atractosteus* (long dashes) and *Lepisosteus* (short dashes), after Wiley (1976).

Family Amiidae – bowfins

There is only one living species, *Amia calva*, a predator widely distributed in lakes and rivers of the eastern USA (Fig. 20.6). We can say something about the historical biogeography of the group thanks to an extensive fossil record. Boreske (1974) gave a thorough review of the American fossils. Figure 20.7A shows the pattern of relationships that emerged from his work, and Figure 20.6 compares the Recent and fossil distribution of *Amia* in North America, giving a good idea of the dominantly mid-western Cretaceous and Tertiary localities from which freshwater fossil fishes are known.

Figure 20.7A shows that *Amia* was previously much more widespread, and that extinction has reduced a Holarctic range to eastern North America. The two species of the genus that were common to Europe and North America in the early Tertiary illustrate the trans-atlantic 'track' mentioned in the discussion of gars, and it is noteworthy that one of these species occurs in Spitsbergen, while Eocene remains of *Amia* (and gars) have recently been found on Ellesmere Island, north-east of Greenland, which was adjacent to Spitsbergen in early Tertiary times. Unlike the gars, *Amia* also occurs as fossils in mainland Asia (*A. mongolensis*). The geographic pattern shown in Figure 20.7A can be simplified to the area cladogram shown in Figure 20.7C: *Amia* implies that North America is more closely related to Europe than to

Asia. This is a hypothesis that can be tested by fishes (and other organisms) represented in the three areas. Apart from *Amia*, the family Amiidae includes a number of late Jurassic and Cretaceous genera from Europe, South America and Asia. Some of these were marine, but so far as we know *Amia* has always been strictly confined to fresh water.

Family Hiodontidae – mooneyes

The family contains only two living species, *Hiodon alosoides* and *H. tergisus*, with largely allopatric distributions in temperate and cold temperate waters of eastern Canada and USA. The family also includes an extinct genus, *Eohiodon*, with three or four species of Palaeocene to Oligocene age from western North America. The closest relatives of the hiodontids are the Lycopteridae, an extinct group from the late Jurassic and early Cretaceous of China and Siberia. Hiodontids are amongst the most archaic living teleosts. They are primitive members of the Osteoglossomorpha, a small group of freshwater fishes with representatives in every continent except Europe, and of interest to biogeographers for that reason. Nelson (1969) published an analysis of osteoglossomorph biogeography which applied to fishes a method developed by entomologists (Brundin, 1966). This can be called the cladistic dispersal method; it is illustrated in Figure 20.8.

The cladistic dispersal method depends on a prior

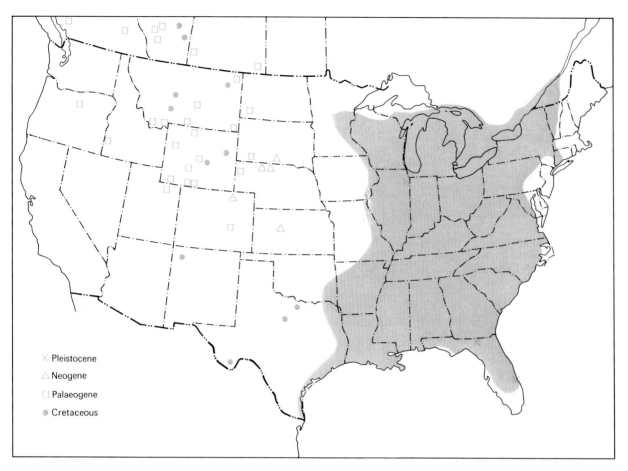

Figure 20.6 Map showing the Recent (shaded) and fossil distribution of *Amia* in North America. Symbols for fossil localities: × = Pleistocene (*Amia calva*), △ = Neogene (Miocene and Pliocene), □ = Palaeogene (Palaeocene, Eocene and Oligocene), ● = Cretaceous. Mainly after Boreske (1974).

phylogenetic analysis of the group under study, set out as a cladogram. One then estimates the distribution of ancestral species (represented by junctions in the cladogram) using the principle of parsimony – accounting for present distribution by the minimum number of dispersals. This is done by working through the cladogram and estimating ancestral distribution at each node by, in Nelson's words, 'combining descendent distributions when they are completely different, and eliminating the unshared element when the descendent distributions are not completely different'. For example, the two living species of *Hiodon* are largely allopatric, but occur together in Manitoba, in the upper reaches of the Mississippi, and south of Hudson's Bay. One could estimate the distribution of their common ancestor, by 'eliminating the unshared element', as the area where they are sympatric. The fossil localities of *Eohiodon* do not overlap that area, so one would add the distribution of the fossil genus (Alberta, British Columbia, Oregon, Montana, Wyoming – no detailed

phylogeny of the species of *Eohiodon* is available, so the total range must be added). The range of *Lycoptera* in Asia is then added, and so on throughout the cladogram (Fig. 20.8). The result of this analysis is the conclusion that hiodontids reached North America by two episodes of dispersal, one from Africa to Asia, and one from Asia to North America.

Nelson, whom I regard as today's leading theorist in phylogenetics and biogeography, no longer accepts the cladistic dispersal method. This is because, as he explained in 1975, the method will resolve dispersal whether or not it actually occurred. In other words, the method was developed within the framework of dispersal biogeography, and followed the basic assumption that the biogeographer's job is to find out when and by what route each taxon dispersed from its centre of origin. Nelson is now an advocate of vicariance biogeography (discussed above, under lepisosteids), and suggests that the areas identified by the cladistic dispersal method do not mark centres of

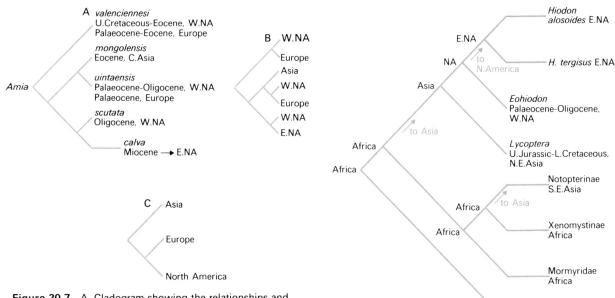

Figure 20.7 A, Cladogram showing the relationships and distribution of *Amia*. The species *valenciennesi* and *uintaensis* are taken in a wide sense; European and American populations are distinguishable, and are usually called separate species. B, Version of A in which the species names are omitted, showing how the areas occupied by them are related. C, Cladogram of the areas occupied by *Amia*, simplified from B.

Figure 20.8 Cladogram showing the relationships and distribution of hiodontids and other osteoglossomorphs. The areas named at the junctions in the diagram are ancestral distributions estimated by the cladistic dispersal method (see text), and arrows mark dispersals postulated by that method. After Nelson (1969).

origin, but the sites of barriers which interrupted widespread ancestral species (Fig. 20.9). Thus, following the vicariance paradigm, he suggests that the fishes were there before the barriers (between Africa and South America, Asia and North America, etc.) developed, not that the fishes crossed those barriers. Ancestral distributions are best estimated, Nelson now suggests, by simply adding descendent distributions (Fig. 20.9).

When Nelson wrote his paper on osteoglossomorph biogeography (1969), he did not know that the Mesozoic lycopterids were related to the hiodontids, and that discovery seemed a significant confirmation of his hypothesis that hiodontids reached North America from Asia. More recently, Chinese palaeontologists have revised the composition of the Lycopteridae, previously thought to be a close-knit group of three or four fossil species. Their conclusions bring out a significant consequence of the cladistic dispersal method. Figure 20.10 is a new cladogram of lycopterid relationships, published in China (Chang & Chou, 1976), and using the cladistic dispersal method. It shows the lycopterids split into four genera, two of which are related to hiodontids, and two to the remaining osteoglossomorphs. Previously (Fig. 20.8), the cladistic dispersal method estimated the area of origin of osteoglossomorphs as Africa, but with the addition of these Chinese fossils that area comes out as east Asia. Thus the addition of a few early and primitive fossils to the diagram gives a new centre of origin, one that coincides with the Chinese fossil localities. Of course, it is possible that this is correct – those localities really are the site of origin of

osteoglossomorphs. But in accepting that, we have to accept an important corollary, that the lack of lycopterid-like fossils in (say) the Americas and Africa is real. In other words, we have to accept that the fossil record is complete. No palaeontologist familiar with the dearth of freshwater Mesozoic fishes from those areas would accept that.

This example brings out the fact that the cladistic dispersal method is inordinately sensitive to fossils, and in a group with a reasonable fossil record it usually gives the same result as the 'oldest fossil' rule of dispersal biogeography – that the site of the oldest fossils marks the centre of origin of a group. Thus both methods depend on the same assumption, that the fossil record is complete. The experience of every vertebrate palaeontologist is against that, and the two methods are suspect for that reason. Remember too the other assumption on which they rest, that the task of biogeography is fixing sites of origin and directions of dispersal.

Superorder Ostariophysi – catfishes, carps and characins

Ostariophysans are characterised by the Weberian apparatus, a set of 'ear ossicles' formed by parts of vertebrae linking the swimbladder and ear. With over 6000 living species, they are by far the largest group of freshwater fishes, and dominate the lakes and rivers of every continent except Australia. Ostariophysans have always been of particular interest to biogeographers, and there are many

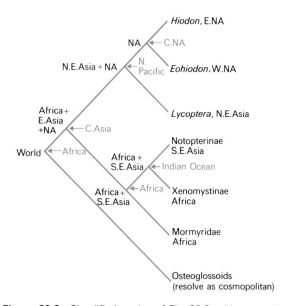

Figure 20.9 Simplified version of Fig. 20.8, with ancestral distributions estimated by adding descendent distributions, and arrows marking areas where postulated vicariance events took place.

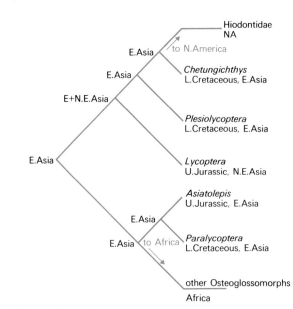

Figure 20.10 Cladogram of hiodontids and other osteoglossomorphs, incorporating new fossil forms from China, with ancestral distributions and dispersals estimated by the cladistic dispersal method.

essays on their history. The following comments summarise a dialogue between Novacek & Marshall (1976) and Briggs (1979), authors who have most recently studied the problem of ostariophysan distribution.

There are three principal groups of ostariophysans, the characoids (characins), cyprinoids (carps and their relatives) and siluroids (catfishes). Characoids are confined to Africa and South and Middle America, except for one species of the South and Middle American genus *Astyanax*, which occurs in Texas and New Mexico. Cyprinoids occur in Europe, Asia and Africa, but are entirely absent from South America. In North America they are represented by two families, the Cyprinidae, a very large group ranging through Eurasia and Africa, and the Catostomidae (suckers), a group of about 60 species which is confined to North America (south to Guatemala) except for one isolated genus in the Yangtse system in China, and one species, *Catostomus catostomus*, which ranges from Siberia to eastern North America. All North American catfishes (siluroids) belong to an endemic family Ictaluridae, with about 35 species.

Traditionally (for instance, Darlington, 1957) the history of the ostariophysans has been interpreted with the help of various rules of thumb. For example, cyprinoids are most numerous and diverse in south-east Asia, and using the 'numbers rule' this has been taken to be the region where they originated, and whence they have dispersed to Eurasia, Africa and North America. The characoids, found only in South America and Africa, and numerous and diverse in both continents, were interpreted as indicating past contact between the two areas, and convinced some ichthyologists of the reality of continental

drift long before it became fashionable. Ironically, within the last ten years fossil characoid teeth have been described from the Palaeocene and Eocene of Europe, too late to alter the opinion of those ichthyologists. The siluroids have been interpreted by a 'relict rule'. In Chile and Argentina there is a very primitive and distinct genus of catfishes, placed in its own family Diplomystidae, which is obviously the sister-group of all other siluroids. There are two ways of interpreting the significance of such a group: first, that primitive, relict species mark the centre of origin, and were left behind there as more advanced or progressive forms dispersed over the globe; and second, that relicts mark the limits of the dispersal of the group, having been 'forced' away from the evolutionary centre by competition with more successful relatives (the same idea is used by some to explain the distribution of beeches – see Chapter 21). It is not obvious which of these alternatives is true, and each has its advocates, ready with examples to back up their choice (for instance, Darlington, 1970, who chose the second, and Brundin, 1972, who chose the first).

Novacek & Marshall (1976) re-assessed the history of ostariophysans using the cladistic dispersal method (discussed above under hiodontids) and the theory of plate tectonics. From a cladogram of ostariophysan interrelationships they estimated that the group originated in the early Cretaceous within Gondwana, in what is now South America, where the most primitive siluroids and, in their view, the most primitive characoids are found (first alternative above). So far as North America is concerned, Novacek & Marshall concluded that its ostariophysans came ultimately from Africa, via Europe or Asia, in Tertiary times.

Briggs (1979) used the theory 'that the region with the more stable ecosystem (the greatest species diversity) would function as the most important evolutionary center and would produce species that could invade adjoining regions but that the more stable region would accept few or no species in return' (second alternative above). Briggs argued that ostariophysans arose in south-east Asia, in the late Jurassic. For North American ostariophysans, he concluded that the siluroids came either from Asia or from Europe in the late Jurassic, and that cyprinoids came from Asia somewhat later.

How might we evaluate or choose between these two different theories? Each is backed up by competent reviews of the taxonomy and fossil record of ostariophysans, and of plate tectonic reconstructions of past continental positions. The only real difference between the two is that Novacek & Marshall locate the centre of origin where primitive forms are now found, and Briggs locates it as far as possible from that point. Such axiomatic differences seem irreconcilable, and the vicariance biogeographer would say that there is no real difference between the two theories: each sees the biogeographer's task as the resolution of centres of origin and directions of dispersal, a task that may be futile.

At least Briggs agrees with Novacek & Marshall about North American ostariophysans – they came from elsewhere, either from Europe or from Asia. But to make any progress with North American ostariophysans we need theories of their relationships, so that for each group we can answer the question 'what is its sister-group, and where does it live?' As yet, that question cannot be answered for North American catfishes (Ictaluridae) or cyprinoids (Cyprinidae and Catostomidae), for the relationships of those groups are poorly understood. Within North America, the interrelationships of catostomids have been studied, most recently by biochemical genetics (Ferris & Whitt, 1978), but we do not know the position of *Myxocyprinus*, the isolated catostomid in China, which might be the sister-group of all American catostomids, or of some American subgroup.

The single North American characin has relatives in Central and South America, and more distantly, in Africa. The only fishes we have met so far with such a pattern are *Atractosteus* gars (Fig. 20.4), but gars are widespread in North America and absent in South America, whereas there are no characins in Cuba, where *Atractosteus tristoechus* is found. *Atractosteus* and characoids occur as fossils in the early Tertiary of Europe.

Thus the characoid and gar patterns are not incongruent, but to assume that they have the same historical cause requires either multiple extinctions, as yet undocumented by fossils (for instance, of characins in Cuba and North America; of gars in South America), or multiple dispersals. At present, the simplest explanation is that the two patterns are not the same, and the single North American characoid has dispersed northwards from Central America.

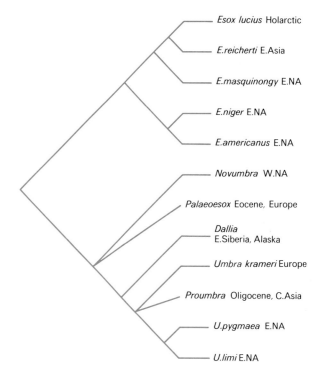

Figure 20.11 Cladogram showing the relationships and distribution of living and certain fossil esocoids, after Nelson (1972).

Families Esocidae and Umbridae – pikes and mudminnows

These two families comprise the suborder Esocoidei, a coldwater Holarctic group found in northern North America, Europe and northern Asia. Each of the families contains only five living species, and four of each are found in North America. The four North American *Esox* (pike) include three endemic (and sympatric) species from the central and eastern regions, and *E. lucius*, which ranges throughout the Holarctic region and so is part of the Holarctic fauna rather than the North American. The fifth *Esox* is found in the Amur Basin, in eastern Asia. The four North American species of Umbridae (mudminnows) are placed in three genera: *Novumbra*, with a single species in the Pacific drainage of Washington state; *Dallia*, with a single arctic species in Alaska, Siberia and islands in the Bering Sea; and *Umbra*, with two allopatric species in central and eastern North America. The one non-American umbrid is an *Umbra* from south-eastern Europe.

There are fossil esocoids in Europe from the Oligocene onwards and in North America from the Palaeocene onwards. Fossil umbrids are found in Asia (Oligocene), Europe (Eocene onwards) and North America (Oligocene onwards).

Figure 20.11 shows the interrelationships of living esocoids and some of the fossils, and Figure 20.12 is a simplified cladogram of their distribution. If the cladistic dispersal method is applied to Recent esocoids, a North

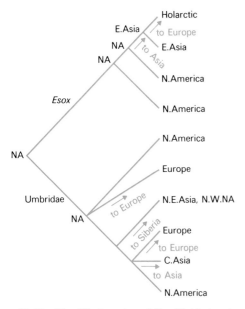

Figure 20.12 Simplified version of Fig. 20.11 showing the areas occupied by esocoids, and a cladistic dispersal interpretation of their areas of origin and dispersals.

American origin is resolved, and three dispersals are necessary: *E. lucius* and *E. reicherti* to Eurasia, *Dallia* to Siberia, and *Umbra* to Europe. But the two extinct umbrid genera complicate this picture (using the 'oldest fossil' rule, *Palaeoesox* has been held to show that the group originated in Europe by Darlington, 1957, p. 34). This is because the fossils do not yield enough information to determine their precise relationships, so that they have to be entered in the cladogram as trichotomies (which say no more than that there are three groups, and give no help in sorting out patterns). So entered, each requires an independent dispersal if the group originated in North America. But it is possible that the two extinct genera are most closely related to *Umbra krameri*, so that one episode of dispersal would account for all three; it is also possible that *Palaeoesox* is the most primitive umbrid, fitting below *Novumbra* in the cladogram. All in all, apart from telling us that esocoids and umbrids have existed since the early Tertiary, the fossils are not yet much help in deciphering the history of the group.

Since esocoids occur in all the sub-regions of the Holarctic, we might expect them to contribute to a more general historical problem, the interrelationships of those sub-regions; for example, do esocoids relate North America to Europe more closely than to Asia, as *Amia* does (Fig. 20.7C)? Figure 20.13 is a cladogram of areas simplified from Figure 20.11, omitting the widespread *E. lucius*, which is uninformative on the interrelationships of areas, and the ambiguous fossils. Here too, esocoids are disappointingly unhelpful; central and eastern North America are related to Europe by *Umbra*, and to eastern Asia by *Esox*, so that no decision is possible. But the umbrids

suggest that central and eastern North America are more closely related to Europe than to the Pacific drainage of North America, a pattern which recalls Mesozoic geography (Fig. 20.2D, E), and suggests a hypothesis that can be tested by other groups.

Family Salmonidae – salmon, trout and whitefish

Like the esocoids, salmonids are a coldwater, Holarctic group. Unlike esocoids, they include some diadromous species, which enter fresh water only to breed. Some of these species have races or subspecies which are restricted to fresh water, like the landlocked salmon, *Salmo salar sebago*, of various north-eastern American lakes. As the name implies, these fishes have become landlocked, unable to return to the sea, by changes in drainage patterns. It is possible that in the more remote past this phenomenon has been responsible for speciation and distribution of salmonids in fresh water.

There are three subfamilies, Salmoninae (salmon and trout), Coregoninae (whitefishes) and Thymallinae (graylings), all three distributed throughout the northern Holarctic. Because of their importance as food and game fishes, the lower level taxonomy of salmonids is highly developed, with many species and subspecies for forms found in different lakes. But I could find virtually nothing useful in print about the interrelationships of salmonid genera and subfamilies. Without such information, nothing can be said about the historical biogeography of the group. Fossils show that salmonids have been in North America since the Eocene.

Order Percopsiformes – troutperches and cavefishes

The percopsiforms are a small group endemic to North America. There are three families: Percopsidae (troutperches), with two species, one in Canada and central and eastern USA and one in the Columbia River, the Pacific drainage of Washington and Oregon; Aphredoderidae (pirateperch), with one species in central and eastern USA; and Amblyopsidae (cavefishes) with five species in swamps and caves of central and south-eastern USA.

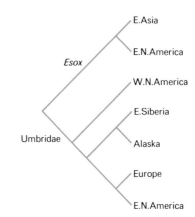

Figure 20.13 Simplified version of Figs 20.11 and 20.12, omitting *Esox lucius* (Holarctic) and the two fossil genera.

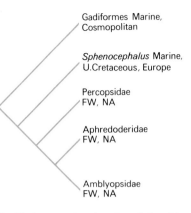

Gadiformes Marine,
Cosmopolitan

Sphenocephalus Marine,
U.Cretaceous, Europe

Percopsidae
FW, NA

Aphredoderidae
FW, NA

Amblyopsidae
FW, NA

Figure 20.14 Cladogram showing the relationships and distribution of percopsiforms.

Percopsiforms have a reasonable fossil record in North America, with three Eocene genera of percopsids and an Oligocene aphredoderid. The only other member of the group is *Sphenocephalus*, a Cretaceous genus from marine beds in Europe.

The relationships of percopsiforms are shown in Figure 20.14. Their sister-group is the order Gadiformes (cods and their relatives), a large, coldwater marine group containing one freshwater species, the burbot (*Lota lota*), which is distributed throughout the Holarctic.

As an endemic North American group with marine relatives, the percopsiforms illustrate the problem of estimating ancestral habitat (freshwater or marine), which is similar to the biogeographic problem of estimating ancestral distribution. As Figure 20.14 shows, both gadiforms and percopsiforms include marine and freshwater forms. In the percopsiforms, estimating the ancestral habitat as marine or freshwater is equally parsimonious: each requires one change in habitat – either *Sphenocephalus* is secondarily marine, or the ancestor of Recent percopsiforms moved from the sea into North American fresh waters. But in Gadiformes, *Lota* is a derived member of one subgroup, and estimating the ancestral habitat of gadiforms as freshwater would be unparsimonious, requiring independent movement into the sea of the ancestors of several subgroups. So parsimony demands that the ancestral gadiform was marine, and the burbot is secondarily in fresh water. From this it follows that the ancestral percopsiform was also marine, and North American percopsiforms are secondarily in fresh water.

This sort of analysis recalls the cladistic dispersal method in biogeography (discussed under hiodontids, above). The alternative to the dispersal method (estimating ancestral distribution by minimising the number of dispersals) is the vicariance method (estimating ancestral distribution by adding descendent distributions together). In estimating ancestral habitat, the assumption that it must be either marine or freshwater is equivalent to the dispersal method. An equivalent of the vicariance method would be to allow a third possibility, that the ancestor of gadiforms

and percopsiforms was euryhaline – capable of life in the sea and fresh water. On that view, *Lota* and the North American percopsiforms represent populations isolated in fresh water, whereas *Sphenocephalus* and the remaining gadiforms represent populations isolated in the sea.

Suborder Cyprinodontoidei – killifishes

Killifishes are small, surface-feeding fishes, found in warmer fresh and brackish water. North American freshwater forms belong to two families (Table 1), Cyprinodontidae and Poeciliidae (viviparous killifishes). Cyprinodontids are also found in South and Central America, Africa, Europe and Asia, whereas poeciliids are confined to the Americas. These distributions will not be discussed in detail here, but the group does illustrate two very different approaches to historical biogeography. Donn Rosen is the authority on American killifishes, and one could hardly find a better example of recent developments in biogeography than to compare two of his papers on the group, published five years apart. The first (1973, especially pp. 257–260) concentrates on cyprinodonts of the eastern seaboard, and discusses their distribution in terms of salt tolerance, dispersal abilities, ecological preferences, ocean currents and so on; in short, in the terms of dispersal biogeography. The second paper (1978) uses two groups of poeciliids as the basis for an exposition of the general question of method in historical biogeography, and of the cladistic vicariance method, in which cladograms of areas occupied by individual taxa are compared, and congruence between them is taken as the pattern which must be explained by the historical biogeographer (see Conclusions, below).

Family Centrarchidae – sunfishes

Centrarchids are spiny-finned fishes, endemic to North America. There are two subfamilies, the Elassominae (pygmy sunfishes), with three species in south-eastern USA, and Centrarchinae, with 26 species in the central and eastern USA and one isolated species in the Pacific drainage of California (Sacramento River). Fossil centrarchids are not surely known before the Miocene. Pliocene fossils from Idaho show that *Archoplites*, now confined to the Sacramento River, formerly had a wider range. The interrelationships of centrarchids have been thoroughly studied, recently by biochemical methods (Avise & Smith, 1977), but the sister-group of the family is unknown, so that centrarchids cannot yet contribute to historical patterns extending outside North America.

Family Percidae – perch and darters

The percids are a Holarctic group, found in Europe, northern Asia and North America. There are two wide gaps in the circumpolar distribution of the family, one across the Atlantic, and one between Asia and North America, since percids are absent in eastern Siberia and in North America west of the Rockies.

There are two subfamilies, Percinae and Luciopercinae,

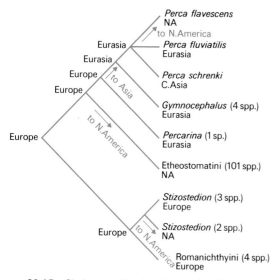

Figure 20.15 Cladogram showing the relationships and distribution of Percidae with their areas of origin and dispersals as postulated by Collette & Bănărescu (1977).

both represented in North America. The Percinae include two tribes, one (Percini) with a single species in North America, one in central Asia, and the remaining six concentrated in Europe, two of them extending into central Asia. The second percine tribe is the Etheostomatini (darters), endemic to North America and with over 100 species there. The Luciopercinae include the North American pikeperches (two species) and six species in Europe and western Asia. There are reputed fossil percids in North America and Europe from the Eocene onwards, but their relationships to the living subgroups are unknown.

Figure 20.15 shows the relationships and history of living percids as deduced by Collette & Bănărescu (1977). They postulated a European origin, and dispersals as shown in the diagram, suggesting that etheostomatins entered North America by the pre-Eocene North Atlantic landbridge, whereas *Perca* and *Stizostedion* entered by the Bering route in Miocene or Pliocene times. Collette & Bănărescu seem to have been using 'numbers clues' and the fossil record to postulate centres of origin and directions of dispersal, but as Fig. 20.15 shows, the cladistic dispersal method gives the same results.

Family Cichlidae – cichlids

Cichlids are a very large (over 600 species) warmwater group, occurring in Central and South America, Africa (including Madagascar), Syria and southern India. One species, *Cichlasoma cyanoguttatum*, is found in North America, in southern Texas. *Cichlasoma* is a Central and northern South American genus, which also occurs in Cuba and in Hispaniola, where there are Miocene fossils. The single North American cichlid, with its relatives in Central and South America, and more distantly in Africa,

recalls the one North American characin, while this pattern and Cuba recalls *Atractosteus* gars. In discussing characins, it was concluded that the characin and gar patterns are not the same, and that the single North American characin arrived from the south. Comparing cichlids and characins, their North American representation is the same, as are their relationships with Central and South America, and more distantly with Africa. But characins do not occur in the Antilles, or in India and Madagascar, whereas cichlids do. As in the gar/characin comparison, the cichlid and characin patterns are not incongruent, but they can only be made coincident by assuming multiple extinctions (of characins in the Antilles, India and Madagascar; of cichlids in Europe) or multiple dispersals. And to render the characin/cichlid pattern coincident with gars requires additional assumptions, such as extinction of cichlids and characins in North America. Cichlid distribution (like that of gars) is usually explained by dispersal through the sea, since a few species are found in brackish water. At present, in the absence of detailed schemes of relationships for Central American cichlids and characins (comparable to Fig. 20.4), it is simplest to follow the traditional interpretation, that the single North American representatives of the two groups arrived from the south.

Conclusions

Having surveyed the groups of freshwater fishes which make up the North American fauna, what general conclusions can be drawn about the development of that fauna? Previous general accounts of North American fish biogeography (for instance, those by Darlington, 1957; Miller, 1959, 1965) were written in the dispersal tradition, aiming to allocate a centre of origin to each group, and using the fossil record and other clues to estimate when and whence the ancestors of each group entered the continent. I have taken a different approach, leaning heavily on phylogenetic systematics (Hennig, 1966) or cladistics, a method which seeks to express relationships as precisely as possible, in dichotomous cladograms. This method forces the biogeographer to ask precise questions about each group – is it strictly monophyletic? If so, what is its sister-group, and where does it live? This approach also leads to a rather different evaluation of the fossil record. Traditionally, biogeographers have given a lot of weight to fossils – 'The best clues, of course, are fossils – the right fossils in the right places' (Darlington, 1957, p. 35). But to the cladist, fossils are often uninformative, because they provide too few characters for their relationships to be determined (paddlefishes, esocoids and percids, above). And 'ancestral' fossils, which may be what Darlington meant by 'the right fossils', are simply fossils which seem to be equally closely related to all their possible descendants; that is, their relationships are vague. Even when fossils can be entered in the cladogram, as in gars (Fig. 20.4), they are often placed on only one or two characters (the fossil gar species in Fig. 20.4 are mostly placed by degree of loss of enamel on skull bones), so that one hesitates in drawing far-reaching conclusions from their relationships.

Given the cladistic approach, there is little to say about several of the potentially most informative groups of North American fishes (salmonids, cyprinids, catostomids, ictalurids, centrarchids, together comprising two-thirds of the species listed in Table 1), because their systematics are still rudimentary at the level we have been asking about, relationships outside North America. Doubtless because of that, previous theories about the historical biogeography of these groups have been vague. Salmonids, cyprinids and ictalurids have been held to show Eurasian relationships of some sort, but the exact nature of those relationships remains ambiguous. Catostomids, with an isolated genus in China and a fossil record since the Palaeocene in North America, are believed to have dispersed from Asia, but their two-area distribution (Yangtse and North America) is not informative (cf. paddlefishes, above).

Those groups for which we do have some cladistic information are the gars (Fig. 20.4), *Amia* (Fig. 20.7), hiodontids (Fig. 20.8), esocoids (Fig. 20.11), percopsiforms (Fig. 20.14) and percids (Fig. 20.15). Three methods of dealing with such information have been summarised or mentioned in the comments on individual groups of fishes: the cladistic dispersal method (see hiodontids and Figs 20.8, 10, 12, 15), the 'track' method (see gars and Fig. 20.5); and the cladistic vicariance method (mentioned under cyprinodonts, and see Figs 20.7, 9, 13).

The cladistic dispersal method is dispersal biogeography married with cladistic systematics. It rests on the same basic assumption as dispersal biogeography, that the biota of a region is built up piecemeal, by chance events, as founder-members of different taxa arrive there, by more-or-less accidental means. Three criticisms of the method have already been mentioned. First, it is inordinately sensitive to fossils, so that fossil localities (or concentrations of palaeontologists) may be resolved as centres of origin (Fig. 20.10). Thus the method depends on the same assumption as the 'oldest fossil' clue – that the fossil record is complete. Second, the method gives much the same results as the 'numbers clues' of dispersal biogeography (Fig. 20.15). And third, the method depends on the assumption that primitive or relict taxa mark the centre of origin (Brundin, 1972), an axiom that is not obviously true.

The 'track' method of vicariance biogeography differs fundamentally from dispersal biogeography in assuming that the dominant factor in historical biogeography is not chance dispersal of individual taxa, but fragmentation of biotas by geological or climatic change. The track method seeks to reconstruct these ancestral biotas by collating the distribution of descendants. The drawback of this method is that it depends on 'raw' similarity. For example, the paddlefishes (*Polyodon* in North America and *Psephurus* in the Yangtse) and the catostomids (*Myxocyprinus* in the

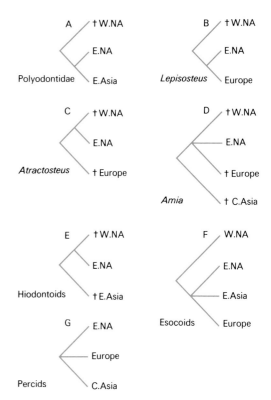

Figure 20.16 Area cladograms derived from Figs 20.3, 4, 7, 9, 13 and 15, omitting areas represented in only one (Cuba, India and Central America in Fig. 20.4; S.E. Asia in Fig. 20.9; Alaska and Siberia in Fig. 20.13) and Africa (found only in Figs 20.4 and 20.9). † = extinct.

Yangtse and the rest in North America) would be two components of a China–North America generalised track. But, as pointed out in discussing paddlefishes, 'similarity', like relationship, is a comparative concept, and a track joining only two points cannot legitimately contribute to comparative biogeography.

The third method, cladistic vicariance, is one developed very recently (Platnick & Nelson, 1978; Rosen, 1978) to bring together vicariance biogeography and cladistic systematics. If, as vicariists assume, biotas have been fragmented by geological and climatic change, how might we decipher the pattern? The problem is similar to that of reconstructing the relationships of organisms (phylogeny), since our theory is that the diversity of life today is the result of successive division and subdivision of ancestral species. With organisms, the elements of the pattern are homologies: a whale is related to mammals, not to the fishes it resembles, because it shares with mammals homologies not found in fishes (four-chambered heart, diaphragm, mammary glands, placenta, and so on). To reconstruct the geographic pattern of life, we need some geographic equivalent of homology. The cladistic vicariance method uses the geographic patterns of individual taxa as the equivalent of individual homologies in systematics. If

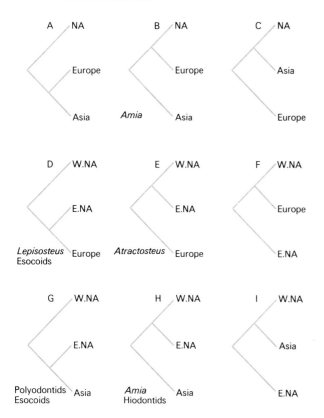

Figure 20.17 Alternative area cladograms for A–C, North America, Europe and Asia; D–F, western and eastern North America and Europe; G–I, western and eastern North America and Asia, with the groups whose patterns (Fig. 20.16) are congruent.

the biotas of areas *A* and *B* share a more recent common ancestry than either does with area *C*, the individual taxa represented in the three areas should show the same pattern – species in *A* and *B* should be related to each other rather than to species in *C*. The data of systematics are most economically expressed in cladograms, which are diagrammatic hierarchies of the homologies shared by species. The data of biogeography can be expressed as cladograms of areas.

To illustrate how the method works, Figure 20.16 summarises the meagre data from North American fishes. Areas occupied by only one taxon in the set (for example,

Cuba and India for gars; Alaska and Siberia for the esocoid *Dallia*; the sea for percopsiforms) are omitted because they cannot contribute to a general pattern. Examples of how the data in Figure 20.16 might be synthesised are shown in Figure 20.17. Given three areas, say North America, Europe, and Asia, they can be related in only three ways (Fig. 20.17 A–C; the fourth way in which they might be related is the pattern shown by percids, Fig. 20.16 G, where all three are equally closely related, but this pattern actually says nothing about their interrelationships, only that there are three areas). Of the taxa shown in Figure 20.16, only *Amia* (Fig. 20.16 D) contributes to the question of the interrelationships of North America, Europe and Asia, relating them as in Figure 20.17 B. The other two groups occurring in all three areas (esocoids, percids) say nothing decisive. As yet, then, fishes hardly contribute to this problem.

If North America is arbitrarily divided into eastern and western sub-regions, we can ask about the relationships of those areas to Europe (Fig. 20.17 D–F) and to Asia (Fig. 20.17 G–I). Europe is related to eastern North America by *Lepisosteus* gars and esocoids, and to North America as a whole by *Atractosteus*. Asia is related to eastern North America by polyodontids and esocoids, and to America as a whole by *Amia* and hiodontids. These conclusions may seem merely contradictory, or due only to the chances of fossilisation, since amongst the six groups just mentioned, all except esocoids are represented in western North America by fossils alone, and the west is where the fossil localities are (Fig. 20.6). But there is a way of deciding whether those results are due only to chance. It is to look for cladograms of other groups – insects, plants or whatever – with representatives in America, Asia and Europe, to see whether they relate the areas in a random way or not. If the relationship is random, all the area cladograms in Figure 20.17 should be equally likely, and one could accept a random explanation, colonisation by chance dispersal. If the relationship were not random, one would seek a non-random explanation, vicariance. The fishes provide too small a sample of cladograms for us to choose between these alternatives, which is perhaps only to say that the title of this article is too restrictive; historical biogeography cannot be studied through the few cladograms available in one group of organisms, but demands integration of cladistic information from all groups of animals and plants.

References

Avise, J. C. & Smith, M. H. 1977. Gene frequency comparisons between sunfish (Centrarchidae) populations at various stages of evolutionary divergence. *Systematic Zoology* **26**: 319–335.

Boreske, J. R. 1974. A review of the North American fossil amiid fishes. *Bulletin of the Museum of Comparative Zoology, Harvard* **146**: 1–87.

Briggs, J. C. 1979. Ostariophysan zoogeography: an alternative hypothesis. *Copeia* **1979**: 111–118.

Brundin, L. 1966. Transantarctic relationships and their significance, as evidence by chironomid midges. *Kungliga Svenska Vetenskapsakademiens Handlingar* (4) **11**, 1: 1–472.

Brundin, L. 1972. Phylogenetics and biogeography. *Systematic Zoology* **21**: 69–79.

Chang, M.-M. & Chou, C.-C. 1976. Discovery of *Plesiolycoptera* in Songhuajiang-Liaoning Basin and origin of Osteoglossomorpha. [In Chinese] *Vertebrata palasiatica* **14**: 146–153.

Collette, B. B. & Bănărescu, P. 1977. Systematics and zoogeography of the fishes of the family Percidae. *Journal of the Fisheries Research Board of Canada* **34**: 1450–1463.

Croizat, L., Nelson, G. & Rosen, D. E. 1974. Centers of origin and related concepts. *Systematic Zoology* **23**: 265–287.

Darlington, P. J. 1957. *Zoogeography: the geographical distribution of animals*. 675 pp. New York: J. Wiley.

Darlington, P. J. 1970. A practical criticism of Hennig-Brundin 'phylogenetic systematics' and Antarctic biogeography. *Systematic Zoology* **19**: 1–18.

Estes, R. 1970. Origin of the Recent North American lower vertebrate fauna: an inquiry into the fossil record. *Forma et Functio* **3**: 139–163.

Ferris, S. D. & Whitt, G. S. 1978. Phylogeny of tetraploid catostomid fishes based on the loss of duplicate gene expression. *Systematic Zoology* **27**: 189–206.

Grande, L. 1979. *Eohiodon falcatus*, a new species of hiodontid (Pisces) from the late early Eocene Green River Formation of Wyoming. *Journal of Paleontology* **53**: 103–111.

Henning, W. 1966. *Phylogenetic systematics*. 263 pp. Urbana: University of Illinois Press.

Hocutt, C. H. 1979. Drainage evolution and fish dispersal in the central Appalachians: summary. *Bulletin of the Geological Society of America* **90**: 129–130.

Hubbs, C. L., Miller, R. R. & Hubbs, L. C. 1974. Hydrographic history and relict fishes of the north-central Great Basin. *Memoirs of the California Academy of Sciences* **7**: 1–257.

Miller, R. R. 1959. Origin and affinities of the freshwater fish fauna of western North America. *Publications of the American Association for the Advancement of Science* **51**: 187–222.

Miller, R. R. 1965. Quaternary freshwater fishes of North America, pp. 569–581 *in* Wright, H. E. & Frey, D. G. (Eds) *The Quaternary of the United States*. Princeton: Princeton University Press.

Miller, R. R. 1966. Geographical distribution of Central American freshwater fishes. *Copeia* **1966**: 773–802.

Myers, G. S. 1966. Derivation of the freshwater fish fauna of Central America. *Copeia* **1966**: 766–773.

Nelson, G. J. 1969. Infraorbital bones and their bearing on the phylogeny and geography of osteoglossomorph fishes. *American Museum Novitates* **2394**: 1–37.

Nelson, G. J. 1972. Cephalic sensory canals, pitlines, and the classification of esocoid fishes, with notes on galaxiids and other teleosts. *American Museum Novitates* **2492**: 1–49.

Nelson, G. 1975. Historical biogeography: an alternative formalization. *Systematic Zoology* **23**: 555–558.

Nelson, G. 1978. From Candolle to Croizat: comments on the history of biogeography. *Journal of the History of Biology* **11**: 269–305.

Novacek, M. J. & Marshall, L. G. 1976. Early biogeographic history of ostariophysan fishes. *Copeia* **1976**: 1–12.

Platnick, N. I. & Nelson, G. J. 1978. A method of analysis for historical biogeography. *Systematic Zoology* **27**: 1–16.

Rosen, D. E. 1973. Suborder Cyprinodontoidei. *Memoirs of the Sears Foundation for Marine Research* **1**, **6**: 229–262.

Rosen, D. E. 1978. Vicariant patterns and historical explanation in biogeography. *Systematic Zoology* **27**: 159–188.

Wiley, E. O. 1976. The phylogeny and biogeography of fossil and Recent gars (Actinopterygii: Lepisosteidae). *Miscellaneous Publications of the Museum of Natural History, University of Kansas* **64**: 1–111.

CHAPTER 21

Biogeographical methods and the southern beeches

C. J. Humphries

'*The key to the history of terrestrial life in the far south may be* Nothofagus'   P. J. Darlington (1965)
'Nothofagus *is uninformative on the interrelationships of southern hemisphere areas*'   C. Patterson (1980)

In the introductory essay to his *Flora Novae Zelandiae* (1853) Sir Joseph Dalton Hooker attempted for the first time to explain why so many different groups of distantly related organisms should show similar, widely disjunct distribution patterns (Fig. 21.1) in the southern hemisphere areas of southern South America, Tasmania, Australia and New Zealand. In an effort to find a general explanation that could account for geographic disjunctions in one hundred or more plant genera, Hooker proposed the then novel theory that all of the different groups originated and subsequently dispersed to the southern hemisphere on a continuous tract of land. Although this was the simplest possible explanation for a major biological problem, Hooker's theory was unacceptable to many of his contempories since fashionable geological theories advocated that continents and intervening oceans are and always had been stable.

Charles Darwin, who was to publish his theory of evolution only six years later, could not agree with Hooker as to how the pattern of plant distribution had developed. Darwin was influenced by theories advocating that the continents had always been in their current positions and suggested that many different groups of organisms had their origins in the northern hemisphere and arrived at various points in the southern hemisphere by way of long distance dispersal.

Even though these extreme viewpoints are some 130 years old, and since that time there have been literally thousands of biogeographical papers published, there are still many differences of opinion amongst contemporary theorists on the origins and subsequent history of austral plants and animals. It is impossible to review all of these papers comprehensively, but certain groups of organisms such as the marsupials and the southern beeches (*Notho-fagus*) have frequently figured as important examples.

This article attempts to review the history of theoretical phytogeography since the middle of the nineteenth century to the present day by examining the different explanations for disjunct distributions of *Nothofagus*. *Nothofagus* has been selected as a review subject for a variety of reasons. First, it has attracted the attention of many biogeographers working from different theoretical frameworks. The quotation given from Darlington is one of many to be found in the writings of classical biogeographers who use contemporary theories on earth history, traditional classifications, fossils and dispersal hypotheses as the basis for explaining disjunct distribution patterns. By contrast, the quotation from Patterson represents the view of a much smaller group of biogeographers who test models of phylogenetic theories of relationship of the organisms against models of earth history in an effort to find general biogeographical explanations for disjunctions. Second, the relationships between *Nothofagus* species and their relationships with other members of the beech family are, on the whole, quite well known. Third, despite the similarity

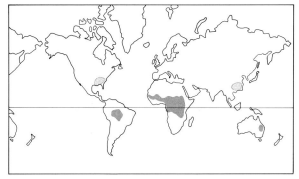

☐ Tulip trees ▉ Lungfishes

Figure 21.1 Disjunct distributions in the tulip trees (*Liriodendron*) and the lungfishes; two examples of widespread taxa which require biogeographical explanations.

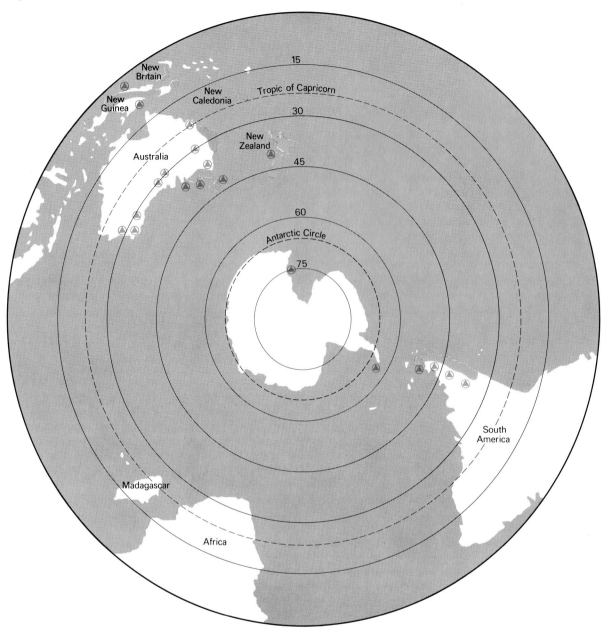

Figure 21.2 A conspectus of the species of *Nothofagus* and a map showing modern (blue tint) and fossil (▲) distributions in the southern hemisphere. Only the Recent species are listed here:

New Caledonia	New Guinea	South America	New Zealand	Australia/ Tasmania
aequilateralis	*perryi*	*alessandri*	*solandri*	*cunninghamii*
balansae	*nuda*	*pumilio*	*cliffortioides*	*moorei*
codonandra	*starkenborghii*	*antarctica*	*truncata*	*gunnii*
baummanniae	(also in	*obliqua*	*fusca*	
discoidea	New Britain)	*glauca*	*menziesii*	
	brassii	*alpina*		
	pullei	*procera*		
	crenata	*betuloides*		
	resinosa	*nitida*		
	pseudoresinosa	*dombeyi*		
	flaviramea			
	carrii			
	grandis			
	discoidea			
	rubra			
	wormersleyi			

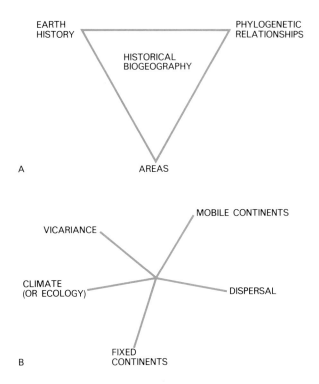

Figure 21.3 A. Historical biogeography receives input from three sources. B. A particular biogeographic analysis is subject to the assumption of two or more of five variables.

of the supporting evidence for each biogeographical explanation the hypotheses and the premises on which they are based are so different that the ways in which the evidence is used can be sharply contrasted.

So this essay is directed towards methodology; it will attempt to explain the varying methods biogeographers have used in trying to solve a common problem – an explanation of the history of the distribution of southern beeches. All previous explanations are found to be wanting in some degree and a new suggestion is put forward. First, we must begin with the data.

Geography of the southern beeches

The 36 species of the southern beech genus *Nothofagus* are the only members of the family Fagaceae to occur in the southern hemisphere (Fig. 21.2). The closest relative is the northern beech genus, *Fagus*, and this is widely distributed in Europe and North America. The remaining genera of the Fagaceae are important trees of the north temperate and south-east Asian forests. The species of *Nothofagus* are almost always tall trees dwarfed only by severe weather conditions at the timber-lines and southern latitude extremities. They are mainly evergreen but several species are deciduous. They have small seeds, which do not travel very far and do not survive in sea water. All species are poorly fitted for dispersal and for this reason alone they

have always raised intriguing questions regarding their wide distribution in the southern hemisphere. They form forests in three principal areas of the southern cold-temperate zone – southern South America, Australia (including Tasmania) and New Zealand and they also occur in the cooler highlands of New Guinea, New Britain and New Caledonia.

There are ten species in southern South America forming almost continuous forest on the tops and along the western slopes of the Andes, variously distributed from latitude 56° S to latitude 33° N. According to McQueen (1976) the southern limit is imposed by the lack of suitable land and the northern limit is due to the aridity that accompanies the 'Mediterranean' climate of central Chile. There are three species in Tasmania and Australia – the deciduous *N. gunni* is endemic to the mountains of Tasmania; *N. moorei* occurs in New South Wales and extends northwards to the border with Queensland. *N. cunninghamii* occurs in Victoria and Tasmania. The five New Zealand species occur in rather isolated patches but extend from the lowlands to the timber-line on both main islands, particularly on their western sides. There are some 18 tropical evergreen species in the mountains of New Guinea, New Britain and New Caledonia. Here, they are important co-dominants of the rain forest, forming extensive but patchy stands, especially in the lower montane areas between about 2300 m and 2800 m.

Narrative theories of *Nothofagus* biogeography

Historical biogeography can be divided into those studies carried out as narrative theories and those under analytical models (Ball, 1976). At this point we consider narrative theories.

In narrative theories the history of the distribution of contemporary species is expressed in terms of the taxonomy of modern species as compared with the distribution and taxonomy of the fossils. Narrative historical biogeography is mostly concerned with providing explanations in terms of five variables (Fig. 21.3). Usually, the fashionable geological or climatological theory determines the biogeographical explanation. For instance, if one assumes a world in which the continents have always occupied the same positions, transoceanic distribution patterns must have come about by long-distance dispersal. On the other hand, if one assumes that the continents have moved relative to one another or are the result of fragmentation of a supercontinent, or both, then the present-day distribution may be explained by dispersal prior to continental break-up or by a vicariance event.

A vicariance event does not assume dispersal. Rather, it assumes that the species was widely distributed over a land mass and that when that land mass fragmented and the fragments moved away from one another the original species was similarly fragmented.

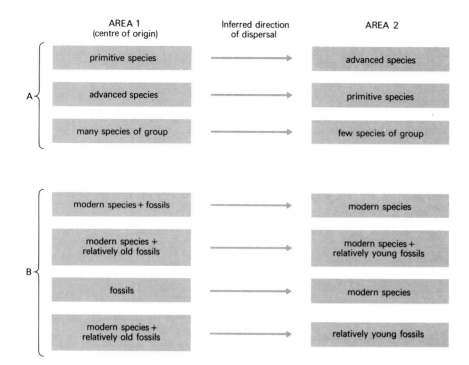

Figure 21.4   In narrative theories of biogeography the centre of origin and, hence, the direction of dispersal is inferred by a number of means. In some cases (A) the centre of origin is decided according to the area occupied by the most primitive or the most specialised members, or the area occupied by the greatest diversity of members of a group. In other cases (B) the issue is decided on the basis of fossils.

Thus, within the context of narrative theories we can recognise those in which dispersal is assumed and those in which dispersal is unneccessary for the explanation. A wide variety of narrative theories, adopting combinations of the above variables, has been proposed for the southern beeches.

Dispersal explanations, of course, assume that there was a 'centre of origin' for the group and that various members moved away from that centre. Several methods have been employed to determine the centre of origin and hence, the direction of subsequent dispersal. These methods are illustrated in Figure 21.4 and explained in the accompanying legend; they are discussed further below.

Explanations involving stable continents and dispersal

In the introduction to this essay it was pointed out that Darwin, who believed in stable continents, thought that austral groups had originated in the northern hemisphere and subsequently dispersed south. The 'monoboreal relict hypothesis' of Schröter (1913) was an early attempt by a botanist to justify Darwin's idea. In its simplest form this hypothesis stated that isolated austral groups were invariably primitive forms, driven southwards from the northern hemisphere through the development of new 'aggressive' groups in the north. The evidence for this hypothesis came from the fact that groups, such as the gymnosperms *Podocarpus* and *Araucaria*, which are largely restricted to the southern hemisphere, have Mesozoic fossil relatives in the northern hemisphere.

Croizat (1952), who was originally working within a stabilist framework, suggested that there were several centres for the origin of the angiosperms. The Beech family (Fagaceae), like several other 'primitive' families, were said to have originated in New Caledonia from a New Caledonian centre. Although no precise directions for dispersal were given, *Nothofagus* was considered to have moved along a circum-antarctic 'track', the other members of the family (for example, the northern beeches) moving through south-east Asia to Europe and North America.

Since it is generally believed that fossils are amongst the best indicators of evolutionary direction the particularly good fossil pollen record of *Nothofagus* has been used extensively in phytogeographical discussion. There are three recognisable pollen morphs among the species of *Nothofagus* (Fig. 21.5 A) and these are named after three species *brassii*, *fusca* and *menziesii* (Hanks & Fairbrothers, 1976). All three pollen types are very distinctive and can easily be distinguished from pollen of other beeches. The fossil types are found from a number of localities and horizons (Fig. 21.5 B). There have been some reports of *Nothofagus* pollen fossils from the northern hemisphere but their identification as belonging to *Nothofagus* is by no means certain. Definite records of *Nothofagus* pollen

B

	Australia b	Australia f	Australia m	New Zealand b	New Zealand f	New Zealand m	Feugia & Patagonia b	Feugia & Patagonia f	Feugia & Patagonia m	New Guinea b	New Guinea f	New Guinea m	New Caledonia b	New Caledonia f	New Caledonia m	Seymour Is. b	Seymour Is. f	Seymour Is. m	McMurdo Sound (Antartica) b	McMurdo Sound (Antartica) f	McMurdo Sound (Antartica) m
Recent + Pleistocene		█	█		█	█		█	█	█			█								
Pliocene	█	█	█	█	█					█											
Upper Miocene	█	█	█	█	█					█											
Lower Miocene	█	█	█	█	█																
Oligocene	█	█	█	█	█	█	█	█	█												
Eocene	█	█	█	█	█		█	█	█										█	█	█
Palaeocene	█	█	█													█	█				
Upper Cretaceous	█	█	█	█	█	█	█														

A

Polar view	Equatorial view
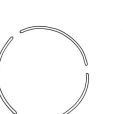	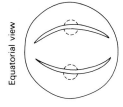

'Fagus'-type
(Fagus sylvatica)

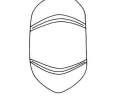

'Brassii'-type
(Nothofagus grandis)

'Fusca'-type
(Nothofagus alessandri)

'Menziesii'-type
(Nothofagus menziesii)

Figure 21.5 A. The pollen of *Nothofagus* species may be one of three types shown here and compared with that of a northern beech (*Fagus*). The terms polar view and equatorial view are conventions adopted by palynologists to facilitate direct comparison between pollen-types. B. Stratigraphical and geographical distribution of the three *Nothofagus* pollen-types.

have been found in areas of present-day *Nothofagus* forest with appreciable extensions of range only in western Australia, Patagonia and Antarctica (Fig. 21. 2).

Amongst the earliest fossil records (Fig. 21.5B) the '*brassii*'-type and '*fusca*'-type occur throughout the Upper Cretaceous of New Zealand. The earliest record for the '*menziesii*'-type is from the Eocene of Australia and South America. It appears from fossil evidence, that *Nothofagus* was reasonably widespread in the Cretaceous becoming more widespread during the Tertiary. Today the genus is absent from Antarctica, the '*brassii*'type is restricted to New Guinea and New Caledonia with only the '*fusca*'-and '*menziesii*'-types occurring in the cooler temperate floras of Australia, New Zealand and South America.

Darlington's (1965) interpretation of the history of *Nothofagus* combined fossil pollen evidence with stable continent geology. Although he acknowledged continental drift as a means of causing disjunction, his dismissal of it rested on the subjective belief that it occurred too early to have affected the history of the angiosperms. Despite a reliable micro-fossil record being present only in the austral continents, Darlington believed that *Nothofagus* could not have originated there because its nearest relative *Fagus* lives in the northern hemisphere. Since the 'centre of diversity' for the Fagaceae is in south-east Asia, *Nothofagus* 'may have originated in Asia primarily in subtropical parts of south-east Asia during the Cretaceous.

It probably was never widespread or dominant in the northern hemisphere. It probably crossed the tropics once, by way of the Indo-Australian archipelago, in the Cretaceous. The *Nothofagus*, of the *brassii* group, now on New Guinea may be descendants of the original, Cretaceous, tropics crossers' (Darlington, 1965, p. 145). He went on to say that an origin in Asia followed by southward dispersal across the Indo-Australian archipelago would bring *Nothofagus* into the southern hemisphere to Australia or New Zealand or both. The '*brassii*' group was considered to be ancestral to the other modern species. He went on to postulate three separate late Cretaceous long-distance dispersal events around the austral land-masses to account for the modern distributions.

Other writers had similar views. Taktajhan (1969), for example, agreed with Darlington that the 'cradle' of the flowering plants was south-east Asia and suggested that *Nothofagus* reached the southern hemisphere by way of Malaysia and Australasia.

Explanations invoking stable continents and land bridges

Toward the end of the nineteenth century the geologist Melchoir Neumayr suggested that, formerly, the major continental land-masses were linked by very large islands. This idea was developed as the land bridge theory which suggested that continents, which had remained in the same positions, were linked by bridges made and broken at various times. The botanist Van Steenis (1962) used this

Figure 21.6 Scheme of morphological interrelationships in the cupule structure in species of *Nothofagus*. The diagram (redrawn from Van Steenis, 1953 and corrected for modern names – see Fig. 21.1) shows the relative number and shape of the cupule lobes, the elaborations of the cupule lamellae, the number of florets and the number of stigmas. For key to abbreviations see Figure 21.8.

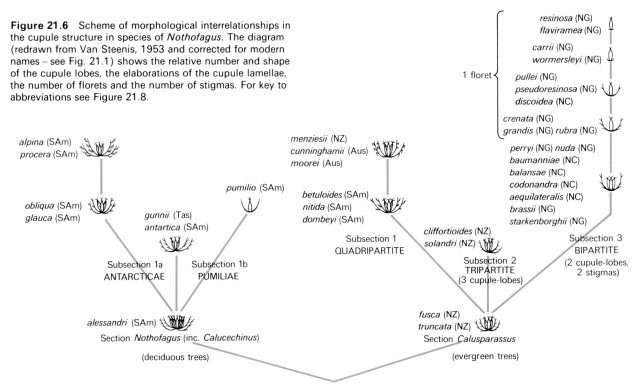

idea and his land bridge theory of plant distribution represents an intermediate position between the stable and mobile continent theories.

Although Van Steenis believed in a steady-state world he also realised that the major plant disjunctions in both tropical and temperate floras could not be accounted for by transoceanic dispersal. In rejecting long-distance dispersal and multiple origins for similar floras on different continents, he developed a land bridge theory. To account for the similarities between the fractions of the austral biota, land bridges were postulated connecting Australia, Tasmania, New Zealand, Antarctica and South America. The presence of submerged continental shelves were used as evidence for a former Mesozoic highway.

The classification of *Nothofagus* proposed by Van Steenis (1953) is based on variations in the leaves, the female flower and the number of florets (Fig. 21.6) and is the one most generally accepted today. The eight deciduous species of South America, New Zealand and Tasmania were considered to be related to one another and grouped together in a section (named as *Calucechinus*). All the evergreen species were placed in another section (*Calusparassus*) which was divided into three subsections on the basis of the number of lobes in the cupule of the female fruit. The first subsection is a grouping of eight South American, Australian and New Zealand species with four lobes in the mature cupule. The second subsection is a small group of two species, *N. cliffortioides* and *N. solandri* from New Zealand with only three lobes in the female cupule, and the third subsection a group of approximately 18 species occurring in New Guinea, New Britain and New Caledonia, characterised by having just two lobes in the female cupule. The 'brassii' pollen type is restricted to the last subsection whilst the 'menziesii'- and 'fusca'-types occur throughout the other groups.

According to Van Steenis the Fagaceae originated in south-east Asia in an area bounded by the Province of Yunnan (China) and Queensland. As *Nothofagus* is the only austral genus of the Fagaceae, and because the New Guinea species are a distinct group with 'brassii' pollen, and because the 'brassii' pollen is considered primitive, a land bridge migration from south-east Asia into the temperate south was postulated to account for present-day distribution patterns. Additionally, since the fossil record (Fig. 21.5B) shows the 'brassii' pollen type to have become extinct in most austral areas during the Pliocene, Van Steenis considers the New Guinea and New Caledonia species as relict survivors of an older, more widespread group.

Explanations invoking mobile continents

Other biologists were not wholly convinced that the world was in a steady state, or that continental drift occurred too early, or that land bridges existed.

Much of the work carried out in the 1970s reflects a general acceptance of plate tectonic theory and the idea of mobile continents as a causal agent of disjunction. Stable continents posed transport problems for such poor dispersers as *Nothofagus*, so former overland dispersal routes seemed more plausible. In terms of biogeographical theory, however, this simply provides an alternative, albeit more plausible, explanation for dispersal but this time with overland routes rather than over water or former land bridges. Most of the major movements of continents deemed to have affected the distribution of the modern biota occurred, according to geologists, some 120 million years ago. The assumption that continent wandering explains the present-day distribution simply had the effect, under narrative theories, of 'moving' the dispersal event back to a time when the continents were closer together.

For *Nothofagus* there are basically two types of narrative explanations invoking mobile continents – those which accept northern origins and those which favour southern origins.

Raven & Axelrod (1972, 1974) using the assumption that the relatives of *Nothofagus* occur in the northern hemisphere, coupled with the fact that various present-day austral gymnosperm groups have Mesozoic and Cenozoic fossil relatives in Europe, suggested that *Nothofagus* 'probably passed between the northern and southern hemispheres by way of Africa and Europe since land connections were absent in Middle America' (Raven & Axelrod, 1972, p. 1382). In the southern hemisphere, *Nothofagus* migrated from Africa into South America and eventually to Australasia. In other words, this idea simply reflects Schröter's (1913) view but tries to define dispersal routes within the context of geological and climatical theories.

In a more detailed essay, Schuster (1976) also accepts a northern origin for *Nothofagus* but gives an alternative migration route because there were land connections between North and South America until the Tertiary. Schuster believes that 'centres of diversity' are equivalent to 'centres of origin' and thus came to the conclusion that North America was the site of origin for our well-travelled genus since the Fagaceae are Eurasian except for *Nothofagus*. The ancestor of *Nothofagus* arrived in Gondwanaland by overland transport from North to South America, along the so-called Marsupial route, possibly by mid-Cretaceous times. This far-fetched explanation continues, completely without evidence, to say that: 'By Upper Cretaceous times, *Nothofagus* had diffused not only to Antarctica but, more than 80 million years ago, across to Australia and New Zealand'. Its entry into New Caledonia and New Zealand 'must have occurred well before the end of the Cretaceous' because of their separation at this time but 'did not reach New Guinea until well into the Tertiary' (Schuster, 1976, p. 120).

Both the Raven & Axelrod and Schuster hypotheses accept the validity of northern hemisphere fossils despite their general rejection by others (for instance, Cranwell, 1963). The main alternative is a southern origin and Moore (1972) completes the narrative picture by giving us the post-tectonic version of Hooker's original observations

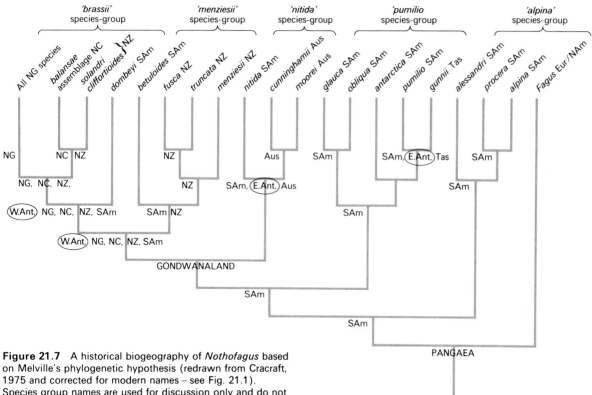

Figure 21.7 A historical biogeography of *Nothofagus* based on Melville's phylogenetic hypothesis (redrawn from Cracraft, 1975 and corrected for modern names – see Fig. 21.1). Species group names are used for discussion only and do not refer to any classifications. Localities from which *Nothofagus* is known only as fossils are circled.

outlined in the introduction. He gives us a starting date for *Nothofagus* by saying that 'at least some of the Palaeo-austral distributions date from the Cretaceous' (Moore, 1972, p. 131). Since the oldest fossils occur in New Zealand and Antarctica and because of the poor dispersal qualities of *Nothofagus* seed 'one is forced to the conclusion that *Nothofagus* achieved migration between New Zealand and Antarctica before their late Cretaceous separation, at a time when its major groups were already differentiated, crossing to Australia and South America while connections were still available during the late Cretaceous–early Tertiary and subsequently moving northwards in the east and west' (p. 131). Hanks & Fairbrothers (1976) (who, incidently, accept northern hemisphere pollen fossil evidence) strengthen Moore's view and provide a new twist by saying that '*Nothofagus* could have developed in a region between New Zealand, Antarctica, and Australia and then migrated through Antarctica into South America; and north through Tasmania, Australia, and New Zealand and eventually arrive in Europe' (Hanks & Fairbrothers, 1976, p. 69), presumably then to become extinct in the northern hemisphere.

To sum up the narrative approach, all we can say is that many possible centres of origin for *Nothofagus* have been proposed involving almost every major continent on the globe; the most favoured places being North America, Europe, south-east Asia, New Caledonia, somewhere

between Yunnan and Queensland or between Antarctica and New Zealand. In all cases the present-day distribution of *Nothofagus* is explained by dispersal in a variety of trans-oceanic or overland routes depending on the acceptance or rejection of continental drift. In each case, there is no attempt to use a refined taxonomy showing relationship at the species level and only vague notions are given as to what are 'centres of origin'. Furthermore, there is little appreciation of the idea that biogeography is more about areas of the globe and their total faunas and floras than about trying to explain how, when and by what route a *particular* group of organisms arrived at its present position.

Analytical theories of *Nothofagus* biogeography

In 1966, Brundin, in his classical investigation of the transantarctic relationships of chironomid midges, said that until such time as there are explicit phylogenetic hypotheses for groups such as *Nothofagus* it will be impossible even to start reconstructing precise distribution patterns. In other words, until phylogenetic schemes can be proposed that postulate relationships between modern species and fossils it is neither possible to provide historical details for particular genera nor to make comparisons, hence biogeographical statements, between distantly re-

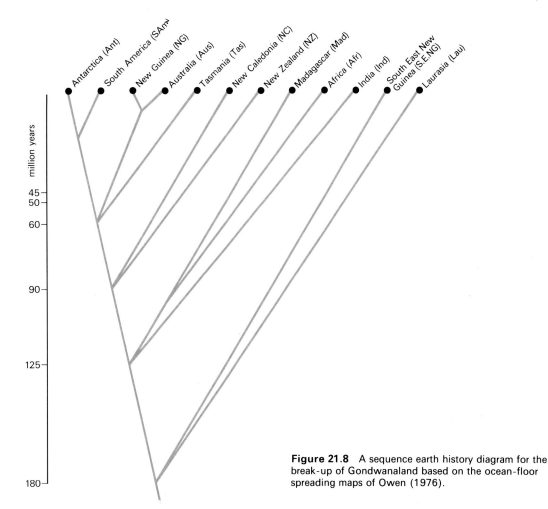

Figure 21.8 A sequence earth history diagram for the break-up of Gondwanaland based on the ocean-floor spreading maps of Owen (1976).

lated groups which occupy similar areas. It is of paramount importance, therefore, that precise interrelationships are known between the species of *Nothofagus*.

Melville (1973) was the first person to reconstruct a fully resolved model of species relationships in *Nothofagus*. His hypothesis is an extension of Van Steenis's (1953) classification and is shown in Figure 21.7. The relationships are based on three evolutionary trends of deciduous to evergreen leaves, a gradual reduction in the parts of the female flower, and an elaboration of the scales on the lobes of the female cupule. The phylogeny is based on Forman's (1966) idea that *Trigonobalanus*, an unusual tree genus from Borneo, has the most primitive female infructescence structure in the Fagaceae. The mature infructescence of *Trigonobalanus* has seven nutlets, provided with a more or less entire cupule derived from nine lobes. The most common condition in *Nothofagus* is a cupule of four lobes, enclosing an infructescence of two lateral triangular nuts and a median, lenticular or flattened one. The most primitive condition is found in the South American species *N. allessandri* which has four lobes in the cupule but seven nutlets in the mature infructescence. Melville considers that reduction from four–three–two cupule

lobes is a continuous transformation which puts Van Steenis's subsections *Tripartite* and *Bipartite* amongst the most derived taxa (Fig. 21.7). Other changes include the development of three lenticular nutlets, nutlet loss and gradual reduction in the cupule scales. The ultimate conditions are seen in the New Guinea *N. grandis* which has one lenticular nutlet and two cupule scales, and *N. pullei* in which the scales are minute and the solitary nutlet virtually naked. Characters used for suggesting relationships between the cool-temperate species include elaborations of the female cupule and modifications of leaf venation.

Cracraft (1975) used Melville's hypothesis of interrelationships to reconstruct the historical biogeography of the modern species of *Nothofagus*. In Figure 21.7 Melville's phylogenetic scheme is drawn with Cracraft's selected species names given to the four transantarctic groups – the '*brassii*' group involves the evergreen *N. dombeyi* from South America and its nearest relatives (the sister group) in New Zealand, New Caledonia and New Guinea; the '*menziesii*' group with *N. betuloides* in South America and its sister group in New Zealand; the '*nitida*' group comprising *N. nitida* in South America and its sister group of two species in Australia; the 'pumilio' group occurring in South America and Tasmania, and finally the 'alpina' group endemic to South America. Information from fossils occurring outside the present range of Nothofagus species is also included in Figure 21.7. The reconstruction was an attempt to determine the centre of origin. The method used to determine the centre of origin was to combine the shared areas and eliminate the unshared or unique areas. The shared areas are thought to represent a continuous land-mass at some time in the past. The unshared areas may or may not represent areas into which Nothofagus has dispersed. The result shows a combination of vicariance and dispersal events. For instance, the ancestors of the 'brassii' group must have been distributed throughout a land-mass comprising South America, western Antarctica, New Zealand, New Caledonia and New Guinea.

Continental drift is used as an explanation for the present disjunct pattern. Separation of the New Zealand and South American continental blocks could account for the isolation of N. dombeyi from its sister group. A different, later geological event isolated the ancestor of N. flaviramea and N. brassii in New Guinea and the ancestor of N. codonandra and N. solandri in New Caledonia and New Zealand. The ancestor of the 'menziesii' group was spread across South America, western Antarctica and New Zealand. Continental break-up could also account for vicariance speciation isolating N. betuloides in South America, leaving the ancestor of N. menziesii and its

relatives in New Zealand. A similar pattern can be reconstructed for the 'nitida' group but over eastern rather than western Antarctica. However, because in the 'pumilio' group the species N. glauca, N. obliqua and N. antarctica all occur in South America and the species N. pumilio and N. gunnii occur in South America and Tasmania, Cracraft suggests that the simplest explanation for N. pumilio and N. gunni is a dispersal event from South America to Tasmania. The 'alpina' group has a wholly South American origin.

Because the taxa in New Zealand, New Guinea, New Caledonia and Australia appear to represent unique events the obvious deduction for the origin of the whole group is in pre-drift west Gondwanaland with subsequent dispersals into other areas. The presence of at least four species groups in pre-drift Gondwanaland means that dispersal to east and west Antarctica and into Australia, New Guinea and New Zealand had occurred by at least 90 million years ago (see Fig. 21.8). According to this model the sister group relationships are not between New Zealand and Australia but across Antarctica to South America, a pattern matching with several groups occurring at the time of continental break-up.

Although Cracraft's analysis is perhaps the most explicit hypothesis to date, its validity is difficult to assess because it is based on dubious taxonomy and incorporated untestable dispersal explanations for some of the ancestral distribution patterns. Since the model is based on general

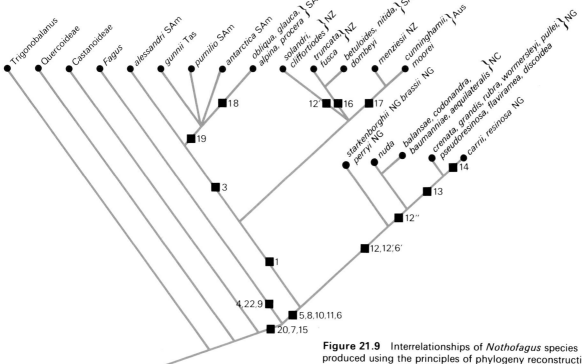

Figure 21.9 Interrelationships of Nothofagus species produced using the principles of phylogeny reconstruction proposed by Hennig (1966). In this method monophyletic groups are identified by the possession of shared derived characters. See Table 1 for key to characters.

Table 1 Characters used for the hypothesis of interrelationships of *Nothofagus* species shown in Figure 21.9

		primitive (−)	derived (+)
1	Tracheids	present	absent
2	Wood fibre	× 1·48 mm	× 0·7–1·16 mm
3	Leaves	evergreen	deciduous
4	Terminal flowers	present	aborted
5	Male dichasia	many flowered	few flowered
6	Male perianth	6-lobed	campanulate or fused tubular (a small group of species are more derived in this feature – 6′)
7	Male inflorescence	long spikes	hanging clusters
8	Pollen aperture number	3	5–8
9	Pollen apertures	colpate	colporate
10	Pollen apertures	long	medium short (– 10′)
11	Ovule integments	2	1
12	Cupules	4-lobed	3-lobed (– 12′) 2-lobed (– 12″)
13	Female flowers per cupule	2–7	1
14	Cupules	reaching nutlet tip	shorter than nutlet
15	Cupules	along rhachis	in leaf axils
16	Cupules	not branched	branched
17	Cupule lamellae	not modified	gland tipped, recurved
18	Cupule lamellae	entire	simple branches, elaborate branches
19	Infructescence	7 fruits	3 fruits
20	Central fruits	trimerous	dimerous (or absent)
21	Lateral fruits	trimerous	dimerous
22	Styles	flattened + long stigmatic surface	cylindrical with short stigmatic surface

interpretations of advancement in some morphological characters rather than on the use of shared derived character states to establish monophyletic groups, it contains several cumbersome and unlikely interpretations. For example, the South American species *N. dombeyi* is, on the basis of cupule and floret morphology, one of the most primitive species in the genus, yet it is placed as the sister species of some, but not all, the taxa with derived cupules. Furthermore, there are a number of species pairings in the diagram that have no evidence at all for their inclusion. For these two reasons, and because various other valuable data on wood anatomy and the male inflorescence were disregarded, another cladogram (phylogenetic diagram) produced on strictly cladistic lines is given in Figure 21.9 based on data given in Table 1. This scheme will be used in subsequent discussions.

The scheme of interrelationship favoured here differs from that proposed by Melville in a number of points. All of the cool-temperate species show a 'specialised' wood anatomy and are assumed to be more closely related to one another than any are to the New Caledonian and New Guinea species. This means that the tripartite cupule in the two New Zealand species is assumed to have evolved independently from the bipartite cupule seen in the tropical taxa. Within the two temperate groups there is insufficient evidence to establish the relative order of some main branching points and so the divisions are given as unresolved trichotomies rather than resolved dichotomies. The net result from the rearrangement is that there are only two transantarctic relationships at the species level rather than four as predicted in the Melville model.

Once an explicit theory of interrelationships of *Nothofagus* species has been produced it can be used to suggest a history of areas and can be tested against phylogenies of other organisms and hypothesised events in earth history (Fig. 21.8). This is the vicariance–cladistic method of biogeographic analysis which needs a brief explanation.

The theory of the phylogenetic relationships of *Nothofagus* (Figs. 21.9, 21.10 A) can be translated into a phylogeny of areas (Fig. 21.10 B). That is to say, the relevant areas are substituted for taxa on a diagram of the relationships of the *Nothofagus* species. Other groups of organisms are then chosen which have representatives in the areas concerned. If several phylogenies match one another and the areas diagram (that is, if they are congruent) then there is reason to suspect that these organisms have a similar history. There is, in effect, a general theory of biogeographic history which should be matched by known events in earth history. The idea is, quite simply, that the pattern of fragmentation of areas will be reflected in the phylogeny of organisms. The vicariance pattern can be explained by the splitting of a former continuous area (Fig. 21.10). There is no reason to assume that dispersal has occurred from one area to another except for those cases of non-congruent phylogenies. From the distribution models produced in this way, Nelson (1975) suggests that within a vicariance paradigm the unshared elements estimate not ancestral distributions but the exact positions where the barriers led to divergence.

By combining the northern and the southern beeches together, *Fagus* and *Nothofagus* form a seven-area pattern of distribution (Fig. 21.10 B). Two of the areas represent North America and Europe in Laurasia, and the other five the principal land-masses of Gondwanaland. The break-up sequence for Gondwanaland is given as a dichotomous diagram in Figure 21.8. Since there are three groups in Figure 21.10 A showing different disjunction patterns for South America–Australia, South America– Tasmania and New Zealand–Australia, they have to be treated in Figure 21.10 B as branching from one point because such a pattern could not have occurred through continental break-up. It is possible to resolve the next branching point because all of the species of the cool-temperate austral areas represent a sister group to the New Guinea–New Caledonia group. Although this model gives only low resolution for a history of *Nothofagus*, a similar area diagram for the Melville/Cracraft model (Fig. 21.10 C) gives even less resolution because the South America–New

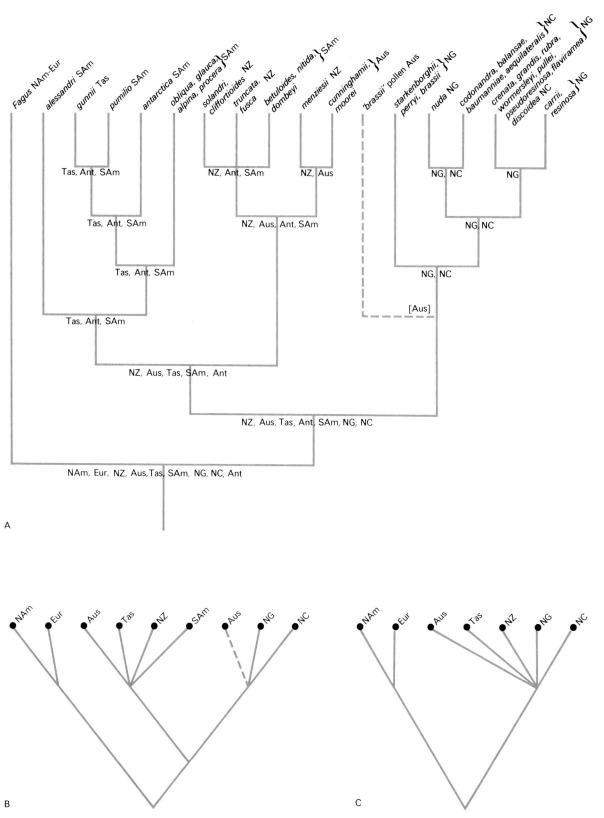

Figure 21.10   A. A historical vicariance biogeography of *Nothofagus* based on Fig. 21.9. B. An area phylogeny based on A. C. An area cladogram based on the phylogeny of *Nothofagus* suggested by Melville (Fig. 21.7).

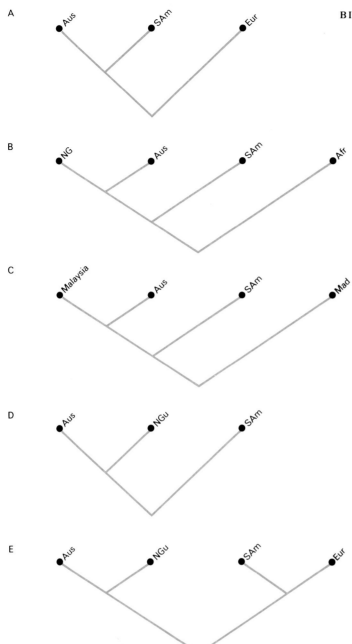

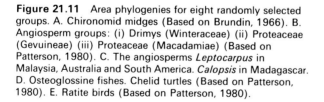

Figure 21.11 Area phylogenies for eight randomly selected groups. A. Chironomid midges (Based on Brundin, 1966). B. Angiosperm groups: (i) Drimys (Winteraceae) (ii) Proteaceae (Gevuineae) (iii) Proteaceae (Macadamiae) (Based on Patterson, 1980). C. The angiosperms *Leptocarpus* in Malaysia, Australia and South America. *Calopsis* in Madagascar. D. Osteoglossine fishes. Chelid turtles (Based on Patterson, 1980). E. Ratite birds (Based on Patterson, 1980).

Guinea/New Caledonia disjunction in the '*brassii*' species group represents yet another pattern (Fig. 21.6).

The only identifiable microfossils that can be associated with particular modern groups are the pollen grains of the '*brassii*' type. Adding these fossils into the area diagram does not extend the areas occupied by *Nothofagus* but does extend the tropical '*brassii*' group into Australia and New Zealand. The net effect is to produce a primary reduced area cladogram identical to the Melville/Cracraft model (Fig. 21.10 C). Nevertheless, fossils do give the minimum age for some of the area relationships. The Australia–South America–New Zealand pattern is at least as old as late Cretaceous and, if the new cladogram in Figure 21.9 is correct, then it must have been preceded by the New Guinea/New Caledonia–Australia/South America/New Zealand dichotomy. The northern hemisphere–Gondwanaland dichotomy for the *Fagus– Nothofagus* split does not have a fossil record at all, but it must predate the Australia/South America/New Zealand split. In other words, the beeches relate northern and southern areas and within the vicariance framework probably diverged at the break-up of Pangaea.

By using area cladograms selected from other taxonomic groups with endemic representatives in at least three of the areas occupied by *Nothofagus* it should be possible to see general area patterns of relationship. From the cladograms in Figure 21.11 it does seem certain that the area relationships are generally similar in all but one situation, the last dichotomy in the genus *Leptocarpus* (Fig. 21.11 C). Since this genus is the only one to occur in Malaysia a parsimonious conclusion demands that, because all the other groups show austral connections, the crossing of Wallace's line must be a dispersal event rather than a case of extinction in Malaysia for *Nothofagus* and the seven other groups.

One thing all the cladograms have in common is the Australia–New Guinea connection. Although most of New Guinea is of very recent origin, the southern part of the island has had two connections with Australia during its geological history – one in the Pleistocene and the other way back in the Jurassic (Fig. 21.8). From the *Nothofagus* cladogram in Figure 21.10 A and from fossil evidence, it would appear that the austral groups are at least as old as the Cretaceous and that the New Guinea connection dates back to the Jurassic. On the other hand, since the Australia–New Guinea dichotomies are the most derived in the cladograms for other organisms (Fig. 21.11 B, D, E) it is likely, in these groups, to represent the more recent Pleistocene connection.

At the first dichotomy the only cladograms to show a similar pattern with *Nothofagus* are those of the chironimid midges and the ratite birds (Fig. 21.11 A, E). The four four-area cladograms for *Drimys*, the two tribes of the Proteaceae and *Leptocarpus/Calopsis* (Fig. 21.11 B, C) do show the same general sequence but with southern African rather than northern hemisphere connections for the first dichotomy; a general area pattern that agrees with the break-up sequence of Gondwanaland (Fig. 21.8).

Conclusion

To sum up the analytical approach, and this chapter, it appears that Patterson's remark given at the beginning is justified. Previous claims that the distributional history of the southern beeches is the key to the general history of life in the far south cannot be substantiated.

What can be said is that the southern beeches do appear to be a wholly southern group whose precise centre of origin is impossible to locate. There were several species groups already extant at the break-up of Gondwanaland and it does appear that continental drift created barriers between some of the modern vicariant species. I have tried to show, by using *Nothofagus* as an example, that narrative theories of biogeography are dependent on many premises, all of which can legitimately be questioned. Earlier narrative theories assumed a world of stable geography, or one in which present land-masses were linked by former land-masses. These assumptions are now unfashionable. Later narrative theories involved continental drift but, like their predecessors, are coupled with the assumption animals and plants dispersed over large tracts of the Earth's surface. This leads to further questions of deciding from where, by what means and by what routes, the particular groups of organisms travelled. It is unlikely that we can solve these questions although attempts to do so have certainly exercised inventive minds. Narrative theories are often based, as in the case of *Nothofagus*, on imprecise ideas of the interrelationships of the organisms and it is not uncommon for the assumed biogeographic history to determine the scheme of relationships.

On the other hand, the analytic approach begins with an attempt to determine, as completely as possible, the phylogenetic interrelationships of the organisms concerned. The distributional data is then added and compared with the results of similar exercises for other groups. If, as the analytical approach assumes, there is a common history, caused by a particular history of geological events, then there should be congruence between the history of many kinds of organisms. In other words, the congruence (or otherwise) provides the test of the hypothesis. Neither the routes and means of dispersal nor the centres of origin are initial assumptions for the analytical approach.

The analytical approach to biogeography is still in its infancy. Phylogenetic interrelationships of very few groups are known with sufficient precision to justify direct comparison with one another. *Nothofagus* may not itself be the key to an understanding of the history of the southern continents. But any general theory of the fauna and flora of the southern hemisphere must explain the distribution of *Nothofagus*. It is my belief that this general theory will be arrived at by means of the analytical approach rather than the narrative approach.

References

Ball, I. R. 1976. Nature and formulation of biogeographical hypotheses. *Systematic Zoology* **24**: 407–430.

Brundin, L. 1966. Transantarctic relationships and their significance, as evidenced by chironomid midges. *Kungliga Svenska VetenskapsAkademien Handlingar* **11**: 1–472.

Cracraft, J. 1975. Historical biogeography and earth history: perspectives for a future synthesis. *Annals of the Missouri Botanical Garden* **62**: 227–250.

Cranwell, L. M. 1963. *Nothofagus*: living and fossil, pp. 387–400 in Gressitt, J. K. (Ed.) *Pacific Basin biogeography symposium* Hawaii: Bishop Museum Press.

Croizat, L. 1952. *Manual of phytogeography: or an account of plant-dispersal throughout the world.* 587 pp. The Hague: W. Junk.

Darlington, P. J. Jr. 1965. *Biogeography of the southern end of the world.* 236 pp., Cambridge, Massachusetts: Harvard University Press.

Forman, L. L. 1966. On the evolution of cupules in the Fagaceae. *Kew Bulletin* **18**: 385–419.

Hanks, S. L. & Fairbrothers, D. E. 1976. Palynotaxonomic investigation of *Fagus* L. and *Nothofagus* Bl.: Light microscopy, scanning electron microscopy and computer analyses. *Botanical Systematics* **1**: 1–141.

Hennig, W. 1966. *Phylogenetic systematics.* 263 pp., Urbana: University of Illinois Press.

Hooker, J. D. 1853. Introductory Essay. pp. i–xxxix in *The botany of the Antarctic voyage of H.M. Discovery ships Erebus and Terror in the years 1853–55. II Flora Novae Zelandiae.* London: Reeve Bros.

McQueen, D. R. 1976. The ecology of *Nothofagus* and associated vegetation in South America. *Tuatara* **22**: 38–68.

Melville, R. 1973. Continental drift and plant distribution, pp. 439–446 in Tarling, D. H. & Runcorn, S. K. (Eds) *Implications of continental drift to the earth sciences.* London: Academic Press.

Moore, D. M. 1972. Connections between cool temperate floras with particular reference to southern South America. pp. 115–138. in Valentine, D. H. (Ed.) *Taxonomy, phytogeography and evolution.* London: Academic Press.

Nelson, G. 1975. Historical biogeography: an alternative formalization. *Systematic Zoology* **23**: 555–558.

Owen, H. G. 1976. Continental displacement and expansion of the earth during the Mesozoic and Cenozoic. *Philosophical Transactions of the Royal Society* (A) **281**: 223–291.

Patterson, C. 1980. Methods of paleobiogeography. pp. 446–500 in Nelson, G. & Rosen, D. E. (Eds) *Vicariance biogeography: a critique.* New York: Columbia University Press.

Raven, P. H. & Axelrod, D. I. 1972. Plate tectonics and Australasian paleobiogeography. *Science* **176**: 1379–1386.

Raven, P. H. & Axelrod, D. I. 1974. Angiosperm biogeography and past continental movements. *Annals of the Missouri Botanical Garden* **61**: 539–673.

Schröter, C. 1913. Geographie der Pflanzen. C. Genetische Pflanzengeographie (Epiontologie). p. 1284 in Teichmann, E. (Ed.) *Handwörterbuch der Naturwissenschaften.* 4. Jena: Gustav Fischer.

Schuster, R. M. 1976. Plate tectonics and its bearing on the geographical origin and dispersal of angiosperms, pp. 48–138 *in* Beck, C. B. (Ed.) *Origin and early evolution of Angiosperms.* New York: Columbia University Press.

Taktajhan, A. 1969. *Flowering plants: origin and dispersal.* 310 pp. Edinburgh: Oliver & Boyd.

Van Steenis, C. G. G. J. 1953. Papuan *Nothofagus. Journal of the Arnold Arboretum* **34**: 301–373.

Van Steenis, C. G. G. J. 1962. The land-bridge theory in Botany with particular reference to tropical plants. *Blumea* **11**: 235–542.

Index of organisms

Subject index

Numbers in *italic* refer to figures. An entry 'natural selection' has not been included since this is coordinate with the subject matter of this book.